BIRKHÄUSER

Publications des Archives Henri-Poincaré
Publications of the Henri Poincaré Archives

Textes et Travaux, Approches Philosophiques en Logique, Mathématiques et Physique autour de 1900
Texts, Studies and Philosophical Insights in Logic, Mathematics and Physics around 1900

Éditeur/Editor: Gerhard Heinzmann, Nancy, France

One Hundred Years of Intuitionism (1907-2007)

The Cerisy Conference

Mark van Atten
Pascal Boldini
Michel Bourdeau
Gerhard Heinzmann
Editors

Birkhäuser
Basel · Boston · Berlin

Editors:

Mark van Atten
IHPST
(CNRS/Paris I/ENS)
13, rue du Four
75006 Paris
France

Pascal Boldini
CAMS-EHESS
Université Paris-Sorbonne
Maison de la Recherche
28, rue Serpente
75006 Paris
France

Michel Bourdeau
IHPST
(CNRS/Paris I/ENS)
13, rue du Four
75006 Paris
France

Gerhard Heinzmann
Archives H. Poincaré
Nancy-Université/CNRS
23, bd. Albert
54015 Nancy CX
France

2000 Mathematical Subject Classification: 01, 03

Library of Congress Control Number: 2008930404

Bibliographic information published by Die Deutsche Bibliothek.
Die Deutsche Bibliothek lists this publication in the Deutsche Nationalbibliografie; detailed bibliographic data is available in the Internet at http://dnb.ddb.de

ISBN 978-3-7643-8652-8 Birkhäuser Verlag AG, Basel - Boston - Berlin

Basel · Boston · Berlin
P.O. Box 133, CH-4010 Basel, Switzerland
Part of Springer Science+Business Media
Printed on acid-free paper produced from chlorine-free pulp. TCF ∞

ISBN 978-3-7643-8652-8

ISBN 978-3-7643-8653-5 (eBook)

9 8 7 6 5 4 3 2 1

www.birkhauser.ch

To

Dirk and Doke van Dalen

in recognition of a shared life that has fostered scholarship on Brouwer and intuitionism through many distinguished publications, the Brouwer Archive, and the encouragement of others

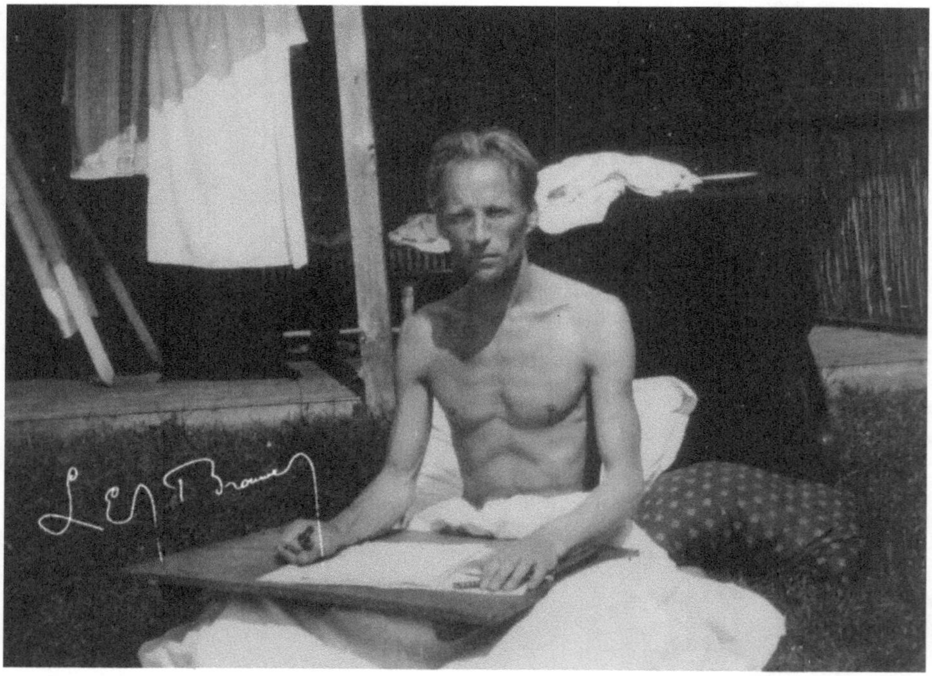

Contents

Contributors

Mohammad Ardeshir, Department of Mathematics, Sharif University of Technology, Tehran.

Mark van Atten, Institut d'Histoire et de Philosophie des Sciences et des Techniques (CNRS/Paris 1/ENS), Paris, France.

Henk Barendregt, The L.E.J. Brouwer Institute, Faculty of Science, Radboud University, Nijmegen, The Netherlands.

Douglas Bridges, Department of Mathematics & Statistics, University of Canterbury, Christchurch, New Zealand.

Dirk van Dalen, Department of Philosophy, Utrecht University, The Netherlands.

Jacques Dubucs, Institut d'Histoire et de Philosophie des Sciences et des Techniques (CNRS/Paris 1/ENS), Paris, France.

Michael Dummett, New College, Oxford University, England.

Jean Fichot, Caen, France.

Marcel Guillaume, Laboratoire de Logique, Algorithmique et Informatique, Université de Clermont-Ferrand 1, France.

Gerhard Heinzmann, Archives Henri Poincaré, Nancy-Université/Nancy 2/CNRS, France.

Mathieu Marion, Département de Philosophie, Université du Québec à Montréal, Canada.

Per Martin-Löf, Department of Mathematics, Stockholm University, Sweden.

Charles McCarty, Department of Philosophy, Indiana University, Bloomington, USA.

Alain Michel, Université de Provence Aix-Marseille 1, France.

Philippe Nabonnand, Archives Henri Poincaré, Nancy-Université/Nancy 2/CNRS, France.

Mitsuhiro Okada, Department of Philosophy, Keio University, Tokyo, Japan.

Carl Posy, Department of Philosophy, The Hebrew University of Jerusalem, Israel.

Giovanni Sambin, Department of Pure and Applied Mathematics, University of Padua, Italy.

Peter Schroeder-Heister, Wilhelm-Schickard-Institut für Informatik, University of Tübingen, Germany.

Anton Setzer, Department of Computer Science, Swansea University, England.

Göran Sundholm, Department of Philosophy, Leiden University, The Netherlands.

Richard Tieszen, Department of Philosophy, San Jose State University, USA.

Wim Veldman, Institute for Mathematics, Astrophysics and Particle Physics, Faculty of Science, Radboud University, Nijmegen, The Netherlands.

Preface

If one takes the year of Brouwer's dissertation, 1907, as the birth year of intuitionism, it has now reached the age of 100. It is in good shape, but has known a turbulent past. After the First World War, the conflict between Brouwer and Hilbert inflamed the mathematical and philosophical communities alike. This 'Grundlagenstreit' came to a violent end in the Annalen-affair. For some fifteen years, Brouwer kept silent; when he started publishing again, after the Second World War, his work reached but a handful of specialists. The future looked bleak. Then a number of developments began that again drew attention to intuitionism and constructivism in general. We mention some examples. In 1958, Gödel devised an interpretation of intuitionistic arithmetic as the required extension of the finitary standpoint to achieve Hilbert's aim, a consistency proof of classical arithmetic. With the posthumous publication of Wittgenstein's *Remarks on the Foundations of Mathematics* in 1956, also Brouwer's ideas again sparked a lively interest among philosophers. Bishop presented and developed constructive mathematics in his *Foundations of Constructive Analysis* from 1967. In 1969, the Curry-Howard isomorphism was introduced, and in 1972, Martin-Löf presented his intuitionistic type theory. Dummett, in his paper 'The philosophical basis of intuitionistic logic' from 1973, argued that the most convincing reasons to reject the principles of classical logic and to adopt those proposed by Brouwer instead, are to be found in a philosophy of language of a Wittgensteinian kind. In combination with the idea that metaphysics has a logical basis, it is intuitionism that led Dummett to his arguments in favour of anti-realism. In the mid-1980s, phenomenological interest in intuitionism saw a revival, which presents an alternative to Wittgensteinian interpretations.

The aim of the meeting *1907–2007: One Hundred Years of Intuitionism*, held in Cerisy, France, from June 5–12, 2007, was twofold. One objective was to complete and review the already rich historical account of intuitionism (in particular in relation to other varieties of constructivism, such as the pre-intuitionism of Poincaré and Borel). The other goal was to provide a forum for continuing philosophical and technical work on intuitionism. As time was limited while the subject is not, not as many topics could be covered as one would have hoped. We regret in particular that we have not been able to give due attention to the Russian contributions to intuitionism starting with Kolmogorov, nor to the school of intuitionistic non-standard analysis in Strasbourg.

The meeting would not have been possible without the enthusiastic and generous support from the following persons and institutes, which is gratefully acknowledged:

- The staff members of the *Centre Culturel International de Cerisy*, whose care and attention made our stay so pleasant;
- The *Evert Willem Beth Stichting*, who chose the Cerisy conference as the host for the 2007 Beth Lecture by Per Martin-Löf;
- The *Centre National de la Recherche Scientifique (CNRS)*;
- The *Archives Henri Poincaré* and the *Conseil Scientifique* of the *Université de Nancy 2*;
- The *Institut d'Histoire et de Philosophie des Sciences et des Techniques (IHPST)* and the *Université Paris 1*;
- The *Centre d'Analyse et de Mathématiques Sociales (CAMS) de l'Ecole des Hautes Etudes en Sciences Sociales (EHESS)*;
- Henk Barendregt, who offered meditation sessions every morning; Yann Kiefer, Wim Veldman, and John Thomas, who performed on the piano and violin.

Finally, we are much obliged to Dirk van Dalen for the frontispiece and for permission to include his full Brouwer bibliography in this volume.

Nancy and Paris
April 2008

Mark van Atten
Pascal Boldini
Michel Bourdeau
Gerhard Heinzmann

Part I

Brouwer and Brouwerian intuitionism

Another look at Brouwer's dissertation

Dirk van Dalen

Dedicated to N.G. de Bruijn for his ninetieth birthday — in admiration and gratitude

Brouwer's dissertation marked the beginning of two separate research activities that played an important role in the mathematics of the twentieth century. The first of these were his first steps in topology and Lie group theory, the second one opened up new directions in the foundations of mathematics. It is with the second one that we are concerned here; for the first one see Freudenthal's comments in (Brouwer 1976).

Even today, one is struck by the extraordinary wealth of ideas in the 182 pages that Brouwer presented to the faculty on February 19, 1907. His aim was to offer directions for a reconstruction of mathematics along constructive lines. In doing so, he was not satisfied with more or less postulating certain objects and operations, but he went down to the ultimate foundation of all of mathematics, as he liked to see it. He posited the Ur-intuition of mathematics as the point of departure for his reconstruction. From that point onwards the development proceeds along natural and well-determined lines. He developed for example by topological and group theoretical means a 'measurable continuum' on the basis of the given intuitive continuum, thus coming to his study of Lie groups in the context of Hilbert's fifth problem.

His foundational program was certainly revolutionary, his contemporaries would even have said 'counter revolutionary'. It contained the germs of his later, full-blown intuitionistic program. The reader familiar with today's constructive mathematics will readily see that some of Brouwer's elaborations do not quite meet the standards that are now generally accepted; there are points where Brouwer reconnoiters the new territory, and where he soon had to revise or strengthen his methods. In that sense the dissertation has a strong programmatic character. The critical discussion of the various foundational doctrines and practices, on the other hand, are completely adequate and pass the test of our day.

The main thrust of Brouwer's foundational program was in the direction of a constructive rebuilding of mathematics along the lines of Kant, Schopenhauer, and others. Mathematics, in his view was a human enterprise; his aim was the revision of existing practice, by stressing that the fundamental objects, and indeed all objects, of mathematics are creations of the human mind. And that the whole edifice of mathematics was to be based on few, but strict principles.

The unfortunate thing is that in setting forth his ideas, he used an almost hermetic language. We are nowadays used to very strict practices of theory formation, with precise definitions, avoidance of colorful suggestive comparisons, and terminological hygiene. Indeed, in Brouwer's days the tradition of suggestive, didactical explanations was on its way out. Authors like Frege, Russell, Hilbert had introduced an austere scientific practice that was gaining influence. As we will see, Brouwer did not shrink from condensed concise formulations, but neither from long-winded suggestive expositions. As the dissertation is extraordinarily rich in ideas, and notions, we will in the present paper restrict ourselves to a few key notions; trace these in Brouwer's published and unpublished notes, in the hope to help readers to get a better grip on Brouwer's intuitionism.

In order to set the stage for discussion of some topics and their treatment, let me say something about the prehistory of the dissertation. In June 1904 Brouwer took the final exam required for admission as a PhD student. At that time he was not only involved in research in mathematics proper, but he was also probing the Dutch philosophical establishment to find a niche for his own views. In the wake of the grand Leyden philosopher Bolland (a Hegelian) he gave a series of lectures on 'Art, Life, and Mysticism' at the Technical University at Delft. From these lectures he emerges as a full-blooded mystic who was not afraid to draw radical consequences from his views. The involvement with mysticism ended, at least in print, in 1905 when the lectures were published. He remained by and large faithful to his mysticism, and it would certainly not go too far to see some of his motivation for his later foundational work in the early monograph.

It appears that Brouwer did not immediately return to his task as a PhD student; in 1906 he published three papers on vector distributions in higher-dimensional spaces, and potential theory in non-Euclidean spaces. From the correspondence that has been preserved we know that he returned to the dissertation in February 1906. The actual writing of the text began in September of that year. The impression that is left from the correspondence between the PhD advisor, Korteweg, and the student, is one of mutual respect. The older man worried about Brouwer's preference for radical mystical and philosophical views, and the younger one used all of his considerable persuasive power to pacify and convince his advisor.

Let me briefly mention some relevant topics and issues. The prime goal of a human being is, in Brouwer's view, the return to the original state of life, where there is no distinction between subject and object, where one is completely shielded from the disturbing influences of the external world.

> And in that merging sea of colours, without separation, without permanence and yet without movement, that chaos without disorder, you know a Direction, which you follow spontaneously, and which you could just as well not follow. You recognize your 'Free Will', in so far as it was free to withdraw itself from the world, in which there was causality, and then remains free, and yet only then has a really determined Direction, which it reversibly follows in freedom. ... The phenomena follow each other in time, bound by causality, because you yourself want, shrouded in clouds, the phenomena in that regularity. (Brouwer 1905A, p.14)

The free will is mentioned here not as a metaphor, but as an actual asset of the human being (from now on to be called 'the subject'). Free will plays a role in the dissertation, and it later appears in full force at the introduction of choice sequences.

One particular consequence of this particular goal in life is that whatever hinders this 'turning into oneself' is to be considered sinful. Particular examples of such obstructions are dealings with mankind and the world at large. One adage in this connection is that one should 'avoid power over nature and over fellow human beings' — do not interfere with nature and leave your fellow man alone. The first chapter of the monograph ends with the pessimistic words

> The life of mankind as a whole, is an arrogant eating away of its nests all over the perfect earth, a meddling with her mothering vegetation, gnawing, spoiling, sterilizing her rich creative powers, until it has gnawed away all life, and the human cancer withers away over the barren earth.

One topic that casts its shadow before it, is Brouwer's view on communication and language. It should of course not come as a surprise that communication by means of language can hardly be expected to be foolproof. But the trend in mathematics had been in the direction of the primacy of language; both logic and formalism had accepted the frailties of daily language, but had at the same time made an exception for logic and mathematics. At this point Brouwer is adamant:

> No two persons will experience exactly the same feeling, and even in the most restricted sciences, logic and mathematics, which can properly speaking not be separated, no two [persons] will think the same thing in the case of the basic notions from which logic and mathematics are built.

A word of warning may be in place here. Normally none would expect people to shun unreliable tools, so one would not be surprised to see Brouwer treating language with disdain; nothing is less true — especially the use of an imperfect tool, that cannot be replaced, calls for extreme care and attention. In his mature mathematical work Brouwer goes to great length to be as precise as possible, and in lectures, and indeed conversation, he exercised great care to perfect his choice of words.

Language carries in itself the danger of sterilizing the true content of our thinking; in his notebooks for the dissertation he noted

> And that killing of what is alive in things resulted from the need to make them the same for different people, and thus to reach an agreement. For originally the world was different for different people.

As to the role of language in a scientific context, he remarked that

> the objectivization of the world in mathematical systems in distinct individuals is kept in mutual connection by the *passionless language*, which causes the erection of the identical mathematical system in both listener and speaker, although the emotional content of that system can be totally different in both.

That much having been said, it is clear that on Brouwer's account it is the internal mathematical construction that counts, and that language is nothing but a useful tool in this context.

1 The Ur-intuition

For Brouwer, and indeed for any subject, the construction or creation of the basic objects of mathematics is of the greatest importance. At this point he had a long way to go before he could, or would, express himself with the desired clarity. The dissertation opens with a brief recapitulation of the by then accepted means of creating the number systems out of the natural numbers. This is followed by the bold step to introduce the basic material of which mathematics is made through the act of intuition.

> In the following chapters we shall go further into the Ur-intuition of mathematics (and of every activity of the intellect) as the substratum, divested of all quality, of any perception of change, a unity of the continuous and the discrete, a possibility of thinking together several entities, connected by a 'between', which is never exhausted by the insertion of new entities. Since continuity and discreteness occur in this Ur-intuition as inseparable complements, both having equal rights and being equally clear, it is impossible to stay clear of one of them as a primitive entity, and then to construct it from the other one, the latter being considered by itself; in fact it is already impossible to consider it by itself. Having recognized the intuition of continuity, the 'flowing' as primitive, as well as the conceiving of several things

> as one, the latter being at the basis of every mathematical structure, we are able to state properties of the continuum as a 'matrix of points joined together in thought'.

This terse introduction of the continuum has the characteristic of a recapitulation of a notion explained to the reader at an earlier occasion. As it stands, it is mystifying rather than clarifying. There is a reason for this; the dissertation originally contained a philosophical introduction (with a definite mystical flavor), which mentioned the intuition of time, the notion of (causal) sequence[1] and the jump from end to means. In this introduction the Ur-intuition is given its place as *intuition of time.*

> But man is endowed with a faculty, that accompanies all his interactions with nature, that is the ability of objectifying the world, seeing in the world recurrences of sequences, seeing in the world causal systems in time.
>
> The ur-phenomenon is the intuition of time by itself, in which iteration, as 'thing in time, and one more thing', is possible, but in which (and this is a phenomenon, which is outside mathematics) also a sensation can fall apart into composing qualities, so that a single moment of life can be lived as a sequence of qualitative distinct things.

The fact that Brouwer, when advised in unmistakable terms by Korteweg, dropped this introduction, made the foundational part of the dissertation harder to grasp. Korteweg's motivation was simply that a text with such strong flavor of mysticism would not contribute to the general appreciation of the faculty members. As Brouwer put it in a letter to Korteweg, the topics of chapter two suddenly appeared in the lime light to take the place of their former leader, 'and it was not possible to dress all of them so that on their own they could together save the show'.

Indeed chapter two contains a few introductory explanations; there is a slightly different formulation of the above quote:

> The ur-phenomenon is the intuition of time by itself, in which iteration, as 'thing in time, and one more thing', is possible, and on the basis of which moments of life fall apart as causal sequences of qualitative distinct things, that subsequently concentrate themselves in the intellect as not sensed, but observed mathematical causal sequences.

The meaning of the Ur-intuition of the continuum (and time) is elaborated, so to speak, in action when the continuum is given its measure and number

[1] Brouwer uses the term 'volgreeks' in his early writings. For lack of a better word, we use 'sequence' as a translation. In his later writings, for example (Brouwer 1929A, Brouwer 1933A1), Brouwer introduces the term 'causal sequence', which denotes the 'equivalence class' of sequences under the identification by the subject.

systems. We will here restrict ourselves to the fundamental discussions that one finds in Brouwer's various publications.

Since Brouwer saw both the continuous and the discrete as inseparable results of the Ur-intuition, we will deal with them simultaneously. In some publications more attention is paid to the one than the other, depending on use to be made, but the twins were never far apart.

In the first real publication (after all, a dissertation was a privately distributed book or booklet) Brouwer briefly returned to the creation of the fundamental structures of mathematics. In his lecture, *Die moeglichen Maechtigkeiten* (The possible cardinalities), at the International Mathematics Conference in Rome (1908), Brouwer discussed the constructively admissible sets. The lecture was an elaboration of the material presented the year before; it returned to the matter of creating mathematical entities and drawing conclusions for the sets mathematics allowed.

> When one investigates how mathematical systems come about, one sees that they are constructed out of the Ur-intuition of two-ity. The intuitions of the continuous and the discrete join here, as [simply] a second [thing] is thought not by itself, but under preservation of the recollection of the first. The first and the second are thus *kept together* and the intuition of the continuous (continere = keeping together) consists of this keeping together. This mathematical Ur-intuition is nothing but the contentless abstraction of the sensation of time. I.e., the sensation of 'fixed' and 'floating' together. Or of 'remaining' and 'changing' together.

Here one sees that the continuous, let us say, the continuum, comes about in the passage from one sensation to the next — it *is* the passage. Hence Brouwer's claim that the two notions are inseparable: you cannot have the one without the other.

Let us now make a jump to the late twenties, when Brouwer, after showing his mathematics capable of treating the central topics of his time,[2] returned to the basic topics of his dissertation. There is no doubt that the presentation of the period of his ripe intuitionism, which is a big step forward in clarity and detail, is indeed a crystallization of his early ideas. The first occasion at which Brouwer spelled out the genesis of mathematics and its objects in print, was in 1928 in Vienna, where Brouwer lectured on a clearly philosophical topic, *Mathematics, Science, and Language.* Here he described the three manifestations of the will to live (*Lebenswille*) of the individual. Only the first two are relevant for us here: (1) the mathematical attention, (2) mathematical abstraction. The mathematical attention consists of two parts: temporal at-

[2] Cf. (Brouwer 1929A, p.164).

tention[3] and *causal attention*[4]. The temporal attention consists of the earlier mentioned 'falling apart of a moment of life', which gives rise to a temporal two-ity. Brouwer went on to say that the act of temporal attention results in separating the split moment of life from the ego and placing it in the so-called 'world of perception'[5]. Note that the latter has no independent existence, it is 'the outside world as created by the subject'. The act of temporal attention can be repeated ad libitum and gives rise to finite sequences of sensations remembered. The next stage in the creation of the subjects 'world' is that of 'causal attention'. This is the identification process that serves to keep the multitude of objects provided by the temporal attention manageable. The subject abstracts, let us say among the n-tuples, those that from the subjects point of view are similar in some specific respect. E.g., he will identify all the n-tuples that he perceives in the morning at a specific place and time as 'his coffee machine'. The ultimate abstraction that respects distinctness is the one that identifies, for example, all two-ities into the *empty two-ity*. This empty two-ity provides the subject with the number 2 which is in Brouwer's approach the starting point of the creation of the mathematical universe.

The sequences, obtained by an abstraction process, are called *causal sequences* by Brouwer, and these causal sequences are the basic ingredients of mathematics and science in general — one might add of the outer world.

Of course, the subject can iterate the act of temporal attention, and thus create strings of arbitrary length. Thus the individual natural numbers are within easy reach. The problem is, does the subject have access to the totality of natural numbers? Whatever may be in doubt, it is clear that he will not get the set of natural numbers with the same status as, say, the set of numbers less than 20. There is no such thing for the subject as a completed totality of natural numbers; the second best for the subject would be the recognition of the totality of the natural numbers as a potential infinite entity. The dissertation does not deal with the matter, but from a letter of Brouwer to the Utrecht mathematics professor J. de Vries (undated draft in the Brouwer archive), in which Brouwer gave a quick survey of the main points of the dissertation, one reads 'I put the "mathematical construction act of complete induction" in the place of "the axiom of complete induction", and show how this is nothing new after the intuition of time.' Thus the act that yields the totality of natural numbers is implicit in the intuition of time.

In the dissertation and various other places, Brouwer uses the term 'the mathematical intuition *and so on*' (discussion of Russell), and at another spot he comments on Dedekind:

[3] German: *Zeitliche Einstellung* (Brouwer 1929A), Dutch: *tijdsgewaarwording* (Brouwer 1933A1).

[4] German: *kausale Einstellung* (Brouwer 1929A), Dutch: *causale aandacht* (Brouwer 1933A1).

[5] German: Anschauungswelt.

> Dedekind's system has no meaning; a logical meaning would require a consistency proof, which Dedekind does not give either, then he would have to appeal to the intuition of 'and so on'.

In later publications Brouwer specifies the generation of the number sequence as caused by the 'self-unfolding of the act of the intellect'.

In his *Intuitionism and Formalism* Brouwer boldly asserts,

> This intuition of two-oneness, the basal intuition of mathematics, creates not only the numbers one and two, but also all finite ordinal numbers, inasmuch as one of the elements of the two-oneness may be thought of as a new two-oneness, which process may be repeated indefinitely; this gives rise to the smallest infinite ordinal ω.

In fact this is the first step towards his version of the second number class.

The first Vienna lecture briefly mentions the matter of the infinite as a mental entity. After the empty two-ity has been introduced, he goes on,

> It is this common substratum of all two-ities, that forms the Ur-intuition of mathematics, the self-unfolding of which introduces among other things the infinite as a mental reality, and in fact first of all the totality of the natural numbers, not to be commented on here; next that of the real numbers, and finally the whole of pure mathematics. (Brouwer 1929A, p.154)

A little later Brouwer gave a lecture in Dutch, *Willen, Weten, Spreken* (Will, Knowledge, Speech) that discussed the matter in some more detail,

> The sensation of time is the fundamental phenomenon of the intellect: the falling apart of a moment of life into two qualitative different things, of which one recedes before the other and nonetheless is retained by memory. The split moment of life is thereby more or less set free from the I and placed in a world of its own, the world of observation. Of the temporal Two-ness or twofold temporal sequence of phenomena resulting from the sensation of time one member can in turn fall apart into two parts, by which the temporal Three-ness is introduced. The continuation of this process, as self unfolding of the fundamental phenomenon of the intellect is the cause of the creation of the temporal sequence of phenomena of arbitrary multiplicity. (Brouwer 1933A1, pp.177–178)/(van Stigt 1990, p.418)

In the paper *Consciousness, Philosophy, and Mathematics*, which sums up his thoughts over the years past, Brouwer states,

> Mathematics comes into being, when the two-ity created by a move of time is divested of all quality by the subject, and when the remaining empty form of the common substratum of all two-ities, as a basic intuition of mathematics is left to an unlimited unfolding, creating new mathematical entities in the shape of *predeterminately or more*

or less freely proceeding infinite sequences of mathematical entities previously acquired, ...

We note that quite early on Brouwer was aware of the role of the natural number sequence. Starting from the 'and-so-on' intuition, he passed on to the more explicit 'self-unfolding' notion. This self-unfolding, so to speak, is part and parcel of the Ur-intuition; it supplements the first acts of mathematical attention that yield the finite sequences. The discussion of Brouwer's views on the natural number sequence should help to alleviate the worries that Brouwer would in fact only be able to create the individual numbers, but not the whole of all natural numbers. If that were the case, Brouwer would not have been able to quantify over the natural numbers.

The axiom of complete induction is dealt with in the dissertation in the list of appended theses. The second one says,

> It is not only impossible to prove the admissibility of the axiom of complete induction, but it ought not to have a place as a separate axiom or as an intuitive truth. Complete induction is an act of mathematical constructing, that has its justification already in the Ur-intuition of mathematics.

Once the totality of natural numbers has been granted by the Ur-intuition, and hence recognized as an entity of mathematics, the step to the justification of the principle of mathematical induction is not that difficult. Given the intuitionistic reading of the logical connectives (according to the 'proof = construction'-interpretation, see below) one immediately gets the *principle of complete induction* as a corollary. Let us have a quick look at the matter. The argument runs roughly as follows: given $A(1)$ and $\forall n(A(n) \rightarrow A(n+1))$, we want to show $\forall n(A(n)$. 'Show' means for a constructivist 'present a proof', where we have to keep in mind that already in 1907 Brouwer was aware that proofs are constructions; he spoke of 'erecting mathematical buildings' and 'fitting buildings into other buildings'. In modern terms this would be read as 'constructing mathematical structures' and 'constructing a structure on the basis of (out of) another structure'. It is quite clear that he knew how proof-constructions for implication, universally and existentially quantified statements were to be made. The cases of conjunction and disjunction were tacitly understood. So — returning to the matter of induction — we may assume that there is a proof a_1 of $A(1)$; notation - $a_1 : A(1)$. Now a proof for $\forall n(A(n) \rightarrow A(n+1))$ is a construction c that for any given n and proof $a : A(n)$ yields a proof $c(n, a) : A(n+1)$.

So $a_2 = c(1, a_1)$ and $a_2 : A(2)$, and $a_3 = c(2, a_2)$ and $a_3 : A(3)$, ... Hence parallel to the construction of the natural numbers we obtain the (potentially infinite) sequence of proofs $a_1, a_2, a_3, \ldots$, i.e., a proof for $\forall n(A(n)$.[6]

[6] In systems with an explicit recursor, one can often write down a term for the proof-construction given by the sequence.

Now let us look at the other twin, the continuum. Already before the Ur-intuition was introduced, Brouwer discussed the continuum in the notebooks where he wrote down his ideas and tested technical details for the dissertation. In the notebooks he accepted the continuum as intuitively given, but without any details. Here are two quotes:

> The intuitive continuum as the opposite of the point; the unknown, about which lack of understanding is impossible.

> About the continuum I see intuitively that there are yet unknown assignments on it, as such it is the matrix of yet unborn points. It exists thus independent from the points to be built on it, is thus different from the set of those points for otherwise its creation would follow that of those points.

He obviously saw that one needed more than just the definable (or lawlike) points. After all, Cantor had made it clear that any attempt to stick to a denumerable continuum was bound to fail; and although Brouwer found a satisfactory solution to the cardinality questions concerning the continuum only after the introduction of choice sequences and spreads, he was aware of the fact that the extra, non-lawlike points had to be incorporated. He called these extra points 'unknown points'. For example,

> One also defines unknown irrationals as limits of unknown series. One assigns the familiar ordering relation, and only afterwards one has to introduce the continuity postulate, in order to carry out the operations on these irrationals.

The 'unknown points' of the continuum are used in the dissertation to refute the well-ordering theorem.

> Now we know that besides the denumerable sets, for which the theorem *certainly holds*, there exists only the continuum, for which the theorem *certainly does not hold*, firstly because the greater part of the elements of the continuum must be considered as unknown, and consequently can never be individually ordered, secondly, because every well-ordered set is denumerable. Thus this question also turns out to be illusory. (Brouwer 1907, p.153)

A similar use is made in the discussion of the Cantor-Bernstein theorem.

2 Choice sequences

It is an anachronism to place the choice sequences in the dissertation; yet, there are certain associations that are hard to ignore. The first question to ask is 'was Brouwer aware of the early discussion of the phenomenon?'. At the time of Brouwer's dissertation the topic of 'choice' was widely discussed in connection with Zermelo's axiom of choice. Émile Borel, in particular, took up the idea of sequences of numbers determined by choice. In the discussion following Zermelo's paper, he explicitly rejected arbitrarily large choice functions, i.e., the possibility of non-denumerably many choices; by default one might conclude that endorsed ω-sequences of choices. The crucial point here is the location of these choice objects, are they to be found in an idealized mathematical world, or inside the human being? Borel took the human side of the issue serious, in his 1908 Rome lecture, *Sur les Principes de la Théorie des Ensembles*, he admitted that denumerably many choices could not be effected in their totality; nonetheless he was willing to allow certain thought experiments on them. But, as the infinite totality of choices (evidently considered as choice sequences *avant la lettre*) is non-denumerable, his conclusion was that it could not be admitted as a mathematical entity for use in mathematical arguments.

Apart from the discussion around the axiom of choice, there was at the time of Brouwer's dissertation only Du Bois-Reymond's book on Function theory, which explicitly discussed 'lawless sequences'. Neither the dissertation, nor the notebooks mention Du Bois-Reymond; hence we may safely assume that Brouwer was not familiar with Du Bois-Reymond's lawless sequences; as Du Bois-Reymond introduced the notion in the context of analyzing the continuum, Brouwer would certainly have seen the relevance, had he been aware of the text. In Brouwer's notebooks there is mention of *kansrij* (chance sequence), and Brouwer uses the term 'prendre au hasard'. Apparently he did not see how to make use of the notion of choice. There is no explicit mention in the dissertation.

In the notebooks, however, there is ample evidence that Brouwer contemplated the choice phenomenon. Whether this was motivated by the discussion around the axiom of choice is debatable. There are two plausible motives for considering choice: the discussion around the axiom of choice, and the non-denumerability of the continuum. Given Brouwer's acceptance of the subjective side of mathematics, the axiom of choice would not be as important as the necessity to 'save' the continuum. In notebook VII the axiom of choice is mentioned:

> (re Zermelo) It is not true, that a law of nature is possible, that picks an element from each set of a partition. One would rather say: [if I] present a living individual a set from a partition, and ask him to choose an element from it, he will mention one (according to his brain structure), but 1. this holds only for definable partitions, and 2. it is not certain that he will mention the same example, if I present the partition in two languages to him.

In the dissertation itself Brouwer mentions the axiom of choice more or less in passing, quoting Borel's remark that anybody introducing such an axiom might just as well adopt the well-ordering theorem straight away. He continues by stating that the latter is illusory, as the continuum, consisting mostly of *unknown elements*, cannot be well-ordered.

Unknown points and choice sequences appear to be closely related, if not identical. There are passages in the notes where both serve the same purpose, i.e., completing the continuum.
Examples:

1. 'If the "prendre au hasard" is projected on a denumerable collection, the it can *only* be through an (empirical) infinite decimal expansion';
2. 'Nothing can be said about the continuum, than with the help of an everywhere dense scale, constructed on it. Hence each subset must after all be expressed by means of such a scale. That can only be done in two ways: 1. directly defined, then the set is denumerable, 2. with the help of an infinite chance sequence. Then the set is of the cardinality of the continuum.'
3. After stating that the 'known' irrationals are obtained as limits of 'known' series, Brouwer goes on to the unknown (i.e., not given by a law or a definition) irrationals, which one defines as limits of unknown sequences. 'One assigns the well-known order relation to it, and only afterwards, to be able to perform the *operations*, one has to introduce the continuity postulate.'

The use of decimals may surprise the reader who is familiar with Brouwer's later paper 'Does every real have a decimal expansion?'. Evidently, Brouwer had not considered the matter at this stage. Nonetheless it is clear that he saw the necessity of admitting non-lawlike sequences. Keeping in mind that already in 1907 he had adopted the position that mathematical objects are creations of the subject, it is no *tour de force* to view the *prendre au hasard* and the unknown sequences as choice sequences.

In the dissertation Brouwer sticks to cautious formulations; there is mention of 'unknown points' (see above) without any elaboration. The reader, familiar with modern intuitionism will find at least one spot in the dissertation that reminds very much of choice sequences — the part that relates to Cantor's fundamental theorem on perfect sets. On page 65 there is a figure that, slightly anachronistically, may be called a 'fan'. And, indeed, the step from one node to a next one is made by *choice*. Hence choice sequences *avant la lettre* do appear in the dissertation, albeit anonymous. The context of Cantor's fundamental theorem makes this unavoidable; sticking to the legitimate sets of Brouwer's classification would deny all sense to the theorem. Indeed, the 'fan' is used to deal with certain subsets of the continuum, and it is just a tool to describe these subsets. Fans and spreads, as objects in their own right, are only introduced in the *Begründungs*-papers in 1918.

There is one more place in the dissertation, where Brouwer introduces spreads with arbitrary branchings, namely in the section where he deals with power sets, p.158. He explicitly accepts $C = M^A$ for finite or countable M and countable A. Of course that does not guarantee that he might not have considered definable branches in the underlying tree, but that would have been altogether out of line. In the 'Addenda and Corrigenda' (Brouwer 1917A2, p.440), the reference to this exponentiation occurs explicitly, here the term 'branching agglomerate' is used for the underlying tree. Finally, in the 'Begründungs'-papers, the basic 'set-construction' device that was to replace all his earlier conventions receives the official name '*Menge*'; perhaps not a very fortunate choice in view of the already firmly established tradition in Cantorian set theory. From Brouwer's point of view, however, the name rightly expressed that these entities were the legitimate totalities that could be grasped by the subject. The name 'spread', which has become the standard one, was only introduced in print in 1948; it was probably already used in Brouwer's Cambridge lectures (Brouwer 1981A).

Within a year after the dissertation, Brouwer returned to subsets of the continuum (Brouwer 1908A). By an analysis as sketched in the dissertation, he showed that these are denumerable or of the cardinality of the continuum. The argument, however, is not constructively acceptable, as Brouwer soon realized. In his own reprint of 'Intuitionism and Formalism' (Brouwer 1913C), he wrote in the margin that the process of weeding out isolated branches was not legitimate, as one cannot in general decide whether a branch will eventually become isolated. The point is elaborated in the subsequent 'Addenda and Corrigenda' (Brouwer 1917A2). By that time Brouwer had come to his mature intuitionistic program with choice sequences and spreads. The notion of 'deconstructible spread', i.e., the spread in which the weeding process can be carried out, returned in the second paper of the *Begründungs* series; after that the notion was not mentioned again, except briefly in Brouwer's correspondence with Fraenkel.[7]

It is quite reasonable, in the light of the existing publications and notes of Brouwer to conclude that the notion of choice sequence was already in his mind at the time of the dissertation. It had to wait, however until he found the key principle for handling functionals — the continuity principle.

3 Denumerably unfinished sets

Whereas Cantor in his treatment of ordinals boldly progressed beyond the bounds of the denumerable, Brouwer considered higher ordinals and cardinals out of bounds, as being unredeemingly beyond human creation. His 1907 set notion allowed him to introduce ω, and in general all ordinals that could be obtained by repeated (finite and denumerable) concatenation of smaller ordinals. Thus he recognized the ordinals of the second number class — but denied ω_1 existence. Nonetheless he was willing to acknowledge that we at least can speak about 'the totality of all countable ordinals', and at the same time refuse to admit it as constructible. For this kind of totality Brouwer coined the

[7] Brouwer to Fraenkel 12.I.1927, see (van Dalen 2000).

term 'denumerably unfinished'. To be precise, he defined X as denumerably unfinished, if (1) only denumerable subsets can be produced, and (2) there is for X a previously defined procedure that for every such denumerable subset A automatically produced an element of X not in A. The denumerable ordinals obviously satisfy this criteria, and so do the definable reals, and the whole of all possible mathematical systems. In the Notes, Brouwer even listed the whole of all mathematical theorems as a denumerably unfinished set. In the absence of a further specification there is not much one could say about the statement. Was it a triviality, or a foreshadowing of Gödel's incompleteness theorem? Whatever it was, it expressed Brouwer's feeling that the totality of all mathematical theorems escaped the bounds of the subject's capacity. The reader will no doubt have recognized a similarity to the notion of 'productive set', so the flavor of 'systematic incompleteness' may be recognized.

The definition seems to allow the intuitive continuum as a denumerably unfinished set, however in the examples it is not mentioned, and in the Notes Brouwer explicitly states that the continuum is not equivalent (under 1-1 mappings) to the second number class. The argument being that if the continuum were built up in a denumerably unfinished way, one would only get the 'known' points, not the 'unknown' ones. Another argument is given in (Brouwer 1913C). So Brouwer apparently saw clause (1) as taking care of the 'building up' in a denumerable way. In view of the fact that ever growing unions of denumerable sets would follow the generation of the second number class, Brouwer concluded that all denumerably unfinished sets have the same cardinality. Here one must be somewhat careful. As the equality relation on the second number class is decidable, the denumerably unfinished set of definable reals with extensional equality provides a counter example.

It is worth noting that Borel was also strongly interested in similar aspects of certain mathematical notions. He was in this sense directly influenced by Du Bois-Reymond.

The notion of denumerably unfinished did not play an important role in the further development of intuitionistic mathematics; it more or less faded away.

4 Logic

We will not go into Brouwer's criticism of contemporary mathematical schools and programs; most of it has been vindicated by developments in the logic of the last century. Instead we will see what can be said about the scarce and scattered remarks in the dissertation and subsequent papers.

The fact that Brouwer apparently had little affinity with logical research, did not mean that he did not see the importance of the subject. The Notes contain a number of remarks that show Brouwer's preoccupation with the role of logic in life and mathematics; some of these have a moral undertone, they portray logic as a lifeless superstructure of the real activities of the subject. We present a few quotations:

> 'Mathematical logic robbed mathematics from all illusion of "truth that touches life", and one finds out to have worked with nothing but a *chimera*, an illusory extract that has been "applied" to reality, but that does not touch it.'
> 'Mathematical logic is like a summarizing hypothesis, like molecular theory and mechanics in the sciences. (A human being *wants* to see fixed laws, and hence sees them).'
> 'That logical formation is only necessary, to prevent that different intuitions of two persons, should they occur, can harm the understanding. In this way the old poorly thought-out Euclidean geometry needed axioms. And that is because the people involved have often not reconstructed the *whole* system themselves, but only some parts, which are kept together by vague impressions; these should indeed be made precise by axioms.'
> 'That mathematical logic can be explained as the naive observation of a person *A who himself cannot construct* who observes *B*, who is naively building.'

The drift of all these quotes is that logic is a secondary activity, which can never replace the intuitive approach to mathematics. The more technical objections to the axiomatic method and to formalism have this same background. One may recall Brouwer's 'consistency does not guarantee existence', which is repeated over and over again in the dissertation. The criticism of the axiomatic and language based approaches to the foundations of mathematics are so well known, that we will move on to a more interesting and surprising part of Brouwer's views, that is the emergence of the 'proof = construction' doctrine. When considering Brouwer's analysis of the basics of logic, in particular the consequence relation, one should bear in mind that the most up-to-date source of logic in the dissertation was Russell's *Principles of Mathematics.* It is fairly evident he had not read Frege's logical works (in mathematical circles Frege was at the time almost exclusively known for his discussion with Hilbert, in which the overwhelming majority took Hilbert's side); Frege is not mentioned at all in the dissertation. In the Notes Frege is mentioned once in connection with the Frege-Hilbert discussion. Not surprisingly, Brouwer appreciates Frege's point:

> (Frege Jahresber. XII cf. in particular p.370, 374) is in so far right against Hilbert, that axiomatic investigations may only be understood as the sub-ordering of an intuitive structure into a more general intuitive structure. And Schopenhauer is right in that every new theorem is nothing but a new 'structure in a structure'.[8]

[8] Brouwer's 'building' is systematically translated by 'structure'.

The basic comments on logic of Brouwer are to be found in the beginning of chapter III, 'Mathematics and Logic'. He sets out to show that mathematics is independent of 'the so-called *laws of logic* (laws of argument or of human thinking)'. The point that is of interest for the history of constructive mathematics today, is that Brouwer had an alternative reading of the logical operations. Mathematical statements describe constructs of the subject, made up of the basic entities of mathematics. In this description also the relations of the various building blocks are incorporated, that is to say the relations are themselves represented as constructions (or construction instructions) in a suitable mathematical way. A successful construction of the required structure then is the proof. Brouwer's terminology is somewhat vague, he speaks of erecting mathematical structures, and fitting structures into other structures. In the letter of January 23, 1907, Brouwer explains to his PhD adviser the difference between 'logical argument' and 'mathematical argument'. The example is the simple theorem from Euclid, 'in an isosceles triangle the angles [opposite the equal sides] are acute'. The logical procedure, he says, is to represent the triangles in the plane by points in R_6, and then observe that the set of points representing the isosceles triangle, is contained in the set of points representing the triangles with (at least) two acute angles. The mathematician, on the other hand, he says, will construct an isosceles triangle and verify that the facing angles are acute.

Forgetting for a moment the actual details of the argument or of the constructions involved, the idea is that the logician simply applies the machinery of logical algebra (or for that matter, the tools of Aristotelian logic), and the mathematician carries out constructions. It may be helpful to think of the basic theorems of primitive arithmetic, where (primitive recursive) constructions are the tools to establish theorems the mathematical way.

In particular, in an implication a structure is constructed and then by a constructive procedure turned into another structure. Neither the dissertation nor the Notes provide any detailed information on the precise reading of the logical connectives; one has, so to speak, to read between the lines. For the conjunction and the disjunction that is fairly evident. In the Unreliability paper, (Brouwer 1908C), Brouwer discusses PEM in some detail. PEM demands that any fitting of a structure into another one, or its blocking, can be constructed. And so the construction for a disjunction has to yield a disjunct that allows a construction. Neither is the reading of an existence quantifier problematic: existence means 'being constructible'. The burden of the constructive interpretation of logic is carried by the implication. The 'fitting procedure' can with justification be seen as the origin of the corresponding formulation in the later 'proof interpretation'. One has to check the meaning of 'implication', so to speak, in use. The place where a logical argument in abstracto is applied, is the paper where Brouwer shows $\neg A \leftrightarrow \neg\neg\neg A$, (Brouwer 1923C1). Keeping in mind that Brouwer defines $\neg A$ as A implies a contradiction, the argument presented by Brouwer indeed fits the proof interpretation. But there it stops, that is to say, there are no elucidations that spell out the underlying argument. Indeed no such elaboration is to be found in Brouwer's writings, perhaps he considered the justification of the daily use of logic obvious, once one understands the basic principles of intuitionism. There happens to be a set of notes of a course of Brouwer in 1925, taken by Van Dantzig, in which the consequence relation is clarified. After listing the axioms for virtual ordering, Brouwer adds the comment

> The axioms *II* through *V* are to be understood in the constructive sense: if the premises of the axiom are satisfied, the virtual ordered set should provide a construction for the order condition in the conclusion.

Here we have to read 'satisfied' of course as 'being established by a construction'.

In the dissertation Brouwer specifically pays attention to the 'hypothetical judgement'; that is to say, he points out that implications $A \to B$ do not assume A to be satisfied, in the sense that one may not be in a position to carry out the construction for the structure belonging to A. Thus the traditional view that in the case of a hypothetical judgement one assumes that the structure for A is realized, offers serious problems to the constructivist. One does not accept the realization of structures without sound evidence. In the beginning of chapter III Brouwer deals with the problem in his own terminology. Unfortunately, it is not easy to fit his explanation into precise modern terms. There are (at least) two readings of Brouwer's text — a strict one, which presents awkward problems, and a more liberal one, which succeeds in justifying the standard 'proof interpretation'. The latter is presented in a penetrating analysis by Mark van Atten, (van Atten 2008); the first can be found in (van Dalen 2004).

Here is Brouwer's original text

> There is a particular case, where the chain of syllogisms has a somewhat different character, that seems closer to the usual logical figures, and which indeed seems to presuppose the hypothetical judgement of logic. This is where a structure inside a structure is defined by some relation, while one does not recognize in it immediately the means to construct it. It seems that in that case one *assumes* that the required structure was constructed, and that one derives from this assumption a chain of hypothetical judgements.
> But this is no more than apparent; what one really does in this case, consists of the following: one starts to construct a system that satisfies part of the required relations, and tries to deduce from these relations other ones in such a way that in the end the deduced relations can be combined with the ones that have so far not been used into a system of conditions, that may serve as a point of departure for the construction of the required system. Only by this construction it is shown that the condition can indeed be satisfied. (Brouwer 1907, pp.126–127)

The crucial point here is the satisfaction of the 'required system', which — in the above reading — has to be the structure for A. But that point of view would have disastrous consequences for the logic. One would have to give up the ex falso principle, and actually much more: even an innocuous implication as $A \to A$ would lose its validity. Roughly speaking, this strict interpretation is asking too much, namely that a construction for the structure of A should be provided. Van Atten's liberal interpretation provides a different reading of the above quote, it turns the question around; instead of asking for a construction for the structure of A and subsequently for B, A and B are considered as conditions on possible constructions of these structures, and the first task thus becomes to modify the conditions on A into conditions on B. The reader is referred for more details to Van Atten's paper, where a refined analysis is provided.

Whatever may be in doubt, it is certain that Brouwer was in possession of a constructive reinterpretation of logic, the one that later became codified by Heyting as the proof interpretation. In this interpretation, that became the standard one for constructive argument, the hypothetical judgement was taken care of by taking the interpretation of the implication seriously — $c : A \to B \Leftrightarrow \forall a(a : A \to c(a) : B)$, where $p : P$ stands for p is a proof of P. This definition has to be taken as a promise (obligation): the subject promises to construct a proof of B for any proof that is provided for A. Obviously $0 = 1 \to A$ is correct under this definition; any construction c will do, as a proof for $0{=}1$ will never be offered. The price to pay for this convention is that we have proofs for $0 = 1 \to A$ that in no way reflect on the meaning of A.

The purpose of the present paper is to concentrate on some points in Brouwer's early foundational work that could easily escape the reader and that, so to speak, live between the lines. It should help to understand the later intuitionistic program in the light of the program outlined in the dissertation, and vice versa. That does by no means suggest that the many points and topics brought up in the dissertation are less interesting, innovative, or rewarding. Even after one hundred years the dissertation has not lost its fascination.

Brouwerian infinity

Carl Posy

1 Introduction

Brouwer believed that we humans build the objects of mathematics, and thus he held that those objects are things that *we* finite beings can intuitively grasp. This was a problem, for mathematics is inherently infinitary (by his time infinite processes, Cantorian higher infinities and a thoroughly infinitary conception of the continuum were already at center stage), but infinite entities and infinite processes exceed our finite grasp. This dilemma — to balance infinity and human intuition — defined Brouwer's intuitionistic career.

I will distinguish three main periods in that career: (i) The period of his Doctoral Dissertation (c. 1907), (ii) the period in which he developed intuitionistic set theory (which takes us from c. 1917 through the late 1920s) and (iii) the period in which he made explicit use of 'creating subject' arguments (1948 and thereafter). In what follows I will show you that these are successive stages in an evolving strategy for solving that dilemma and that that evolution was philosophical as well as mathematical. In particular I'll point out that the first stage — couched though it be in modern terms — was fairly traditional mathematically and philosophically. The second, however, involved a profound mathematical revolution in the notions of infinity and the continuum, and I'll show that there was a corresponding adjustment in Brouwer's philosophical theory about human intuition. Then I'll suggest that the third stage was driven by Brouwer's desire to resolve some internal mathematical problems, but that this time the corresponding philosophical adjustment took him to a notion of intuition that was once again traditional (indeed, Kantian) in an important sense.

2 The Early Period

2.1 Mathematical

Brouwer's dissertation confronted head-on the three-fold use of infinity in Cantorian set theory. There is the *infinitely small*: individual real numbers given by convergent infinite sequences of rational numbers; the *infinitely large*, Cantor's second number class and higher: and the *continuum*, which is really three layers deep into infinity: an infinite set of infinite equivalence classes of co-convergent infinite sequences. Its cardinality, topology and algebra are derived from this reduction to individual real numbers. Brouwer accepted the first of these three notions of infinity — in his own way — but rejected the latter two, for his own special reasons.

He adopts the first, because, constructivist though he is, he admits ω-sequences (and of course convergent ones) as legitimate mathematical objects.[1] To be sure, for him these infinite sequences must be 'constructively given'. But having admitted such sequences, Brouwer points out that you can go very far with this tool, especially when you embed one such process in another.

But these are the only infinite objects that he does accept.[2] So in particular Brouwer rejects Cantor's 'second number class' and any whiff of a hierarchy of increasing cardinalities. Indeed, he speaks of the set of real numbers as 'denumerably unfinished'. By this he means that given a denumerable subset, we can straightaway find an element of the continuum that is not in the given subset. But this is a strictly negative result, and there is no positive existence claim to accompany it.

And as for the continuum: The Cantorian continuum — the continuum as the collection of all the individual real numbers — has a topological nature that Brouwer denies and a reductive character that he rejects as well. Topologically, the Cantorian continuum is what I call 'brittle'; it can be split into disjoint sets.

$$\Re = \{x|x < {}^1\!/_2\} \cup \{x|x \geq {}^1\!/_2\}$$

And Brouwer by contrast favors what I call a viscous continuum, one which cannot be thus split. Or more generally Brouwer held that there is no pair of disjoint non-empty sets A and B such that:

---

[1] An ω-sequence is a sequence (of discrete elements) that is generated by a countably ordered process.

[2] Here is a quote from the Dissertation on this: 'In chapter I we have seen that there exist no other sets than finite and denumerably infinite sets and continua; this has been shown on the basis of the intuitively clear fact that in mathematics we can create only finite sequences, further by means of the clearly conceived "and so on" the order type ω, but only consisting of equal elements ...' (Brouwer 1907, pp.142–143)/(Brouwer 1975, p.80)

$$\Re = A \cup B$$

By 'reductive' character, I mean that the continuum is itself the sum of individual independently generable points. And this too Brouwer denies: Though you can impose various points and constructions from points on the continuum, you cannot construct the continuum out of (or equivalently reduce it to) a set of independently given points. No set of points can exhaust the continuum. It is a unity in its own right.

> [W]e are able to state properties of the continuum as a 'matrix of points to be thought of as a whole'.
> (Brouwer 1907, p.9)/(Brouwer 1975, p.17)

> However ... a construction for it, an action that would create ... 'all' its points as individuals, is inconceivable and impossible. (Brouwer 1907, p.62)/(Brouwer 1975, p.45)

From these considerations Brouwer proposed his own alternative to the Cantorian hierarchy (Brouwer 1907, p.149)/(Brouwer 1975, p.83):

1. the different finite numbers;
2. the denumerably infinite;
3. the denumerably unfinished;
4. the continuous.

And he discussed how various algebraic and measure properties can be derived from it.

2.2 Philosophical

But what are the constructivist philosophical doctrines behind these mathematical views? Some people use the metaphor of 'inside' versus 'outside' perspective here. The forbidden infinities presuppose an 'outside' point of view, they say.

This is a nice metaphor, but as I said we want to tease out a more specific philosophical doctrine about the appropriate notion of 'intuitive grasp'. And indeed each of Brouwer's mathematical positions is in fact underwritten by a corresponding philosophical principle about the nature of intuitive grasp.

Constructively given sequences: Brouwer's notion of a constructively given sequence is quite precise: It is what we now call the intuitionistic reading of the $\forall\exists$ quantifier combination in $\forall x \exists y \varphi(x, y)$ where φ is a linear (perhaps discrete) order. According to this reading, $\forall x \exists y \varphi(x, y)$ means that given x we know how to find (construct) a y such that $\varphi(x, y)$ holds. So a constructively given infinite sequence is one in which $\forall n \exists y \varphi(n, y)$ holds, where n ranges over the natural numbers.

Rejection of the higher infinite: But this is not enough to deny existence to the second number class and other putative large objects. In order to get that negative result, Brouwer needs to assume an additional doctrine. Brouwer hints at the doctrine I have in mind when he suggests that our mention of large infinite classes is in general a mere *façon de parler*.[3] In saying this, Brouwer is expressing the important philosophical distinction between an *object* on the one hand and a '*mere collection*' a '*mere aggregate*' on the other.

This is a distinction that goes back at least to Leibniz, who spoke of entities which we might like to refer to as objects, but which in reality are not objects at all, but mere phenomena or 'well founded fictions'. Leibniz considered his 'monads' to be real objects and physical bodies to be fictions. Brouwer however is using this device to say that finite numbers and constructively given denumerable sets are objects, while the (Cantorian) collection of all real numbers is no more than a convenient fiction.

But, why is that? Well, Brouwer says, that latter set is 'unfinished'. Clearly the former are 'finished' or 'finish-able'. We've seen the mathematical content of his notion of unfinished: Given any subset that you can denumerably construct, you can straightaway get a part that is not in that subset.

So it must be that a finished set is one in which you have some legitimate grasp of all of its parts. Not necessarily all at once; for, some infinite sets are graspable. So what Brouwer must intend is that there is a *recognizable process* that will deliver the entity with *all its parts*.[4] (Hence I spoke of 'finish-able'.)

So Brouwer's underlying notion of intuitive grasp is: a grasp of something which provides in turn a grasp of each of the parts (or at least a means of attaining a grasp of each of the parts). We can say that the parts are 'determined' or 'determinable' by the initial grasp. There is a straightforward duality here: A legitimate object is intuitively graspable, an intuitive grasp presents you with a legitimate object.

An unfinished collection is one that we cannot grasp in a way that suffices to determine all the parts. A collection which is indeterminate in this way is not a legitimate object. And that is what the second number class is like. We might speak in the singular about such a collection — I just did, didn't I — but this, as I said, is for Brouwer no more than a *façon de parler*, 'an expressed intention'. The Cantorian continuum, the second number class and beyond, these are not intuitively graspable and for that reason they are not legitimate objects.

The Intuition of the Continuum: But of course Brouwer thinks that we do intuitively grasp the continuum as a whole. And he believes that this grasp is the basis for the mathematics of the continuum. Thus he is led to posit a separate and primitive form of intuition, which delivers the continuum as a whole and which generates the mathematics of the continuum. This is sometimes called an '*Ur-intuition*'.

---

[3] 'We can introduce these words as an expression for a known intention.' (Brouwer 1907, p.148)/(Brouwer 1975, p.82)

[4] I say 'parts' here, and not 'elements', intentionally.

Moreover, we grasp the continuum as 'viscous', non-separable. And Brouwer here uses the notion of 'fluidity' to represent what I have called viscosity. Indeed here is the complete version of the Dissertation passage with which I ended the preceding section:

> Having recognized that the intuition of 'fluidity' is as primitive as that of several things conceived as forming a unit together, the latter being at the basis of every mathematical construction, we are able to state properties of the continuum as a 'matrix of points to be thought of as a whole'. (Brouwer 1907, pp.8–9)/(Brouwer 1975, p.17)

> However the *continuum as a whole* was given to us by intuition; a construction for it, an action that would create from the mathematical intuition 'all' its points as individuals, is inconceivable and impossible. (Brouwer 1907, p.62)/(Brouwer 1975, p.45)

So in sum in the Dissertation we find Brouwer embracing four main mathematical views: the constructive notion of convergent sequence, the denial of higher infinities, the viscous topology for the continuum, and the rejection for any reductive (set-theoretic) construction of the continuum. And philosophically parallel to these we find his idea of the constructive grasp of an infinite process, his claim that an intuitive grasp must completely determine its object, and his positing of the separate Ur-intuition of fluidity in order to account for the mathematics of the continuum.

Let me add, as I mentioned at the outset, that the idea of an irreducible, viscous continuum, that is given by a separate intuition of 'flowing' — this too is a venerable package of doctrines. Here I would cite Kant, rather than Leibniz, as the forerunner. Thus:

> Points and instants are only limits, that is, mere positions which limit space and time. But positions always presuppose the intuitions which they limit or are intended to limit; and out of mere positions, viewed as constituents capable of being given prior to space or time, neither space nor time can be constructed. Such magnitudes may also be called *flowing*, since the synthesis of productive imagination involved in their production is a progression in time, and the continuity of time is ordinarily designated by the term flowing or flowing away. (Kant 1965, A169–70/B211–12)

3 Choice sequences

3.1 Mathematical

Brouwer's second — often called mature — intuitionistic stage starts from publications from 1917. (Though there were relevant lectures that took place a bit before that.) To be sure, Brouwer still clung to three of his four main mathematical doctrines: the active, constructive grasp of denumerable infinity; the rejection of higher infinities; and the topological viscosity of the continuum. The striking mathematical change is that Brouwer now adopts the reductive conception of the continuum: He is prepared to allow that the continuum is indeed composed of independently generated real numbers. And I emphasize that he makes this change without abandoning his commitment to the 'viscous' topology of the continuum. This is a mathematical achievement of the first rank, involving a radical new way constructively to construe infinite sequences and infinite sets of these.

Brouwer's main mathematical moves in achieving this *coup* — choice sequences, spreads and the uniform continuity theorem — are by now familiar notions, well studied in the literature on intuitionism: The theory of *choice sequences* allows infinite sequences whose elements may be randomly generated or even freely chosen; so long as we are guaranteed that the sequence of elements can be continued without stop. The general notion of a *spread* of such sequences — given by a spread law and a complementary law — is well known. And so too is the representation of the unit continuum $[0,1]$ as a finitary spread (a fan) consisting of canonical real number generators (e.g., decimal expansions).[5] And Brouwer's *Uniform Continuity Theorem* ('Every total function on $[0,1]$ is uniformly continuous') is a famous — if not notorious — consequence of these basic notions.

Of course, the viscosity (un-split-ability) of the continuum follows immediately from the continuity theorem: For if we could perform a clean split of

[5] Here are some of the technical notions:

- A *choice sequence* α is given by setting out some preset finite initial segment $\langle\alpha(1),\alpha(2),\ldots,\alpha(n)\rangle$ together with a growth rule which, given $\langle\alpha(0),\alpha(1),\ldots,\alpha(n),\ldots,\alpha(k)\rangle$ determines the range of possible choices for $\alpha(k+1)$.
- A *spread* is a set of rules for admissible finite sequences of natural numbers such that each such admissible finite sequence $\langle n_1,\ldots,n_k\rangle$ has at least one admissible successor $\langle n_1,\ldots,n_k,n_{k+1}\rangle$. These rules constitute what is called the *spread law.*
- A finitary spread (fan) is a spread such that for each admissible sequence there are only finitely many admissible continuations.
- An infinite sequence of natural numbers will belong to the spread if each finite subsequence does.
- We get infinite sequences whose elements are more sophisticated objects by assigning such an object to each admissible finite sequence. The complementary law gives the means of making these assignments.
- An infinite sequence whose elements form a convergent sequence of rational numbers is a real number generator.
- The unit continuum $[0,1]$ can be represented as a *fan* of real number generators such that each generator consists of a sequence of rational numbers whose denominators are powers of 2.

Heyting's book (Heyting 1971) is still an excellent source for learning the basic notions of intuitionistic mathematics.

the continuum, say at $1/2$, then we could equally well define a discontinuous total function: $f(x) = 3$ for $x < 1/2$, $f(x) = 4$ for $x \geq 1/2$.

To see the thrust of Brouwer's new mathematical thinking here try to define f at a real number, r, whose decimal expansion is generated in the following way: Suppose that the first $n + 2$ places in the decimal expansion are $0, 4, 9, \ldots, 9$ (so $r = 0.499\ldots$); and then suppose from that point onward we throw a 10-sided die to determine the subsequent digits. (Or if you don't like that,[6] then simply say that we freely choose the digits in that decimal expansion. Towards the end, I'll say something about whether this 'free choice' option helps.) In any event, we do know that this sequence generates a real number, r; and we know that r lies in the interval $[0, 1]$. We can say that r is an element of the decimal (= ten-branching) finitary spread that expresses this interval. At the time that we define r, we don't know where in the interval it lies (indeed, we may *never* know *exactly* where it lies). We may think of r as a generic (or 'relatively lawless') element of the spread. So Brouwer will claim that $f(r)$ is undefined. And that scotches your clean split as well.

Actually, since any option other than 9 for any one of the digits after the decimal point already determines that $r < 1/2$, we can make do with the simpler binary spread of Figure 1. (In this case the solid circles are shorthand for an un-branching straight line.)

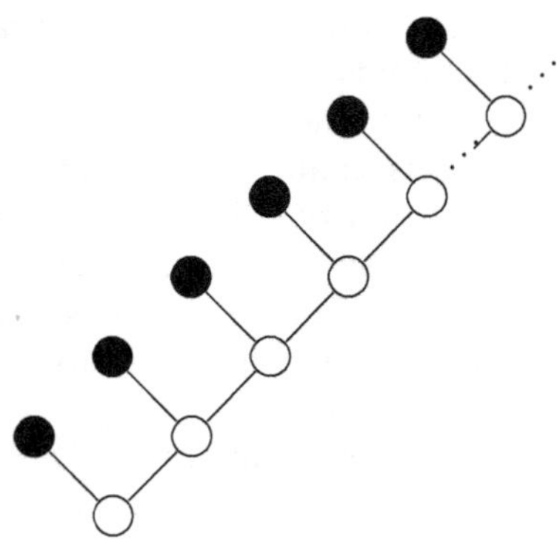

FIGURE 1

But in any case, this is a nice sense of viscosity. Speaking *mathematically*, this is Brouwer's coup: a viscous continuum which is nonetheless reducible to individually generated real numbers.

[6] According to Hans Freudenthal (Brouwer 1981A, p.xi), Brouwer himself objected to the die-throwing image. Freudenthal thought that Brouwer's objection here stemmed from a discomfort with the empirical aspect of die throwing.

3.2 Speaking philosophically

Let me say that Brouwer no longer appeals to the notion of a special intuition of 'flowing' to underly the mathematics of the continuum. He no longer needs to.[7] That is a radical, but readily explainable change. But there are more subtle, but no less important realignments in his other philosophical doctrines as well.

Constructive denumerable infinity: Here we have to note that if Brouwer intends his constructive reading of $\forall n\exists y\varphi(n, y)$ in particular (and thus $\forall x\exists y\varphi(x, y)$ in general) to cover choice sequences, then he must now employ a broader conception of existence. Specifically, up till now we have assumed (and Brouwer has assumed) that in order to assert the existence of something, you have either to have the thing in hand, or at least you have to have a mathematical method for constructing the thing in question. This is what I called a 'recognizable process' in section 2.2 above. But this is too strong. We might well not have a recognizable mathematical method for determining the elements of the choice sequence that generates r. So Brouwer now needs to weaken that requirement. Instead of calling for having an actual mathematical construction of the entity whose existence is being asserted, he now must say that it suffices that we have a guarantee that we can eventually produce such a construction. Nowadays we call this a mild assertabilism: In order to assert $\exists xP(x)$ it suffices to have a guarantee that we can eventually find an object c such that $P(c)$ holds.

There is a subtle point here: r is indeterminate in many ways: Most prominently we don't know whether it is smaller than or equal to $1/2$; we don't even know whether we will ever find that out. But still, r counts as a legitimate object. And I should add that this weaker requirement in no way diminishes the ontological status of the things whose existence is thus established. Thus, for instance, r and its ilk can be arguments (and even values) of functions. They are embeddable, full fledged objects.[8]

Determinacy and Intuitive Grasp: Brouwer's mode of thinking here, leads us to say that infinity is the real source of indeterminacy for him. $f(r)$ and its ilk make this precise. To say that the function f is not defined at the argument r, is to say that $f(r)$ is not (or not yet) a legitimate object at all. And that of course is because we do not have a fully determinate grasp of it with all its parts. (Indeed it is the first element that is so far missing.) But, given Brouwer's mild assertabilism, he can only maintain this negative attitude

[7] In some sense Brouwer's notion of the Ur-intuition may well remain. (See in particular (van Atten 2007, pp.34–35).) But at this point one must separate his allegiance to a viscous continuum, on the one hand, which might indeed be based on phenomenological grounds, from any technically mathematical use of that phenomenology on the other. My point is that there is no longer a need for such a source within his mathematics proper.

[8] Spreads are embeddable too — though Brouwer ultimately finds this fact superfluous. See (Brouwer 1942C).

towards $f(r)$ in circumstances which require a possibly infinite process. If $f(r)$ were to depend upon a recognizably finite process, then we could simply wait around until that process is completed and after that see what there is to see about it. It would be no different than determining the parts of r itself.

So we see on the philosophical side that in his mature period Brouwer dropped his appeal to the primordial intuition of flowing — a deep philosophical change attendant to his new conception of the continuum — and he also refined his philosophical doctrine about construction and infinity as part of the same realignment as well.

4 Theory of the Creating Subject

Now in 1948 we find yet another overt change in Brouwer's published work: The introduction of his theory of the 'creating subject' into his mathematical arguments. This is a method of generating infinite sequences, tied to the stages of solving unsolved mathematical problems. Though Brouwer thought about this method earlier in his career,[9] from 1948 onward he explicitly introduced it into his mathematical arguments.

Now what led Brouwer suddenly to introduce this method into his published works? I will suggest that he was led to this in order to address a particular difficulty about his mature notion of infinity and choice sequences.

4.1 A Complication in Brouwer's Thought

The real number r that I described above was defined in order to help prove that the continuum is unsplittable (viscous, as I have put it); or equivalently in the proof of his 'Negative Continuity Theorem' it is to show that $f(r)$ *cannot be intuitively grasped*. I.e., it is contradictory to assume that f is defined at r.

But notice that r, as I have described it, seems actually inadequate to this task. For, it remains open whether any digit other than 9 will occur in the decimal expansion of r. If that happens, of course, then $r < 1/2$, and $f(r)$ is clearly defined.

This is not the complication I have in mind. For, Brouwer apparently took this into account in the 1927 paper in which he gives a formal proof of the 'Negative Continuity Theorem' (Brouwer 1927B). That paper contains a more complicated definition of a real number, r^*, a number that is designed to do the job. r^* is defined as follows:[10]

> $r^* = 0.4d_2d_3 \ldots d_n \ldots$ The remaining digits after the 4, $\{d_{i+1}\}_i$, are chosen ... by means of an unlimited sequence of choices such that (a)

[9] See (Posy 1980), which suggests that the method of the creating subject was tacitly at work in Brouwer's thought at least as early as 1927.

[10] I have modified the quotation to make it consistent with the treatment I am giving here. The changes preserve the sense of Brouwer's definition.

> we temporarily choose for every k that we have already considered $d_2 = d_3 = \cdots = d_k = 9$. But (b) we reserve the right *(die Freiheit vorbehalten)* to determine at any time after the first, second, ... $(j-1)$st, and j-th choice have been made the choice of all further elements [i.e., $(j+1)$st, $(j+2)$nd, and so on] in such a way that either (i) $d_k = 9$ or (ii) $d_k < 9$.

r^*, like r, is clearly in the interval $[0, 1]$, but Brouwer is adding further conditions to r^* designed to *guarantee* that $f(r^*)$ is not intuitively graspable, and hence undefined.

But the definition is unclear. Let me tell you about two prominent interpretations of the passage. One makes the guarantee too strong (and thus self defeating), the other is too weak and disingenuous. It also will conflict with other things that Brouwer says.[11]

The first reading: In some of my own papers I have said that Brouwer wants to *force* r^* to be even more indeterminate than r; so that we can *never* say whether r^* is equal to $1/2$ or smaller than $1/2$.[12] This is what is called a second-order restriction: you must never make a choice which restricts further freedom of choice. And this is, as I said, self-defeating. It means that only 9 can be chosen. (For, once again, if you were to choose any thing else as a digit of r^*, then you would know that $r^* < 1/2$, and $f(r^*) = 3$.)

Effectively this puts r^* on the (spinal) rightmost path of the spread in Figure 1, and guarantees that $f(r^*) = 4$, and thus is intuitively grasped. If you like the metaphor of 'inside versus outside' points of view, then you can say that the 'outside point of view' is the one in which one knows whether or not a digit other than 9 will be chosen. The problem with this definition then is that in forcing the indeterminacy, it improperly mixes the inside and the outside points of view.[13]

The second reading: This reading of the definition of r^* uses the notion of 'provisional' second-order restriction, that has been developed by Dirk van Dalen and Mark van Atten:[14] The idea is that you start with the second-order restriction I mentioned just now ('don't break the indeterminacy') but *you reserve the right to remove it at any later point.* In this case, you are not caught in the self-defeating bind.

---

[11] There are other interpretations of this argument of Brouwer's with even more problems.

[12] See, for instance, (Posy 2000).

[13] I point out in (Posy 2000) that this reasoning is actually a version of the Hangman paradox. I show there that there is a fuller theory of forced indeterminacy at work here. In particular it is worth noting that the problem of forced indeterminacy occurs even in finite contexts.

[14] See (van Atten and van Dalen 2002). See also (van Atten 2007, pp.108–109). Van Atten has informed me that though his reading of this passage in Brouwer does indeed involve a second-order provisional restriction, it does not quite coincide with the second reading of the passage that I am suggesting here. We will explore this matter further on another occasion.

But this seems, as I said, too weak a definition. For, it remains possible that we will make choices — second-order[15] or first-order choices — that fix the value of $f(r^*)$ and again undermine the counter-example. We can say that we *don't currently have a grasp of* $f(r^*)$; we cannot say that we *can't have such a grasp.* What's more, this is introspectively disingenuous: We know full well that we want $f(r^*)$ to be undefined, and won't forget that fact in later 2nd and 1st order choices about r^*.

Actually, let me point out that if in these weaker circumstances you really do want to say that the function *cannot be defined at* r^*, then you need to adopt the following strict definition of what it means for a function to be defined at a point x:

A function, g, is defined at x only if for each k there is an initial part of the decimal expansion of x (say of length j) such that one can calculate in a finite time the first k elements in the decimal expansion of $g(x)$ using the first j digits in the decimal expansion of x, and the initial descriptions of g and x.

The point is that the value $g(x)$ must be deducible from a finite amount of information about x and from the definition of g alone. However in a subsequent research note, Brouwer quite explicitly loosens this notion of definability for functions: He allows real-valued functions which themselves 'develop freely' (Brouwer 1942A). His example is an infinite sum the sign of whose summands is freely chosen.

$h(x) = \sum \pm \frac{x^n}{n!}$ where for each n the sign preceding the expression $\frac{x^n}{n!}$ is freely chosen.[16]

Once we allow functions like this to be defined we can't require that in order for $f(x)$ to be defined its value must be deducible from the definition of f alone.

Here then is a more usable condition for the definability of a function:

> *A function, g, is defined at x* only if for each k there is an initial part of the decimal expansion of x (say of length j) such that one can calculate in a finite time the first k elements in the decimal expansion of $g(x)$ using the first j digits in the decimal expansion of x, the initial descriptions of g and x, plus additional information that is given within a finite time but which may be independent of the formal definition of g.

But if this is Brouwer's general conception of definability, then once again, he will need a strong reading of r^* to serve as the counter-example for which the function f is not defined. And, once again, the second reading could not work.

[15] We might, in particular, lift the provisional restriction.

[16] Moreover, in that 1942 paper Brouwer goes on to say that though in his published work he had until then always confined himself to examples of law-given functions, nevertheless in his lectures and discussions he had made it clear that he believed the class of definable functions must include those that are freely developed.

This is the deep complication to which I alluded above. On the one hand, we want to assure a non-trivial degree of indeterminacy, but, on the other hand, we cannot explicitly force it. I have emphasized the 'Negative Continuity Theorem' and the definition of r^*, because this difficulty is especially prominent there. But this is a general problem facing second-order restrictions on freedom. Indeed in his later work Brouwer explicitly disparages any use of second-order restrictions:

> In former publications I have sometimes admitted restrictions of freedom with regard also to future restrictions of freedom. However this admission is not justified by close introspection and moreover would endanger the simplicity and rigour of further developments. (Brouwer 1952B, p.142)

4.2 The Creating Subject

As I said, I think that Brouwer's notion of the 'creating subject' serves as his solution to this difficulty. Though we know that he thought about it before then, Brouwer introduced this notion explicitly only in 1948. In this theory Brouwer defines a choice sequence according to the stages of solving a mathematical problem, e.g., proving Goldbach's conjecture. The 'creating subject' stands for an idealized mathematician engaged in solving that problem. (For that matter, we could think of the creating subject as a suitably idealized version of the entire mathematical community, and focus simply on efforts throughout that community to solve this problem.) The main point is that we regularly check the progress of finding the proof, and based on the outcome of these checks we define a real number, $r^{\#}$ as follows:

$r^{\#} = 0.499..d_n, d_{n+1}, \ldots$, where

a. The n-th digit in the decimal expansion of $r^{\#}$ is 9 if by the n-th stage, neither a proof nor a refutation of the conjecture has been found.
b. The n-th digit in the decimal expansion of $r^{\#}$ is 0 if $n \geq k$ and at stage k, the problem has been solved. (I.e., at stage k either a proof or a refutation of the conjecture has been found.)

Symbolically:

$$\neg \vdash^G_n G \vee \neg G \quad \Rightarrow d_n = 9$$

$$\vdash^G_n G \vee \neg G \;\&\; n \geq k \Rightarrow d_n = 0$$

(where $\vdash^G_n P$ holds when P is known at the n-th stage in the progress of the Creating Subject who is working on proving or refuting proposition G)

This method of defining sequences has many mathematical advantages. Brouwer, for instance, uses it to derive outright contradictions from classically valid but intuitionistically invalid principles, and Van Dalen has gotten some nice results about the fine-structure of the intuitionistic continuum using aspects of the theory of the creating subject. But I want now to focus on just three aspects of this new way of defining indeterminate real numbers:

First: The method of the creating subject will give Brouwer a reasonable means to state his 'Negative Continuity Theorem' and reconcile it with his later research note: To say that f is fully defined is to say that for any given x the descriptions of f and of x suffice to guarantee that $\exists y(f(x) = y)$. Or, put formally it is to say:

$$\vdash_0^f \exists y(f(r^{\#}) = y)$$

even with the broader notion of when a function is defined that we find in Brouwer's later note. Then even for plain old r, we can say

$$\neg \vdash_0^f \exists y(f(r^{\#}) = y)$$

And that is a strong negation.

Second: This method now gives a deeper insight into Brouwer's mature notion of infinity. I'll show you this in two stages:

a. Notice there is a series of which we cannot say that it is infinite or that it is finite: the series of indices in the decimal expansion of $r^{\#}$ at which a 9 occurs. We cannot say that it is finite, because we don't know that the problem will be solved. And we can't say that it is infinite — we don't know that it won't be solved. Indeed, if we add the principle of 'Christian charity',

$$P \Rightarrow \neg\neg\exists n \vdash_n^P P$$

which says that there are no true propositions which cannot be proved, then we make the even stronger claim: the series of 9-indices in the decimal expansion of $r^{\#}$ is *not* infinite: For we can safely say that we will *never* know that this series *is* infinite. (We can never assert $\forall n(d_n = 9)$: for, to assert that is to say that the problem will never be solved. But that means we know that Goldbach's conjecture is false, and thus have solved the problem.) So, again, the series of 9-indices in the decimal expansion of $r^{\#}$ is not infinite; but we cannot say that this series is finite.

b. Of course it could work out that we do find out that the series is finite. (We might in fact solve the problem). But here's an elaboration: Suppose that we do come to solve the particular problem at hand, and arrive at a 0 in the decimal expansion of $r^{\#}$; let us in this situation supply a new unsolved problem and continue the process further. And suppose we continue on in this way, turning to a new unsolved problem every time we happen to solve one. We are generating a new real number $r^{\#\#}$, and we get a diagram like Figure 2.

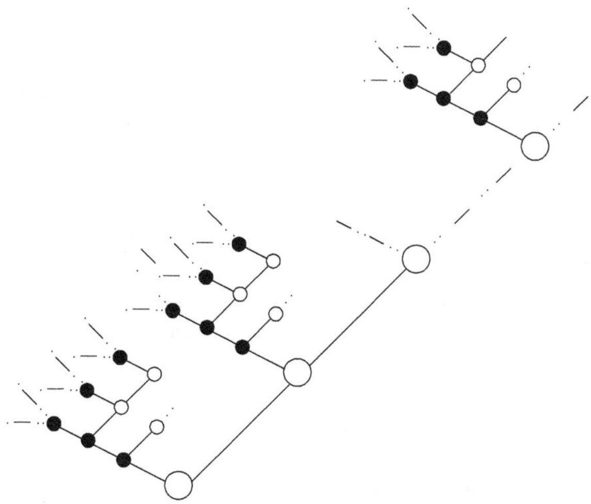

FIGURE 2

$r^{\#\#}$ is a path in this diagram. Now consider the series of 0-indices in the decimal expansion of $r^{\#\#}$. This series *cannot* be infinite. For, to say that it is infinite — i.e., to say that $\forall n \exists k > n(d_k = 0)$ — is to say that we know in advance that every problem we encounter will eventually be solved. This is something that Brouwer would vehemently deny. Indeed, early on and constantly he opposed Hilbert's optimistic claim that every mathematical problem must be solvable.

But on the other hand, this series of 0-indices in the decimal expansion of $r^{\#\#}$ *cannot* be finite either. For at no point can we assert that $\exists n \forall k > n(d_k = 9)$. This is because, as above, at every point we hold open the possibility that the problem on which our subject is working at that point will be solved.

So with respect to $r^{\#\#}$ we can say quite strongly: The set of 0-indices in $r^{\#\#}$'s decimal expansion is a set that *cannot be finite*, and it also *cannot be infinite.* Indeed, if you look at Figure 2 as a Kripke model, and number the solid nodes by their distance from the origin, then you can see that the model validates: $\neg \exists x \forall y \neq x(Fxy)$ and also $\neg \forall y \exists x F(xy)$.

But, formal invalidation aside, we must ask why something like Figure 2 can be a legitimate model to trace our knowledge. The model rests, after all, on Brouwer's controversial anti-optimism. And my *third* point is that his theory of the Creating Subject actually can explain that anti-optimism in an historical context. For this theory emphasizes that it is not only infinity that leads to indeterminacy: it is infinity plus our *dependency.* I can put myself in a position to see whether I am on a finite path in Figure 1 — to see whether the problem will be solved — but I can do no more than that. I have to wait and see if the problem gets solved, even though it is my own progress that I am tracking.

And the historical context for this is Kant once again. Kant emphasized that this 'dependency' — he called it the *receptivity* of human intuition — is the true source of indeterminacy in our knowledge. It is precisely when what he calls 'the key' to answering a question does not lie within us — when we must wait to see whether an answer will come — it is precisely in this situation that we must consider the possibility of an unending wait. And that is the source of indeterminacy.[17] To be sure, Kant, himself, endorsed the Hilbertian optimism in mathematics; for, he thought that in mathematics the 'key' to each problem is indeed within us. For him mathematical knowledge is not 'receptive' or 'dependent' at all.[18] But Brouwer here emphasizes that even when we are tracking our own *a priori* research, still we are dependent, we have to wait for the outcome! So to return to my earlier point about the 'dice throwing' model of choice sequences: The method of the creating subject may well eliminate the *physical* contingency that this model suggests, but it also shows that in the end we cannot eliminate the contingency altogether (or as I have called it 'dependency') that is expressed by the 'dice throwing' image.

So in mathematics for Brouwer, just as in empirical research for Kant, because our knowledge is 'receptive' we cannot expect every question we ask to be answerable. And thus, as I said, Brouwer's most *avant garde* conception of infinity rests ultimately on a traditional notion of receptive intuition.

One final remark: I am not claiming that Brouwer was directly influenced by Kant on this business of receptivity. The stages I have described in Brouwer's developing thought may be no more than an internal dialog and evolution. It is likely that these two thinkers, Kant and Brouwer — each in his own way — came to see the connection between receptivity and indeterminacy on their own. This is I believe tells us that this connection is profound

[17] 'In natural science ... there is endless conjecture, and certainty is not to be counted upon. For the natural appearances are objects which are given to us independently of our concepts, and the key to them lies not in us and our pure thinking, but outside us; and therefore in many cases, since the key is not to be found, an assured solution is not to be expected.' (Kant 1965, A480–481/B508–509)

[18] '[T]here are sciences the very nature of which requires that every question arising within their domain should be completely answerable in terms of what is known, inasmuch as the answer must issue from the same sources from which the question proceeds. In these sciences it is not permissible to plead unavoidable ignorance; the solution can be demanded.' (Kant 1965, A476/B504); and 'It is not so extraordinary as at first seems the case, that a science should be in a position to demand and expect none but assured answers to all the questions within its domain (*quaestiones domesticae*) although up to the present they have perhaps not been found. In addition to transcendental philosophy, there are two pure rational sciences, ..., namely, *pure mathematics* and *pure ethics*.' (Kant 1965, A480/B508)

and will emerge in any deep thought about the balance between intuition and infinity.[19]

[19] I am grateful to the editors for their patience and support. And I am particularly indebted to Mark van Atten for a number of very helpful comments on an earlier draft of this paper.

The new intuitionism

Charles McCarty

1 On Denoting

I confer the title 'intuitionism' and its cognates, here and throughout, upon intuitionistic mathematics solely, the mathematics inaugurated officially by L.E.J. Brouwer but prefigured in the writings of Paul du Bois-Reymond and Émile Borel. I do not use that word to designate any philosophy or philosophies of mathematics that might, at one time or another, have danced attendance upon the mathematics. Nor do I use 'intuitionist' as Felix Klein once did, in his Evanston lectures of 1893 (Klein 1911), to denote a mentality prompting certain mathematicians to favor geometric visualization over algorithmic calculation or logical analysis. Even with that delimitation, I freely admit that the definite article in my title is inappropriate. In the proclamation of another 'new' intuitionism, there is little new. Since the day Klein bestowed the appellation, some mathematicians and philosophers, among them Brouwer, have trumpeted the presumptive advantages of the very latest, the very 'newest' intuitionism. Surveying the foundational trends of the past forty years, one espies a number of developments, each modish in its day, each then meriting the designation 'new intuitionism:' the proof-theoretic intuitionism of Georg Kreisel, the category-theoretic intuitionism of William Lawvere, the intuitionistic type theories of Per Martin-Löf, the constructivism of Errett Bishop and that of Paul Lorenzen.

In what you are about to read, there is no new mathematics. You will find no novel theorems or proofs. I doubt that any specific mathematical result, work or research project now marching to the intuitionistic tune will be elbowed out of bounds by this writing. You will find instead a take on the subject, an order of affairs, an attitude. You will also find a new demand — or at least a demand insufficiently pressed of late: that we intuitionists take theorems Brouwer proved, especially his proofs that the *tertium non datur* is invalid, very seriously, not simply as specimens of antiquarian interest, but as valves and chambers in the heart enlivening a body of real mathematical truth.

As for that new order of affairs, I wish to honor one of Brouwer's most fecund insights into intuitionism. According to Brouwer, full engagement with intuitionism calls for two foundational acts or *Handlungen* (Brouwer 1992). The first act is, in the terms of my own idiolect, the recursivization of the mathematics and logic of certain discrete domains, domains with decidable equality. This first act incorporates, but is not exhausted by, the elementary study of the natural, integral and rational numbers. Upon this landscape Kronecker would have smiled. I suggest that, as a theory, the first of Brouwer's acts is blood-brother to Markovian constructivism with (perhaps a weak form of) Church's Thesis, but without Markov's Principle, i.e., without the assertion that Σ^0_1 numerical predicates are stable. The second of the two acts has its apotheosis in Brouwer's characteristically intuitionistic approach to higher-order domains without decidable equality, exemplified by his celebrated handling of real-valued functions and culminating in the Uniform Continuity Theorem. In (Brouwer 1992, p.23), Brouwer linked the second act to *Mengenkonstruktion*, the construction of intuitionistic sets or *spreads*. In the vernacular, I would describe the second act as a non-standard extension of a strictly constructivistic or Markovian universe. Incidentally, Brouwer's own marginalium to the manuscript of his Cambridge lectures, preserved as a footnote to page 93 of (Brouwer 1981A), seems to recommend that the second act be viewed as an elaboration of the first. This note, if rightly interpreted here, does not disconfirm the rough-and-ready account of the two acts just sketched.

For good foundational reasons, I choose to pay tribute to Brouwer's acts by reversing the order he imposed on them, by which the decidable comes first, the undecidable second. Of course, you could retain his original ordering as mirroring the timeline of historical discovery or the evolution in his own thinking. But his order is not thereby privileged: it is not the royal road to intuitionism. I deem the opposite order, starting with the undecidable, most appropriate to theoretical progress and scientific explanation. This means that I begin my own mathematical work not with elementary logic and arithmetic, but with higher-order logic and set theory. Accordingly, sets or properties get formed freely and impredicatively. The investigation of full powersets like the set of all truth-values (or propositions) is not just permitted, it is encouraged. This rich mathematical country of mine lies well beyond the lands of the strictly predicative, beloved of Martin-Löf and those resident on the old Stanford-Munich meridian. An intuitionist has now stolen into a domain once ruled by Dedekind.

2 Some Main Problems of Philosophy

In public lectures, when I mention that I conceive of Brouwerian intuitionism primarily as mathematics and only secondarily, if at all, as philosophy, critics respond — at times indignantly — 'We all know that intuitionism is a classic "ism". So, it is a philosophy of mathematics, and hence philosophy, not mathematics'. I reply that cobordism is a famous 'ism', but hardly a philosophy. Furthermore, if what the critics say is true, they should tell me precisely *which* philosophy or philosophies, of the several distinct possibilities, intuitionism is. Is it the idiosyncratic idealism of Brouwer? Or is it the eclectic mix of philosophies favored by his more pragmatic student, Arend Heyting? Or is it anti-realism in one of its several popular versions? Or is it none of these? Certainly, I would not count every philosopher an 'intuitionist' who held, once upon a time, that we know some mathematical proposition thanks to an exercise of an intuitive faculty. Nor do I so classify every thinker, Aristotle and Paul du Bois-Reymond among them, with the gumption to dispute the law of the excluded third.

Savants agree that intuitionism is not always well served by its philosophical handmaids. Brouwer, Heyting and Dummett were wrong, and not in some ancillary fashion, about intuitionism. It is not merely that the murky idiom of Brouwer's writings cast an intellectual pall over intuitionism so dark and lingering that its prospects still seem dim to many. (A prominent logician — a man who should know better — once leaned over the dinner table and whispered to me, 'The real problem with intuitionism is that your jokes — your theorems — take too long to tell!' I replied, also *sotto voce*, 'Not at all. We intuitionists learn to laugh more slowly.') The predicative strictures Brouwer imposed on *Mengenkonstruktion* once veiled the intuitionistic vision. Few would concur with Heyting's insistence that '$26+38=64$', issuing assertorically from my mouth, is an empirical claim about me and the arithmetical doings of my past, viz., that I have calculated the sum of 26 and 38, and confirmed it to be 64. For I cannot decide to make the false conditional sentence

If $26 + 38 = 64$, then $0 = 1$

true by neglecting to carry out personally the addition specified in its antecedent.

As for the contemporary anti-realist, I would be intrigued to see, coming from her pen, a non-circular, strictly mathematical proof, from acceptably anti-realistic principles, of the invalidity of the law of the excluded third. There was once a deal of buzz in the halls of philosophy departments about the power of anti-realistic semantics to overturn conventional logical laws. I would love to have a knockdown proof that it does so. Simple metamathematics establishes that such proof will not be forthcoming from intuitionistic set theory plus the Axiom of Choice plus the Continuum Hypothesis, for these together are compatible with the *tertium non datur*, should ZF set theory in classical logic be itself consistent.

Thanks to Gödel, Kreisel, Leivant and others, logicians know that the soundness and completeness of intuitionistic predicate logic with respect to Tarski's or Kripke's semantics implies Markov's Principle (Kreisel 1962a). From this, some conclude that there is no hope for a completeness proof for formal intuitionistic predicate logic from purely intuitionistic assumptions, as anti-realists once demanded (Dummett 1978). In reply, I would point first to the striking completeness theorems, admittedly with respect to Kripke models amended to force logical contradictions, obtained by Veldman (Veldman 1976) and De Swart (de Swart 1976). Second, I would argue that it is now much too early in the intuitionism game to pronounce with any finality upon the truth of Markov's Principle. Markov's Principle is consistent with intuitionistic set theory plus Extended Church's Thesis, the Uniformity Principle, the Presentation Axiom (Aczel and Rathjen 2001) and Brouwer's Continuity Theorem in the form that every total, real-valued function of a real variable is continuous. Ergo, I would now be loath to spurn a successful anti-realistic foundation for all that — where the 'all that' properly includes Bishop-style constructive mathematics — if we discovered that it calls for a bit of Markov's Principle on the side. Were the project of antirealism proven successful and seen to require Markov's Principle for that success, I should count those facts a persuasive argument *prima facie* for adopting Markov's Principle and for re-examining those aspects of Brouwer's thought that are often construed as inconsistent with it. Third, the main business of (McCarty 2006a) pinches the shady underside of antirealism more smartly than does the Gödel-Kreisel Theorem: strong soundness and completeness for intuitionistic predicate logic imply that the goals of antirealism, when it comes to the *tertium non datur*, are unattainable. Were predicate logic strongly sound and complete with respect to substitutions or for Tarski, Beth or Kripke models, weak counterexamples would be out of bounds logically.

3 The Meaning of 'Meaning'

What would it take to prove mathematically, in a guaranteed fashion, that the law of the excluded third is invalid, this being a key element in Brouwer's intuitionism? One could avail oneself of the Continuity Theorem, or a form of intuitionistic Church's Thesis such as Weak Church's Thesis. The former asserts that every function taking natural numbers to natural numbers is Turing computable. The latter inserts a double negation into the consequent: that every natural number function is not not Turing computable. The Uniformity Principle, the statement that every natural-number-valued relation over any powerset is uniformized by a constant function, would also do the trick. Any of these entail, in straightforward fashion, that the *tertium non datur* is wrong. I must wonder therefore why a(n admittedly rather mathematical) philosophy of intuitionism does not begin by asking for good reasons for thinking one, two or all of these principles true? I am aware of significant work on the mathematical infrastructure beneath the Continuity Theorem (van Atten and van Dalen 2002) and I applaud it. Scholars are, rightly in my view, beginning to think of Church's Thesis as a claim admitting of strict mathematical proof (McCarty 2006b). This is a trend I also salute. More attention could be devoted to analyzing the Uniformity Principle, given the neat way, first revealed in a theorem of A.S. Troelstra (Troelstra 1980), that it separates mathematical from logical domains within a single universe for intuitionistic set theory.

I reject arguments on behalf of Church's Thesis, Continuity and Uniformity, as well as arguments against the law of the excluded third that rely heavily upon a contention that the intuitionistic logical signs bear meanings distinct from those of the homonymous conventional signs. There really are no special intuitionistic or constructive meanings. When I say 'exists' in my intuitionistic mode, I mean 'exists' and no more. Similarly for 'all', 'and', 'or', 'if ... then' and 'not', *mutatis mutandis.* These signs denote and connote exactly what a conventional mathematician denotes and connotes by them. To see why I believe this, please read the following paragraph.

> 'Weak' mathematical induction is a true principle. Consequently, for any natural number $n > 3$, $2^n < n!$. The inequality holds of 4, since $2^4 = 16 < 24 = 4!$. Assuming that it holds of k, we see easily that it also holds of $k+1$, because $2^{k+1} = 2 \times 2^k < (k+1) \times k! = (k+1)!$ for $k > 3$. However, least-number induction, also known as the well-ordering principle, is false. Let S be any mathematical statement whatsoever. Consider the set A^S that definitely contains the number 1 but contains 0 if and only if S is true. In familiar notation, A^S is $\{n \mid n = 0 \wedge S\} \cup \{1\}$. A^S has a member, for 1 belongs to it no matter what. For the sake of argument, suppose that it has a smallest member. As you can see from the definition of A^S, the putative least member has to be either 0 or 1, as required by the meaning of the union symbol '$\cup$'. On the one hand, if the least member is 0, then 0 belongs to A^S and S has to be true. On the other hand, if the least member of A^S is 1, 0 must fail to belong to A^S. There's only one way that can happen: S is false. So, if A^S has a smallest member, then S is either true or false. This result is generalizable: if every set of natural numbers that has a member also has a least member, then, for any statement S, either S is true or it is false. This last claim, an expression of the classical law of bivalence or *tertium non datur*, is itself false. Therefore, the least-number principle is false as well.

So endeth my little paragraph. Whether you are a practicing intuitionist or not, I wager that you understood perfectly the paragraph you just read, every word of it, precisely as meant. Thanks to that understanding, you know that its topic is natural number induction in two of its familiar forms. Moreover, you followed every inferential move and recognized every step in the reasoning as valid. You are aware that, with the opening four statements, I proved that 'weak' induction, which is true, implies the simple arithmetical verity that $2^n < n!$ provided that $n > 3$. Immediately thereafter, I continued to communicate intelligibly and to reason correctly about the same natural numbers, but about a different form of induction, least-number induction. I showed that the latter entails the principle of bivalence or the *tertium non datur.* If you did not know it already, you learned — thanks again to your solid grasp of the senses of my everyday words and their modes of combination — that I, an intuitionist and writing as such, take the *tertium non datur* to be logically invalid and least-number induction to be false.

If you are not an intuitionist, these biographical facts about me may cause you some head-scratching. Nevertheless, they are facts you, or any non-intuitionist, could discern by comprehending the entire paragraph in its ordinary, straightforward, widely-shared, old-fashioned meaning, as I intended you should. You may think the business of my little paragraph trivial or curious or both. But you wouldn't have gotten as far as thinking my brief mathematical message trivial or curious, as far as demanding reasons for my beliefs, unless you had understood perfectly well the ideas I wished to convey by the words and sentences displayed above.

Please note that my mathematical paragraph falls neatly and naturally into two parts. The first, ending with the words 'because $2^{k+1} = 2 \times 2^k < (k+1) \times k! = (k+1)!$ for $k > 3$', although written by an intuitionist, features conclusions acceptable to everyday mathematicians. The second part, taking up the remainder of the paragraph, aids and abets the contention that the least-number principle is specious, something a conventional mathematician would reject. I find it hard to grant, as some would have it, that, in those first few sentences dealing with weak induction, my words — both logical and mathematical — have one kind of meaning attached to them but that, when I swing effortlessly into the second, distinctively intuitionistic part of the paragraph, I suddenly write or read in a quite different tongue, using a quite different set of meanings, 'intuitionistic meanings', all the while employing words many of which are syntactically identical to words that I used a moment before. I find it harder to believe that, when I leave the mathematical paragraph behind and return to the non-mathematical prose immediately on its heels, I fill my head — and the heads of my readers — with other, ordinary senses, different again from the 'intuitionistic senses' I entertained a millisecond prior to that.

Some isotopes of the 'weird meanings' view of intuitionism, such as that Quine propounded in his monograph *Philosophy of Logic* (Quine 1970), would have it that intuitionists not only speak a bizarre language when doing mathematics, but also refer to mathematically bizarre entities. The idea was that we intuitionists, in using, say, the familiar term 'natural number', do not refer to the counting numbers, as understood by grade school kids, but to some proprietary intuitionistic mathematical objects, perhaps some natural-number *Doppelgänger* loitering in a Kleene realizability universe. I refuse to believe that, when intuitionists talk about natural numbers, we are not speaking intelligibly about the very same entities as our classical colleagues. I refuse to believe that we are not, for good reason, disagreeing vehemently with the classical mathematicians over the properties of those very same entities. Please keep in mind that, in large part, the intuitionistic set or number theorist employs, when conducting mathematical business, a language syntactically and phonetically indistinguishable from that of his or her classical brethren.

Critics may be moved to object, 'In the intuitionism of Brouwer, which you claim to defend, the concept "choice sequence" loomed large. For some of his latter-day apostles (Troelstra 1977), that concept and its varied elaborations were to establish the footings of a specifically Brouwerian analysis. Yet "choice sequence" is a concept that enters nowhere into classical analysis or function theory. Does its proper study not require that appropriately non-classical meanings be attached to the intuitionist's words?' (The introduction to (Tait 2005) seems to entertain such a view.) First, intuitionists joyfully pursue investigations into objects and structures not only absent from conventional mathematics but inconsistent with it as well. In that regard, there is nothing special about choice sequences. Some models of intuitionistic set theory contain choice sequences, even lawless sequences. Others contain non-trivial sets S that stand in bijective correspondence with their full (and not just Scott-continuous) function spaces $[S \Rightarrow S]$ (McCarty 1984). Others exhibit non-trivial sets that model the polymorphic lambda calculus beloved of functional-programming mavens (Pitts 1987). The existence of these resplendent anticlassical structures no more demonstrates that the words of intuitionists bear special meanings than the fact that Germany boasts some superior varieties of beer — not vouchsafed for regular consumption in the United States — proves that 'Bier' does not mean 'beer'. The intuitionistic structures just listed contravene the narrow strictures of classical logic and set theory. None obliges us to invent a special-purpose language. Indeed, it is via our usual language only that I can convey their extraordinary mathematical characters — as I just attempted to do. Second, it is historically false that choice sequences enter nowhere into classical analysis. What may well have been their first appearance on the mathematical stage was in a treatise on that very subject, the intriguing *Die allgemeine Functionentheorie* of Paul du Bois-Reymond (Du Bois-Reymond 1882). Later, Émile Borel encouraged their use in classical probability theory:

> For my part I think it is possible to pose problems about probability for decimal numbers which are obtained by choosing the digits either randomly or by imposing certain restrictions on the choice-restrictions leaving some randomness to the choice. (Borel 1972b, p.2130)

Third, the notion of choice sequence, in Brouwer or in Troelstra, is perfectly intelligible to a classical mathematician speaking conventional mathese. It makes good classical sense — as in the above quotation from Borel — to imagine a sequence of numbers emerging over time such that one is, at any stage, apprized of its first few terms and of any explicit restrictions imposed upon its further development. What strikes the conventional mathematician (and, I must say, this intuitionist, too) as dubious is the further suggestion that, by admitting such a concept, we have thereby credited Brouwer's Principle for Numbers (Troelstra and van Dalen 1988, vol.1,p.209) and undermined the *tertium non datur*. You cannot disprove the latter by staring long and hard into the heart of an idea like that of choice sequence. You must argue cogently from a platform of true principles.

When I lecture the graduate students on intuitionistic mathematics, I start out with the mathematics, be it number theory, algebra, set theory, or logic and with principles governing those subjects, and, on that basis, plunge right into proving theorems. I emphasize that, at no point in a process of graduate training familiar to all mathematics professors, need I stop to explain special meanings that I have covertly assigned to the connectives and quantifiers I constantly employ. The students already know what those words mean, and they act accordingly. Occasionally, graduate students run up against obstacles to knowledge, but those are uniformly of a mathematical nature. They never fall into a state one might reasonably construe as deeply-rooted linguistic confusion.

Now, my benevolent critics may inquire, 'If that is so, how do your students ever learn that certain laws of conventional logic are not intuitionistically valid?' My answer is that they learn it in precisely the way they learn that such schemes as $\forall x \exists y.\psi \vdash \exists y \forall x.\psi$ are not conventionally valid: *by proof.* I demonstrate from intuitionistic first principles that the law of the excluded third, double negation-elimination, negation-quantifier exchange and other schemes are invalid. Here is one of the proofs I offer them; it is short and sweet. It presupposes familiarity with the Uniformity Principle or UP, with p ranging over truth-values, i.e., subsets of $\{0\}$, and R a suitable binary relation: $\forall p\ \exists n \in \mathbb{N}\ R(p,n) \rightarrow \exists n \in \mathbb{N}\ \forall p\ R(p,n)$. UP tells us that every total number-valued relation over the truth-values is uniformized by a constant function. Casual readers, if any, should be warned that the collection of truth-values is far more extensive here than dreamt of in classical mathematics. For one thing, another argument from UP shows that the subsets of $\{0\}$ comprise an uncountable set.

Now, I define a binary relation S to hold between truth-values and natural numbers so that $S(p,n)$ is true just in case

$$\text{either } p \text{ holds and } n = 0, \text{ or } p \text{ fails and } n = 1.$$

By assuming — for the sake of argument — the *tertium non datur* in the form that every truth-value is either true or false, ones sees that $\forall p\ \exists n \in \mathbb{N}\ S(p,n)$. From UP, it follows that $\exists n \in \mathbb{N}\ \forall p\ S(p,n)$. But this claim must be wrong. The empty truth-value F exists and 1 is the only number m such that $S(\mathrm{F}, m)$. The full truth-value T exists; 0 is the only number m such that $S(\mathrm{T}, m)$. Therefore, our assumption was mistaken, and the law of the excluded third is seen to be invalid.

Incidentally, it is not a theorem of intuitionistic set theory that validity in propositional logic is decidable (McCarty 1991). One must distinguish, even extensionally, between validity for propositional formulae and derivability in Heyting's formalization of that notion. The latter is decidable. Obviously, it is false that, for every logical scheme underivable in Heyting's predicate logic, there is a proof in an intuitionistic set theory including assorted intuitionistic principles like UP that the scheme is invalid. Were that fact otherwise, Heyting's predicate logic would be decidable. Such limitations erect no serious barriers for advocates of intuitionism. On the contrary, they confirm nicely the early pronouncements of Brouwer and Heyting on the incompletability of intuitionism. Even in logic, there is *Ignorabimus* — and by the truckload.

In my lecture courses on intuitionism, the first principles include familiar set-theoretic axioms, such as the Axiom of Unions or Powerset. There are also truths proprietary to intuitionistic mathematics, like UP. When I introduce these in the lecture hall, I give rough-and-ready reasons for them, I illustrate them and I appeal to students' intuitions, in the everyday sense of that term. I inform them that researchers are now hard at work, and have been for some time, in varied endeavors to reduce these true intuitionistic precepts to other true intuitionistic precepts arguably more fundamental. And I leave things at that. In this, I am no worse off, as an intuitionist, than I would be as a conventional mathematician who lectures on axiomatic set theory. He or she tries to show the students, but does not prove to them, that the axioms of ordinary set theory, like the Axiom of Replacement, are true, and appeals to their (ordinary) mathematical intuitions in aid of the effort.

Undaunted, critics may demand to know, 'If intuitionistic results don't have special intuitionistic meanings, why don't classical mathematicians employ the intuitionists' theorems in their own work? Were you to agree that intuitionists use signs with meanings distinct from those of the syntactically identical signs of classical mathematics, you would have a ready answer to this question. For example, a classical mathematician who concluded from Brouwer's Continuity Theorem, without further interpretational ado, that a specific, classically-defined real-valued function is continuous would be committing the fallacy of equivocation.' In my view, words common to both forms of mathematics usually carry common meanings. With those words, the intuitionist may be asserting something the classical mathematician thinks to be false, e.g., that every total, real-valued function of a real variable is continuous. A conventional mathematician does not normally utilize Brouwer's Continuity Theorem because he or she does not believe it to be true. More generally, classical mathematicians do not customarily put the results of intuitionists to work in their vineyards because, when those results deviate from conventional theorems, classical mathematicians deem them false, and, therefore, ruinous to proofs devoted to uncovering the truth. By a similar token, an advocate of the classical Axiom of Choice will not exploit the Axiom of Determinacy (AD) in proofs that are not hypothetical: he takes AD to be false.

'But what about all the interpretations of formalized intuitionistic mathematics: realizability, topological models, topoi? Do these yield us no insight into the special meanings of the signs at work in the corresponding informal mathematics?' I do not know how we would have lit upon such principles as Extended Church's Thesis and Uniformity had they not played major roles in Kleene's realizability and other interpretational variants on that recursive theme. We learned and continue to learn a great deal about the character and extent of informal intuitionistic mathematics from research conducted via interpretations in proof and model theory. That said, I am in no way obliged to conclude from important facts of metamathematics that the connectives and quantifiers I use when explaining, teaching, and practicing intuitionistic mathematics informally have the meanings the formal correlates of those signs seem to take on, say, under a realizability reading of set theory. What the symbolic logician calls 'interpretation' is hardly what the man or the scholar on the street means by it. Standardly, an interpretation, when given, conveys and illuminates either the original meaning of what is interpreted or some special aspect of it. By 'interpretation', the logician is often thinking of a wholesale redrafting of formal symbols so that all the system's theorems come out true under the so-called interpretation, the original meanings of the symbols be damned. In classical metamathematics, interpretations are constrained to preserve the meanings of the logical signs only and, for intuitionistic formal systems, one need not do that. The endless profusion of logicians' interpretations is no more embarrassing for the intuitionistic mathematician than it is for her classical cousin. After all, there are lots and lots of Boolean-valued or measure-theoretic interpretations of formalized classical mathematics, but few would maintain that the standard meanings of the classical connectives and quantifiers are uniquely captured by such interpretations, revealing as they may be for model-theoretic purposes.

Remember that the Löwenheim-Skolem Theorem and its attendant 'paradox' are about the logicians' idea of interpretation for set theory, interpretations generally having little or nothing to do with the original meanings of set-theoretic terms. The Löwenheim-Skolem Theorem informs us that, if transfinite set theory is consistent and if there are an infinite number of disposable razors, there is an 'interpretation' of set theory on which all the theorems are true but speak only of disposable razors, hardly an interpretation of set theory in any quotidian sense of the word. Imagine announcing to a conference of philologists your new proof of the consistency of Dante's *Inferno* and your interpretation of that darkly magnificent poem with reference exclusively to an infinity of disposable razors. Besides, I think we intuitionists should worry that constant, ill-defined talk of interpretation or meaning distracts us from the brilliance and audacity of those postulates sufficient to prove outright, as Brouwer did, the invalidity of the *tertium non datur.*

4 The Philosophical Basis of Intuitionistic Logic

In my lectures on intuitionism, I do not avail myself of Heyting's famous explanations of the logical signs for at least two reasons. These are reasons well known to philosophers. First, those explanations, as they appear in (Troelstra and van Dalen 1988, vol.1,p.9), make what seems essential reference to entities named 'proofs'. I presume that these are proofs in some familiar sense of the term, everyday proofs of ordinary mathematical statements. Under the heading 'Heyting's Explanations', one is likely to find such assertions as these:

A proof of $A \wedge B$ is given by presenting an ordered pair consisting of proofs, the first of A and the second of B.

A proof of $A \vee B$ is given by presenting either a proof of A or a proof of B.

A proof of $A \rightarrow B$ is a construction that, applied to any proof of A, converts it into a proof of B.

A proof of $\neg A$ is a construction that converts any hypothetical proof of A into a proof of a standard contradiction. There are no proofs of standard contradictions.

A proof of $\exists n.A(n)$ is given by providing an $m \in \mathbb{N}$ and a proof of $A(m)$.

A proof of $\forall X.A(X)$ is a proof of $A(X)$ for any set X.

Now, whatever else they are, proofs of the sort just delineated cannot be classical or conventional, but seem to be intuitionistic. It is plainly false that, as Heyting's explanations would demand, a classical proof of $A \vee B$ always yields us immediately or derivatively either a proof of A or one of B. As scholars of the subject have remarked, the statement $G \vee \neg G$, where G is a Gödel sentence for ZF set theory, is a consequence of a law of conventional logic. But there is no conventional proof either of G or of $\neg G$.

Some may object that, for the example to succeed, the negation in $\neg G$ should be intuitionistic rather than classical. This raises the question, 'If words in the mouths of intuitionists have special meanings, might there be conventional proofs of intuitionistic statements, mathematical sentences with those special meanings attached?' I am disposed to think that, if there cannot be classical proofs of intuitionistic statements, then the idea of special intuitionistic meanings is indeed all wet. Were there such meanings, and were, with those meanings, intuitionistic claims to express intelligible mathematical statements, there would seem to be no prohibition on classical mathematicians attempting to prove those claims in a conventional fashion. For example, some aver that an intuitionistic statement S is true when there exists a proof of S, as explained above. If so, a classical mathematician might be able to show, using methods intuitionists would not adopt, that there exists a proof of S and, hence, that S — as understood by intuitionists — is true.

Second, Heyting's explanations, when deployed as such, suffer some of the same drawbacks as attempted explanations of meaning that adapt for their purposes definitions of truth from conventional mathematics. The signs one is desperately trying to explain via the definition feature prominently and essentially in the definiens itself. Heyting tells us that a proof of $A \wedge B$ contains a proof of A *and* one of B. But how do students know the supposedly distinctive meanings of the intuitionistic signs before they start learning intuitionistic mathematics? When I give the opening lecture in my graduate course on intuitionism, the students have little or no idea of what an intuitionistic proof is. So, a pronouncement of Heyting's explanations would be to them an empty gesture. Once the lecture course is well under way, and the students have a much better understanding of intuitionistic proof, they seem to have no need for Heyting's explanations. At that point, they know perfectly well what the mathematical and logical signs in my lectures mean. Of course, to my thinking, they knew the meanings of those signs all along, for those meanings are the customary ones.

It is still worth asking after the strictly mathematical wallop carried by Heyting's explanations. With the foregoing, I did not mean to suggest that they have no real power or that they are not, suitably understood, true. I call 'constructions' those mathematical objects that are apposite for consideration as proofs in the sense of Heyting. Hidden under the label 'Heyting's Explanations' is, at least, a tripartite assumption: first, that there is an algebra of constructions closed under such abstract operations as pairing and application, second, that there is a canonical relation, named 'is a proof of', holding between constructions and sentences as laid out by Heyting, and, third, that Convention T holds. By 'Convention T', I mean the claim, for open and closed statements A, that

> A if and only if there exists a construction that is a Heyting-style proof of A.

In the meld of these three, there is some mathematical power. Let the forcing sign, $\Vdash$, denote the 'is a proof of' relation holding between constructions and statements, so that $q \Vdash A$ means that q is a Heyting proof of A. Now, assume the antecedent of the Uniformity Principle UP: $\forall p\, \exists n \in \mathbb{N}\ R(p, n)$. By Convention T, it follows that there is a suitable construction q such that $q \Vdash \forall p\, \exists n \in \mathbb{N}\ R(p, n)$. By the clause in the display *supra* pertaining to universal set quantifications $\forall X.A(X)$, q commutes with the universal set quantifier, giving us $\forall p.\ q \Vdash \exists n \in \mathbb{N}\ R(p, n)$. Hence, from the clause governing $\exists$ for natural numbers and Convention T, we see that q_0 is a natural number and $\forall p\ R(p, q_0)$. Therefore, $\exists n \in \mathbb{N}\ \forall p\ R(p, n)$, which is the consequent of UP.

Therefore, from a rendering of Heyting's explanations, the anticlassical principle UP is derivable. Versions of intuitionistic Church's Thesis (or Countable Choice) and the Presentation Axiom can also be derived. Heyting's explanations need not be read as an incantation conferring upon the logical signs arcane meanings to which only intuitionists and die-hard constructivists are privy. If true and justified, Heyting's work supplies the intuitionist with cogent mathematical reasons (but not the only cogent reasons) for thinking the *tertium non datur* invalid.

5 On Saying That

The axioms and theorems of intuitionistic mathematics should be taken at their semantical face values. When an intuitionist asserts that a number specified by a linguistic expression ρ exists or that a function defined by ϕ is continuous, the intuitionist is often making the very same statement as a classical mathematician who asserts that ρ exists or that ϕ is continuous. According to this account, intuitionists can disagree with their conventional colleagues and, in so doing, dispute with them not meanings but truths. Intuitionists should be allowed their straightforward, mathematical say, without re-interpretative interference. Each contender offers a persuasive depiction of the primary objects of mathematics: the natural numbers, the rational numbers, the universe of cumulative sets, and the intuitive continuum — not a mock-up continuum in some model or other of set theory.

A view here put forward and defended, that the logical signs of the working intuitionist mean just what those signs mean in other mathematical contexts, has the great advantage that, when intuitionists use those signs, they are doing precisely what they say they are doing, for example, proving that the law of the excluded middle is invalid. They are not, except by implication, proving some other scheme invalid, some mysterious or hidden scheme bristling with modal operators, double negatives or indices for Turing machines. They are neither equivocating nor refuting a statement bearing only a superficial resemblance to a canon of classical logic. In overturning supposed logical laws on purely mathematical grounds, the intuitionists have toppled a central pillar beneath conventional mathematics. Likewise, when intuitionists prove that every real-valued, total function of a real variable is continuous, they are not quasi-proving that every quasi-real-valued, quasi-total quasi-function is quasi-continuous. They aim into the very heart of analysis. This is what Brouwer did; this is what Brouwer proved. Were it not so, we would have little reason, as intuitionists, to celebrate his name. What, then, is my new intuitionism? In full allegiance to truth, construction, and proof, it is

Back to Brouwer!

6 Acknowledgements

My heartfelt thanks go to Henk Barendregt, Bernd Buldt, Dirk van Dalen, Per Martin-Löf, Göran Sundholm, and other participants in the conference *1907–2007: One Hundred Years of Intuitionism* for their comments and suggestions on an ancestor of the present article. I am grateful as well to the Centre Culturel de Cerisy, the Beth Foundation, and the Institute for the History and Philosophy of Sciences and Technology, in addition to the conference organizers: Mark van Atten, Pascal Boldini, Michel Bourdeau and Gerhard Heinzmann.

Truth and experience of truth

Jacques Dubucs

Fifty years ago (Bourbaki 1960, p.46) wrote of intuitionism as 'a school that will probably be remembered only as a historical blip'. Of course, this very prophecy looks today like an historical oddity because, as a matter of fact, intuitionism has survived. Nevertheless, the sense in which it has survived needs clarification. The most striking feature of the present situation of intuitionism is that some sort of peaceful coexistence with classicism has been eventually reached. Times where controversy was raging are disappearing from collective memory, and the whole intuitionistic enterprise nowadays tends to be merely considered one place in the whole landscape of logic, mathematics, and philosophy. This place can be specified in various ways.

1 Five ways of situating intuitionism

1.1 Labelling

In the most unpretentious and neutral way, mere *labelling* is used to signal proofs that use non-intuitionistic principles of reasoning or theorems for which intuitionistic proofs are not available:

> Our results are marked with the symbol '°' in every case when the demonstration we give is not valid for both [classical and intuitionistic] systems, but only for the classical (and no demonstration which the reader is expected to discover for himself is available for the intuitionistic system). (Kleene 1952, p.101)

At this basic level, classicism and intuitionism seem simply to provide different ways of proving theorems, where the intuitionistic way sometimes does not succeed but, when it does, gives more information than the classical way about the proposition that is proved: an intuitionistic proof of $A \vee B$ provides a proof of A or provides a proof of B, and an intuitionistic proof of $\exists x \phi x$ provides a proof of ϕa for some a.

1.2 Specializing

A second way of considering intuitionism is to see it as a *specialization*. Starting from Kleene's seminal paper on realizability (Kleene 1945), there is an important tradition that thinks of intuitionism as the logic appropriate to computation and theoretical computer science. This computational reading of intuitionism, which culminates in the propositions-as-types paradigm, equates the constructive content of intuitionistic proofs with computer specifications that provide, when carried out, objects that bear 'witness' to the proved assertions. For example, from an intuitionistic proof of $\forall x \exists y \psi(x, y)$ one can extract a program that, given a value for x, will compute the corresponding value for y. Intuitionistic logic is in this way considered as the specific logic of how computing machines proceed.

1.3 Representing

The third way of situating intuitionism in the whole logical landscape consists in giving a mathematical *representation* of its logic, namely in proposing specific structures that could play for intuitionism the same representing role that is played for classicism by the familiar Boolean algebras. Starting from (McKinsey and Tarski 1948), this trend suggests that intuitionistic semantics could be viewed as a generalization of the classical one, with truth-values ranging over the elements of a Heyting algebra (as the collection of the open sets of a topological space) instead of a Boolean one. Whatever the interest of this strategy could be to the mathematical study of intuitionistic logic, it is plain that such a representation does not do justice to the basic philosophical tenet of intuitionism, namely that truth should be epistemically constrained: the fact that a truth is known or unknown, knowable or unknowable, cannot be equated with, or explained by, the fact that an infinite number of values can be attributed to it.

1.4 Relativizing

The fourth strategy does try to take into account the main motivation of intuitionism, namely the idea that the truth-conditions of a sentence should not be defined independently of the means of its verification. To do that, it proposes a semantics where the truth-value of a sentence is allowed to vary according to the nodes (states of knowledge) at which the sentence is evaluated, it being understood that the logical behavior at each node is classical, but that the value of a complex sentence at some node may depend on the value of its constituents at other nodes (Kripke 1965). In this strategy, the epistemic component of intuitionism is expressed by the relational structure of the nodes (for example, the partial ordering of the nodes reflects the intuitionist thesis of the temporal persistence of knowledge).

1.5 Uncovering

The leading idea of the final strategy is to enlarge the language of logic by means of a modal operator intended to express in an explicit way the epistemic component that is hidden in the intuitionistic logical constants, assuming that the non-modal part of the resulting system behaves classically. This 'modal-logical route' towards intuitionism was initiated by (Gödel 1933b), who considered the modal operator of provability, not in a definite formal system, but generally and absolutely speaking, as a suitable way of uncovering this epistemic component. Indeed, according to Brouwer, intuitionistic truth amounts to having a proof (in a suitably idealized sense):

> A statement is true if we have a proof of it, and false if we can show that the assumption that there is a proof for the statement leads to a contradiction. (Troelstra and van Dalen 1988, p.4)

2 Intuitionism as a conflicting doctrine

To varying degrees, the five ways above succeed in making intuitionism a *presentable* doctrine. Its theorems can be clearly delimited, its use in theoretical computer science grants it a high level of respectability in an increasingly computerized world, its logic gives rise to interesting mathematical inquiries, its semantics can be treated by means of the possible worlds machinery as some sort of localized truth-conditional semantics, and its most formidable difficulty to the newcomer (how could 'A' be equated to 'A is known', rather than 'A is true'?) can be disentangled by separating the ontic and the epistemic parts of its assertions: who could ask for more?

Actually, the efforts that have been made to make intuitionistic *logic* intelligible, acceptable, or even advantageous to classically-minded people may have been quite successful (there is no need to be a Brouwerian zealot to understand that 'the object a satisfies ϕ' is more informative than '$\exists x \phi x$'). But if that is all, then only one half of the job has been done, for intuitionism is both the doctrine that intuitionistic reasoning is intelligible, acceptable or advantageous, and the doctrine that classical reasoning is obscure, inadmissible and specious. The rest of the present paper is devoted to possible vindications of this conflicting aspect of intuitionism.

2.1 Grounds for revisionism

One of the salient difficulties in the dialectical situation of intuitionism lies in the fact that there does not seem to exist any possibility of an intuitionistic justification for the existence of the very situations that are its main motivation. Intuitionists declare themselves in agreement with classicists when decidable situations are at stake, as when the issue is to evaluate a universal sentence about a finite domain, and claim that the disagreement with their opponents only starts when undecidable questions appear (Brouwer 1924N, pp.1–2). Now, according to the intuitionistic meaning of negation, one is justified to assert $\neg A$ as soon as one is justified to assert that one will never be justified to assert A (formally: a proof of $\neg A$ is a construction that transforms any (putative) proof of A into a proof of $\bot$). Thus, there can be no proposition A such that it is provable that neither A nor $\neg A$ is provable.[1] There is therefore no possibility of an intuitionistic justification of the undecidability of any proposition A, for if one could justify that A were unjustifiable for ever, then one would get, by that very same reason, a justification for $\neg A$. Thus, it cannot ever be justified in intuitionistic propositional logic that the *tertium non datur* suffers from counterexamples.

Unsolvability

The difficulty above raises another vexing issue. One could wonder whether the rejection of the excluded middle rests on the unproblematic observation that there are hitherto unresolved problems, or whether it is only consistent with the more controversial idea that some problems are absolutely unsolvable. None of these alternatives seems satisfactory. On the one hand, the thesis of absolute unsolvability fits well with the idea, favored by some scholars, that intuitionism is, in the end, some sort of 'anguished branch of the philosophy of mathematics' (Gray 2000, p.165): what could one imagine to be farther away from Hilbertian 'epistemological optimism' than the conviction that there are unsolvable questions whose unsolvability could not even be conclusively recognized?

Nevertheless, little in Brouwer's writings supports the view that there is much we cannot know. Brouwer insists that genuine knowledge differs from what one usually designates by this name, and he emphasizes that one has no guarantee of being able to communicate what one knows, but that does not at all mean that there are mathematical questions that are forever beyond the reach of the human mind. On the other hand, if we reject the idea of absolutely unsolvable problems, and if we interpret 'unsolvable' more prosaically as 'unsolved', or as 'unsolvable by the means that are currently available', then our interpretation is not consonant with the Brouwerian conception of negation. To be in a position to deny a proposition amounts, from the intuitionistic point of view, to being justified in asserting that *no* extension of our current state of information will *ever* support the assertion of the proposition in question, or to have at hand a method capable of transforming *any* putative proof of that proposition into a proof of an absurdity. In other words, the weak interpretation of unsolvability does not fit well with the unrestricted generality

[1] Taking the modal route explained above, this fact is expressible by McKinsey's formula $\Box\Diamond A \rightarrow \Diamond\Box A$, which is trivially equivalent to $\neg(\Box\neg\Box A \;\&\; \Box\neg\Box\neg A)$

that underlies the meaning of the intuitionistic negation. As regards a third possible way out, namely the idea that any well-formed sentence possesses a proof or a refutation, but that some of these proofs and refutations will never be discovered, this suggests a 'platonic' image of the realm of proofs that is in itself problematic, as we will try to show at the end of this paper.

Indeterminacy

The less implausible conclusion of the previous considerations is that the invocation of hitherto unsolved or forever unsolvable mathematical problems is not the best way of grounding the intuitionistic rejection of excluded middle. A more convincing way might be to refer to the heterodox theory of the continuum based on choice sequences that (Brouwer 1918B) presented. To deny a certain property of an infinite sequence conclusively, we must be able to derive a contradiction from the assumption that the sequence has the property, together with the information about the generation of the sequence that is already available at the moment when we reason. In this view, a situation in which we are unable either to attribute or to deny a property to a sequence should be a situation in which the information we have about the generation of the sequence at some stage of its development is definitely not enough to predict its future behavior. This is exactly the case for 'free choice sequences', which are procedures in which the creating subject is not limited by a law but proceeds freely in the generation of arbitrary elements of the sequence. As these choice sequences are open-ended at any time, nothing can be said in advance about the way they will continue on the basis of the information currently available, and hence the principle of excluded middle generally does not apply to propositions that refer to these objects. This is the reason why it has been argued by (Placek 1999, pp.120–121) that the free growth of mathematical entities as well as the resulting indeterminacy of their properties provide a better ground than unsolvability, and probably the best ground *tout court*, for the intuitionistic rejection of the unrestricted validity of the principle of the excluded middle. Nevertheless, there are two difficulties to overcome with this kind of grounding for intuitionistic revisionism.

The first one is that an ontology of 'becoming entities' is not compulsory and can therefore not be used as a *knock-down* argument against classical logic.

The second difficulty, perhaps greater, is that this solution depends on considerations relating specifically to a mathematical domain, while nothing, in the very idea that truth does not transcend the possibility to recognize it, prevents an extension of the intuitionistic principles to arbitrary domains. One can very well admit that lawless sequences represent a paradigmatic case where *tertium non datur* does not apply, while continuing to maintain the universal extendability of the principles that underly the intuitionistic enterprise. Now, no *grounding* of the rejection of excluded middle that rests on the putative indeterminacy of some ontological domain will help us to achieve the intuitionistic objective at a higher level of generality. Last but not least, there is the non-negligible risk that, by grounding the rejection of the excluded middle on the specific indeterminacy of some domain, intuitionism gets the agreement of its opponent for irrelevant and insidious reasons, namely that intuitionism is no longer a *distinctive* position. As a matter of fact, one could perfectly agree with the suspension of some classical principles in very specific situations, as when the propositions at stake refer to contingent futures or to the quantum universe, while maintaining that, in doing so, one is just applying in a sensible way the principles of truth-conditional semantics. Grounding intuitionistic logical revisionism on intuitionistic mathematical revisionism is therefore no more satisfactory than invoking unsolvability considerations.

2.2 Intuitionistic justification

Epistemic immanence

Intuitionism is often said to propose the replacement of the notion of truth by the notion of justified assertability. While this characterization is, of course, roughly correct, it raises the difficulty that the very notion of truth is involved in the notion of justification, in such a way that the whole proposal has a certain air of absurdity. To disentangle the problem, the best strategy is to start by distinguishing, as it is currently and classically done, between two requirements for the relation that should hold between a situation s and a proposition A in order for the relation to deserve the name of a justification of A by s. The first requirement, let me call it the *ontic* requirement, is that of *reliability*: if s obtains, then A must be true. The second requirement is that of *epistemic immanence*: if s obtains, then it must be possible to know that fact — the fact that s, as a reliable indicator of A, obtains, or, somewhat pedantically, the fact that s obtains and the fact that s cannot obtain without A obtaining as well. Justification ensures both truth (the ontic requirement) and recognizability of truth (the epistemic requirement). Both requirements are compulsory. The necessity of the first one is self-evident and the necessity of the second one results from the absurdity of the hypothesis that the very situation by which one recognizes, when it obtains, that a proposition is true, could itself obtain in an undetectable way. As an absolute minimum, epistemic immanence requires therefore that any situation that justifies the assertion of a given sentence is such that an individual who is suitably related to that situation should be in a position to recognize that it justifies the assertion of the sentence in question:

(EI) $\mathit{Justif}(s, A) \rightarrow \forall a[\mathit{Rel}(a, s) \rightarrow \Diamond K_a(A)]$

(EI), when mathematical sentences are concerned, simply says that, if s is a proof of A, then everybody who reads s should be able to recognize that A is correct. Moreover, **(EI)** applies also beyond mathematics, as it should do if, as

I have argued above, intuitionistic principles should be conceived as *general*. To take the favorite example of (Bolzano 1987, §162, p.341), the fact that the thermometer reads higher in summer than in winter justifies the proposition that it is warmer in summer than in winter, because the thermometer cannot read higher except if it is warmer, and that everybody who is in a position to check the proposition about the thermometer (by standing in front of it at suitable moments) should be able to recognize that the proposition about the temperature is true.[2] While **(EI)** looks like a platitude, it is here that the dispute really begins.

Reference capacities

While the classicist generally tends to overlook epistemic elements and to concentrate on the purely alethic part of justification, it would be unfair to omit that she requires that any justification makes really knowable the proposition it justifies. She insists that a proof, to deserve its name, should be clear and detailed enough to remove any doubt about the truth of the theorem, she agrees that the proof should make this truth recognizable, and hence she shows some real interest in the issue of epistemic immanence. Nevertheless, she considers that this immanence, namely the possibility, for a, to recognize the truth of A by means of s, should be understood as strictly subjected to the accessibility condition $Rel(a, s)$. To her, a proof of A justifies A because *if* somebody could read it, this person would recognize the correctness of A. While (at least I imagine so) the presence of molybdenum on Aldebaran has never been proved or disproved until now, she will consider that either its presence or its absence is justifiable, because we have already at hand perfect molybdenum detectors that we use on Earth. To sum up, the classicist inclines to make recognizability of the truth strictly subject to the hypothesis that we have access to the relevant justification. In absence of such access, it remains nevertheless a justification, provided that, if we would have access, we could not fail to recognize the truth.

Quite at the opposite, the intuitionist considers the accessibility condition as superfluous and hence she puts forward a much more demanding concept of justification. To her, the possibility for the cognizers to be suitably related to the item that justifies the assertion should be conceived as immediately flowing from the very nature of the justification. The epistemic immanence of a justification lies not only, not even principally, in the fact that, when equipped with it we are able to recognize the correctness of the justified propositions, but also and above all in the fact that they are actually available. Indeed, to be in a suitable relation to s could cost extensively more efforts than required simply to recognize that s authorizes the assertion of A (for a non-mathematical example, it could be an absolute physiological impossibility for us to move

[2] Bolzano's notion of justification is discussed in detail in (Dubucs and Lapointe 2006).

towards that location where we would however never fail to realize, should we ever find ourselves there, that A is correct). For this reason, she considers that the graspability of the truth-value of the proposition must *unhypothetically* follow from the existence of a justification.

This difference between intuitionism and classicism can be expressed by reference to the general notion of justifiability. While the classical logician is happy with the *weak* notion of justifiability

(WJ) A is justifiable iff $\exists s Justif(s, A)$

the intuitionist wants the stronger notion

(SJ) A is justifiable iff $\exists s[Justif(s, A) \,\&\, \Diamond Rel(a, s)]$

Thus, the strong reading of justifiability is required, if intuitionism should be a *distinctive* position. At this point, two difficulties appear, a generic one about the notion of strong justification, and a specific one about the sufficiency of the appeal to strong justification to make intuitionism a distinctive position.

Strong justifiability

Justifiability, as ordinarily conceived, is a monadic property of assertions. True, one often says, in loose terms, that A is justifiable to a, while not justifiable to b, to express that a thinks that A is justifiable, while b disagrees. Of course, there are domains like, arguably, aesthetics, where the belief *that* A can be considered to be at least partially constitutive of the fact *that* A. But mathematics cannot be considered, anyway I assume so, as such a domain: a mathematical proposition is, or is not, justifiable, in a way that does not depend on opinion. The weak notion of justifiability meets this requirement, but the strong one might not do so. Strong justifiability, as defined by **(SJ)**, seems to be a *relationship* between propositions and agents, which only obtains if a given agent is able to enter into the suitable relation to a justification of the proposition. There is a risk that the integration into logic of such a notion makes logic collapse into psychology, and it is often debated if intuitionism, which rests on that notion, can escape this threat. There is little doubt that some statements by intuitionists are a bit adventurous in this respect, as when (Heyting 1958a, p.103) writes that 'the only justification for intuitionistic mathematics lies in the immediate conviction of the self-evidence of ... propositions'. Nevertheless, what is meant here is neither that mathematical assertions are justifiable according to, and by virtue of, the strength of our convictions, nor that the belief in their correctness to any degree makes them correct. Rather, they are justified by the successful performance of mental *constructions* satisfying determinate requirements. Heyting calls this the 'principle of positivity' (Heyting 1958a, p.108). Thus, to put himself in a relevant relation to the item that justifies a proposition is nothing else than to *carry out* a construction that corresponds to the proposition.

In this respect, the question whether A is strongly justifiable, far from being a matter of appreciation or opinion, is an objective question: did a perform the construction that is required to justify the proposition at stake, or did he not? The answer is not a choice open to free will. Justifiability may well depend on mathematicians, but not on their opinions, just on their performances and on their constructive acts. Is it possible to go further and to remove this last dependence? In one sense, yes, and in another sense, no. On the one hand, if one admits some sort of structural similarity between different minds, namely if one is ready to consider that a mental construction is something reasonably generic that may be realized in different minds, then there is no doubt that when I tell you that the sum of the interior angles of Euclidean triangles amounts to two right angles, I am *predicting* the result you will get by performing in your mind constructions similar to those I have performed in mine. In that sense, individual idiosyncrasies are not the last word of intuitionism. But, on the other hand, that prediction, as well as its linguistic expression in the words I am uttering to you, cannot be taken as a substitute for the construction that it is your duty to perform. My prediction of what would happen if you performed the relevant construction in your mind is not to be taken as having, on you, the same epistemic effect as *your* performance would have. In particular, to hear my prediction, even with confidence, or to read, even attentively, a written account by me of what construction I performed, is in no way sufficient to make you access the situation that justifies the proposition at stake. To be recognized as such, truth should be *experienced.* Hence, 'truth is only in reality, i.e., in the present and past experiences of consciousness', and 'there are no non-experienced truths' (Brouwer 1949C, p.1243). To believe that the mere reading of the text that describes the constructions others have performed is enough to put you in the same epistemic state as mine, is to believe something illusory. Inert reading is just a transition from the experienced to the unexperienced.

According to (Brouwer 1908C), the default of classical logic is the same. The principles of that logic are not 'reliable', not in the sense that they could lead from true premises to false conclusions, but because they lead their followers to the impression of possessing knowledge that they actually do not possess. From transformations that were performed in accordance with the principles of classical logic, results arise that we are actually unable to have genuine experience of. In particular, one can, according to these principles, draw from sentences that have a justification in perceptual experience other ones that go beyond the domain of our potential perceptions. To Brouwer, the general remedy to the defects of the usual logic consists therefore in only keeping the principles that are just dispensable, that is, in never admitting a principle which is such that we could get, owing to it, something that we would hardly get without it: a correct principle of assertability is a principle that limits itself to describing acts we can perform in order to experience truth, not a principle which could, in some cases, serve as a substitute when there is no experience to be had. This Cartesian perspective on the lack of utility of any correct logic is definitely the element that prevents any absorption of intuitionism into the contemporary landscape.

The proper explanation of intuitionistic logic: on Brouwer's demonstration of the Bar Theorem

Göran Sundholm and Mark van Atten

Der ... geführte Beweis scheint mir aber trotzdem wegen der in seinem Gedankengange enthaltenen Aussagen Interesse zu besitzen.

(Brouwer 1927B, n.7)[1]

Brouwer's demonstration of his Bar Theorem gives rise to provocative questions regarding the proper explanation of the logical connectives within intuitionistic and constructivist frameworks, respectively, and, more generally, regarding the role of logic within intuitionism. It is the purpose of the present note to discuss a number of these issues, both from an historical, as well as a systematic point of view.

1 Background

The Bar Theorem is a theorem in intuitionistic mathematics about trees.[2] A bar is a set of nodes in a tree such that every infinite path through the tree intersects it; every infinite path is barred by that set of nodes. The question arises whether a bar admits of a direct, well-ordered construction. For the development of intuitionistic analysis it turns out to be crucial that such a well-ordered construction be possible. The content of the Bar Theorem is that this is indeed the case: it states that if a tree contains a bar, then it contains a well-ordered bar.[3]

[1] 'The demonstration given ... nevertheless seems to me to be of interest, owing to the statements that are contained in its train of thought.'

[2] Brouwer's reasoning has been carefully scrutinized in the literature, beginning with (Kleene and Vesley 1965). We refer the reader to (Dummett 1977, section 3.4), (Martino and Giaretta 1981), (Dummett 2000a, section 3.4), and (van Atten 2004a, ch.4).

[3] As Kleene showed (Kleene and Vesley 1965, pp.87–88), a condition is that the bar is decidable. Brouwer does not make this condition explicit, but in his proofs

Instantiated by a particular tree t, the Bar Theorem takes the form of the judgement: If P(t) has been demonstrated, then Q(t) is demonstrable. What is required for an intuitionistic demonstration of the Bar Theorem, then, is a method that transforms any demonstration[4] that P(t) is true (the tree contains a decidable bar) into one that is Q(t) is true (the tree contains a well-ordered bar). Brouwer's demonstration depends on a general observation. Intuitionistically, to give a demonstration of a mathematical theorem is not to produce a certain linguistic object, but to produce a mental mathematical construction (or a method to obtain one, which method is of course also a mental mathematical construction) that makes the corresponding proposition true. Therefore, the requirement, for a demonstration that the consequence

$$\text{A is true} \Rightarrow \text{B is true}$$

holds, of a method that transforms any demonstration that A is true into one that B is true, is really the requirement of a method that transforms any mathematical construction that makes A true into one that makes B true. Any mathematical means will do, and not just the ones that are so general as to correspond to logical laws.[5]

Towards the end of his dissertation, Brouwer (Brouwer 1907, pp.173–174) gave a lucid distinction between 'mathematical language' and 'mathematics of the second order' that in modern terms correspond to an 'object language' void of content and its (mathematical) study in a contentual 'meta-theory'. Brouwer told Hilbert about his conception during Hilbert's stay at Scheveningen in 1909 and, as he himself notes (Brouwer 1928A2, p.375n.2), it seems fair to regard this distinction as the (unacknowledged) source for Hilbert's metamathematics. Whether Hilbert ever read Brouwer's dissertation, or indeed any of Brouwer's writings on foundational matters, in a careful way, or at all, is a moot point;[6] obviously the writings of Weyl were both his prime source of information concerning intuitionism, as well as his main bone of contention.

---

of 1924 (Brouwer 1924D2) and 1927 (Brouwer 1927B), it is satisfied, because there the bar is defined by an application of the continuity principle for choice sequences. The proof from 1954 (Brouwer 1954A), however, in which the bar is neither defined so as to be decidable, nor explicitly required to be, is incorrect. This seems to have been an oversight on Brouwer's part, rather than an overgeneralization, as the method he gives to construct the well-ordering depends on the decidability of the bar. See on this point also (Veldman 2006b).

[4] Here we note and accept an ambiguity. In the primary sense, a demonstration is the act by and in which knowledge is acquired; this corresponds to Brouwer's 'Beweis'. But we also use 'demonstration' to translate Brouwer's 'Beweisführung', which is not an act but an object, namely the act objectified.

[5] Considered intuitionistically, logical theorems are nothing but very general mathematical theorems (Heyting 1931, p.114), (Heyting 1954, p.78), (Heyting 1956, p.6).

[6] Paul Bernays told Dirk van Dalen in 1977 that Hilbert had never even looked at Brouwer's writings: 'Er hat die Sachen überhaupt nicht gelesen' ('He has not read these things at all') (van Dalen 2005, p.637).

In the opposite direction, however, Brouwer read Hilbert's papers very carefully, and it may well be that Hilbert's lectures (Hilbert 1922, 1923) served as a direct source of inspiration for Brouwer's work on the Bar Theorem (Brouwer 1924D2).[7] Regarding the question of how to establish that a formal system is free from contradiction, Hilbert notes that

> [d]er allgemeine Grundgedanke wie ein solcher Nachweis geschieht ist stets der folgende: wir nehmen an *es liege ein Beweis konkret als Figur ... vor*; ... Sodann zeigen wir durch eine inhaltlich finite Betrachtungsweise, daß dies kein unseren Anforderungen genügender Beweis sein kann. ... Das Verfahren besteht darin, daß wir den als vorliegend angenommenen Beweis sukzessive abändern ... . (Hilbert 1923, p.157, our italics)[8]

Brouwer's fullest account of the Bar Theorem uses the very same term *vorliegen* (Brouwer 1927B, p.63) with respect to a concrete demonstration that we have at hand. Also, his way of proceeding is very aptly described as 'performing successive changes on an assumed concrete proof', but now as a part of, not metamathematics, but mathematics.[9] That Hilbert's work was an important source of inspiration for Brouwer gains further support from Brouwer's famous claim (Brouwer 1927B, p.64n.8) that some mathematical demonstrations are infinite mental constructions. It is expressly directed against Hilbert's metamathematics and, in fact, the possibly infinitary structure of a mathematical demonstration constituted his foremost argument against Hilbert. Whether, on the other hand, Hilbert's later incorporation of the infinitary 'omega-rule' (Hilbert 1931a,b) was a response to Brouwer's views on infinite proofs and his strictures against metamathematics, we cannot say.

A few years after Brouwer's work on the Bar Theorem, his student Arend Heyting contributed to the foundational debate in a two-fold way: first, he presented formal languages, as well as *formalen Regeln* (formal rules), for intuitionistic logic and mathematics (Heyting 1930a,b,c), and secondly, somewhat later, he provided a suitably constructive notion of proposition for intuitionism, by replacing the classical use of truth-conditions by 'proof-conditions', and gave concomitant meaning-explanations for the connectives and quantifiers of his formal languages (Heyting 1930d, 1931, 1934). The added preci-

---

[7] We owe this suggestion to Per Martin-Löf. Already Heyting (Heyting 1962, p.104) notes that in Brouwer's work on the Bar Theorem, mathematical demonstrations are themselves considered as objects of mathematical study and in this way the method resembles Hilbert's.

[8] 'The general line of thought as to how such a demonstration takes place is always the following: we assume that a proof is at hand as a concrete figure ... ; Then, from a finitist contentual point of view, we show that this cannot be a proof that meets our conditions. ... The procedure consists in performing successive changes on the assumed concrete proof ... .'

[9] See (Heyting 1956, p.45). From an intuitionistic point of view there is no special metamathematics.

sion was particularly welcome in the often confused debates on intuitionism: now also mathematicians and philosophers outside the intuitionistic fold could get a grasp of basic intuitionistic notions. From the vantage point of today, Brouwer's work on the Bar Theorem is often viewed in the light of Heyting's proof-explanations, and perhaps naturally so, but the temporal sequence was the opposite: Brouwer's work served as a major inspiration for Heyting.[10] This decisive intuitionistic contribution was not the only inspiration for Heyting's work, though. Formalism and logicism contributed in equal measure: formalism with the idea of a formal system (Heyting's system for propositional logic was in fact culled from *Principia Mathematica*, striking out constructively unacceptable propositions (van Atten 2005, p.129)) and (late) logicism with the idea of systematic meaning-explanations after the fashion of the truth tables.

It is perhaps worth mentioning that Heyting's semantic work antedates that of Tarski (Tarski 1935) by half a decade; it should be stressed though, that Heyting's work belongs to the pre-metamathematical period. The formal system is not a mere object of study, but a tool for gaining mathematical insights.[11] His formulae are meaningful propositions, rather than mere elements of a *semiotisches Quadrupel* as in (Hermes 1938). Accordingly, for the variables in the system should be substituted only meaningful propositions (Heyting 1930a, p.3), and the turnstile serves as a Fregean assertion-sign, rather than as a Kleene-Rosser metamathematical theorem-predicate.[12] In the light of the temporal mismatch and the encroachment of themes possibly alien to Brouwer's intuitionism, such as the use of (formal) languages and semantic meaning-explanations, caution is called for when applying the notions of Heyting to Brouwer's work.

Elsewhere we have dealt extensively with Heyting's constructive notion of a proposition and the matching meaning-explanations for the logical connectives, for instance (van Atten 2004b, ch.2) and (Sundholm 1983, 1997), and we see no need to repeat that discussion here. In particular, we assume familiarity with the general form of the BHK explanation of the logical constants.

[10] Brouwer applied the Proof Interpretation *avant la lettre* in a lecture on ordering from 1925; see the quotation in Dirk van Dalen's paper on p.19 of the present volume, and (van Dalen 2005, p.676). In a personal communication, Dirk van Dalen told us that Heyting was probably not in the audience then, but that he very likely heard Brouwer make the remark on another occasion.

[11] It is possible to divest the well-formed formulae of their content and study the objects thus obtained by mathematical means, as Heyting himself points out and also practised, with his Bernays-style independence proofs. But: '[D]iese Betrachtungsweise tritt für uns hinter die vorige [inhaltliche] zurück' ('This approach for us takes second place to the former [contentual] one.') See (Heyting 1930a, p.43 and appendix).

[12] See (Kleene 1952, pp.88,526).

2 Brouwer's strategy: closure under rules versus truth of an implication

Gentzen's techniques of Natural Deduction have become very well known, and are now part of the standard logic curriculum both for philosophers and mathematicians. However, Brouwer's hypothesis in the demonstration of the Bar Theorem from an assumed proof is not of the familiar Natural Deduction kind, where a well-formed formula is put forward as an assumption and the derivation is continued by applying the rules of inference irrespective of whether the premises have been demonstrated or not. In the demonstration of the Bar Theorem, what is given is a concrete *Beweisführung* for the truth of the antecedent, and on this entity various transformation steps are performed in such a way that one obtains a demonstration also of the truth of the consequent. In Natural Deduction terms, Brouwer's assumption does not correspond to a customary 'assumption of A towards deriving B'. Instead, it is an assumption that one has a closed derivation of the antecedent formula A, that is, that formula A is a derived *theorem* and that we possess its (closed) derivation. By performing suitable operations on this derivation we might obtain closed derivations for other theorems and have then established a 'closure property' of the system in question. Reflection on the easy proof of the Herbrand-Tarski Deduction Theorem for Hilbert-style Propositional Calculus (HPC) may put the matter into proper relief. Here one is given a concrete derivation D of the consequence A, $\Gamma \Rightarrow$ B, whence D is a finite tree with end-formula B, with either (propositional) axioms, assumptions in Γ, or the assumption A, as top formulae, and further regulated by applications of Modus Ponens, that is, the sole derivation-rule in HPC. On this tree D one then performs an 'A $\supset$-transformation' by adding the sign-combination A $\supset$ in front of every formula that occurs in D. This transformation obviously destroys inferential links and the rest of the demonstration consists in checking that the transformed tree A$\supset$-D can be rebuilt into a derivation from assumptions in Γ only that establishes the consequence $\Gamma \Rightarrow$ A $\supset$ B. Brouwer's reasoning makes use of a certain canonical form of the original *Beweisführung* and it is on the latter that the transformations are performed that yield a demonstration of the conclusion.

Note that it is only a closed proof-object a for proposition A (notation: a:Proof(A)) that can be evaluated to canonical form, the particular of which is determined by the meaning explanation of A; in a hypothetical judgement of the form b:proof(B) given that x:proof(A), or in other notation, the conditional judgement

x:proof(A) $\Rightarrow$ b:proof(B)

neither x nor b admit such evaluation. In the first edition of his *Elements of intuitionism*, Michael Dummett (Dummett 1977, pp.103,400) has suggested that 'in order to exploit fully the intuitionistic meaning' of $\supset$ we need to be able to cite, for each statement A, an axiom of the form

A $\supset$ there exists a proof of A of such-and-such a kind

'where a proof will be a suitable dressed spread, and the axiom will state restrictions on the form of the proof, depending upon the structure of the statement A'. But this is mistaken; as Dummett himself explains, in the second edition of *Elements*:

> We cannot, however, hope to generate, for any but the simplest type of mathematical statement A, an axiom with antecedent A and consequent an existential statement specifying the form a canonical proof of A must take. We cannot do so because, if A is a conditional or a universally quantified statement, we cannot circumscribe the effective operations that might serve as a proof of it. Such an operation might be recognized as efficacious only in the light of various known mathematical results, or of some intricate reasoning. (Dummett 2000b, p.273)

'Canonizability' does not apply to conditional constructions, but applies only to closed proof-objects. A given proof(-object) a of A can be evaluated to canonical form; to demand that this evaluation should take the form of a(n axiomatic or) demonstrable implication

(*) A $\supset \exists$p(p is a canonical proof of A)

asks too much. It asks too much, because it requires that the existence of a canonical proof can be obtained from A by the highly general means of logical inferences; whereas such a canonical proof may well depend on specific details of constructions for A, and in Brouwer's demonstration of the Bar Theorem this is indeed the case.[13] In the case of the Deduction Theorem there is no need to perform a 'canonization' of the original derivation; Brouwer went from his given *Beweisführung* to one in canonical form, because his method of transformation needs control over the modes of inference that can occur in the deduction to which the transformation is applied. In the canonical form that is the case, since then the *Beweisführung* is built up from very basic inference steps only. This difference is due to the fact that in a formal system all the axioms and rules of inference are known from the outset. Brouwer, of course,

[13] Dummett's alleged axiom (*) from 1977 is closely related to his Principle K (Dummett 1976, p.99): 'If a statement is true, then it must be in principle possible to know that it is true', as well as to the formula p$\supset \Diamond$Kp that plays a crucial role in the so-called Fitch's paradox. The matter is dealt with by Martin-Löf (Martin-Löf 1998) in considerable detail and for further discussion we refer to his treatment. The alleged axiom (*) also raises some other hard questions: what is the domain of quantification? If it is the whole 'universe of constructions', the worries concerning impredicativity come back with a vengeance. Furthermore, what is the status of the proof-relation? If it is a propositional function, over what domain? (The answer to this question would seem to take us back to the first worry.) And how, on pain of an infinite regress, can its meaning-explanation be cast in terms of proof-conditions using that very proof-relation itself?

does not establish the closure of any formal system; his transformations apply to real *Beweisführungen* and not to their *simulacra* in formal systems.

Thus what Brouwer demonstrates, is not the truth of the implication

$$P(t) \supset Q(t)$$

nor the holding of the equi-assertible consequence

$$P(t) \text{ is true} \Rightarrow Q(t) \text{ is true}$$

Instead, what is actually done in his reasoning is an attempt to demonstrate the closure of all of mathematics under the Bar Rule, that is, the rule

From the premise: P(t) is true, *conclude*: Q(t) is true[14]

3 Brouwer and the variety of proofs

A quirk of the English language makes it extra difficult to treat of Brouwer's work in English: Brouwer's Germanic languages deploy *Beweis* (German) and *bewijs* (Dutch) and their cognates where today's mathematico-philosophical English uses *proof. Proof* stems from Latin *probare* whence, at least etymologically, proving and probing (testing) are quite close; *demonstration* is a more fitting translation of the terms used by Brouwer and in the sequel this will be adhered to. Clearly, for Brouwer the most important aspect of a mathematical demonstration is the deed, or act, whereby a theorem is established. This fits very well with one of the meanings offered by the *Oxford English Dictionary*:

> 3. a. The action or process of demonstrating or making evident by reasoning; the action of proving beyond the possibility of doubt by a process of argument or logical deduction or by practical proof; clear or indubitable proof; also (with pl.) an argument or series of propositions proving an asserted conclusion.

In this primary sense the (act of) demonstration has only a limited endurance. After its completion it is no more and only the asserted theorem remains. What are called demonstrations, or proofs, in mathematical texts clearly are not acts; rather, they are blueprints for acts of demonstration and can be used by mathematicians for carrying out such acts. Brouwer famously spoke about demonstrations as mental constructions (Brouwer 1927B, p.64n.8); this is ambiguous between, on the one hand, the demonstrative deed and, on the other, a mental objectification thereof,

---

[14] Similar remarks can be made about the principle of Christian Charity: 'If a statement is true, then there is a moment t at which the creating subject proves it'. As an implication this is highly problematic, but as a principle about demonstrations it is clear: given a demonstration by the subject of a statement, the subject can indicate the moment t at which it arrived at that demonstration.

of which the written proof (demonstration) attempts a description. In connection with Brouwer's deviant rendering of the Law of Excluded Middle in the dissertation, Dirk van Dalen (van Dalen 1999, p.106) has plausibly suggested that Brouwer was familiar with the traditional logic course given by Bellaar-Spruyt at Amsterdam, where a similar use can be found. Also his views on the act of demonstration, its mental products, and external linguistic signs, are strongly reminiscent of a scheme that is well known from traditional philosophy:[15]

THE TRADITIONAL STRUCTURE OF LOGIC:

	OPERATION OF THE INTELLECT	(MENTAL) PRODUCT	(EXTERNAL) SIGN
1	*Simple Apprehension*	Concept, Idea, (Mental) Term	(Written/spoken) Term
2	*Judging*, Composition/Division of two terms	Judgement, (Mental) Proposition: *S* is *P*.	Assertion, (Written/spoken) Proposition
3	*Reasoning*, Inferring	(Mental) Inference	(Written/spoken) Inference, Reasoning

A(n act of) demonstration is a mediate act pieced together out of several applications of Judging and Reasoning, with a matching (mental) product, and a written demonstration (proof) as external sign.

Brouwer's terminology in the original publications respects this scheme to a surprising extent. Thus, the title in the 1924 version of the Bar Theorem uses *bewijs* in Dutch (Brouwer 1924D1) and *Beweis* in German (Brouwer 1924D2), that is, *demonstration*, whereas for the mental entity to which the transformation is applied Brouwer uses *Beweisführung* in German in 1924 and 1927, and *redeneering* in Dutch (Brouwer 1924D1) in the spelling of the time, which latter term perhaps is best translated with *reasoning*. In his English presentations of the Bar Theorem, Brouwer (Brouwer 1954A, Brouwer 1981A) uses the verb 'to prove'. However, at crucial places he deliberately seems to avoid the use of the noun 'proof'. The term *Beweisführung* is rendered as 'constructive mathematical argument' and 'mathematical deduction', respectively.[16]

[15] The diagram draws on a similar one in (Maritain 1946, p.6), but is reasonably standard. Maritain's source, and also that of virtually all other Neo-Thomists, is the splendid *Ars Logica* by John of St. Thomas. Bellaar-Spruyt's lectures (Bellaar-Spruyt 1903) had the traditional tripartite structure: Of terms; Of propositions; Of reasoning or syllogisms.

[16] Traditional influence on Brouwer's thinking regarding logic and language is visible throughout his career and not only in the 1920s. Thus, for instance, in 1947 he gave

4 Heyting and the proof of implications

Reflection upon the construction 5+3 shows that there is a fourfold sense of 'construction' involved here:[17]

1. the composite temporal act of generating the numbers 5 and 3 and then adding them together;
2. the object obtained as a product of this act, namely 5+3;
3. the act objectified, as a description of the steps effected by the particular subject at a particular time; and
4. by abstracting of the objectified act with respect to subject and time, a prescription or blueprint for construction acts yielding 5+3.

Heyting's semantical explanations are cast in terms of proof conditions. But:

> Ein Beweis für eine Aussage ist eine mathematische Konstruktion, welche selbst wieder mathematisch betrachtet werden kann. (Heyting 1931, p.114)[18]

Accordingly, propositions are regarded as conditions on constructions, where the relevant sense of construction is here that of 2; these constructions are the proof-objects that above we notated as 'a:Proof(A)'. This notion of proof at the level of propositional content, rather than that of asserted theorem, is novel with intuitionism. Complex conditions can be built up using dependency relations.

Of Heyting's clauses, only that for implication has proved controversial. There are two worries here. One is that Heyting's notion of implication appears to be impredicative, the other that it seems to require the introduction of hypothetical entities into intuitionistic mathematics (Griss).

The worry of impredicativity is that in an application of modus ponens, the proof construction for A⊃B is applied to a proof construction for A, which may itself use applications of modus ponens with antecedent A. This is readily accounted for by insisting that Heyting's clauses do not explain propositions in terms of any proofs, but only in terms of canonical proofs. Thus, a canonical proof for a conjunction A&B is an ordered pair ⟨a,b⟩ where a is a (not necessarily canonical) proof for A, and similar for b. The sole

---

the following comment on his spread definition of 1919: 'Because mathematics is independent of language, the word *symbol* (*Zeichen*) and in particular the words *complex of digits* (*Ziffernkomplex*) must be understood in this definition in the sense of *mental symbols*, consisting in previously obtained mathematical concepts.' (Brouwer 1947). Brouwer's views here are remarkably similar to certain medieval views on 'mental terms' due to William of Ockham, e.g., (Ockham 1974, §3, pp.52–54).

[17] We here draw upon (Sundholm 1983) and (Sundholm 1993).

[18] 'A proof for a proposition is a mathematical construction, which in turn can itself be considered mathematically.'

condition on a non-canonical proof is that it be evaluable to a canonical one. Applying this insight to implication, we see that ultimately, no matter how a proof of A is given, it will have to be equal to a proof of A in canonical form, whence the impredicativity issue does not arise.

In natural deduction, a derivation commonly begins with assuming that a proposition A is true. One then shows that under this assumption, certain propositions are also true, for instance B, and may continue by concluding that the proposition A⊃B is true outright, discharging the assumption that A is true. In terms of proof constructions, it might seem (and it did to Griss) as if we had to start this derivation by introducing a hypothetical construction for A. This semblance, however, is spurious, as a careful analysis reveals that all constructions involved are actual.

Comparison with the case of function is illuminating. In the old sense, a function is an analytical expression in a variable, obtaining a value upon substitution of suitable arguments. For example, $x + 2 \in \mathbb{N}$ given that $x \in \mathbb{N}$, as the following simple construction shows:

$$\frac{\dfrac{x \in \mathbb{N}}{\mathrm{S}x \in \mathbb{N}}}{\mathrm{SS}x \in \mathbb{N}}$$

Conveniently summarizing, we may write $(x + 2) \in \mathbb{N}(x \in \mathbb{N})$, as an expression of the relation of dependency. Functions in the old-fashioned sense are dependent objects of lowest level. Such dependent objects are not hypothetical, but enjoy a perfectly actual existence. A function in the modern sense of a mapping, that is, an independent object of higher level, is readily obtained from this by λ-abstraction: $\lambda x.(x + 2) \in \mathbb{N} \to \mathbb{N}$.

In the premise of an application of implication-introduction, we have a proposition B whose truth depends on that of A. In the light of the above, its proof object will be a dependent one of the form b:Proof(B) (x:Proof(A)). Moreover, this dependent object is actual and not hypothetical, which meets Griss' objection. From this, by implication-introduction, we readily obtain a suitable canonical proof-object for A⊃B:

$$\frac{\mathrm{b{:}Proof(B)}\ (x{:}\mathrm{Proof(A)})}{\supset\mathrm{I}(\mathrm{A},\mathrm{B},\lambda x.\mathrm{b}){:}\mathrm{Proof}(\mathrm{A}\supset\mathrm{B})}$$

Consequences are verified by functions in the old sense, whereas the canonical proof-object for an implication draws upon a function in the modern sense.

This way of elaborating the matter has been perfected in the constructive type theory of Per Martin-Löf (Martin-Löf 1994). Note that it generates the proof objects inductively, also for implication, thereby blocking impredicativity, and respects the parallel between canonical proofs and those of Gentzen introductory form.[19]

A proposition A is given in terms of a proof-condition Proof(A) that is explained in terms of how a canonical proof for A may be put together out of parts, depending on the structure and meaning of A. Propositional truth is explained in terms of proof-conditions: A is true $=_{\text{def}}$ Proof(A) exists. This only defines the truth-conditions for A; whether this condition is fulfillable, that is whether the proposition really is true, is of course not a matter of definition. This truth-condition may be fulfillable and then a proof-object can be found; the proposition can then be known to be true, whereas when it has been found the proposition is known to be true. The proof-condition for an implication A⊃B requires a relation between the proof-conditions for the propositions A and B; neither the condition itself nor its fulfillability presupposes any information concerning the fulfillability of the conditions for A and B, that is, whether these propositions really are true. In order to understand, and even to know that an implication is true, it is only necessary to know the conditions for the truth of A and B, but not whether these conditions are, or can be, fulfilled.

Brouwer was not overly concerned about implications where such knowledge is lacking. His referee report from around 1936 for *Compositio Mathematica* on the discussion between Freudenthal and Heyting[20] contained merely this:

> Interessante Diskussion über den Sinn der Implikation eines Satzes durch einen anderen, wenn über die Richtigkeit des letzteren Satzes nichts bekannt ist.[21]

And in a later statement in 'Points and Spaces', one finds the likely reason for this brevity:

> [T]he wording of a mathematical theorem has no sense unless it indicates the construction either of an actual mathematical entity or of an incompatibility (e.g., the identity of the empty two-ity with an empty unity) out of some constructional condition imposed on a hypothetical mathematical system. (Brouwer 1954A, p.3)

---

[19] As Michael Dummett observed, the usual kind of formulation, for instance 'A proof of a A⊃B is a method, or function, f, taking a proof a:A into a proof f(a):B' does not respect this and is rather an elimination rule (Dummett 1994, pp.321–322).

[20] (Freudenthal 1936), (Heyting 1936).

[21] 'Interesting discussion on the sense of one sentence being implied by another, when nothing is known about the correctness of the latter sentence.' Brouwer Archive, Utrecht. Kolmogorov's lucid review in the *Zentralblatt* deserves to be better known; it also provides, in brief compass, a perfect illustration of the interpretative possibilities that we have been concerned to explore, whence we give it as an appendix.

We read 'no sense' as 'no mathematical sense': the wording of the theorem may still make sense as a logical implication. Brouwer's point is that in that case, the implication is not directly informative about how things are, or are not, in the mathematical universe, and therefore is not descriptive of it. This brings us to our final theme.

5 Ontological Descriptivism versus Meaning Theory

With respect to ontology, intuitionism is a form of idealism: its denizens are mental constructions. However, this should not prevent us from seeing that epistemologically, it shares a realist stance with platonism: the correctness of a knowledge claim is in both cases ultimately reduced to matters of ontology, namely, in the intuitionistic case, to features of our mental constructions, and in the platonist case, to what states of affairs obtain in the platonist realm of mathematics. Both intuitionism and platonism therefore are versions of ontological descriptivism. In intuitionism, the ontology is idealist, in the sense that its objects are mental constructions, whereas in platonism the ontology is realist, in the sense that the objects are not of our making but self-subsistent.

This ontological descriptivism is clearly present in Brouwer, who attempts to exploit freely, without constraints, all four kinds of constructions (p.68 above) for mathematical purposes.[22] His demonstration of the Bar Theorem pivots around the issue of how to transform mathematical constructions. The properties of such transformations clearly depend on what kinds of constructions are being transformed.

In the meaning-theoretical tradition that started with Heyting and was continued by Martin-Löf making essential use of the Curry-Howard insight, so far only constructions of the second kind have played a role. It is difficult to see how that could be otherwise, in view of meaning-theoretical uniformity constraints.

One such constraint is that of compositionality. As we have noted above, Brouwer's *Beweisführungen*, that is, constructions of kind 3 or 4, in natural deduction terms correspond to closed derivations, and thus an assumption that we possess such a *Beweisführung* amounts to an assumption not of A, but of $\vdash$A. Such assumptions cannot take antecedent place in an implication, on pain of begging Geach's so-called 'Frege Point' (Geach 1965). This seems to put a tension on Arend Heyting: in his mathematical practice, he was undoubtedly a Brouwerian intuitionist, but in his semantical theorizing a constraint such as compositionality would seem to rule out accounting for that mathematical practice.[23] From the vantage point of today, with the benefit

[22] For an ontological descriptivist justification of the weak continuity principle for choice sequences (exploiting aspects of constructions in sense 1), see (van Atten and van Dalen 2002) and (van Atten 2007, ch.7).

[23] See (Sundholm 1985, section 3). Also Brouwer might have noticed this tension in Heyting: in the letter (July 17, 1928) in which he suggests to Heyting to submit his

of hindsight, these two different tendencies can be discerned within Heyting's work from the outset. Research after Heyting has commonly taken his views on proofs and propositions in a meaning-theoretical sense, and in the light of some of his formulations not unreasonably so. Thus Kreisel (Kreisel 1962b, p.198) wanted to 'give a formal semantic foundation for intuitionistic formal systems in terms of an abstract theory of constructions' by means of converting Heyting's informal explanations even into statable theorems. Prominent expositors of intuitionism, for instance Troelstra, Van Dalen, and Beeson, have continued this meaning-theoretical tendency,[24] which from the early 1970s gained strength and cohesion from the Curry-Howard insight, as well as formal codification in the emerging constructive type theory of Per Martin-Löf. Support for this reading can be found in Heyting's early writings (Heyting 1930d, p.958), (Heyting 1931, p.114), which jointly yield that the construction demanded by the proposition 'Euler's constant C is rational' is a pair $\langle a,b \rangle$ of natural numbers such that $C = a/b$. Here, certainly, it is natural to view the construction demanded for the truth of a mathematical proposition as a mathematical object, rather than as a demonstration(-act or -trace). The demonstration that the proposition in question is true would consist in the production of a suitable construction-object. In later works, for instance (Heyting 1960, p.180), Heyting listed construction methods that square perfectly with the way proof-objects are built, namely (i) construction of a natural number; (ii) hypothetical construction; (iii) general method of construction; (iv) contradiction, and commented upon the use of these methods with great insight (Heyting 1958b). Also Heyting's very early — in fact prior to their appearance in print — adoption (Heyting 1935, pp.78–79) of Gentzen's natural deduction methods, together with the Curry-Howard insight, serve to make the proof-object reading natural.

Of course, in the literature one finds routine references — no names, no pack-drill! —to (Heyting 1956, ch.7) as a convenient source of the proof-explanations; it is therefore somewhat surprising to discover that these explanations cannot be found there. On the contrary, the explanations of logic are at the level of assertions and not at the level of propositions, e.g., p&q can be asserted if and only if both p and q can be asserted. The justification of an asserted theorem is always via a demonstration: what we have here are assertion- and demonstration-conditions rather than proof-(object-)conditions.[25]

formalization of intuitionistic logic and analysis to the *Mathematische Annalen*, he advises: 'En misschien zou zich ook (met het oog op §13) het begrip "Gesetz" laten formaliseeren.' — 'And, with an eye on §13, perhaps also the notion of "law" can be formalized.'

[24] (Troelstra 1969, 2.1–2.3), (van Dalen 1973, 2.1), (Troelstra 1977, 2.1–2.2), (Beeson 1985, II:4–II:6), (Troelstra and van Dalen 1988, 3.1–3.3).

[25] Also formulations in the early expository text *Grundlagenforschung* seem to point in this direction, e.g., '[J]ede Aussage steht ... für die Intention auf eine mathematische Konstruktion die gewisse Bedingungen genügen soll. Ein Beweis für eine Aussage besteht in die Verwirklichung der in ihr geforderte Konstruktion.'

Another meaning-theoretical uniformity constraint is that of subject-independence. Constructions in sense 1 and 3 are subject-dependent, and it would therefore not seem possible to exploit them in meaning explanations that should be objectively valid. An ontological descriptivist who accepts a Husserlian notion of transcendental subjectivity can have it both ways: in that case, the relevant aspects of subjectivity in mathematics can be exploited in such a way that it gives the same result for all subjects. For a discussion of the matter, and a defense of the claim that Brouwer's notion of the subject is, as far as mathematics is concerned, indeed best understood as that of the transcendental subject in Husserl's sense, see (van Atten 2004a, ch.6); for a discussion of intuitionism and intersubjectivity from a different perspective, see (Placek 1999).[26]

In the meaning-theoretical tradition, the Bar Theorem so far has been proved only from the axiom of Bar Induction, whereas Brouwer's alternative demonstration has not been subsumed under that tradition's paradigm; on account of the above considerations on *Beweisführungen*, we suggest this is no coincidence. In this connection, one should observe however, that in the opposite direction it seems most likely that everything justifiable within the meaning-theoretical tradition would be accepted by Brouwer,[27] with the likely exception of ex falso.[28] Brouwer also accepted the demonstration of the Bar Theorem from the axiom of Bar Induction, indeed it was he who proposed it, but he held that the alternative ontological descriptivist route taken in the main text of his paper from 1927 was of independent interest:

(Heyting 1934, p.14) — 'Every mathematical proposition stands for the intention towards a mathematical construction that should meet certain conditions. A demonstration of a proposition consists in the actualization [realization] of the construction it demands.' Here the *Verwirklichung* clearly is an act. On the other hand, it is not said that this realization-act is the construction demanded, and hence an ambiguity remains. We leave the matter for a later occasion.

[26] Husserl's influence on Heyting, via Becker, and more generally the influence of the phenomenological movement on Heyting, does not seem to have gone further than the (important) conception of a proposition as an expression of an intention, to be fulfilled by giving a demonstration (Heyting 1931). In particular, Heyting's later reflections on solipsism (Franchella 1995) do not draw on phenomenology. See also (Franchella 2007).

[27] One indication is the fact that Brouwer's reasoning for the equivalence of $\neg A$ and $\neg\neg\neg A$ (Brouwer 1923C1) is presented unconditionally; in particular, it is not ruled out that A has already been demonstrated.

[28] Also in meaning-theoretical explanations, the grounds on which ex falso is accepted, if it is, are not quite of a kind with those that justify the other principles: in the former case, instead of providing a concrete transformation of conditions on constructions, one relies on the fact that constructions for $\bot$ are impossible. See (van Atten 2008) for a careful examination of the issue and a defense of the view that Brouwer, for whom this distinction is important, does not accept ex falso.

> Intuitionistisch durchdacht, ist diese Versicherbarkeit nichts anderes als diejenige Eigenschaft, welche dadurch definiert ist, daß sie für jedes Element von μ_1 und für jedes Element von μ besteht, und daß sie für ein beliebiges $F_{sn_1\ldots n_r}$ besteht, sobald sie für jedes ν für $F_{sn_1\ldots n_r\nu}$ erfüllt ist [formulation of Bar Induction]. Diese Bemerkung zieht die Wohlordnungseigenschaft eines beliebigen $F_{sn_1\ldots n_r}$ sofort nach sich. Der im Texte für die letztere Eigenschaft geführte Beweis scheint mir aber trotzdem wegen der in seinem Gedankengange enthaltenen Aussagen Interesse zu besitzen. (Brouwer 1927B, p.63n.7)[29]

6 Brouwer's Demonstration of the Bar Theorem as a Transcendental Argument

We recall that for Brouwer, an argument for the Bar Theorem requires one to show that the consequent is demonstrable given that the antecedent has been demonstrated. We further saw that, on account of his ontological descriptivism, also specifically mental aspects of mathematical constructions may be exploited in building the required transformation method. And this, indeed, is the path his reasoning took. In acts of reflection, he observed what types of mental acts are available to us in constructing bars and determining their properties. On the basis of that (in effect, phenomenological) analysis, Brouwer claimed that any demonstration that a tree is barred, when analyzed into sufficient detail, can be decomposed into elementary mental steps that come in only a few different kinds. This analyzed demonstration Brouwer called the canonical form of the demonstration that the tree is barred.[30] In effect, then, Brouwer's reflections led him to formulate a necessary condition: nothing can be a demonstration that a tree is barred unless it is analyzable into a mental demonstration of the specified canonical form.[31] As we

[29] 'When properly thought out intuitionistically, this securability is seen to be nothing but the property defined by the stipulation that it shall hold for every element of μ_1 and for every inhibited element of μ, and that it shall hold for an arbitrary $F_{sn_1\ldots n_r}$ as soon as it is satisfied, for every ν, for $F_{sn_1\ldots n_r\nu}$. This remark immediately implies the well-ordering property for an arbitrary $F_{sn_1\ldots n_r}$. The demonstration given in the text for the latter property nevertheless seems to me to be of interest, owing to the statements that are contained in its train of thought.' (van Heijenoort 1967, p.460n.7, modified)

[30] Or rather, 'a canonical form'; for Brouwer left open whether the canonical form that he proposed can be decomposed even further (Brouwer 1954A, p.13). For his purposes it suffices to have just the one he suggests.

[31] We stress that *canonical* in Brouwer's sense is not the same as *cut-free* or *normal*; a canonical demonstration may contain ζ-inferences. In fact, the difference between the 1924 and 1927 versions of his demonstrations of the Bar Theorem is that in the former, all ζ-inferences are eliminated, while in the 1927 presentation Brouwer realizes that for his purposes normalization with respect to ζ-inferences is superfluous (Brouwer 1927B, p.65n.9).

remarked above, the device of canonization serves to reduce the potentially great variety in demonstrations that the antecedent is true to the one specific structure of canonical demonstrations; and the specificity of this structure enabled Brouwer, in the second part of his demonstration of the Bar Theorem, to devise a construction method that transforms a canonical demonstration into a well-ordered construction of a bar.

That every demonstration of the antecedent can be rendered in such a simple and narrowly circumscribed canonical form is not proved by Brouwer but put forward as an insight or Thesis. As Heyting (Heyting 1958b, p.342) noted, Church's Thesis is closely parallel to Brouwer's way of proceeding: also there we have the situation that a function is given with a calculation method. The Church-Turing thesis then says that the simple and narrowly defined rules for Turing machines can compute the function in question.

The first part of Brouwer's argument, then, has the following structure:

- I possess a demonstration that the tree is barred;
- (Insight or Thesis) It is a necessary condition of the possibility of possessing such a demonstration that it be analyzable into a mental demonstration of the specified canonical form;
- Therefore, the demonstration is analyzable into a mental demonstration of the specified canonical form.

As for Brouwer, to possess a demonstration is to have a certain mental experience, this argument is an instance of the following well-known type of transcendental argument:

- I have mental experience E;
- It is a necessary condition of the possibility of having mental experience E that C;
- Therefore, C.

where the necessity invoked in the second premise is not merely logical or conceptual necessity.

We claim, therefore, that Brouwer's demonstration of the Bar Theorem is based on a transcendental argument. Brouwer nowhere uses the term 'transcendental' to characterize arguments, but he explicitly recognized the mechanism and its role in intuitionistic mathematics. In a late, prominent statement of his philosophical position from 1948, he said:

> [T]heorems holding in intuitionism, but not in classical mathematics, often originate from the circumstance that for mathematical entities belonging to a certain species, the possession of a certain property imposes a special character on their way of development from the basic intuition [i.e., on the way they are constructed in the mind], and that from this special character of their way of development from the basic intuition, properties ensue which for classical mathematics are false. A striking example is the intuitionist theorem that a full function

of the unit continuum, i.e. a function assigning a real number to every non-negative real number not exceeding unity, is necessarily uniformly continuous. (Brouwer 1949C, p.1244)[32]

The reference to the uniform continuity theorem of course is an implicit reference to the Bar Theorem as well, for Brouwer used the latter as a lemma for the former. One might of course wish to contest that the crucial claim at which Brouwer arrived by reflection on the mental is evident or at least correct.[33] We do not propose to discuss the question of correctness here; what interests us here is the availability to Brouwer of this particular strategy. For Brouwer, it is quite natural to devise transcendental arguments in mathematics, because he is an ontological descriptivist for whom all mathematics is mental to begin with. In doing mathematics, this mental nature can often be left implicit, but one is free to exploit it.

For contrast, one may think of Leibniz. Leibniz held that, ontologically, the objects of geometry are objects in God's mind, but he also said that this ontological fact does not make it impossible for an atheist to be a geometer (*Théodicée*, §184). What Leibniz seems to imply is that in doing geometry, the specifically ontological properties of the objects are of no interest. In particular, Leibniz nowhere suggests that the specifically ontological properties of mathematical objects can be exploited to demonstrate mathematical truths about them. However, it is not clear that for Leibniz that would have been impossible in principle.

Acknowledgement

We are grateful to Dirk van Dalen, Director of the Brouwer Archive, Utrecht, for permission to publish the items on page 70 and in footnote 23.

Appendix: Kolmogorov's report on (Freudenthal 1936) and (Heyting 1936), Zbl 0015.24202

Bekanntlich besteht vom intuitionistischen Standpunkte aus ein mathematischer B e w e i s immer in der Durchführung einer K o n s t r u k t i o n. Wird ein S a t z hypothetisch ohne Beweis ausgesprochen, so ist es vielmehr nur eine I n t e n t i o n oder eine A u f g a b e — die entsprechende Konstruktion zu finden. F r e u d e n t h a l lehnt die Betrachtung solcher Intentionen, oder Aufgaben, ab und behauptet: Ein mathematischer Satz läßt sich nur mit seinem Beweis formulieren. Folglich kann man, nach F r e u d e n t h a l, eine Implikation $a \supset b$ nur dann formulieren, wenn a schon bewiesen ist. Eine solche Deutung der Implikation ist, offenbar, praktisch inhaltslos. F r e u d e n t h a l versucht deswegen $a \supset b$ im Sinne des Prädikatenkalküls zu deuten. H e y t i n g unterstreicht in seinen Bemerkungen die Notwendigkeit und

[32] Compare (Brouwer 1981A, p.94).

[33] See, for example, Charles Parsons' remarks on p.451 of his introduction to Brouwer's 1927 paper in (van Heijenoort 1967).

Fruchtbarkeit, die Intentionen (oder Aufgaben) als Bestandteile einer Implikation zu betrachten. *A. Kolmogoroff.*

As is familiar, from the intuitionistic point of view a mathematical demonstration always consists in the execution of a construction. When a Satz [proposition, theorem] is stated hypothetically without demonstration, it is rather a mere intention or task — to find the corresponding construction. Freudenthal rejects the consideration of such intentions, or tasks, and claims: a mathematical *Satz* [proposition, theorem] can only be formulated with its *Beweis* [demonstration, proof]. Consequently, according to Freudenthal, one can only formulate an implication a⊃b when *a* has already been demonstrated. Such an interpretation of the implication is, obviously, practically void of content. Freudenthal therefore attempts to interpret a⊃b in the sense of the predicate calculus. In his remarks, Heyting emphasizes the necessity and fecundity of regarding intentions (or tasks) as components of an implication.

A. Kolmogorov.

The intersection of intuitionism (Brouwer) and phenomenology (Husserl)

Richard Tieszen

1 Brief Historical Background on the Intersection of Intuitionism and Phenomenology

There are important and interesting conceptual links between Brouwer's intuitionism and Husserl's transcendental phenomenology. This is true independently of whether or not there had ever been a historical intersection of the two approaches to mathematics and logic. But of course there was a historical intersection. It occurred primarily in the nineteen twenties and early thirties, in the middle of the foundational debate between Brouwer's intuitionism and Hilbert's formalism. During this period there is a fascinating development and exchange of ideas on intuitionism, phenomenology, and formalism in the work of Hermann Weyl, Oskar Becker, Felix Kaufmann, and Arend Heyting, which is set against the background of the ideas of L.E.J. Brouwer, Edmund Husserl and David Hilbert.[1]

Philosophically, the fact that intuitionism and transcendental phenomenology should have been associated more or less strongly during this period is not surprising, since both Brouwer and Husserl wanted to take consciousness seriously, without objectifying it, attempting to reduce it to something else, or ignoring it. There are many details in their views that are involved in the association. In this paper I will only be able to scratch the surface. To see what some of these details are, I will start in §2 with an overview of ideas

[1] See, e.g., (Becker 1923), 1927, (Heyting 1931), (Kaufmann 1930), (Weyl 1918, 1921, 1925, 1926, 1928). See also the recent archival and scholarly research in, e.g., (Mancosu and Ryckman 2002), (van Atten 2005), and (van Dalen 1984). Philosophical and technical work in this tradition can be found, e.g., in (Martin-Löf 1984, 1987), (Sundholm 1983), (Tieszen 1989, 1992, 1994, 2000a,b), (van Atten et al. 2002), (van Atten 2004a, 2007). By the way, Brouwer and Husserl did actually meet once, in April 1928, in Amsterdam. We do not know what they talked about but their correspondence with other people indicates that they were evidently both pleased with the meeting.

in Brouwer's work that would resonate with those interested in Husserlian transcendental phenomenology. A certain subset of ideas in transcendental phenomenology is especially relevant. Once we see what these ideas are in §3 it will be very natural, I think, to speak of an intuitionistic (transcendental) phenomenology. I would like to make it clear from the outset, however, that I want to distinguish what might be called 'intuitionistic phenomenology' from traditional Husserlian transcendental phenomenology. It is the *intersection* of some important ideas in intuitionism and phenomenology that I want to focus on in this paper. There are also some important differences but I do not have space to discuss them here. In general, mathematics and logic enjoy a robust ideality, objectivity, and transcendence in Husserl's work after 1900, to an extent that is arguably incompatible with intuitionism.[2]

2 Some Central Philosophical Ideas in Brouwer's Intuitionism

In discussing Brouwer's Intuitionism in this paper I will refer to several periods in his development although, on the whole, I think that what is retained and developed in his later work preferable. In two manuscripts from 1951 (van Stigt 1990, pp.447–458), and also (Brouwer 1955), Brouwer says that for many years mathematics was considered to be the science of space and time, and that logic was used in mathematics in these circumstances, not only when space and time were believed to exist independently of human experience but even after they were taken to be forms of perception of the human mind to be studied in an interior world. The conviction that a mathematical assertion is either true of false whether anyone knows it or not continued to carry weight, as did the view that mathematical truths, like the laws of nature, will survive after the extinction of humankind.

Mathematics was detached from the external world of experience in several steps. Brouwer says that this first happened in the philosophy of Kant. The semi-intuitionism or pre-intuitionism of Poincaré, Borel, and Lebesgue followed in this vein but Brouwer argues that these developments were defective in several respects. Intuitionism then entered the scene. With intuitionism the question whether a meaningful mathematical assertion is correct or incorrect is freed from any recourse to forces independent of the thinking subject, such as the external world, mutual understanding, the use of language, or axioms that are not supported by inner experience. This question now becomes exclusively a matter of individual consciousness of the subject. Correctness of an assertion now has no meaning other than that its content appears in the consciousness of the subject. The criterion of truth or falsehood of a mathematical assertion is confined to mathematical activity itself, without appeal to logic or to a hypothetical omniscient being. An immediate consequence

[2] See, e,g., (Tieszen, forthcoming).

is that in mathematics no truths can be recognized that have not been experienced. Intuitionism rejects all non-experienced truths. This is Brouwer's idealism, which I will consider below in relation to Husserl's transcendental phenomenological idealism. As a result of this idealism, Brouwer says, the two cases of truth and falsehood for a mathematical assertion P, formerly exclusively admitted, are replaced by the following three:

1. P has been proved to be true.
2. P has been proved to be false.
3. P has neither been proved to be true nor to be absurd, nor do we know a method (Brouwer sometimes says 'finite algorithm') leading to the statement that either P is true or that P is absurd. If we do know a method leading to either P is true or P is absurd, then this third case is reducible to the first or second cases. This applies in particular to assertions of possibility of some construction of bounded finite character in some finite mathematical system, because such a construction can be attempted only in a finite number of ways and each attempt proves successful or abortive in a finite number of steps.

Brouwer sometimes uses slightly different language to set out his threefold categorization. He says, for example, that we need to distinguish the 'true', the 'impossible now and ever', and the 'at present neither true nor impossible'.

In contrast with the 'perpetual' character of the cases where P has been proved to be true, P has been proved to be absurd, or we have a method leading to P or the absurdity of P, the case in which we do not presently have such a method is said by Brouwer to be 'temporary', in the sense that it is possible that one day a method might be found. Thus, an assertion in case 3 can pass over into one of the other two because we go on thinking and this may generate P or ¬P or a method leading to one of these, but also because in intuitionistic mathematics a mathematical entity (e.g., a choice sequence) is not necessarily predeterminate and may in its state of free growth acquire at some time a property that it did not possess before. The mathematical entities underlying the assertions, Brouwer says, are often in a state of becoming.

Now this 'time' of which Brouwer speaks in describing intuitionistic constructions is 'internal time'. It is the time that is a form of human consciousness, in line with the Kantian turn mentioned earlier. It is the time of 'internal sense', not of 'external sense'. In his 1907 thesis Brouwer distinguishes 'intuitive' time from 'scientific' time, and says that the only *a priori* element in science is time in the 'intuitive' sense (Brouwer 1907, p.99n.2)/(Brouwer 1975, p.61n.2).

The so-called 'first act of intuitionism', with which Brouwer intervened in the development of foundations, refers to internal time in an essential way. This first act, he says, separates mathematics from mathematical language and, in particular from the phenomena of language described by theoretical logic. One can read Brouwer's remarks here as cautionary. The reason for wanting this 'separation' is that the use of language in mathematics and in logic in particular can outrun our mental mathematical constructions. Furthermore, the separation allows us to avoid or to at least be clearly aware of the Hilbertian move of making mathematical language itself the object of study in mathematics (see Brouwer's characterization below of the 'genetic order' of stages in the logico-linguistic treatment of mathematics). It is also a way of acknowledging the imperfections of language in communication. Thus, Brouwer describes mathematics as a languageless activity of the mind that has its origin in the basic phenomenon of the perception of a *move of time*. The notion of 'origin' here is important, since Husserl also seeks the origins of the various sciences, including mathematics and logic, and in Husserl's lectures on internal time ranging from 1893 to 1917 (Husserl 1966) we see that all activities of consciousness have their origin in the structures of internal time. In Brouwer's description the perception of the move of time is the 'falling apart of a life moment into two distinct things, one of which gives way to the other, but is retained in *memory*.' Brouwer says that a two-ity emerges in this manner and, upon being divested of all quality, it is the common substratum of all two-ities that remains, the mental creation of the *empty two-ity*. It is this common substratum, this empty *form*, which is the basic intuition of mathematics (see (Brouwer 1952B) and (Brouwer 1954A)). In 'Intuitionism and Formalism' (Brouwer 1912A) Brouwer says that the falling apart of moments of life into qualitatively different parts, to be reunited only while remaining separated by time, is the fundamental phenomenon of the human intellect. When we *abstract* from its emotional or qualitative content we obtain the fundamental phenomenon of mathematical thinking, the intuition of bare two-oneness.

Brouwer's idea here does not seem to me to be mysterious. If there is consciousness or experience at all then it has a temporal character. Now suppose that something stands out in your sensory experience. You take note of some stimulus. Then at a later point something stands out in your sensory experience. This is already a structural description, since I am saying that *something* stands out but I am not saying *what* the content of the sensory experience is. If I diagram what I have just said as

then at the later stage the earlier stage is still retained in memory. It is not just forgotten or no longer in mind. It is retained in an appropriately modified way in the later awareness. On the horizontal axis we have a multiplicity, two, separated by time. If we drop a vertical axis below the rightmost stroke to show what is retained in memory at that time, then we have the present content and the past content together as a unity in consciousness, not separated by time, at that stage. We have a 'oneness', thanks to the retention. It is when we are aware of this 'two-oneness' as a form, irrespective of the qualitative or sensory content given at the stages, that we have Brouwer's 'basic intuition of mathematics'. There is an abstraction from the sensory contents of the experience (see also (Tieszen 1989, pp.99–111)).

A similar situation would arise if you were listening to me speak. As I speak, the words of my utterance would be slipping back into the past but you would be (passively) retaining them and, as retained, they figure into your present understanding of my utterance. The earlier parts of the utterance are retained in an appropriately modified manner. A multiplicity of parts of the utterance, separated by time, are held together at a later stage as a whole in order for your understanding of the utterance to be possible. Brouwer's two-oneness is just a very simple phenomenon of the same type.

From the point of view of Husserl's phenomenology, many further interesting questions can be asked about Brouwer's analysis of the origins of the concept of number.[3] I cannot go into the matter here, except to say that I think that Brouwer can at least be viewed as describing a process or structure at the deepest level of constitution — the level of the constitution of all objects in internal time — that founds or is a realization of the concept of number, whereas this will presumably not be possible for all mathematical concepts. It seems that for many concepts of mathematics there will not be such a correlation or founding basis in the most primitive cognitive processes of human beings in sense experience.

Brouwer of course says that this intuition of two-oneness creates not only the numbers one and two, but also all the finite *ordinal* numbers, inasmuch as one of the elements of the two-oneness may be thought of as a new two-oneness, which process may be repeated indefinitely. This is a process we can continue to carry out. If we write down a trace of the two-oneness, we have $|(|)$. The parentheses are used to indicate what comes later in time. Let us always replace the rightmost stroke with a two-ity. Then we have $|(|(|))$. And so on. Brouwer says it is 'introspectively realized' how the basic operation of thinking of one element of the two-oneness as a new two-oneness, continually displaying *unaltered* retention by memory, successively generates each natural number. The operation generates the infinitely proceeding sequence of the natural numbers, which is how the infinite ordinal number ω is to be understood. It also generates arbitrary finite sequences, and infinitely proceeding sequences of mathematical systems previously acquired.

Brouwer also says in 'Intuitionism and Formalism' that this basal intuition of mathematics in which the connected and the separate, the continuous and the discrete are united, gives rise immediately to the intuition of the linear continuum; i.e, to the 'between', which is not exhaustible by the interposition

[3] For example, Robert Tragesser has argued in conversation that Brouwer's description of the move of time does not found the concept of natural number but, rather, it presupposes the concept of natural number. Space limitations prohibit further discussion here, but I think one can at least argue for founding in the sense indicated in this paragraph. Tragesser is not singling out Brouwer here. He also objects to a major part of Husserl's work: the analysis of the origins of concepts of mathematics and logic.

of new units and which can therefore never be thought of as a mere collection of units. I will compare this below with Husserl's views on the essential continuity of the stream of consciousness. This claim is already in Brouwer's 1907 thesis (Brouwer 1907, p.17), where he argues that the discrete and the continuous each require the other. They are inseparable and each depends on the other. Neither is eliminable in favor of the other.

Continuing his analysis of origins, Brouwer says that all mathematical sets of units, properly so-called, can be developed out of the basal intuition. Then any previously constructed objects (including previously constructed sets of units or previously constructed sequences) can be taken as a unit. In all of this, language is supposed to have no other role than serving as a non-mathematical auxiliary to assist memory or to enable different individuals to build up the same objects (Brouwer 1913C, p.86). Brouwer also says that the intuitionist can never feel assured of the exactness of a mathematical theory by such guarantees as a proof of its being consistent (Hilbert), the possibility of defining its concepts by a finite number of words (Poincaré), or the practical certainty that it will never lead to a misunderstanding in human relations (Borel).

Suppose that an intuitionistic mathematical construction has been carefully described in words and the introspective or inner character of the mathematical construction is ignored for a moment. If its linguistic description is considered by itself and submitted to a linguistic application of a principle of classical logic, then is it always possible to perform a (languageless) mathematical construction corresponding to the linguistic application of the principle of classical logic? The answer is 'yes' for the principles of contradiction and syllogism but 'no', except in special cases, for the principle of the excluded third. If each linguistic application of the principle of the excluded third in a mathematical argument were to accompany some intuitionist mathematical construction, then each intuitionistic assertion (i.e., assignment of a property to an intuitionistic mathematical entity) could be *judged*, i.e., either be proved or reduced to absurdity. Every construction of a bounded finite character in a finite mathematical system can be attempted in only a finite number of ways, and each attempt can either be carried through to completion or continued until further progress is impossible. Hence, as mentioned earlier, every assertion of possibility of a construction of a bounded finite character in a finite mathematical system can be judged. In this special case the application of the principle of the excluded middle is permissible. In the case of infinite systems, however, Brouwer argues that we must abandon this principle. Many intuitionistic counterexamples to the principle of the excluded third can of course be constructed. Brouwer says that the belief in the universal validity of this principle should be considered a phenomenon of the history of civilization of the same order as the former belief in the rationality of π or in the rotation of the firmament around the earth. Why would the dogma of its universal validity have persisted for so long? Perhaps because (i) within a given domain of mathematical entities previously acquired a single application of the principle is non-contradictory; and (ii) the principle stands the test of its application to an extensive group of everyday phenomena of the external world.

Now how does Brouwer's account of origins lead us up to real analysis? Based on what has been said thus far one might fear that there is no place for analysis. This fear presupposes, however, that the infinite sequences generated by the self-unfolding of the basic intuition would have to be predeterminate infinite sequences which proceed in such a way that from the beginning the n-th term is fixed for each n. The 'second act of intuitionism' recognizes the possibility of generating new mathematical entities in the form of infinitely proceeding sequences whose terms are chosen more or less freely from mathematical entities previously acquired. Brouwer says in some of his writings that in these choice sequences the freedom existing at perhaps the first choice may be subjected again and again to progressive restrictions at subsequent choices, and that all these restricting interventions, as well as the choices themselves, may at any stage be made to depend on possible future mathematical experiences of the creating subject. (In a footnote in (Brouwer 1952B), p.142, Brouwer says that in former publications he has sometimes admitted restrictions of freedom with regard to future restrictions of freedom, but that this is not justified by close introspection and would also endanger the simplicity and rigor of further developments.) The 'second act of intuitionism' also recognizes the possibility of generating new mathematical entities in the form of species, that is, in the form of properties supposable for mathematical entities previously acquired, and satisfying the condition that if they hold for a certain mathematical entity, then they also hold for all mathematical entities that have been defined to be equal to it (where relations of equality are to be symmetric, reflexive and transitive). The mathematical elements for which the property holds are called the elements of the species. Brouwer sometimes indicates explicitly that a species can be an element of another species but never an element of itself (see (Brouwer 1952B), but this condition is not included in (Brouwer 1954A)).

With the second act of intuitionism it is possible to introduce the intuitionistic continuum as the species of more or less freely proceeding convergent infinite sequences of rational numbers. Brouwer introduces his definition of the species which above all others 'fertilizes intuitionistic mathematics', the *spread*. Already in a paper of 1918 (Brouwer 1918B) he had presented his continuity principle and used it to show the non-denumerability of the continuum. With additional reflection on the nature of choice sequences Brouwer was able to prove his uniform continuity theorem: all total real-valued functions are uniformly continuous. The Bar Theorem, the Fan Theorem, and other distinctive elements in Brouwer's treatment of analysis are then developed. These distinctive features of intuitionistic real analysis are thus supposed to have their origins in the 'second act of intuitionism'.

If this is how we are to build up intuitionistic mathematics from origins then how far can we go? What are the limits? In his critique of Cantorian set theory, already in his 1907 thesis Brouwer argues that Cantor's second number class does not exist (Brouwer 1907, p.83). The reasoning that allegedly leads to the second number class, with the meaning that Cantor puts into it, 'makes no sense'. The same thing will of course have to be said of all of higher set theory. Although one could go into many fine points about Brouwer's views on the infinite, I will just note for now that in his thesis Brouwer says that for sets we have the following cardinal numbers, in order of magnitude: the different finite numbers, the denumerably infinite, the denumerably unfinished, and the continuous. Brouwer's critical and also positive analyses of parts of mathematics that go beyond arithmetic, elementary geometry, and very elementary set theory are much more detailed and fine-grained than anything we find in Husserl's work.

Although I do not have space to consider the matter here, it is also very interesting to compare Brouwer's critique of Hilbert's formalism with ideas in Husserl's transcendental phenomenology. Brouwer's 1907 enumeration of the 'genetic order' of the stages in the logico-linguistic treatment of mathematics is, for example, quite remarkable. These stages, Brouwer claims, are confused in the 'logical' treatment of mathematics: (1) The construction of pure mental mathematical systems; (2) The linguistic description of mathematics; (3) The mathematical study of (2), including the logical structure; (4) Abstraction from the meaning of the expressions of (3), and of the logical operations. This creates a 'second-order' mathematical system, as in the work of the logicists, but without the precautions taken by Hilbert to insure consistency. (5) A focus on the language of (4); (6) The mathematical study of (5), which is essential to Hilbert's work but is not found in the work of Peano or Russell. (7) Abstraction from or forgetting the meaning of the expressions of (6), thus creating a mathematical system of 'third-order'; and (8) The language of (7), which is the last stage considered by Hilbert. (See (Brouwer 1907, pp.94–95), and (Troelstra and van Dalen 1988, vol.1,p.23).)

Let us now turn to some of the most relevant ideas in Husserl's transcendental phenomenology.

3 On the Way to Intuitionistic (Transcendental) Phenomenology

We can now make comparisons with some ideas in Husserl's transcendental phenomenology in order to see how intuitionism and phenomenology came to be closely associated with one another in some of the work of Weyl and Becker, and also to some extent in the work of Heyting and Kaufmann.

One of the key ingredients in this association was a certain way of reading the transcendental idealism that Husserl began to develop around 1907. In the *Idea of Phenomenology* from 1907 (Husserl 1999) and *Ideas I* from (1913) (Husserl 1983) we are told that the effect of the phenomenological reduction (*epoché*), the method of bracketing or suspension which is necessary to do phenomenology, is to restrict ourselves to what is immanent. It is a disengagement from the natural attitude, the ordinary non-philosophical attitude in which we take ourselves to be experiencing transcendent objects. In the phenomenological attitude, obtained by the reduction, we experience, on the basis of reflection, the immanent. It is a method for inducing in us, in the first instance, an awareness of the internal. This would be an awareness of the internal of the sort that is crucial to Brouwer's work. Husserl then says that the immanent is what is absolute, while the transcendent is not. What is transcendent is always relative to consciousness. As a result of the transcendental phenomenological reduction Husserl argues, in §46 of *Ideas I* for example, that any physical thing that is given 'in person' can be non-existent but that no mental process which is given 'in person' can be non-existent. The non-existence of the world is conceivable but the existence of what is immanent — the absolute being of mental processes — would in no respect be altered thereby.

Whatever things are, Husserl says, they are as experienceable things. It is experience alone that prescribes their sense. We must not let ourselves be deceived by speaking of the thing naively as something that transcends consciousness and exists in itself, apart from any possible relation to consciousness. The genuine concept of transcendence can only be derived from the contents of our experience itself. '*An object existing in itself is never one with which consciousness or the ego pertaining to consciousness has nothing to do.*' (*Ideas I*, §47). In §49 Husserl says that the whole spatiotemporal world and each of its constituents is thus, according to its sense, a merely intentional being. It is a being posited by consciousness in experience. Each constituent of the world, of essential necessity, can be determined and intuited only as something identical through motivated multiplicities of appearances. It is something invariant for consciousness through a manifold of appearances. Beyond that it is nothing.

It is not that we cannot speak of objects (of consciousness) from this perspective. Rather, we will find that we can speak of objects from within the *epoché*, thanks to the intentionality of consciousness, for intentionality is just object-directedness.

This sphere of absolute consciousness that remains as a residuum after the conceivable annihilation of the world is what provides the subject matter for pure phenomenology. It is quite natural to suppose that this is just the sphere in which Brouwer is working, the sphere of consciousness which, for mathematics in particular, is Brouwer's sphere of 'introvert mathematics'. From the point of view of this sphere of absolute consciousness, Husserl says, we are to think of all reality as existent by virtue of a *sense-bestowing* consciousness which, for its part, exists absolutely and not by virtue of another sense-bestowal.

Transcendental phenomenological idealism is therefore the view that it is only in our own experience that things are 'there' for us, given as what they are, with the whole content and mode of being that experience attributes to these things. In Husserl's *Formal and Transcendental Logic* from 1929 (Husserl 1969) this is put as follows: 'nothing exists for me otherwise than by virtue of the actual and potential performance of my own consciousness' (*FTL* §94). Whatever I encounter as an existing object is something that has received its whole sense of being from my intentionality.

In *FTL* and other places Husserl notes that the claim that every existent is *constituted* in the subjectivity of consciousness is, in the first instance, the claim that it is from *me* that everything receives its sense of being. Each person must start from his or her own subjectivity. In his investigations of intersubjectivity in the *Cartesian Meditations* from 1931 (Husserl 1973b) Husserl in fact starts with a reduction to what he calls the 'sphere of ownness' in order to determine how both intersubjectivity and the sense of an objective world are constituted. We are supposed to analyze the manner in which these phenomena are constituted. There is a place in phenomenology, he says, for a transcendental egology. Brouwer also seems to start with an egology and, depending on how he is read, perhaps he never gets beyond this sphere. It is quite interesting to study Brouwer's views on solipsism, intersubjectivity, and language from this phenomenological perspective, but I cannot go into the matter in more detail here.

Another effect of the transcendental phenomenological reduction which is central to transcendental idealism is that consciousness loses its attachment, so to speak, to material reality naively understood, and thus its incorporation into space and its place in transcendent or 'external' time. What we are left with is internal time, the time of mental processes with their modes of givenness of now, before, and after, and their modally determined simultaneity and succession. Internal time cannot be measured by clocks, positions of the sun, or any other external means. Internal time is the unitary form of all mental processes within one stream of consciousness.

Anything that can be an object for us is constituted as such (or the consciousness of it is constituted as such) in the stream of consciousness, including all internal objects, such as mental processes themselves, and all external objects. This must also be true for what Husserl calls 'categorial' objects, such as mathematical objects, which are neither spatial nor temporal. By 1917 or so Husserl says that mathematical objects are 'omnitemporal'. This means that they are not outside of time but, rather, that they have being at all times, in the sense that correlated with them at any time — past, present, or future — are possible acts of consciousness of them.

The flow of consciousness in internal time is continuous. Husserl's reflections on the continuity involved are quite interesting. This continuum is not composed of discrete points. The phases of consciousness, the 'parts' of the stream of consciousness, are evidently non-independent. Furthermore, in the language of Investigation III of the *Logical Investigations* from 1901 (Husserl 1973a), they interpenetrate one another. We do not actually experience a fixed, durationless time-point in the stream of consciousness. There are no discrete parts of the temporal stream itself, no exact now points. To speak of such a point is already an abstraction or idealization. Rather, what we actually have is a 'specious' or 'extended' present, with its backward-looking retention and its forward-looking protention. Using metaphorical language, Husserl describes this specious present as a 'halo' or a 'comet's tail'. It's extension is indeterminate and shades off continuously.

As Brouwer says, in the flow of internal time one can be aware of the falling apart of a life moment into two distinct things, one of which gives way to the other but is retained in memory. One can discern a before and after, or a now and a later, but there is always that which is 'between' these two points, and this 'between' is not exhaustible by the interposition of new units. This 'between' can never be thought of as a mere collection of units.

Influenced by his study of phenomenology, Hermann Weyl suggested in his book *Das Kontinuum* of 1918 that we take the flow of internal time itself to be the most fundamental continuum (see (Weyl 1994, ch.2, §6)). If we do this, however, then we cannot think of the continuum as composed of a set of durationless points. To resolve the continuum into such a set is precisely to break up the flow, to misconstrue the fluid paste of the intuitive continuum. We would have an atomistic and static view of the continuum. Could there possibly be a mathematics or logic of the continuum of internal time itself? By 1921, Weyl had turned to Brouwer's mathematics of the intuitive continuum on this issue (Weyl 1921).

As noted above, Brouwer, like Husserl, distinguishes internal time from external 'scientific' time. On Brouwer's view the continuum is also not composed of idealized discrete, durationless points. Rather, a 'point' is something that is itself becoming as it is generated through time. Such 'points' are given by free choice sequences (see (van Atten et al. 2002), and (Tieszen 2000b)). Brouwer says that the 'synthetic construction' of sets, which consists of combining discrete elements into a set, is entirely inappropriate for generating the continuum. Rather, the characteristic construction of the continuum is 'analysis', i.e., decomposition into homogeneous parts. 'Analysis' in this sense splits the continuum or interval into two parts, subintervals, homogeneous to each other and to the whole interval. Each subinterval is an interval in its own right. The act of insertion and the natural order of the continuum generate a relation of inclusion and an order relation between subintervals. Brouwer holds that the relation of inclusion or subinterval, which is a part-whole relation, is the fundamental relation of the continuum, not the set-element relation. The order relation between disjoint subintervals is the natural order of the continuum, abstracted from the progression of time. Much more can be said about all of this (as in (van Atten et al. 2002)). Some principles of classical logic (e.g., excluded middle), with their built-in assumptions about the exactness or discreteness of objects, would need to be abandoned if we take such an intuitive continuum seriously.

For Husserl, one of the most important features of consciousness that we find if we attend to the sphere of absolute consciousness given to us after the phenomenological reduction is intentionality. Intentionality is the 'aboutness' or object-directedness of the consciousness of the (transcendental) subject. Consciousness is consciousness of something. Now in Husserl's *Logical Investigations*, parts of which clearly influenced Weyl, Becker, and Heyting to associate phenomenology with intuitionism, we are told that consciousness is directed toward objects or states of affairs by way of (in the language of *LI*) the 'meaning-intentions' of our acts of consciousness. An act of consciousness can be directed toward an object or state of affairs by virtue of its meaning-intention even if there is no object or state of affairs.

We only have *knowledge* of an object or state of affairs toward which we are directed, however, when our meaning-intentions are fulfilled. In Husserl's analysis of knowledge in the *LI*, especially Investigation VI, the distinction between empty meaning-intention and meaning-fulfillment plays a key role. A meaning-intention is fulfilled when the object toward which we are directed by the meaning-intention of our act is present in our experience, and is fulfillable when the object could be made present to consciousness through sequences of acts in which it comes to be experienced. It is only when this happens that we have *evidence* for the object or state of affairs. Short of that we have only the empty meaning-intention. To actually experience the object is just to *intuit* it. Fulfillment, in other words, is another way of speaking of intuition, and fulfillability another of speaking of intuitability. The meaning-intention/meaning-fulfillment distinction is something like the conception/intuition distinction. One might meaningfully conceive of certain objects or states of affairs but we might also be able to intuit or experience the objects of some of these conceptions.

There can be either a static or a dynamic fulfillment of a meaning-intention. The fulfillment is static when the object is immediately present to consciousness, and it is dynamic when a sequence of acts in time is required in order to experience the object, or at least to experience it completely.

The meaning-intention/meaning-fulfillment distinction is applied by Husserl in ordinary perception, which he calls 'straightforward intuition', but also in the case of mathematical consciousness. It is argued that mathematical meaning-intentions, which can be expressed in mathematical propositions or judgments, may also have corresponding meaning-fulfillments. In this case we are no longer dealing with ordinary sensory perception but rather with what Husserl calls 'categorial intuition', the intuition of categorial objects. Categorial intuition is founded intuition. It is founded on underlying acts of straightforward perception. It has its *origins* in such underlying acts. Categorial intuition, as founded intuition, requires acts of abstraction and reflection.

It is natural to interpret Brouwer's form of two-ity in terms of some of the elements of this scheme. First there is an awareness of the falling apart of a life moment into two distinct things, one of which gives way to the other but is retained appropriately in memory. Reflection is already required in order for there to be a perception of a move of time in the first place. Then there is an abstraction from the qualitative content of the experience in which the *form* of the two-ity is retained, the common substratum of all two-ities. This is a formal abstraction, as it were, from the content of the experience in which the two-ity is noticed. The form could be instantiated in many different contents. It is in this manner that we have what Brouwer calls the mental creation of the empty (in the sense of formal) two-ity. The basic intuition of mathematics, in Brouwer's sense, is a founded, formal intuition in Husserl's sense.[4] Fulfillment of a meaning-intention involving a natural number just amounts to iterating the two-ity until we obtain the particular number intended. This would typically be a dynamic fulfillment. To be presented with the number requires a sequence of acts through time.

Now we just think of Brouwer's view of the build-up of mathematics from this origin in terms of founding. There are the founding acts and then there are the higher-level founded acts in which the previously constructed objects can be taken as the data for new constructions. From this point of view, Brouwer is investigating the mathematical meaning-intentions for which there are meaning-fulfillments. He is asking, in effect, about the meaning-intentions for which we in fact have *evidence* (in intuition): what is the extent of mathematical knowledge, at least insofar as it is knowledge of particular mathematical objects? What we have here is an account of mathematical intuition, since fulfillment just is fulfillment in intuition. One can see why Weyl and Becker, who were both quite aware of Husserl's view of founded intuition and categorial intuition, would be tempted to identify Brouwerian mathematical intuition with Husserlian categorial intuition of mathematical objects. It is another question, as indicated in footnote 4, whether such an identification holds in general or not. It seems safe to say that Brouwerian mathematical intuition is a founded form of intuition, but that it is not categorial intuition of unchanging, exact objects in Husserl's sense.

Thus far I have focused on how the fulfillment of mathematical meaning-intentions is to be understood. Meaning-intentions may also, however, be frustrated. If we consider the phenomenology of knowledge as it is described in the VIth *Logical Investigation* then we are told that a meaning-intention may be either fulfilled, frustrated, or neither. A meaning-intention is frustrated when it is negated by what we are presented with in experience, in intuition. To take a simple example from ordinary perception, I might believe that the ball I am perceiving is red all over and then find in a subsequent course of expe-

---

[4] It might be argued that Brouwer is in this manner speaking of a founded form of inner intuition but not yet of categorial intuition in Husserl's sense. That is, the form or structure of a mental phenomenon is not yet a categorial or ideal object in Husserl's sense. Categorial objects, for Husserl, are not supposed to be mental entities nor are they supposed to depend in an essential way for their existence on mental phenomena. Rather, in the case of mathematics they are supposed to be non-mental, 'ideal' objects. In Husserl's view, it appears that the subject can in a non-arbitrary way constitute a very robust sense of objectivity and mind-independence in the case of mathematics and logic. See, e.g., (Tieszen, forthcoming).

rience or intuition of the ball that part of it is green. The meaning-intention expressed by the sentence 'The ball is red all over' is thereby negated. It is *not* the case that the ball is red all over. My experience of it as red all over is frustrated. Indeed, Husserl says that it is in ordinary perceptual experience of this sort that we find the origins of negation. Meaning-intentions that are frustrated are said to be 'empty', and intentions that are neither fulfilled nor frustrated may also be regarded as 'presently empty'. Now if we apply this categorization in mathematics then we have a remarkable parallel with the categorization given by Brouwer above, for now a mathematical judgment P may be either

1. fulfilled,
2. frustrated,
3. neither fulfilled nor frustrated, nor do we have a method leading to either the fulfillment or frustration of P.

In mathematics, the parallel, in Husserl's language, would be to hold in case 1 that the judgment is 'perpetually' fulfilled, and in case 2 that the judgment is perpetually empty. Why perpetually? The idea is that once a *mathematical* judgment is fulfilled or frustrated we are to regard it as fulfilled or frustrated for all times thereafter. (See also Kripke of Beth semantics for intuitionistic logic.) One might say that this marks an 'idealization' that is involved in mathematical proof but that is not involved in empirical confirmation. A perceptual judgment, such as 'This room is painted white', might be fulfilled now but at some later time it might be frustrated. The difference between the mathematical and perceptual cases would thus mark at least the possibility of obtaining a kind of necessity or certainty in mathematics and mathematical knowledge. It would mark an *a priori* character, in the sense that, ideally, no future course of experience should overturn a mathematical proof. Those mathematical judgments which really are either fulfilled or frustrated are not defeasible. In case 3 the judgment is 'temporarily empty', in the sense that one day it might be either fulfilled or frustrated. As we noted above, Brouwer says that this is because we go on thinking and this may generate P or $\neg$P, but also because mathematical entities such as choice sequence are not necessarily predeterminate.[5]

It is quite remarkable that Brouwer and Husserl both arrived at these three possibilities for judgments, evidently quite independently of one another. As

[5] Husserl does not comment explicitly on choice sequences. It appears, however, that for Husserl choice sequences could not count as ideal, objective, exact *mathematical* objects. Even if this is so, it does not follow that there are not mental phenomena answering to Brouwer's conception of choice sequences. I think there are in fact such phenomena. Van Atten also argues that there are mental phenomena answering to the conception of choice sequences, that the conception is mathematically coherent, that these objects are formal in Husserl's sense (although dependent on time), and that they preserve intersubjectivity and monotonicity of mathematical truth. See especially section 6.3 of (van Atten 2007).

I said, this categorization is clearly present in Husserl's *Logical Investigations* (Husserl 1973a), especially Investigation VI. If we combine the categorization with a transcendental phenomenological idealism according to which there are no non-experienced and non-mental objects and truths, then we arrive at a convergence of views.[6] Correctness of a judgment now has no other meaning than that its fulfillment appears in the consciousness of the subject.

Oskar Becker, in his 1927 monograph *Mathematische Existenz* (Becker 1927), was the first person to notice this parallel between Husserl's judgments that are fulfilled, frustrated, or neither fulfilled nor frustrated and Brouwer's assertions that are either proved true, proved false, or neither proved true nor proved false. His work on this in *ME* influenced Heyting (see (Heyting 1931) and (van Atten 2005)), who used the idea that constructions (= intuitionistic proofs) are fulfilled mathematical intentions in order to interpret the intuitionistic logic constants. In Becker's own work in *ME* there is a mistake in interpreting the third case in this triad, the case in which a judgment is neither presently fulfilled nor frustrated. In effect, Becker represents this case symbolically as $\neg(P \vee \neg P)$. In other words, he has the three cases as $(P \vee \neg P \vee \neg(P \vee \neg P))$, which is intuitionistically invalid. The problem, from an intuitionistic point of view, is that $\neg(P \vee \neg P)$ is not what it would mean to say that a meaning-intention is temporarily (presently) empty. Becker does not correctly render either Husserl's phenomenological view of this situation or Brouwer's view. $\neg(P \vee \neg P)$ is just $(P \vee \neg P) \rightarrow \perp$, which is not intuitionistically acceptable. If G is Goldbach's conjecture then we can say that $(G \vee \neg G)$ has not been experienced but that we also do not know which one is true. We are certainly not in a position to say that on the basis of this meaning-intention we can derive an absurdity. Nonetheless, Husserl's categorization, by way of Becker's work, played a role in the history of what is now called the BHK (Brouwer-Heyting-Kolmogorov) interpretation of the logical constants (see also (Kolmogorov 1932)).[7] This is typically either not known or forgotten, but it should not be forgotten because it leads to what is still a perfectly relevant theory of mathematical intentionality and, thereby, to an interesting philosophy of mind for mathematics.

---

[6] Can the (transcendental) subject non-arbitrarily constitute non-mental, ideal and abstract transcendent objects and truths? What I have referred to as 'intuitionistic phenomenology' would evidently hold that this is still dubious epistemology and metaphysics, whereas Husserl's original transcendental phenomenological idealism, applied to mathematics, seems to be of just this type.

[7] It is worth noting that the BHK interpretation does not by itself force an intuitionistic understanding of the logical constants. If one interprets 'construction' (= fulfillment), 'operation', or 'function' in a classical manner, then the clauses of the BHK interpretation can be used to justify the principles of classical two-valued logic. See (Troelstra and van Dalen 1988, vol.1,p.9 and p.33, exercise 1.3.4). Thus, it is the understanding of the notion of construction (or fulfillment) that is crucial. In intuitionistic phenomenology it is the set of features summarized briefly in §3 below that provide this understanding.

As I said above, dynamic fulfillment of a meaning-intention requires a sequence of acts, and any sequence of cognitive acts for a human being will be finite. The sequence of acts in which we generate the natural numbers from the basic two-ity is completely determinate. The future course of experience here is fixed. In Husserl's language we could say that the '*horizon*' of possibilities of experience associated with the meaning-intentions that direct us toward particular natural numbers is fixed. It is 'lawlike'. The choice sequences introduced by Brouwer's second act of intuitionism can also be understood in terms of the sequences of acts by virtue of which certain kinds of meaning-intentions are fulfilled. In this case we are dealing with meaning-intentions that direct us toward particular real numbers. Choice sequences may be lawlike but need not be. The transcendental subject in this case is free to choose. There is a range of possibilities, with lawlessness providing the other limiting case. Choice sequences are still sequences carried out in time by a transcendental subject, and the subject will complete only a finite initial segment of any choice sequence, but in this case the future course of experience need not be fixed. There can be degrees of freedom/determinacy. The horizons associated with the meaning-intentions in this case are variable. A kind of free variation is possible here, which the transcendental subject can adjust and restrict.

Now the distinctive mathematics of the intuitive continuum that can be built on the basis of choice sequences is entirely Brouwer's creation. There is nothing like it in Husserl's work, but I want to suggest that one should continue to explore and deepen the phenomenological interpretation, underpinning, and development of this mathematical work of Brouwer.

We have already seen a number of similarities between the work of Brouwer and that of Husserl. Before concluding, I would like to mention one more similarity. I think it is evident that Husserl, like Brouwer, does not wish to conflate formal consistency and truth. This is already apparent in Husserl's *LI*, and is even clearer in *FTL*. In *FTL* Husserl distinguishes three levels of logic: pure grammar, consequence (or consistency) logic, and truth logic. Expressions must be grammatical in order to be consistent and they must be consistent in order to be true. Consequence logic is the level concerned with purely formal or symbolic mathematics and in order for such purely symbolic theories to be true of anything they must be consistent. Formal consistency, however, is not taken to guarantee truth, for otherwise there would not be a need for the level of truth-logic. Husserl does occasionally say in *FTL* that in consequence logic 'truth' is identified with formal derivability from axioms (provided the formal system is consistent), but this is a conception of truth that is still confined to the second level of logic. Truth-logic itself is concerned with evidence, intuition, and the fulfillability of mathematical intentions. Intuition and evidence are required for truth if truth is to amount to anything more than formal derivability in some purely symbolic theory.

Becker was working closely with Husserl during the period just before *FTL* was published. He was Husserl's assistant already in 1923. In *ME* (Becker 1927, p.509) Becker associates Hilbert's formalism with what Husserl calls 'consequence logic' or 'consistency logic', and he associates Brouwer's intuitionism with what Husserl calls 'truth logic'. He in fact frames the matter in terminology that is found later in *FTL* (1929):

> The basic distinction consists in the fact that the logic of truth is related to states of affairs of formal-*ontological* nature and strives to intuit (of course, categorially) these facts themselves insightfully, i.e., to grasp them in their original givenness ... By contrast, the 'logic of consequence' (and with it also Hilbert's mathematics — but not his metamathematics) is not directed, strictly speaking, at objects (which indeed have to be somehow accessible, in some way phenomena) but rather to mere 'posits' (*Gesetzheiten*), which in their inner structure are impenetrable. (Becker 1927, p.509)

Becker says that a metamathematical consistency proof might show the consistency of the 'posits' formulated in axioms involving transfinite objects but that objects such as actually infinite sets cannot be explicitly given to consciousness. Intentions directed toward such objects are in principle not fulfillable in categorial intuition. The transfinite is obtained by analogy with the laws holding for finite sets but, when interpreted contentually, transfinite axioms presuppose the givenness of actually infinite sets, and actually infinite sets are not accessible (Becker 1927, pp.515–516). Of course Becker does try at a later point in *ME* to provide a phenomenological account of the non-denumerable infinite in terms of potential infinity. This led to a significant rift with Weyl. Indeed, Weyl thought that such work 'would discredit the name of phenomenology among the concrete sciences'.[8]

4 Rejection of Classical Logic and Mathematics?

Now why would it be held, on the grounds of the particular subset of phenomenological ideas about the mind that we have considered, that classical logic and mathematics should be rejected in favor of intuitionistic logic and mathematics? The short and simple answer is this: it is on the basis of (i) a form of transcendental idealism according to which there are no non-experienced non-mental objects, and (ii) the view that *knowledge* of objects and truths in mathematics requires the founded *intuition* (i.e., fulfillment of mathematical meaning-intentions) through time by limited (transcendental) subjects that we must regard classical mathematics and logic as transcending our knowledge, as postulating objects and truths that are beyond all possible experience. We cannot in principle be conscious of the objects and truths of which classical mathematics and logic speak. In his later writings Brouwer

[8] See (Mancosu and Ryckman 2002), which includes the correspondence between Weyl and Becker that chronicles this rift.

already has the idealized subject, which we can think of as a transcendental subject, but classical mathematics and logic would go beyond even the capacities of an idealized mathematician.

That, to put it briefly, would be the reason why classical logic and mathematics should be rejected in favor of intuitionistic logic and mathematics. The grounds for this rejection, moreover, are found in a philosophical psychology. The concept of intentionality has been at the center of the philosophy of mind for the past thirty years or so. The case for the rejection of classical mathematics and logic is made on the basis of arguments about a form of transcendental idealism, the intentionality of human consciousness, and the possibilities of intuition, that is, of fulfillment or fulfillability of meaning-intentions, without a detour through the philosophy of language.

Much more needs to be said about the issue of the rejection of classical mathematics and logic, the sense in which it could be rejected, what the nature of the rejection would be, and so on. Furthermore, I have said very little in this paper about ideas in the work of Brouwer and Husserl that lie outside of the intersection of intuitionism and phenomenology. I certainly regard this as an important task, but it will have to be the subject of a different paper.[9]

[9] I would like to thank the editors of the present volume, and especially Mark van Atten, for helpful comments. I also thank the members of the audience at the Cerisy meeting *1907–2007: One Hundred Years of Intuitionism* for their comments on the lecture on which this paper is based. An earlier version of the lecture was presented in the philosophy of mathematics seminar at the Institut d'histoire et de philosophie des sciences et des techniques (IHPST), Paris, April 2007. Thanks are due to the audience members of that seminar for their comments. I hereby gratefully acknowledge that work on this paper was partially supported by a National Endowment for the Humanities (NEH) fellowship.

Brouwer on 'hypotheses' and the middle Wittgenstein

Mathieu Marion

Ich werde scheinbar, wider meinen Willen,
auf die Arithmetik zurückgeworfen.

L. Wittgenstein, January 1929[1]

Wittgenstein returned to Cambridge in January 1929. The earliest manuscripts of that period which we possess, MS 105, date from that month, opening on a few personal remarks, including a comment on his conversations with Ramsey, followed by remarks where we see Wittgenstein exploring new ideas about topics *not* covered in the *Tractatus*, such as the nature of irrational numbers or the contrast between a physical and phenomenological description of visual space. Whence these new topics and ideas? It is, exegetically speaking, natural to look for an answer in the prehistory of MS 105. Alas, any earlier manuscript, if any, must be assumed either destroyed or lost. An obvious starting point is Brouwer's lectures in Vienna in March 1928, 'Mathematik, Wissenschaft und Sprache' on the 10th (Brouwer 1929A) and 'Die Struktur des Kontinuums' on the 14th (Brouwer 1930A).[2] It appears that Wittgenstein only attended the first one, but it was reported by a witness, Herbert Feigl, that it spurred him into coming back to philosophy. According to Feigl:

> When the Dutch mathematician Luitzen Egbertus Jan Brouwer was scheduled to lecture on intuitionism in mathematics in Vienna, Waismann and I managed to coax Wittgenstein, after much resistance, to join us in attending the lecture. When, afterwards, Wittgenstein went to a café with us, a great event took place. Suddenly and very volubly

[1] 'It seems that I am thrown back, against my will, to arithmetic.'

[2] English translations under the titles 'Mathematics, Science, and Language' and 'The Structure of the Continuum' are available in (Ewald 1996, vol.2, pp.1175–1185,1186–1197) and in (Mancosu 1998, pp.45–53,54–63); I am using the latter, refer to 'Mathematics, Science, and Language' by '*MSL*', and to the original German by '*MWS*'.

> Wittgenstein began talking philosophy — at great length. Perhaps this was the turning point, for ever since that time, 1929, when he moved to Cambridge University Wittgenstein was a philosopher again. (Feigl 1981, p.64)

Perhaps one might find some clues in the text of that lecture.[3] However, although Wittgenstein's philosophy of mathematics was perceived by those who discussed it with him (an impressive list that includes, in roughly chronological order from the 1920s to the 1940s: F.P. Ramsey, B. Russell, R.L. Goodstein, A. Watson, A.M. Turing, G. Kreisel) as 'constructivist' and close to Brouwer's, this is not today's 'received view'.

The fact that Wittgenstein's views after 1929 were perceived as close in spirit to Brouwer's intuitionism is evidenced by a passage in Russell's report to the Council of Trinity College, in May 1930, written after reading part of the typescript of the *Philosophical Remarks* and discussing it for five days with Wittgenstein. Russell's sketch is brief, incomplete, and necessarily vague, but not, I contend, wholly inaccurate as an outline. The passage is worth quoting in full, although for the moment only the second paragraph is important (I shall come back to the first paragraph in section 2):

> According to Wittgenstein, when anything is the case there are certain other things that might have been the case in regard, so to speak, to that particular region of fact. Suppose, for example, a certain patch of wall is blue; it might have been red, or green, or &c. To say that it is any of these colours is false, but not meaningless. On the other hand to say that it is loud, or shrill, or to apply to it any other adjective appropriate to a sound, would be to talk nonsense. There is thus a collection of possibilities of a certain kind which is concerned in any fact. Such a collection of possibilities Wittgenstein calls a 'space'. Thus there is a 'space' of colours, and a 'space' of sounds. There are various relations among colours which constitute the geometry of that 'space'. ... Wittgenstein uses the word 'grammar' to cover what corresponds in language to the existence of these various 'spaces'.
>
> A considerable part of Wittgenstein's work is concerned with the interpretation of mathematics, He considers it false to say that math-

[3] One should note that this was not Wittgenstein's first encounter with Brouwer's ideas, since we know that he had read Ramsey's 1925 paper 'The Foundations of Mathematics' (Ramsey 1990, pp.164–224), and that it spurred him into an exchange with Ramsey in 1927 on identity, the gist of which is an attempt to undermine Ramsey's purported rescue of Russell's theory of types by using some of his ideas. For a discussion, see (Marion 1996) or (Marion 1998, pp.55–72). However, Ramsey mentions Brouwer only twice almost *en passant* (Ramsey 1990, p.166, p.219), and does not provide a detailed discussion. Incidentally, one of these two remarks is the well-known comment on 'the Bolshevik menace of Brouwer and Weyl' (Ramsey 1990, p.219), to which Wittgenstein replied (in 1931) by calling Ramsey a 'bourgeois thinker'.

> ematics is logic or consists of tautologies. He discusses 'infinity' at considerable length and links it with the conception of possibility that he has developed in connection with his various 'spaces'. He believes in 'infinite possibility', as he calls it, but not in actual 'infinite classes' or 'infinite series'. What he says about infinity tends, obviously against his will, to have a certain resemblance to what has been said by Brouwer. I think perhaps the resemblance is not so close as it appears at first sight. (Russell 1998, pp.439–440)

It is worth paying attention for a moment to Russell's wording in this last paragraph: the alleged resemblance between Wittgenstein's and Brouwer's views is presented as being *against Wittgenstein's will*. It seems obvious, therefore, that Russell raised the issue with Wittgenstein, who disagreed strongly, so that Russell felt hesitant: 'perhaps the resemblance is not so close as it appears at first sight'. However, this does not prove that Wittgenstein agreed with adversaries of Brouwer, as it is coherent with the view only that Wittgenstein's own views, although in the same ball park, nevertheless differ from Brouwer's — this is quite normal, after all there is a variety of viewpoints that the 'plain man' would *ordinarily* qualify as 'constructivist'. And this implies that Russell's report gives no support to the view that Wittgenstein did not get anything from listening to Brouwer in 1929; it would be a *non sequitur*. At all events, what I am after is not the claim that Wittgenstein held Brouwer's views, but that he found in his Viennese lecture inspiration for his later thoughts.

I have already published a paper (Marion 2003) where, in an attempt at gaining an *Übersicht* of the relation of Wittgenstein's philosophy to Brouwer's intuitionism, I dealt at length with Brouwer's lecture and its influence on the middle Wittgenstein's philosophy of mathematics, and I shall not repeat my analyses here. I wish instead to add two new elements. The first one concerns the historical events themselves and their significance. On the basis of new evidence, I argue in the next section that there are no reasons to believe that Wittgenstein's reaction to Brouwer's lecture was negative, on the contrary. Secondly, I shall argue that early sections of Brouwer's Viennese lecture, hitherto overlooked by commentators, contain one of the key elements for the explanation of Wittgenstein's early moves away from the *Tractatus*. If so, then Brouwer's influence on Wittgenstein is far more important than previously thought, even by sympathetic readers, and on a point not directly pertaining to the philosophy of mathematics, but concerning rather Wittgenstein's own conception of 'analysis'.

1 Some Historical Facts and Their Meaning

Wishing to shed light on a controversial topic such as the philosophical relationship between Wittgenstein and Brouwer, one's first task should be, to use

Collingwood's expression, to study the 'history of historical thought about it' (Collingwood 1939, 132). As we saw from Russell's account, Wittgenstein was perceived by those who knew him as holding views about mathematics that were not identical but close to Brouwer's. In a recent paper, Brian McGuinness quoted a letter from Ramsey to Abraham Fraenkel, which gives further support to this claim:

> Ramsey reports in a letter to Fraenkel in 1928 that [in 1924] Wittgenstein did not accept his (Ramsey's) solution which avoided the need for the axiom of reducibility but rejected all those parts of mathematics that depended on it: 'his conclusions were nearly those of the moderate intuitionists. What he now thinks I do not know'. (McGuinness 2006, p.25)[4]

From reviews by Paul Bernays, Sir Michael Dummett and Georg Kreisel,[5] the general agreement at the time of the posthumous publication, in 1956, of the *Remarks on the Foundations of Mathematics*, was, however, that Wittgenstein had actually been defending views that were neither antithetical nor akin to Brouwer's, i.e., what actually amounts to a radical form of constructivism called 'strict finitism'.

The first mention in the secondary literature that Wittgenstein actually attended Brouwer's lecture dates from 1964, in a footnote to a book by George Pitcher, *The Philosophy of Ludwig Wittgenstein*:

> Professor Herbert Feigl has kindly informed me, in correspondence, that he and F. Waismann spent a few hours with Wittgenstein in a café after the lecture, and that 'it was fascinating to behold the change that had come over W[ittgenstein] that evening.' Whereas previously he had been reluctant to discuss philosophy, and had even had to be persuaded by Waismann and Feigl to attend the lecture at all, now 'he became extremely voluble and began sketching ideas that were the beginning of his later writings.' In Feigl's opinion, 'that evening marked the return of W[ittgenstein] to strong philosophical interests and activities.' (Pitcher 1964, 8n.)

(Herbert Feigl's account, quoted above, was actually published much later, in 1981.) In this context, Wittgenstein's attendance and reaction to Brouwer's

[4] The easy way out here is to dismiss Ramsey (as well as Russell, as quoted above) as being clueless about Wittgenstein's views, an hypothesis that I am not ready to entertain myself, especially when voiced by people who are themselves clueless about issues pertaining to mathematical logic and the foundations of mathematics. After all, Ramsey was a first-rate philosopher on his own, who was the only person *at the time* to have had extensive discussions *with Wittgenstein*, and his critical study of the *Tractatus* shows that he understood it very well (at least better than some second-rate, know-it-all commentators who never met Wittgenstein).

[5] Respectively, (Bernays 1959), (Dummett 1959), and (Kreisel 1958).

lecture were perceived as a significant event. For someone like Robert Fogelin, who wished to defend an intuitionistic reading of Wittgenstein (Fogelin 1968), this was proof enough, but he did not bother tracing the connections between the content of Brouwer's lectures and that of the *Remarks on the Foundations of Mathematics.* More importantly, this was taken as evidence supporting a reading of the later Wittgenstein along lines laid out by Sir Michael Dummett in his review of that book and later developed in, e.g., the first edition of Hacker's *Insight and Illusion* (Hacker 1972) or Crispin Wright's *Wittgenstein on the Foundations of Mathematics* (Wright 1980). Kripke's controversial account of 'rule-following' (Kripke 1982) can also be seen as genetically related to this Dummettian approach.

Since then, however, the whole topic appears to have fallen victim to what is largely a parochial dispute, within the confines of Oxford. Repudiating his earlier views expressed in the first edition of *Insight and Illusion* (Hacker 1972, p.104), Peter Hacker stated in the revised second edition of that book that:

> [Wittgenstein] viewed intuitionism as an aberration, a perversion in mathematics that stands in need of philosophical therapy, not as a source of inspiration in philosophy of mathematics, let alone as involving an insight that can be generalized to the whole domain of philosophical logic and philosophy of language. (Hacker 1986, pp.120–121)

Furthermore, since the 1980s, under the successive influence of Gordon Baker and Peter Hacker in Britain as well as, more recently, the 'New Wittgenstein' in America, the 'received view' has it that *Wittgenstein held no particular view on mathematics* but was critical of all possible views, and if one can pinpoint a 'thesis' apparently held by him, this must be only a stage in a dialectical refutation of some other 'thesis', etc. In general, Brouwer's views are usually described in terms that are both anathema to the vast majority of analytical philosophers, and obviously anti-Wittgensteinian: 'mysticism', 'solipsism', 'psychologism', etc. Now, one might as well add 'constructivism' to this list; a safe bet, since most mathematicians profess not to be constructivists. In view of its current unquestioned popularity — indeed, not to be a conformist on this point is to elicit serious suspicions of incompetence — this may indeed be called the 'received view', although those who hold it fancy themselves as swimming against the current.

Within this context, it appears obvious that Wittgenstein could not have found any line of thought that he could have made his own in Brouwer's lecture. However, if Wittgenstein indeed viewed intuitionism as an 'aberration' and a 'perversion', then his reaction to Brouwer's lecture must have been entirely negative, so Feigl's report appears at first sight to stand in the way of this reading. But neither Feigl's report nor Menger's[6] say anything specific about the nature of Wittgenstein's reaction, so it is easy to assume that

[6] Karl Menger's evocative description was published only in 1994: 'Motionless from beginning to end, Wittgenstein looked at the speaker first with a slightly startled

Wittgenstein's excitement after the lecture was caused by his strong *disagreement* with Brouwer. This is how Ray Monk tentatively re-described the event in his biography:

> It would be wrong to infer from Feigl's report that Wittgenstein underwent a sudden conversion to Brouwerian intuitionism — although there can be no doubt that hearing Brouwer was a tremendous stimulus and may well have planted a seed that developed during the following years. ... It is likely, however, that Wittgenstein's excitement after the lecture had as much to do with his disagreements with Brouwer as with his agreement. (Monk 1990, p.249)

However, since eyewitness accounts do not go either way, they do not provide firm support to this claim either. I don't think, however, that this settles the issue, as there is further evidence that supports the claim that his reaction was not negative.

Before looking at it, I would like to comment on this parochial affair. It looks very much as if Hacker initially tried (in the first edition of his book) to develop an interpretation of Wittgenstein along lines set out by Dummett but then realized not only that this was not possible, but also that, when properly understood, Wittgenstein gave us reasons to reject Dummett's own philosophical enterprise to begin with. Hence the second edition of the book and, especially, the necessity for Hacker to distance himself on this very point (Wittgenstein's reaction to Brouwer's lecture) with the excessive language used in the passage quoted above. That Hacker's remarks are not based on a careful reading of Brouwer's lecture is one thing. But the irony in all this is that Dummett's own version of intuitionism, e.g., in *Elements of Intuitionism* (Dummett 2000a), builds on Heyting's introduction of an intuitionistic semantics for logical connectives, the well-known 'Brouwer-Heyting-Kolmogorov' or BHK semantics (Troelstra and van Dalen 1988, vol.1, pp.9, 24, 32–33). Heyting's semantics was introduced, however, in disregard of Brouwer's original philosophy, which was based on a radical divorce between the language (to which logic belongs) in which the mathematician expresses his thoughts and the mathematician's thoughts themselves, which are occurring in a medium outside language. Therefore the attempt was after all to link Wittgenstein not with Brouwer but with Heyting the semanticist, in order to make him serviceable to Dummett's own intellectual project of providing new philosophical foundations to intuitionism.[7] What I propose in what follows is therefore not part of a vindication of Dummett's project (for which I have otherwise much admiration), but an analysis of the impact of the real Brouwer on the real Wittgenstein.

---

expression which later gave way to a faint smile of enjoyment' (Menger 1994, p.92).

[7] Dummett's peculiar reading of Brouwer comes out in (Dummett 1980). Göran Sundholm had already warned against confusing Brouwer with Heyting in (Sundholm 1985, pp.274–276).

The further evidence alluded to consists of reports that Wittgenstein met with Brouwer in Vienna in the days that followed his first lecture, a fact that is not reported in Monk's biography. The evidence is scant, but cannot be overlooked. There is first a footnote in Henry Le Roy Finch's book, *Wittgenstein — The Later Philosophy*:

> Copies of [Brouwer's lectures in Vienna in March1928] were given to the present writer by friends of Brouwer who told him that Brouwer and Wittgenstein subsequently met two or three times privately. (Finch 1977, p.260n.)

This remark is in print since 1977, but it is constantly overlooked. It should not, as there is the further testimony of the logician F.B. Fitch, in an hitherto unpublished letter to Dirk van Dalen:

> I recall meeting Brouwer at the Gutkind apartment on Riverside Drive just after he [had] been at the Institute of Advanced Studies at Princeton. ... Brouwer also told us (when the conversation got around to Wittgenstein) that he had had a private all-day meeting with Wittgenstein on an island, during which they had discussed Brouwer's lecture which had made such an impression on Wittgenstein.' (Letter of F.B. Fitch to D. van Dalen, December 10, 1990)[8]

I should like to make brief comments on these pieces of historical evidence, from the point of view of historiography. First, Finch, who died in 1997, does not cite his sources. Who were these 'friends'? In his letter, Fitch mentioned Eric and Lucia Gutkind, who were indeed close friends of Brouwer,[9] and who lived in New York — as Finch did —, so it is possible that these two were the 'friends of Brouwer' alluded to. And the same goes for Fitch himself, so it is possible that we have here two reports with only one source. Secondly, there is a minor inconsistency in the reports, between meeting 'two or three times privately' and 'a private all-day meeting ... on an island' — the latter being most probably the *Donauinsel*, where not only the Prater gardens but numerous cafés and restaurants are located; an ideal spot for long conversations. At all events, this minor inconsistency does not undermine the principal claim one can extract from these letters, namely that Wittgenstein and Brouwer met privately for a long discussion. That it was 'two or three times' or 'all-day', it sounds as if this was not just a courtesy call.

One reason to take this evidence seriously is a matter of chronology: Fitch is referring in his letter to a meeting with Brouwer after his stay at Princeton, and that would be in the Autumn of 1953.[10] One should thus note that the

---

[8] Van Dalen does quote the last sentence of that letter, but, in what can only be an unfortunate oversight, he confuses Fitch with Finch (van Dalen 2005, p.566).

[9] Gutkind was involved with Brouwer in the Significs movement in the Netherlands, before the war.

[10] See (van Dalen 2005, pp.869–872).

first published account of Wittgenstein's reaction to his lecture dates from 1964, eleven years later. One can therefore not suspect Brouwer of trying to mislead anyone about this.[11] Actually, Brouwer died two years later, in 1966, and there is no reason to believe that he even knew about the footnote in Pitcher's book. One must thus raise the following question against the 'received view': if he thought that intuitionism was an 'aberration' and a 'perversion', why would Wittgenstein bother meeting Brouwer for a day-long conversation? If the hypothesis is that, on the contrary, Wittgenstein found Brouwer's views intriguing, to say the least, it would be normal that he took the opportunity further to discuss them with him. With this, I rest my case.

2 Brouwer on 'Hypotheses' as a Source for the Middle Wittgenstein

What could Wittgenstein have found of interest in Brouwer's lecture? What could have been the topic of their lengthy discussion afterwards? Any answer is purely a matter of conjecture since, as we saw, the eyewitness accounts provide no clue. However, one finds many elements of Brouwer's intuitionism, such as his arguments against the universal applicability of the Law of Excluded Middle or his 'pendulum number', that are critically discussed by Wittgenstein early on in 1929, in particular in MS 105, and there is no reason to believe that Wittgenstein's source for these was not Brouwer's lecture. As I said, I have discussed these in my earlier paper (Marion 2003) and I do not wish to come back to these topics until I find reasons to doubt the correctness of my views. I wish to present a new element that Wittgenstein seems to have taken up, namely Brouwer's concept of 'hypothesis', which is introduced in section 2 of the text of the lecture, i.e., amidst those very considerations that were rejected by both sides in our parochial dispute, as expression of 'solipsism', 'psychologism', etc.[12] Let me thus begin by presenting, with the minimum of reconstruction necessary, Brouwer's train of thought in this section.

[11] At all events, Brouwer was clearly not the sort of person likely to mislead others, and, being one of the great mathematicians of the century, it would be *hubris* on the part of Wittgenstein's admirers to think that he felt he needed that sort of self-aggrandizing 'publicity'.

[12] Brouwer's first Viennese lecture contains in his first sections one of the more detailed expositions of his *philosophical* views in the 1920s. As it turns out, Brouwer remained mostly silent on these issues throughout the 1930s and early 1940s, the 'silent years' as Walter van Stigt called them (van Stigt 1990, p.103). There was, however, a paper published in 1933, 'Willen, Weten, Spreken' (Brouwer 1933A1), incompletely translated in (Brouwer 1975, pp.443–446); for an English translation of the whole text, see (van Stigt 1990, pp.418–431). In it, most of the text of the first Viennese lecture is reproduced, often verbatim, and the passage on 'hypotheses' occurs again in (van Stigt 1990, p.420). Although Brouwer's philosophical views did not change in the intervening years, his post-war presentations differ, and, as far as I can tell, the notion

According to Brouwer, 'mathematical attention' comes in two phases: 'time awareness' and 'causal attention'. The former is defined as the awareness of the 'falling apart of a moment of life into two qualitatively different things of which one is experienced as giving way to the other and yet is retained by an act of memory' (*MWS*, p.153)/(*MSL*, p.45). This is the fundamental 'two-ity'; one such 'two-ity' can become in turn an element of a new 'two-ity', thus creating a 'three-ity', and by iteration of this process one obtains a 'time sequence of phenomena of arbitrary multiplicity' (*MWS*, p.153)/(*MSL*, p.45). By mathematical abstraction, the 'two-ity' can be stripped of its material content: an empty form remains, a 'common substratum' that Brouwer calls the '*Ur-intuition*' or 'primordial intuition' of all mathematics. From it, Brouwer claims, one gets the natural numbers, the real numbers and 'the whole of pure mathematics' (*MWS*, pp.154–155)/(*MSL*, p.46). These are further described as 'the more extended pure mathematical systems' (*MWS*, p.155)/(*MSL*, p.46).

As is well known, this is the key to Brouwer's philosophical basis for intuitionistic mathematics, and it has been the focus of much scholarly work. As I argued in my paper, this is where Wittgenstein could not follow Brouwer, because there is no clear indication how one really gets the natural number sequence from the iteration of the 'two-ity'. But it is the other phase of 'mathematical attention', namely 'causal attention', on which I would like to focus here. According to Brouwer, it allows for the stabilization of the perceptual world: in causal attention, one identifies different 'time sequences of phenomena', extending from the past to the future — an important point as it turns out — , therefore bringing into being a 'common substratum' that Brouwer describes as a 'causal sequence' (*MWS*, p.153)/(*MSL*, p.45). This is not explicitly said, but these 'causal sequences' form what Brouwer calls the 'more restricted system' of the perceptual world (*MWS*, p.155)/(*MSL*, p.46). Using here a strong 'idealist' language, Brouwer claims further that one 'special case of causal attention' is

> the construction in thought of *objects*, that is, of persistent, permanent things (simple or compound) of the perceptional world, so that at the same time the perceptional world becomes stabilized. (*MWS*, p.153)/(*MSL*, p.45)

Brouwer does not quote any source. Leaving aside the 'idealist' tone of the remark — its least interesting aspect here — , and assuming that the members of the 'time sequences' are in some way or another 'phenomenal', one should note at first blush a resemblance with the analysis of sensation or perception,

of 'hypothesis' is briefly mentioned only once, in 'Consciousness, Philosophy, and Mathematics', dated 1948 (Brouwer 1949C, p.1237)/(Brouwer 1975, p.482). In the other direction, so to speak, the earliest mention of 'hypothesis', a very brief one, is found in chapter II of Brouwer's doctoral thesis, in 1907 (Brouwer 1907, p.83n.1)/(Brouwer 1975, p.54n.1).

which was common at the time, in particular since Ernst Mach's *The Analysis of Sensations* (Mach 1959). This is not yet Mach's view, as it is hard to see how the remark about the 'causal sequence' as 'common substratum' fits with Mach's idea of a 'physical phenomenology' (Mach 1943, p.250), i.e., with his attempted construal of physics as dealing solely with observable/measurable variables, without any recourse to theoretical entities, such as atoms.

Of particular interest are remarks immediately following, on the relation between the 'more extended' mathematical systems constructed on the basis of the 'primordial intuition' of 'two-ity' and the 'more restricted system' of the perceptual world obtained by 'causal attention'. The passage is worth quoting in full:

> The effectiveness of mathematical abstraction is based on the fact that many causal sequences are much easier to control if they are *projected* on subsystems of pure mathematics, that is, if their empty abstractions are embedded in more extended mathematical systems. In this way the relations existing within the extended system can also be used to provide an overview of the more restricted system, which often results in a drastic simplification of the latter. This way *scientific theories* comes into being, where not only the elements of the causal system appear as a '*hypothesis*' but also the pure mathematical system that played a centralizing role in the overview. Certain scientific theories are earmarked as *exact scientific theories.* These theories first of all relate to especially stable causal sequences (perceived as natural laws or called into being artificially as technical facts); second, their hypotheses produce a considerable simplification, and third, the causal sequences in their case correspond to special values of numerical parameters whose full value domain belongs to the mathematical system on which they are projected. It is especially in these exact sciences that the phenomenon of *the heuristic character of scientific hypotheses* presents itself; this consists in the discovery afterwards that in the case of sequences that were originally inserted as hypotheses there are corresponding real causal sequences in the perceptional world. (*MWS*, p.155)/(*MSL*, pp.46–47)

In this passage, Brouwer makes four points: (1) the causal sequences of the 'more restricted system' are projected on 'subsystems of pure mathematics', and (2) this gives rise to scientific theories. So, presumably, physics arises from the projection of 'causal sequences' in the 'more restricted system' of the perceptual world, into 'subsystems of pure mathematics'. (3) By doing so, one achieves great simplification. (4) In scientific theories, the heuristic aspect of 'hypotheses' is brought to the fore: they allow one to derive new aspects, which are then verified by 'corresponding real causal sequences in the perceptional world'. To this I should add the obvious, namely that it is a constructivist conception of infinity, i.e., an understanding of infinity in terms of the possibility of carrying out a process without ever completing it, that underlies all of this.

Nevertheless, it is not clear at first blush if there is some surface element here, that could have caught Wittgenstein's attention. In order to see what may have interested him, we may look at his own background. One should therefore recall that Wittgenstein was a great admirer of Boltzmann, who is well known for his contributions to kinetics and for his concomitant opposition to Mach's idea that one should do away with theoretical entities in physics, and that he was trained by Russell, at a time when the latter was developing his own philosophy of science. A quick look at Russell is, as usual, helpful. We find him discussing these matters in 'The Relation of Sense-Data to Physics', which opens with a statement of the problem:

> What can we learn by observation and experiment?
>
> Nothing, so far as physics is concerned, except immediate data of sense: certain patches of colour, sounds, tastes, smells, etc., with certain spatio-temporal relations.
>
> The supposed content of the physical world are *prima facie* very different from these: molecules have no colour, atoms make no noise, electrons have no taste, and corpuscles do not even smell.
>
> If such objects are to be verified, it must be solely through their relation to sense-data: they must have some kind of correlation with sense-data, and must be verifiable through their correlation alone. (Russell 2004, p.139)

To speak of a *correlation* between the theoretical entities of physics and the (observable) data of the sciences sounds already like Brouwer, when he speaks of *projection* of 'causal sequences' of the perceptual world, into 'subsystems of pure mathematics'.[13] Now Russell's solution was to define the objects of physics as functions of sense-data:

> We may succeed in actually defining the objects of physics as functions of sense-data. Just in so far as physics leads to expectations, this *must* be possible, since we can only *expect* what can be experienced. And in so far as the physical state of affairs is inferred from sense-data, it must be capable of expression as a function of sense-data. (Russell 2004, p.140)

Accordingly, Russell wrote in *Our Knowledge of the External World* that

> starting from a world of helter-skelter sense-data, we wish to collect them into series, each of which can be regarded as consisting of the successive appearances of one 'thing'. (Russell 1926, p.114)

[13] That Russell employs here the peculiar notion of 'sense-data' and the related, more general notion of 'sensibilia' brings about its own set of problems, but these can be left aside here.

And he defined a 'thing' as a 'series of aspects which obey the laws of physics' (Russell 1926, pp.115–116). One should note further that those 'aspects' or 'appearances' are collected from different *perspectives*, that are part of what Russell calls a 'perspective-space' (Russell 1926, chap.III). Each perspective will give us a set of 'aspects' or 'appearances', and a 'thing' will be defined by collecting them, i.e., by selecting them one by one, from the sets given with the various perspectives:

> Given an object in one perspective, for the system of all the objects correlated with it in all perspectives ... Thus an aspect of a 'thing' is a member of the system of aspects which is the 'thing' at that moment. (Russell 1926, p.96)

So Russell's definition of a 'thing' turns out to be, as he presented it, highly non-constructive, since it involves nothing less than (a weak version) of axiom of choice, which he called the 'Multiplicative Axiom'.[14] Indeed, a 'thing' is defined in terms of a choice set:

> The 'thing' being defined as the class of its appearances, if κ is the class of perspectives in which a certain thing θ appears, then θ is a member of the multiplicative class of κ, κ being a class of mutually exclusive classes of 'sensibilia'. And similarly a perspective is a member of the multiplicative class of the things which appear in it. (Russell 2004, p.153)

This was certainly not to please Wittgenstein, who rejected this version of the axiom of choice because he could not countenance the possibility of an infinite selection from an infinite class of classes, without possession of a law for selecting (Wittgenstein 1975, §146). With this, we are getting rather close to the content of Brouwer's lecture, as he offers a constructive alternative from which Wittgenstein will be able to criticize Russell — see (Wittgenstein 1979a, p.257), quoted at the very end of this section.

The key idea for our understanding of Wittgenstein seems to me to be hidden within (3) and (4) above. Brouwer's remark on the heuristic character of scientific hypotheses refers implicitly to a debate within philosophy of science, concerning Mach's understanding of scientific laws. As is well known, Mach often insisted, e.g., in 'The Economical Nature of Physical Inquiry' (Mach 1943, pp.186–213), on the fact that laws help us to describe nature economically: the 'economy of thought' comes from subsuming numerous particular, already observed, instances under a general law. But there was an obvious gap in Mach's conception, concerning future, yet unobserved instances. A law cannot be said, strictly speaking, to describe these and this is why it has the character of an 'hypothesis'.[15] So Brouwer's point seems a perfectly good one — no 'mysticism', 'psychologism', etc. is involved here.

[14] See, e.g., (Russell 1903, 308–309).

[15] The point was, of course, already made before Brouwer, e.g., by Poincaré (Poincaré 1952, p.150); this is a rather standard conception by now. See, e.g.,

What has any of this to do with Wittgenstein? In 1932, Carnap sent an offprint of his papers, 'Die physikalische Sprache als Universalsprache der Wissenschaft' (Carnap 1934), to Wittgenstein, eliciting an angry reaction, as Wittgenstein accused Carnap of nothing short of *plagiarism* in an exchange of letters, with Schlick acting as the middleman. This incident has been discussed at length in the secondary literature and this is not the place for a detailed analysis.[16] I merely wish to point out that one of the items in Wittgenstein's list of ideas supposedly stolen by Carnap was his notion of 'hypothesis':

> Carnap has got his conception of hypotheses from me and again I have found this out from Waismann. Neither Poincaré nor Reichenbach could have the same conception, because they do not share my conception of propositions and grammar. (Quoted in (Hintikka 1996, p.140).)

This is a telling comment. Poincaré and Reichenbach were both against Mach on the nature of 'hypotheses', and so were a lot of other people, including Carnap, so there is little, if any, ground for the accusation of plagiarism. However, Wittgenstein cites as the crucial difference his 'conception of propositions and grammar'. The conception of 'grammar' that he refers to is, in a nutshell, that presented in Russell's report to the Council of Trinity College, quoted above, whose key phrase is the remark according to which Wittgenstein

> discusses 'infinity' at considerable length and links it with the conception of possibility that he has developed in connection with his various 'spaces'.

It is clear that the notion of 'hypothesis' was the link between his thoughts on mathematics, which were definitely constructivist, and his 'grammatical' thinking about the 'space' of colour, etc. As for the conception of 'propositions' alluded to, it seems fitting to quote here from Schlick, in 1931 (note that Wittgenstein did not mind the reference to his views in this passage):

> It has often been noted, indeed, that we can really never speak of the absolute verification of a law ... If I may say a few words in passing about the logical situation, the circumstance just mentioned means that at bottom a law of nature does not even have the logical character of an 'assertion', but represents, rather, a 'prescription for the making of assertions'. (I owe this idea and terminology to Ludwig Wittgenstein.) (Schlick 1979, p.188)

In my book *Wittgenstein, Finitism and the Foundations of Mathematics*, I have argued on the basis of textual evidence that Wittgenstein's expression 'prescription' or 'law for constructing propositions' originates from

---

(Popper 1968, p.247) and, for a critique of Mach on this very point, (Braithwaite 1953, p.348).

[16] See (Marion 2002).

Weyl's 1921 paper 'On the New Foundational Crisis in Mathematics', where he construed statements of the form $\forall x F x$ intuitionistically as '*Anweisung für Urteile*' or 'prescriptions for the construction of judgments' (Mancosu 1998, p.98); Wittgenstein read Weyl's paper and he had plenty of conversations with Ramsey, whose turn towards intuitionism in 1929 was the result of the influence of Weyl on that very point.[17] It is fitting to notice here that we are somehow back to Brouwer, who commented on Weyl's '*Urteilsanweisungen*' that

> This is only a matter of name and certainly does not reflect any lacking insight on my part. (Mancosu 1998, p.122)

This is all the more obvious when one considers that these 'prescriptions' do not have the status of assertions or statements, and could very well be seen by Brouwer as non-linguistic.

I was never able to find a source for Wittgenstein's peculiar use in this context of the word 'hypothesis'. I now think that it is right here, in the above passage from Brouwer's lecture. I shall now substantiate this claim by drawing, over and above what has already been said, a further five points of analogy, with illustrations taken mainly from two typescripts representative of the early middle period 1929–1931, *Philosophical Remarks* and Waismann's 'Theses', reproduced in *Wittgenstein and the Vienna Circle* (Wittgenstein 1979a, pp.233–261). (A list of all the relevant passages in Wittgenstein would be pointless here, but some textual evidence needs to be quoted.)

The first point is mostly terminological, but it is worth fleshing it out. Wittgenstein followed previous philosophers of science such as Hertz in distinguishing between a 'primary' and a 'secondary' system. Indeed, according to Hertz:

> A material system is said to be a dynamical model of a second system when the connections of the first can be expressed by such coordinates as to satisfy the [condition that] the number of coordinates of the first system is equal to the number of the second. (Hertz 1899, § 418)

Wittgenstein had already referred to this very passage at 4.04 of the *Tractatus*, and the middle Wittgenstein often speaks of a 'first' and a 'second' system and, in conjunction with these, he also spoke of a 'primary' and a 'secondary' language. Furthermore, Wittgenstein also called the former 'phenomenological', as he described the latter as the 'ordinary, physical language' (Wittgenstein 1975, §157). As I did in my *Wittgenstein, Finitism, and the Foundations of Mathematics* (Marion 1998, p.129), one can thus establish the following equivalences:

Secondary language	$\leftrightarrow$	Physical language
Primary language	$\leftrightarrow$	Phenomenological language

[17] See (Marion 1995) and (Marion 1998, chapter 4).

One should thus notice the parallel with Brouwer's notions:

Secondary (physical) system $\leftrightarrow$ Extended (mathematical) system
Primary (phenomenological) system $\leftrightarrow$ Restricted (perceptual) system

The fit is not perfect, especially since Brouwer does not consider either his 'extended' or 'restricted' systems as having anything to do with language, and this is far from being the case with Wittgenstein, for whom they obviously have (see, e.g., (Wittgenstein 1980, p.53), quoted below). But one should nevertheless understand that Brouwer had in mind the use of a (constructive) mathematical system for the analysis of perception, and this is the same for Wittgenstein inasmuch as the primary system will contain descriptions of immediate experience.

Secondly, the 'projection' of the 'restricted' system into the 'extended' system is said by Brouwer to give rise to scientific theories, in which elements of the 'restricted' system become 'hypotheses'. In my book (Marion 1998, section 5.2), I have argued on the basis of textual evidence in Wittgenstein's writings that 'hypotheses' play a crucial role at the level of the secondary, physical language, so that the further equivalences obtain:

Secondary $\leftrightarrow$ Physical $\leftrightarrow$ Hypothesis
Primary $\leftrightarrow$ Phenomenological $\leftrightarrow$ Assertion

One should note in this typical passage the connection with mathematical induction, an essential aspect for both Brouwer and Wittgenstein:

> It is mathematical induction that occurs in an hypothesis. Connection with the system of space-time, with mathematics.
>
> ...
>
> Physics constructs a system of hypotheses represented as a system of equations. The equations of physics can be neither true nor false. It is only the findings in the course of a verification, i.e., phen<omenological> statements, that are true or false. (Wittgenstein 1979a, p.101)

Thirdly, Brouwer makes a connection between *objects* and *hypotheses.* This is also a point made by Wittgenstein:

> The truth of the matter is that the concept of an object is connected with induction. Induction appears in the form of hypotheses. By an hypothesis we here mean not a statement but rather a law for constructing statements. (Wittgenstein 1979a, p.255)

> The concept of an object involves an hypothesis, for we assume as an hypothesis that the particular aspects we perceive are connected in a law-governed manner. ... Aspects are spatially and temporally connected. An object is the way aspects are connected. An object is a connection of aspects represented by an hypothesis. (Wittgenstein 1979a, p.256)

One should note an essential difference, however, with Brouwer, over which one should not gloss. According to Wittgenstein, 'hypotheses' do not appear just within scientific theories, they are already in ordinary language:

> 'Propositions' about physical objects and most of the things we talk about in ordinary life are always really hypotheses. (Wittgenstein 1980, p.53)

Fourthly, according to Brouwer, the introduction of an hypothesis in the extended system allows for a 'simplification' of the restricted system. This is again a point made by Wittgenstein:

> When I say, 'All the different pictures I see belong to one object, say, to a table', that means that I connect the seen pictures by means of an hypothetically assumed law. On the basis of that law from given pictures I can derive a new picture. If I wanted to describe the particular aspects, that would be tremendously complicated. The structuring achieved by our language consists, therefore, in assembling all those innumerable aspects in a hypothetically assumed connection. (Wittgenstein 1979a, p.256)

> Induction is a process based on a principle of economy. (Wittgenstein 1975, §227)

> Describing phenomena by means of the hypothesis of a world of material objects is unavoidable in view of its simplicity when compared with the unmanageably complicated phenomenological description. If I can see different discrete parts of a circle, it's perhaps impossible to give precise direct description of them, but the statement that they're parts of a circle ... is simple. (Wittgenstein 1975, §230)

Finally, Brouwer claimed that through 'hypotheses' one could derive new 'aspects', to use Wittgenstein's expression, and find out that there are corresponding real aspects in the perceptional world.

> *Physics is not history.* It prophesies. If you tried to conceive of physics as a mere report on the facts observed to date, it would be lacking its most essential element, its relation to the future. (Wittgenstein 1979a, p.101)

> What is essential to an hypothesis is, I believe, that it arouses an expectation by admitting of future confirmation. (Wittgenstein 1975, §228)

Not only is this one of the key features of Wittgenstein's notion of hypothesis, it is also at the basis of Wittgenstein's objection to Russell's already-mentioned definition of a 'thing' as a 'series of aspects which obey the laws of physics'. So Wittgenstein, if we are to trust Waismann's 'Theses', can turn his newly developed constructivist views against Russell:

> Russell does not represent the nature of objects correctly when he conceives of an object as a class. For a class does not help us at all to obtain a statement about a further aspect. A class has nothing to do with induction — an object, however, is essentially connected with induction. ... It is not an accidental fact that our description of an object always remains unfinished. The possibility of such a description must already be contained in the nature of an object — in the form of an hypothesis. (Wittgenstein 1979a, p.257)

3 Conclusion

I believe that these parallels are sufficient grounds to justify our regarding Brouwer's notion of 'hypothesis' as one of, if not *the* source of Wittgenstein's notion. It remains indeed an open possibility that Wittgenstein came up with this notion of 'hypothesis' on his own or even that he got the idea from somewhere else, as the topic was so widely discussed in the first decades of the last century, and not a lot of people agreed with Mach.[18] Nevertheless, the above affinities are there and we know that Wittgenstein heard Brouwer's lecture and discussed its content with him afterwards. Therefore, although this conclusion must remain provisional, it seems fair to conclude, in absence of evidence to the contrary, that Brouwer's lecture is a source for Wittgenstein's notion of 'hypothesis'.

The role that this notion played in the writings of the 'middle Wittgenstein' from 1929 to 1934 is crucial.[19] I should like to conclude by simply pointing out how the above relates to the first major step away from the *Tractatus*, the critique of the logical independence of elementary propositions. Although he had sharply separated mathematics from logic in the *Tractatus*, Wittgenstein nevertheless believed that the calculus of truth-functions was sufficient for the analysis of the perceptual world. What happened in 1929 is that he came to realize that this could not be the case. In the *Tractatus*, analysis was conceived as decomposition from complex to simple and, since complex propositions are concatenations of elementary propositions by means of truth-functional operators, to say that 'statements of degree' are analyzable meant therefore that they are logical products of elementary propositions. At an early stage in 1929, Wittgenstein realized that it makes no sense to analyze statements about degrees in terms of logical product of elementary propositions:

> And different degrees of red are incompatible with one another. Someone might perhaps imagine this being explained by supposing that certain small quantities of red added together would yield a specific degree of red. But in that case what does it mean if we say, for example, that five of these quantities of red are present? It cannot, of

[18] I owe this point to David Stern.

[19] See (Marion 1995) and (Marion 1998, chs. 4 and 5).

> course, be a logical product of quantity no. 1 being present, and quantity no. 2, etc., up to 5; for how would these be distinguished from another? Thus the proposition that 5 degrees of red are present can't be analysed like this. Neither can I have a concluding proposition that this is all the red that is present in this colour: for there is no sense in saying that no more red is needed, since I can't add quantities of red with the 'and' of logic. (Wittgenstein 1975, §76)

Thus, Wittgenstein realized that there must therefore be 'constructions' in language outside the scope of truth-functional logic:

> That makes it look as if a construction might be possible within the elementary proposition. That is to say, as if there were a construction in logic which didn't work by means of truth functions. What's more, it also seems that these constructions have an effect on one proposition's following logically from another. (Wittgenstein 1975, §76)

One should thus notice here the *priority* of a non-logical construction. This is strangely reminiscent of Brouwer's claim that the restricted system of causal sequences is built prior to the intervention of language and, therefore, of logic. Of course, Wittgenstein did not follow Brouwer along this road, but it is clear that he must have been impressed by the fact that some mathematical constructions are not only independent from logic but also must be assumed to have precedence over the calculus of truth-functions in the analyses of perception. And that may be the legacy of Brouwer's lecture on his thinking. If it is the case, then it is clear that changes in Wittgenstein's conception of analysis may indeed have something to do with the content of Brouwer's lecture.

The notion of 'hypothesis' played also a crucial role as the link between the 'grammar' of 'spaces' and the constructivist view on infinity that Russell was already talking about in his report, which was clearly the focus of Wittgenstein's thinking as early as January 1929. But to discuss this would lead us far afield, as my purpose was merely to establish the *prima facie* case for the link with Brouwer. I hope to have succeeded in the process, to some extent at least, to rehabilitate him as a philosopher, as well as to have shown that his little-known remarks on 'hypotheses' are worth paying attention to.

Acknowledgement

I would like to thank Pascal Boldini, Michel Bourdeau, Gerhard Heinzmann and Mark van Atten for their kind invitation to the international colloquium *1907–2007: One Hundred Years of Intuitionism*, at the Château de Cerisy-la-Salle (France) in June 2007, as well as to João Virgílio Gallerani Cuter and Bento Prado Neto for their kind invitation to the international colloquium *The Middle Wittgenstein*, Federal University of São Paulo, São Carlos (Brazil), in

October 2007. The present paper is based on my lectures to these colloquia, especially the latter. I have benefited from discussions with participants on both occasions. And many thanks to Dirk van Dalen for showing me the letter from F.B. Fitch discussed in section 1.

Brouwer's notion of intuition and theory of knowledge by presence

Mohammad Ardeshir

1 Introduction

The notion of 'intuition' has a long history in philosophy, and in particular in the philosophy of mathematics. Almost every philosopher of mathematics, in one way or another, admits the role of intuition in the epistemology of mathematical objects and / or mathematical knowledge. There are two main questions about the notion of 'intuition':

1. what is the *nature* of intuition? and
2. how vast is the extension and the domain of its application?

A survey of the investigations that have been made so far on both of these questions in intuitionism will clearly demonstrate that the answers are still unsatisfactory and obscure.

L.E.J. Brouwer (1886–1966) is the founder of mathematical *intuition*ism. However, the notion of 'intuition' remained unclear in Brouwer for a century, 1907–2007. In his writings, Brouwer uses two versions of the notion, 'ur-[basic, primordial] intuition' and 'intuition'. It seems to me that the notion of 'Ur-intuition' in Brouwer is a *ground* notion, in the sense that it is the *only* a priori notion sufficient to construct the whole of mathematics.

> Mathematics is created by a free action independent of experience; it develops from a single aprioristic basic intuition [Ur-intuition], which may be called *invariance in change* as well as *unity in multitude.* (Brouwer 1907, p.179)/(Brouwer 1975, p.97)

On the other hand, Brouwer uses the notion of 'intuition' as a *global* one, in the sense that it is the notion that would control and permit which constructions are allowed.

> It will be explained why no mathematics can exist which has not been intuitively built up in this way, why consequently the only possible foundation of mathematics must be sought in this construction under

> the obligation carefully to watch which constructions intuition allows and which not, and why any other attempt at such a foundation is condemned to failure. (Brouwer 1907, p.77)/(Brouwer 1975, p.52)

The common feature of both versions of the notion of intuition is the notion of *construction.*

> but *to exist* in mathematics means: *to be constructed by intuition;* and the question whether a corresponding language is consistent, is not only unimportant in itself, it is also not a test for mathematical existence. (Brouwer 1907, p.177)/(Brouwer 1975, p.96)

That means the *epistemological* notion of intuition is entwined with the *ontological* notion of construction. So to have *intuition of* X, or *to know* X means X is *constructible*, which means X *exists* and vice versa.

There are two more notions in Brouwerian intuitionism, which we believe are closely related to the notion of 'intuition'. The first one is the fact that 'mathematics is essentially *languageless*':

> Because mathematics is independent of language, ... (Brouwer 1947)

Or[1]

> Strictly speaking the construction of intuitive mathematics in itself is an *action* and not a *science*; it only becomes a science ... in a mathematics of second order, which consists of the *mathematical consideration of mathematics* or *of the language of mathematics.* (Brouwer 1975, p.98n.1)/(Brouwer 1975, p.61n.1)

The other one is the notion of 'truth' in mathematics:

> Truth is only in reality ... There are no non-experienced truths. (Brouwer 1949C, p.1243)

We believe that any successful theory of intuition in mathematical intuitionism must be able to explain the notions of *existence*, *construction*, *languageless* or *first-order* mathematics in the sense of Brouwer, and *truth.*

There have been some efforts to clarify the Brouwerian notion of intuition. For example, C. Parsons in (Parsons 1993) and (Parsons 1986), tries to explain the notion of intuition in mathematics. However, in (Parsons 1986), he relies on (Parsons 1996), 'which owes more to Kant and Hilbert than to Brouwer', (Parsons 1986, p.214). In (van Atten 2004b), our question on the nature of the Brouwerian notion is not asked, but it is argued that Brouwer would have agreed with Gödel that although intuition requires more care than formal proofs, it is epistemologically more fundamental (p.8). For a genealogy

[1] It is interesting to observe that even well-known intuitionists hardly believe that mathematics has the property of being *languageless*, see (Troelstra and van Dalen 1988, vol.2,pp.831–835).

of the Brouwerian notion of intuition, readers may consult (van Stigt 1990). Among the different interpretations that have been introduced so far, our understanding of Brouwer's notion of intuition is very close to the one explained and elaborated in (Detlefsen 1998).

In this paper, we will present another new interpretation of Brouwer's notion of intuition based on our *expanded* theory of 'knowledge by presence', which is explained in the next section. The theory of knowledge by presence was introduced first by Sŭhrawardī(1155-1191).[2] We expand Sŭhrawardī's theory by the *a priori* notion of *time* which was argued by Abū al-Barakāt al-Baghdādī (1050-1140)[3] to be a consequence of the prime mode of knowledge by presence, i.e., *self-knowledge*.

In his critique of the peripatetic Ibn Sīnā (980-1037),[4] Sŭhrawardī, a Persian philosopher, founded his illuminationist philosophy (*Ḥikmat al-ishrāq*). The title of this philosophy is from the Arabic root *SHARQ*, meaning 'rising', in particular, 'rising of the sun'. The term is also linked to the Arabic word for 'East', which represents a specifically Eastern form of philosophical thought,

[2] Shīhāb al-Dīn Yahyā al-Sŭhrawardī, known as al-Maqtŭl (the Slain) in reference to his execution, and usually referred to as Shaykh al-Ishraq after the Illuminationist philosophy which he espoused, was born in AH 549/AD 1154 in the village of Sŭhraward in northwest Iran. He fell foul of the religious authorities, and was executed in AH 587/AD 1191 on the orders of Salāh al-Dīn, in circumstances which remain unclear but which involved charges of corrupting the religion and allegations of claims to prophecy, and may also have had a political dimension. See the *Routledge Encyclopedia of Philosophy*, for more details.

[3] Hībat Allāh Alī ibn Malkā Abu al-Barakāt al-Baghdādī was an idiosyncratic, highly original philosopher who flourished in the first half of the twelfth century. Precise biographical information is unavailable. The appellation awhad al-zamān, 'the singular [personage] of his time', probably reflects his medical rather than philosophical achievements. His formal teaching seems to have been limited to medicine, in which he had a number of students. His involvement in philosophy seems to have been informal (even by the standards of the time) and tentative. See *Routledge Encyclopedia of Philosophy*, for more details.

[4] Ibn Sīnā was born in AH 370/AD 980 near Bŭkhārā in Central Asia, where his father governed a village in one of the royal estates. At thirteen, Ibn Sīnā began a study of medicine. His medical expertise brought him to the attention of the Sŭltan of Bŭkhārā, Nŭh ibn Mansŭr, whom he treated successfully; as a result he was given permission to use the sultan's library and its rare manuscripts, allowing him to continue his research into modes of knowledge. He composed his major philosophical treatise, Kitāb al-shifā (*Book of Healing*), a comprehensive account of learning that ranges from logic and mathematics to metaphysics and the afterlife. His other major works include Kitāb al-najāt (*Book of Salvation*), two major works on logic: one, al-Mantiq, translated as The Propositional Logic of Ibn Sīnā, was a commentary on Aristotle's Prior Analytics and forms part of al-Shifā; the other, al-Ishārāt wa al-tanbīhāt (*Remarks and Admonitions*). More details may be found, e.g., in the *Routledge Encyclopedia of Philosophy*.

a form of thought in contrast with cognitive reason, based on intuitive and immediate knowledge.

We argue that the theory of intuition in Brouwer's intuitionism can be interpreted in the *expanded* theory of 'knowledge by presence'. As every philosophical interpretation, our interpretation of Brouwer's notion of intuition in the expanded theory of knowledge by presence is not a *faithful one* in the logical sense. We are able to explain some philosophical concepts in intuitionism in terms of the expanded theory of knowledge by presence, and unfortunately some other main concepts remain unexplained or become more difficult to understand.

2 Theory of knowledge by presence

2.1 A brief history

Even if the full-fledged theory of knowledge by presence was developed by Sŭhrawardī, historically, we may find a trace of this theory in Ibn Sīnā. Ibn Sīnā's theory of self-knowledge is exemplified in his notion of 'suspended man', (Sīnā 1960). He conceives a man who is suspended in an absolutely temperate air. That means the soul in this condition does not sense anything extraneous to itself. In this condition the soul is oblivious of everything, except its *pure reality* and the fact that *it exists.* Later, a maverick philosopher, Abū al-Barakāt al-Baghdādī, in his elaboration and critique of Ibn Sīnā's theory of self-awareness, developed his striking innovative theory of *time.* Dissatisfied with the well-known approach, which considered *time* as the *measure of movement,* al-Baghdadi inferred the conclusion that time is an entity whose conception is *a priori* and almost as general as 'being', encompassing the sensible and non-sensible, that which moves and that which is at rest (Pines 1979). He thus transferred the notion of *time* as a topic in *physics* (al Baghdādī 1357, vol.2, pp.35–41), to a topic in *metaphysics* (al Baghdādī 1357, vol.3, pp.69–80). According to al-Baghdadi, the soul has three primary apperceptions anterior to all others: 1. the apperception of its own *self,* 2. that of *existence,* 3. that of *time* (Pines 1979, p.289). Then Sŭhrawardī arrived on the scene and developed his theory of knowledge, which is a culmination of Ibn Sīnā and Abū al-Barakāt al-Baghdādī. It is argued that the influence of Abū al-Barakāt al-Baghdādī on the making of Sŭhrawardī's illuminationist philosophy is crucial (Pines 1979, p.254). However, the problem is why Sŭhrawardī did not come to accept the Abū al-Barakāt al-Baghdādī position on the apriority of *time* in his theory of knowledge. As a matter of fact, he retained the same position as Ibn Sīnā on *time, as measure of movement, not as measure of being.* By *expanded* theory of knowledge by presence, I mean Sŭhrawardī's theory of knowledge by presence augmented by al-Baghdadi's *a priori* notion of time.

2.2 Some examples of 'knowledge by presence'

- The most familiar example of knowledge by presence is *sense perception.* In sense perception, the object is present for the subject. In this empirical case of knowledge by presence, it is qualified by the fact *at the time of immediately experiencing our sensation and feeling.* That means that we *are* in the act of experiencing our sensation, not 'reflecting' on our experience. This means that we have a *presentation* of the thing perceived, and not a *representation* of something we are already acquainted with.
- A better example of knowledge by presence in the case of sense perception is when one feels pain or pleasure. As Wittgenstein argued in (Wittgenstein 1958, p.246):

  > In what sense are my sensations *private*? ... It cannot be said of me at all (except perhaps as a joke) that I *know* I am in pain. What is it supposed to mean - except perhaps that I am in pain.

  That means that the sentence

  I know that I am in pain.

  really means

  I am in pain.

  We may conclude that *to know* to be in pain means *to be* in pain.[5]
- The *prime* mode of knowledge by presence is *self-knowledge.* In the process of knowing, there is an *agent* which stands for the performative 'I', that is established by its nature of authority to *act.* Its most familiar statement is: *I know myself.* In this mode of knowledge by presence, *knowing* means exactly the *being* of myself. In the case of self-knowledge, *to know means to exist and to exist means to know.* That is the meaning of the self-objectivity of knowledge by presence (Ha'iri Yazdi 1982, pp.141–143).
- Our knowledge of our body, powers of our imagination and our phantasms, are acquired through knowledge by presence. The main characteristic of all these is their *particularity.* They are particular in the sense that they are *mine.*
- Our knowledge of our mental forms or ideas are acquired through knowledge by presence. They are qualia of our self, and are united with 'I'.

2.3 The main characteristics of knowledge by presence

> Knowledge is the *presence* of the thing itself within the self, free of materiality. Or, in other words, the non-absence of something within the self, free from materiality, and this definition is not considered to be the most complete. The reason for this is that it encompasses

[5] It should be noted that according to Wittgenstein, this kind of experience is *not* knowledge, since there is no meaning in negating it (Wittgenstein 1979b, items 435–37).

the perception of the self, and the perception of others. (Sŭhrawardī 2002b, p.24,italics mine)

Sŭhrawardī criticizes Ibn Sīnā's theory of knowledge, essentially inherited from Aristotle, in which *definitions* play a great role. In his critiques of Ibn Sīnā's theory of definition, he explains that *formal* characterization of an object is *not* sufficient to define the object, and moreover, it does not lead us to *know* it. He then argues that *to know* an object we need to found it on certain bases. These bases are what Sŭhrawardī called 'innate knowledge', and they have their origin in self awareness (Sŭhrawardī 2002a, p.18). So he believes that in the process of defining any object, some kind of 'innate knowledge' is prior to any formal characterization. This very important observation on the theory of definitions led him to establish his general theory of knowledge. In his theory of knowledge, Sŭhrawardī believes that 'experience' of the self is the most valid type of knowledge. He believes that this experience is the basis of any philosophical reasoning.

In a primary division, knowledge can be divided into two types, *knowledge by presence* or *immediate knowledge* (al-ilm al-ḥuḍūrī), or *non-representational knowledge* and *knowledge by correspondence*, or *knowledge by representation* (al-ilm al-ḥuṣūlī).[6] In 'knowledge by presence', as it stands, the object is *present* to the subject, and that is not the case for 'knowledge by correspondence'. Let us elaborate on this literal meaning of 'knowledge by presence'.

In any 'act of knowing', there is a complex unity of *three* elements, i.e., *subject, object* and *the knowing relation*. The 'subject' as the knower, the 'object' as the known and the 'relation' as the act of knowing constitutes the whole of the intentional act of knowledge. Since knowing in itself is an intentional relation, both sides of the relation are characterized by this intentionality and immanence. As far as the object is concerned, there should be an immanent object which is essential to the very structure of knowledge. This immanent object is different from the object which is outside of our mind and is independent of us. The immanent object is called the *subjective object* and the independent object is called the *objective object* or *transitive object* (Ha'iri Yazdi 1982, pp.45–48).

Sŭhrawardī's theory of knowledge by presence may be characterized by *self-objectivity*, which means the essential nature of this knowledge is that *the reality of awareness* and *that of which the self is aware* are *existentially* one and the same thing. So, the self is aware of itself without any interposition of a 'representation'. So knowledge by presence is a non-representational knowl-

[6] A similar distinction is found in Russell. According to Russell, knowledge is either by acquaintance or by description. See (Russell 1997) for more details. It is worth mentioning that Russell's knowledge by acquaintance has some similarities with Suhrawardi's knowledge by presence, but they are not the same (Ha'iri Yazdi 1982, pp.112–115).

edge. This implies that the self and awareness of the self are individually and numerically a simple single unity.[7]

The main question that arises now is which kind of objects are involved in any act of knowing? The answer is 'the subjective object'. Let me explain what I mean by this claim. When I say that I know the object X, in fact I know a representation of X. The object must be present to me, if I know it. The real object outside of my mind is clearly absent to me, and I don't have any epistemological access to it except through its representation. That is what I meant when I claimed that in every act of knowing, the knower knows the subjective object. I will return to this point in a moment.

We can resume the main characteristics of knowledge by presence with the following clauses:

1. The first main characteristic of knowledge by presence is that the subjective object and the objective object are the same.
2. The second main characteristic of knowledge by presence is its freedom from dualism of truth and falsehood. The essence of the dualism of truth and falsehood is some sort of 'correspondence' between the subjective object and the objective objects, or in other words, between a statement and its objective reference. This dualism is not applicable to knowledge by presence, since there is no objective object in knowledge by presence to be corresponded by its dual.[8]
3. The third main characteristic of knowledge by presence is its freedom from any kind of knowledge by 'conception' or knowledge by 'belief'. The reality of knowledge by presence does not involve any sense of conceptualization and representation. In contrast to knowledge by correspondence, which is based on conceptualization and representation, knowledge by presence belongs to the order of 'being'. In fact, knowledge by presence is a sort of instantiation of the object in the mind.

In contrast to knowledge by presence, knowledge by correspondence comprises two senses of 'objectivity', i.e., subjective object and objective object. In illuminative philosophy, the former is called the *present* object and the latter is called the *absent* object. It is clear that 'presence' and 'absence' are measured relative to the knowing subject. While the subjective object is present to the mind of the subject, the objective object is absent from the mind. As a

---

[7] There is an etymological connection between the two words *existence, being* (*wŭjŭd*) and *consciousness* (*wijdān*), both of which come from the Arabic root *w-j-d*, meaning to 'find' and 'to be found'.

[8] There is, however, a sense of truth in illuminationist philosophy which is applicable to knowledge by presence. This sense of truth is involved in the *experience* of the presence of object in mind. Since knowledge by presence is a sort of 'instantiation' of the *reality* of the object in mind, we may apply such an *existential sense of truth* to the reality of knowledge by presence, (Ha'iri Yazdi 1982, pp.79–80). In this sense, 'truth is only in reality ... [T]here are no non-experienced truths' (Brouwer 1949C, p.1243).

matter of fact, the role played by the subjective object is as an 'intermediary' representation in the act of knowing. This means that the subjective object *represents* the (external) objective object before the mind of the knowing subject via conceptualization. In this setting, 'correspondence' makes sense. That is correspondence between the subjective object as representative and the objective object (external object) as the real object outside of the mind.

So knowledge by correspondence is a type of knowledge in which (a) there are two kinds of objects: internal object and external object. This means that both subjective object and objective object are already involved in any act of knowing. And (b) there is a correspondence relation between these two objects.

> In the course of human knowledge man must first make his inquiry into his awareness of himself (*'ilmihi bi ḏātih*), then proceed from this step to what is above and beyond himself [the external world].
> Thus we say: the fact that our souls apprehend the reality of themselves (*ḏātiha*) does not imply that apprehension has come to them by a representation.
> This is because of the following:
> First: The representation that is used which appears in the mind is not exactly the mind as it is in itself (*hiya hiya*). But that which is aware of itself is supposedly aware of what its objective 'I-ness' (*li 'ayān mā bihi anā'iyatuh*) consists of, rather than that with which this objective 'I-ness' is in conformity. Any representation taking place in the mind of the knower is in fact something added to his reality which, in comparison with that reality, serves as an 'it' (*huwa*) and never as an 'I' (*ana*).
> Second: Suppose that self-apprehension is by representation. Now every representation existing in the self as an intellect is universal [in the sense that it is not impossible to predicate it of many]. Even if the complete sum of the universals referring altogether to one single individual among others have been gathered together in a unitary complex, it still cannot make the representation cease to be a universal. But the fact is that everyone's apprehension of his own reality is with such strict individuality that it can have nothing in common with another. Thus one's understanding of one's own individual reality can never be admitted as being by means of any representation at all.
> Furthermore, the self does in fact apprehend its body as well as its imagination and its phantasm. The supposition that these things are apprehended through a representation inherent in the reality of the self, while the representation as such is universal, would imply that the self is the mover of a universal body using universal mental powers, and has no apprehension of its particular body, nor acquaintance with the powers which belong to itself. This is obviously not right.

((Sŭhrawardī 2002c, p.485), translated in (Ha'iri Yazdi 1982, pp.169–170).)

We can summarize the above passage in two main claims of Sŭhrawardī (Ha'iri Yazdi 1982, p.171):

1. The apprehension of one's own body and all its mental powers and activities are like self awareness, and are forms of knowledge by presence.
2. There is an *epistemic* criterion for the universality and particularity of knowledge.

2.4 The relation between knowledge by presence and knowledge by correspondence

One of the main questions in illuminationist philosophy is that of the nature of the 'relation' between knowledge by presence and knowledge by correspondence. Unlike the Aristotelian category of relation, this illuminative relation is *not* a relation between two sides binding them together in one unity. This 'illuminative relation' is a basic concept in illuminationist philosophy, and justifies some important philosophical claims made by Sŭhrawardī. He called this particular relation *al-iḍāfat al-ishrāqiyah*. This relation, by nature, is an existential monadic relation in which the relation itself and what are related are one and the same thing. We may say that the relationship of knowledge by presence to knowledge by correspondence should be taken as a *cause and effect* relationship in terms of *illumination* and *emanation*. To distinguish this cause and effect type relation from the physical one, it is called *illuminative relation*.

In Sŭhrawardī's terminology, the illuminative relation is like *light*; it has a source, where the light comes from, and an object, upon which the light hits. The main source or the first source of light is called 'light of lights', and it is the top, it is the God. The passage of light from an upper level to a lower level is called *illumination*, and in the context of perception, it is achieved through 'knowledge by presence'. According to this picture of the illuminative relation, we, as human beings, are 'gods' with respect to our creations.

However, it is also possible for a lower level to perceive an upper level existant. In illuminationist philosophy, this type of perception is called 'intuition'. In this sense, it seems that the notion of intuition has nothing to do with creation or construction in Brouwer's sense. It is only a tool to discover existants which already exist.

The following dialogue would make clear what Sŭhrawardī had in mind about the illuminative relation. Before quoting this dialogue, it is worth mentioning that according to Sŭhrawardī this dialogue has really happened between him and the First Teacher [Aristotle] in a dream, after long meditation and introspection on the nature of knowledge.

Q.: How can we ever have knowledge at all?

> A.: Think introvertively of yourself. If you do so, you will certainly find out what truly answers your question.
> Q.: But how?
> A.: If I introvertively consider myself, I will find in all certainty that I am truly aware of myself in such a way that I can never miss myself. This state of self-certainty convinces me that my awareness of myself does not mean anything but the awareness of 'myself', 'by' myself, not by anyone or anything else. If I were aware of myself 'by' anyone or anything else, it would obviously mean that the awareness of myself belonged to another active power which is not myself. In this case there would be a knowing subject operating in myself in knowing myself. Thus it would not be myself that knows myself. But it has been assumed that it is the very performative 'I' as the subject reality of myself who knows myself. ((Sŭhrawardī 2002a, p.70), translated in (Ha'iri Yazdi 1982, pp.88–89).)

We can consider the above argument proceeding in two ways. One way will lead to the conclusion that in the case of self-knowledge, the *self* as the *subject* of knowledge and, at the same time, as the *object* of knowledge, are absolutely identical. This is the primary characterization of knowledge by presence. The second way leads to the point that in any instance of knowledge by correspondence, the 'I', the 'self' *already knows itself* by presence and knows its (external) object by correspondence. In other words, a knowing 'I' is known to itself by presence, and acts like an active intellect to construct in itself the forms of its objects, so that it can know them by correspondence. In this sense, we say that knowledge by presence has *creative* priority over knowledge by correspondence.

2.5 Knowledge by presence vs. knowledge by correspondence

We may summarize the main differences between knowledge by presence and knowledge by correspondence as follows:

1. Knowledge by presence is *immediate*, in the sense that there is no medium in which or by which the act of knowing appears. In knowledge by presence, there is a simple *unity* of *subject* and *object*. In contrast to this state of knowing, knowledge by correspondence is formed by the aid of 'something'.
2. In knowledge by presence, there is no 'correspondence', 'truth' and 'falsity'. These concepts, implicitly or explicitly, are based in one way or another, on a dichotomy of the subject and object.[9]
3. Knowledge by presence is *infallible*, since the subject is object and vice versa. On the other hand, knowledge by correspondence is fallible, since it is always possible that the subjective object does not 'match' the object perceived.

[9] See footnote 8.

4. The object of knowledge by presence is *being* (of the object), but the objects of knowledge by correspondence are *concepts*.
5. Knowledge by presence is *languageless* in its nature, but knowledge by correspondence needs some forms of 'language' to be processed.
6. Knowledge by presence is not divided into 'representation' and 'judgement', but knowledge by correspondence is acquired through these 'linguistic' models of mental processing.
7. Knowledge by presence is not subject to logical rules.
8. Knowledge by presence is 'being' itself, and so it does not fall under any *categories*, but knowledge by correspondence is under the category of 'quality of soul' (*kaif nafsānī*).
9. There is no abstraction in knowledge by presence.
10. In knowledge by presence there is no distinction between what is called 'known essentially' (*ma'lum bi-ḏat*) and what is called 'accidentally known' (*ma'lum bi-laraḍ*).

Acquiring knowledge by correspondence is through representations (in the mind). We may consider these representations in two aspects: *representations in themselves* and *representations for themselves*. By representations for themselves, I mean to consider representations as independent entities, or to put it in another way, to consider them not as *means* to achieve correspondence with external objects. By contrast, by representations in themselves, I mean representations as *means* to acquire knowledge by correspondence. If we ignore the act of acquiring knowledge of external objects, i.e., if we look only at representations for themselves, we may consider *knowledge* as *presence* only. In this way, the division of knowledge into knowledge by presence and knowledge by correspondence is only a *relational* division. We may find similar philosophical conceptual divisions. Look at the concepts of 'plurality' and 'unity'. The concept of 'plurality' is clearly a 'relational' property of the concept of 'unity', and there is no contradiction here. In fact, 'unity' is accompanied with 'existent', but we may see it as a 'plurality'. Similarly, we may consider division of 'existence' into 'potential' and 'actual', etc.

Mental existence is *mental* whenever it is in opposition with external existence, otherwise it is *external*. In other words, mental representation is *really* external existence, but if we compare it with the outer world, it becomes mental existence. So, externality of representation is its essence, and its mental existence is an additional and relational property.

3 The notion of intuition in Brouwer

The notion of 'intuition' is very involved and opaque in Brouwer. On the one hand, it is argued that the Brouwerian notion of *intuition* is *not* exactly what is called intuition by Bergson, Borel, Kant, Poincaré or Russell, (van Stigt

1990).[10] From the arguments of C. Parsons in (Parsons 1996), we may infer that it is not Gödel's intuition either. On the other hand, the notion of intuition has not been clarified enough to the extent that the label 'intuitionism' merits. It seems to us that in Brouwer's writings, we find two notions of 'intuition': the *ground* notion and the *global* notion.

The first one is called by different names in the English translation of his works: Ur-intuition, primordial intuition, basic intuition, etc. This 'Ur-intuition' itself does not refer to a unique concept in Brouwer. It is called *invariance in change* as well as *unity in multitude*:

> Mathematics is created by a free action independent of experience; it develops from a single aprioristic basic intuition [Ur-intuition], which may be called *invariance in change* as well as *unity in multitude.* (Brouwer 1907, p.179)/(Brouwer 1975, p.97)

Here we have 'two' examples of Ur-intuition. These two examples should be taken as extensions of what is called the 'Ur-intuition', not as 'Ur-intuition' itself. In what sense does Brouwer consider these two notions as a single one? 'Invariance in change' applies to an entity which preserves its identity during any change that may happen to it. So, it implicitly carries an element of temporality in itself. Contrary to that is the notion of 'unity in multitude', which in itself may not carry any element of temporality. There is one way to conceive a possible solution to how a 'multitude' may be 'united', but I am not sure that is what Brouwer had in mind. The 'binding' element of a 'multitude' may be argued to be the movement of 'time'. This justification does not seem very plausible at first. Considering the notion of 'unity in multitude' as a basic concept of mathematical building has a long history. We can find it in Plato's *Philebus* (Plato 1993), Cantor's definition of 'set', (Cantor 1883) and Gödel's writings (Wang 1996). As a final remark on this subject, let me say that the notions of 'invariance' and 'change' are *contradictory* if we consider them separately. The same thing is true of the notions 'unity' and 'multitude'. Thus it seems that according to Brouwer, we are able to intuit contradictory notions in a single *action.*

Let us go further and explore more examples of the notion of 'Ur-intuition'. Another one is the intuition of *continuous* and *discrete* which are irreducible to each other.

> In the following chapters we will examine further the basic intuition [Ur-intuition] of mathematics (and of every intellectual activity) as the substratum, divested of all quality, of any perception of change, a unity of continuity and discreteness, a possibility of thinking several entities, connected by a 'between', which is never exhausted by the insertion of new entities. Since continuity and discreteness occur as

[10] R. Tieszen believes that Brouwer's notion of intuition is different also from Husserl's notion (see, e.g., (Levinas 1995)), but we may find some overlap between the two notions (private conversation at the Cerisy Conference.)

> inseparable complements, both having equal rights and being equally clear, it is impossible to avoid one of them as a primitive entity, trying to construe it from the other one, the latter being put forward as self-sufficient; in fact it is impossible to consider it as self-sufficient. (Brouwer 1907, p.8)/(Brouwer 1975, p.17)

It is important to know that according to Brouwer, the perception of 'continuity' *and* 'discreteness' together in a 'unity', is an Ur-intuition. That means that we cannot perceive any one of the pair continuity - discreteness without the other. Again, the two notions of 'continuity' and 'discreteness' would be contradictory if we consider them separately. What makes them united is the *act of intuition.*

In his PhD dissertation in 1907, Brouwer believed that the notion of *continuum* is also intuitive:

> Having recognized that the intuition of continuity, of 'fluidity', is as primitive as that of several things conceived as forming together a unit, the latter being at the basis of every mathematical construction, we are able to state properties of the continuum as a 'matrix of points to be thought of as a whole'. (Brouwer 1907, pp.8–9)/(Brouwer 1975, p.17)

In his genealogy of the notion of intuition, Brouwer founded it on the most crucial one, namely the so-called intuition of the *movement of time*:

> [T]he first act of intuitionism completely separates mathematics from mathematical language, in particular from the phenomena of language, which are described by theoretical logic, and recognizes that intuitionist mathematics is an essentially languageless activity of the mind, having its origin in the perception of a *move of time*, i.e., of the falling apart of a life moment into two distinct things, one of which gives way to the other, but is retained by memory. If the two-ity thus born is divested of all quality, there remains the *empty form of the common substratum of all two-ities.* It is this common substratum, this empty form, which is the basic intuition of mathematics. (Brouwer 1952B, pp.140–141)

The aprioristic character of the intuition of time has been repeated many times in Brouwer's writings. For example in the following passage, he goes further to explain how the concept of number will arise from the intuition of time:

> This neo-intuitionism considers the falling apart of moments of life into qualitatively different parts, to be reunited only while remaining separated by time as the fundamental phenomenon of the human intellect, passing by abstraction from its emotional content into the fundamental phenomenon of mathematical thinking, the intuition of the bare two-oneness. (Brouwer 1913C, p.85)

Until now, we have encountered at least five different Ur-intuitions, 1. *time*, 2. *invariance in change*, 3. *unity in multitude*, 4. *continuous and discrete*, 5. *continuum*. One way to resolve this, as Brouwer himself has done, is to reduce these different Ur-intuitions to the most basic one. Brouwer reduces all notions of Ur-intuition to the Ur-intuition of *time*. The reduction of one mode of intuition to another one is an eligible action in Brouwer's point of view:

> Finally this basal intuition [Ur-intuition] of mathematics, in which the connected and the separate, the continuous and the discrete are united, *gives rise* immediately to the intuition of the linear continuum, i.e., of the 'between', which is not exhaustible by the interposition of new units and which therefore can never be thought of as a mere collection of units. (Brouwer 1913C, p.86) [italics mine]

And he explicitly admits:

> And since in this intuition we become conscious of time as change per se, we can say: *The only aprioristic element in science is time.* (Brouwer 1907, p.99)/(Brouwer 1975, p.61)

But that raises a very important question in Brouwer's epistemology: *how does an intuition* give rise *to another intuition?* The problem is connected to the basic fact that *intuitions are free of any logic*. That means we should look for another phenomenal experience to justify speaking of *transforming* one mode of intuition to another one. My answer to the above question is that *the act of construction* is what makes the 'bridge' between two different intuitions. The reason that 'construction' is an action that does not change the intuitive quality of the object constructed is that it acts *in* intuition. In fact, 'construction' is a tool of exhibition of mathematical objects in intuition.

All that we said was about Ur-intuition, which I call the *ground* notion of intuition in intuitionism. There is a second notion which we call the *global* notion of intuition. This second notion burdens intuitionism and we need to say more to make it plausible. According to Brouwer, 'no mathematics can exist which has not been *intuitively* built up' (Brouwer 1907, p.77)/(Brouwer 1975, p.52). Moreover, it is the 'intuition' which allows a construction to be an eligible construction. In its global sense, intuition is responsible for any eligible mathematical construction at every stage. Restricting Brouwer's notion of intuition only to the early stage of consciousness, i.e., Ur-intuition, will cause the neglect of the fundamental unifying and justifying principles which underlie Brouwer's philosophy of mathematics, (van Stigt 1990, p.158). We should pay attention to the notion of 'primordial' intuition, in the sense that it is not some prehistoric event in consciousness whose only effect was to form some basic constructs such as number. In the broader sense, 'primordial' must be taken as *metaphysical a priori*, which means that it is *certain* and is *the base* for any further mathematical reasoning. In fact, it is a regulative power for any mathematical reasoning. This 'regulator' property of intuition is manifested in the act of 'introspection', re-living and reflecting on the primordial happening, 'a regression which is hard to realize' and requires 'detachment concentration' (Brouwer 1949C, p.1238).

In fact there is something more fundamental and more general to intuition in Brouwer's ontology of mathematical objects, in which *to exist* in mathematics means *to be constructed by intuition* (Brouwer 1907, p.177)/ (Brouwer 1975, p.96).

Combining this with the well-known passage of Brouwer which says 'truth is only in reality; i.e., in the present and the past experiences of consciousness' (Brouwer 1949C, p.1243), we come up with the meaning of intuition in Brouwer.

'Intuition' is a kind of perception in which I, as the subject, or in Brouwer's terminology, as an *idealized mathematician* or *creating subject*, construct my objects freely. The *act* of constructing mathematical objects itself, 'I' as the *actor* and the *objects constructed* all take place in intuition as a whole phenomenon. So 'intuition' is some kind of *self-knowledge*. It is a certain knowledge, since I know myself without any representations. To *know* myself is exactly the same as *I am*. That is the meaning of the *present* experience of consciousness. It is languageless, and first-order mathematics in the sense of Brouwer. In this sense, mathematical objects are *present* to me, and my knowledge of them is just their *presence* to me. Constructions are actions which *preserve* intuitions, in the sense that they are either Ur-intuitions or kinds of actions on Ur-intuitions which are closed under intuition. In this interpretation, 'intuition' and 'construction' are identified. That means that 'intuition without construction' is empty and is not a perception at all, and 'construction is an action only in intuition'. Having these in mind and the fact that both 'intuition' and 'construction' are mental entities and brought about in consciousness, we have a better understanding of 'truth is only in reality'.

Any mathematical judgement or proposition is some kind of *report* of what has been constructed in my intuition. In reporting what has been done in my intuition, I may make errors in reading it from my memory.

4 Similarities and dissimilarities

- In our expanded theory of knowledge by presence, *time* is an *a priori* notion, which is perceived by presence, like the notion of *existence* and *self* (al Baghdādī 1357, vol.3, pp.69–80). As Brouwer said, *time is the only a priori* intuition of mathematics.

- As the list 1–10 in section 2.5 shows, we can explain the key notions of *existence*, *construction*, *languageless* or *first-order* mathematics and *truth*[11] in the Brouwerian intuitionism in terms of 'knowledge by presence'.
- Knowledge by presence is not subject to logical rules (item 7 in section 2.5). This may be considered as one point where our interpretation does not work well. However, the place of logic in intuitionistic mathematics is determined by Brouwer. It does not belong to first-order mathematics in the sense of Brouwer. As we have already seen, language can have no foundational role in intuitionistic mathematics, in particular logic cannot (for more details see (van Atten 2004b), section 1.6).
- We have already noted that there is no *abstraction* in knowledge by presence (item 9 in section 2.5). The act of abstraction is a complicated process in the mind, in which a particular objective object *transforms* itself into a more general subjective object. We have noted that this subjective object is *present* to the knower. That means that we, as the subject of knowing, know the subjective object by knowledge by presence. However, it is not clear to us if the action of abstraction, in the sense of Brouwer, takes place in intuition.
  For example, the concept of two-ity in Brouwer is an *abstraction* of the notion of *invariance in change* or *unity in multitude*. These two notions are clearly perceived through knowledge by presence, and as we have already noted, they are also examples of Ur-intuition.
- It is worth mentioning that the notion of 'intuition' in Sŭhrawardī is different from what is called 'intuition' in Brouwer. We may say that in illuminationist philosophy of Sŭhrawardī, the notion of 'intuition' is a *special case* of knowledge by presence. In contrast with the Brouwerian notion of 'intuition', in illuminationist philosophy, it is a *power of discovery* of the essence of objects, not construction of objects.

Acknowledgement

The author would like to thank M.S. Adib Soltani, H. Masoumi and S. Shahshahani for reading the draft of this paper and suggesting very helpful comments. The corrections and comments by the referees were much appreciated.

---

[11] In the sense of the existential truth, explained in footnote 8. Also compare item 3 in section 2.5 on the *infallibility of knowledge by presence* with this passage in Brouwer:

> [T]he languageless constructions originating in the self-unfolding of the primordial intuition, on the basis of their being presence in memory alone, are *exact and correct*, that, however, the human power of memory, which has to oversee these constructions, by its very nature is limited and fallible. (Brouwer 1933A2, p.58)/(van Stigt 1990, p.427) [italics mine].

Buddhist models of the mind and the common core thesis on mysticism

Henk Barendregt

Dedicated to N.G. (Dick) de Bruijn for his 90th birthday

1 Mystical experience

Mystics have experiences that for most people do not occur often in their daily life. This holds in many cases also for the mystics themselves. There is certainly an important component of euphoria in the experience. But it seems impossible to express the experience in ordinary language. Nevertheless, more than one third of the human population is reported to have had a mystical experience, see (Hood et al. 2003).

When the mystics try to convey their experience, it is given different conceptual descriptions. These include feelings of unity (with the divine or all beings), alterations in time and space perception, loss of control, to mention a few. Early interest in psychology to understand these experiences include (James 1902) and (Leuba 1925).

The *common core thesis* on mysticism holds that all forms are in the end different ways to express the same thing. This thesis is mentioned under this name in (van Atten and Tragesser 2003), but also in (Staal 1975), who added that the 'metaphysical underpinning may be different'. Studying the mysticism of Brouwer[1] and Gödel,[2] Van Atten and Tragesser argue that if these two mathematicians-mystics are right, then the CCT cannot be valid: for Brouwer mathematics conceals the mystical experience and for Gödel mathematics reveals it (van Atten and Tragesser 2003).

---

[1] 1881-1966. See (van Dalen 1999, 2005) for his biography.
[2] 1906-1978. See (Wang 1987) and (Wang 1996) for his biography.

2 Forms of meditation and the cover-up model

This model is based on personal experience with insight meditation (also called *vipassana*), see (Barendregt 1988) and (Barendregt 1996).

Two forms of meditation

First we explain what meditation and in particular vipassana is. Consciousness is considered as having a *content* and a *type*. Seeing, for example, starts with the experience of colours. This is an object of consciousness. This seeing can be accompanied by a type. Pleasure, sadness, desire are possible types of consciousness. Usually people are concerned with the object of consciousness. One often tries to optimize the contents of our consciousness. For example a man may strive for a nice house, a beautiful wife and a cool car. [These are male values and in women there are usually other ones.] This optimalization of content is done with the often implicit assumption that one obtains as a consequence happiness and well-being, a positive type of consciousness. But the house may have a high mortgage, the wife may be opinionated and the car uses a lot of gas. One therefore becomes dependent on conditions.

In meditation one is not concerned with the content of consciousness, but with its type directly. There are many forms of meditation that can be divided into two main families. In *concentration meditation* one strives for improvement of the positive types of consciousness, like joy, loving kindness and compassion. In *insight meditation* one strives for the diminishing and eventual eradication of the negative types of consciousness, like hatred, fear and restlessness. It turns out that it is impossible to guarantee permanently positive types of consciousness, as they require mental energy, while radical elimination of negative types is possible once and for always.

Concentration meditation

The high point of concentration meditation is one-pointedness, in which there is a serene bliss and equilibrium. This is a mystical state. But one may fall back to ordinary consciousness with its usual unpleasant types, including existential fear. The high point of insight meditation is the consecutive elimination of the negative types, first of all this existential fear. Therefore mystics like John of the Cross adhort us to go beyond indulging in bliss. But — as these mystics did experience — then one needs to go through *the dark night of the soul.*

Insight meditation

We have actual and potential consciousness. Seeing this sentence being written on a laptop is at this moment my actual consciousness. Having an itch in my toe is potential consciousness, as I do not experience it at this moment. Potential consciousness may become actual.

The cover-up model states the following.

1. In our potential consciousness there is a *process* having three features.
   a) Perception is fluctuating, chaotic;
   b) It is unbearable, nauseating;
   c) We have no control over these phenomena.

   Traditionally these three features are called the *three (fundamental) characteristics* [of consciousness]. These are *anicca, dukkha, anatta*: non-permanence, suffering, selflessness. This means that there is an actual experience of chaos, nausea and uncontrollability. We would rather commit a crime than have to experience this process. The process is most probably the dark night of the soul as experienced by the mystics. Actually, everyone has his or her personal version of the process, depending on their personality (or karma as the Buddhists would say).
2. Usually we hide *(cover-up)* this process by our feelings and thoughts. Feelings of greed (wanting to get something), aversion (wanting to get rid of something), and attachment (wanting to keep something) have a temporary stabilizing effect on our consciousness. As we do not like the process of the three characteristics, we are addicted to this symptomatic cover-up and the necessary behaviour to cause these feelings and thoughts.
3. Using *discipline, concentration*, and notably *mindfulness* one can purify the process so that it disappears. This so-called *cessation* is the goal of insight meditation.

Mindfulness consists of intended attention on the content and type of consciousness, together with the right distance. This distance should be there not in order to suppress the content or type, but in order to be able to observe them, without being sucked away by or into them. Mindfulness is a type of mind that can be trained. Similar to the regular training of a jogger, in which the muscles and stamina are being exercised, the insight meditator trains mindfulness and at the same time the concentration to keep it being present in the type of consciousness. *Concentration* is the skill to have some kind of consciousness (content and type) and to keep it. It may be compared to making a photograph of the content and type. Mindfulness on the other hand is making a movie of the various contents and types. For this filming one does need some concentration in order to keep doing it: carefully observing and not getting involved. One starts observing the breathing movements of the abdomen and keeps following them. If there is a hindrance, like boredom, that cannot be avoided, then one learns to switch mindfulness towards the hindrance ('O, there is boredom.'). In this way one is mindful again. One develops during 10 day intensive retreats the capacity to be mindful for extended periods. Then one develops equanimity, calmness and bliss, even in the presence of the three characteristics. One surrenders and 'cessation' may happen. For more details on the path of insight meditation, see my two Buddhist Phenomenology papers (Barendregt 1988, 1996).

3 The AM

We will introduce the AM, a model of conscious cognition inspired by the *Abhidhamma* and *Abhidharma*. The Abhidhamma ('Higher Teachings') is a canonical text from the Buddhist Theravada Pali Canon, dealing with what may be called psychology. The work consists of seven volumes comprising more than 5000 pages. It is quite technical and consists of many lists of concepts related to consciousness in all of its forms. Because of the concise form most people study the commentaries or subcommentaries (i.e., a commentary of a commentary). See, e.g., (Bodhi 2000) or (Barendregt 2006). In the later Mahayana Buddhism several additions have been made. These later works we refer to as the Abhidharma, the Sanskrit translation of the Pali word Abhidhamma.

Abhidhamma

The following aspects come from the Abhidhamma.

- Consciousness consists of a linear sequence of short consciousness moments called *cetas*. Each *ceta* lasts a short time interval, certainly much less than a second (the exact range of durations is to be investigated by science).
- Each *ceta* has a *content* (or *object*) and a *type*. The content of a *ceta* is related to what is seen, heard, smelled, tasted, felt, or pops up from memory. The type is related to how this is done; for example with joy or with anger, with an urge to act or with calm equanimity. Also intended actions are seen as contents of consciousness.
- Each *ceta* influences the next or a future *ceta*. The objects in the cetas are considered as side conditions. One speaks about causes and conditions, the cetas types being the causes and the objects the conditions. (Being hungry relates to a *ceta*-type; seeing food relates to the available objects. The next *ceta* in which one comes to act and eat is the influenced *ceta*. In daily life, on the other hand, one often considers the seeing of the food as the cause of eating. In the Abidhamma, hunger is considered as the cause, and food as the condition of the resulting eating.) *Kamma* (karma) is the name of the 'law of influence' cetas have on future cetas.
- The contents of consciousness are divided into two classes: *rupa* (body; what is meant is perceived awareness of the physical senses) and *nama* (mind; for example coming from memory), depending on the sense where they come from: eye, ear, nose, tongue and touch-sense provide the rupa contents; the mind provides the nama content, see Figure 1.

- The type of a *ceta* is determined by a set of subtypes called *cetasikas* acting in parallel. There are unwholesome (increasing suffering), neutral and wholesome (decreasing suffering) *cetasikas*. The unwholesome and wholesome are mutually exclusive: they cannot occur together.

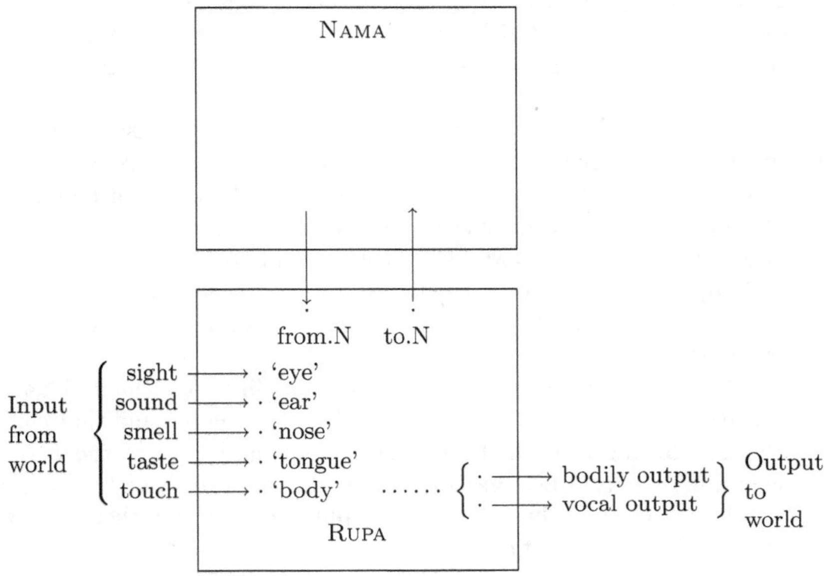

FIGURE 1.Nama-Rupa. Here 'from.N' stands for input from Nama and 'to.N' for output to Nama. The Nama part in this figure should not be considered as a homunculus. It is the process of the stream of cetas that run according to strict rules of cause and effect, i.e., deterministically. The Nama cannot exist if there is no object.

- Some cetas provide stability in the *ceta* stream. This can be based on desire, negative feelings or neutral feelings. Indeed if we fall in love this gives a stable type of consciousness. The same effect is provided by anger. An example of a neutral feeling is concentration (not yet accompanied by joy). All these types of consciousness are called 'rooted cetas'. Also consciousness with mindfulness is rooted.
- Meditation improves the types of the cetas. Samadhi (concentration meditation) is directed towards more occurrences of wholesome *ceta* types; vipassana (insight meditation) based on mindfulness is directed towards less occurrences of unwholesome *ceta* types.
- Both strong concentration and mindfulness are 'beautiful' *cetasikas.* Combined they can give temporary relief from mental and physical hindrances. Concentration is close to bliss and rapture. Mindfulness implies equanimity: not affected by fear and desire, without becoming indifferent, i.e., closing one's eyes to what happens.
- Freedom (enlightenment) consists of the permanent elimination of some or all (at the final step) unwholesome *ceta* types.
- The cetas with a nama or rupa content form sequences of cognitive units, called *vithis.* A rupa vithi consists of 17 cetas (8 cetas for preparation, 7

equal cetas (the *javana*) forming the awareness with its conditioning force to the future, and 2 for finishing off), while a nama vithi consists of just 12 cetas (this time just 3 cetas for preparation, as the content already is in the mind).

- Cognition consists of a sequence of four consecutive vithis: the first for input (both for nama and rupa), the second for obtaining information from memory (where was that observed before?), the third for naming the input and the fourth for getting the meaning of the input.
- The output of our body-mind system consists of bodily action, speech, and mental objects (being input of the mind). Only such actions with intention are recognized as output. These intentions are considered as contents of the rupa or nama cetas.
- Many of the cetas belong to the so-called baseline, *bhavanga*. This is a sequence of cetas of which the content and type are equal and that cannot be changed during one's lifetime. The content and type of the baseline cetas is different from person to person and constitutes partly his or her personality. Each time there is sensory input and cognition there is a start of vithis with different cetas.

Abhidharma

One remarkable aspect of the Abhidhamma model is that cetas with their types can influence other cetas even in the distant future. How is this action at distance possible? This question was addressed satisfactorily in the later Abhidharma (compiled during the period of Mahayana Buddhism after 100 AD), where a few essential additions were made, by Nagarjuna, (Asanga 2001) and (Vasubandhu 1988). The main one is the notion of *alaya* or memory. This concept does not occur in the Abhidhamma, because that theory is concerned only with pure phenomenology, consciousness as it actually appears. Memory on the other hand is potential consciousness. Using memory a *ceta* can influence a *ceta* in the distant future by modifying the memory state. This lingers until the circumstances are ready and the effect takes place in the future. Also Brouwer's analysis of our fundamental mathematical intuition using memory and time shows the importance of this addition of the notion *alaya*.

A second addition in the Abhidharma is the notion of *paratantra*. In the old Abhidhamma one considers only a single sequence of cetas (ceta line) to describe someone's life. In the Abhidharma one considers the mutual interaction of several *ceta* lines. The laws of the single *ceta* lines remain valid. Interaction can be understood from the laws of input and output and *ceta* transitions. But a more comfortable level of abstraction is to recognize that there are other *ceta* lines (of 'other persons') and that there is a new world in which the interaction takes place.[3]

[3] The addition of *paratantra* may be compared to the transition of computer science before and after 1969. In that year the internet was invented (consisting of

A final important addition in the Abhidharma is called *manas* and is concerned with the mental capability to view a process as an entity, to make a reification.

We will not present evidence for this AM model of the mind, neither from phenomenology, nor from neuroscience. We do, however, give two of its plausible consequences.

Consequences of the AM

In traditional psychoanalysis one speaks about the neurotic core of the mind. Later psychoanalysts even speak about its psychotic core, see, e.g., (Eigen 1986) and (Suler 1993). We will see that both neuroses and psychoses are natural side effects of the 'operating system of the mind' as stated in the AM.

Neurotic core

All *cetas* contain *cetasikas* of feeling and other components of the type. In a *vithi* this is enlarged by the 7 *ceta* occurrences in the *javana*. On the other hand several *vithis* together, as a long train of *cetas*, constitute the cognitive emotional process. The feeling present in the single *cetas* may prefer things different from the thoughts in the *vithis*. In this way an inner conflict may result. This is an interpretation of Freud's description of neuroses as a conflict between the super-ego, at the level of cognition, and the *id* that directs our drives. Freud made a good point when he noticed the unscrupulous power of the *id* doing things that morality does not allow. This is an example of strong kamma that cannot be avoided. The organization of consciousness, as described in the Abhidhamma model, shows that we have a basic neurotic core in our mind.

Some forms of psychotherapy are directed to address the cognitive level of our mind by trying to modify one's thinking, i.e., one's *vithis*. One may learn that one should not desire some things that do not belong to us. But this does not help if there is inside us a *ceta* with strong *cetasika* that desires that very object. Working directly with the *cetas* and *cetasikas* may be a welcome and powerful addition to the mentioned forms of psychotherapy.

We now turn to the neurotic core of the mind. If the *cetasikas* determine the type of mind, including feelings, while our thinking occurs in the cognitive process, then we may have processes with conflicting goals. For example, desire for an object of consciousness in the *cetasikas* and a thought that this desire

4 nodes only). Before the internet, computers were rather 'autistic', having as a typical task to compute $\sqrt{2}$ in many decimals. In the internet era, computers are interacting with other computers and new theories in computer science have arisen, such as the Calculus of Communicating Systems (CCS) and Process Algebra. These theories extended existing computer science.

is not allowed in the *vithi*. It is well known that in many cases the desire wins over the moral.[4]

Psychotic core

If consciousness is built up from discrete units smoothly blended together, then the integration mechanism may fail. Some psychoanalysts speak about the basic *psychotic core* of the mind. This can be explained in a natural way from the AM. Falling apart from the apparent continuous consciousness into its fundamental form of *ceta*-based *vithis* can be seen as the underlying mechanism. This also explains why psychotic episodes tend to reoccur. If one has seen things as they are, then that insight remains.

Psychosis can be linked to phenomena of dissociation. In a light and innocent form a dissociation occurs in children (and older humans as well) doing the 'repeated word game'. Taking a word like 'yellow', they speak it out repeatedly. After a while they notice that the word loses its meaning. In psychology one speaks about a 'semantic fatigue' or a '*jamais vu*'. Usually there is a close connection between a word as sound and the meaning of a word. These two come in consecutive *vithis*. After repeating the word many times, the sound function gets emphasized and may occur without an immediately following meaning, which may be absent for a short moment. The child has succeeded in separating the sound and the meaning that are both associated to a concept.

In stronger forms of dissociation the cognitive integration may not be functioning as usual, by being temporarily dysfunctioning or even being completely absent. From the Abhidhamma model this is quite understandable: cognition has to be formed by a process and this may temporarily not be active. For the common sense intuition the phenomenon can be quite startling. One loses one's grip, as the usual models of the world and of oneself slide away. As a reaction several mental alarms may start to ring. It is the presence of these alarms that causes the dissociation to be felt as an extremely painful experience. Indeed, the alteration of perception of our body, our person or our world (desomatization, depersonalization, or derealization) may be utterly shocking. This 'disidentification' often is accompanied by a powerful feeling of lack of control.

In psychiatry the cause of psychoses can be a genetic predisposition (less strength in the *manas*), a developmental deficit or a side effect of an illness (e.g., caused by encephalitis). According to the AM a psychosis may be simply the insight that our consciousness is discrete. The meditative purification found via insight meditation consists in taking out the existential fear of the dissociation. One becomes like a parachute jumper or astronaut no longer

[4] In Christianity one says: 'The spirit is willing, but the flesh is weak.' A Buddhist would rather say the opposite: 'The body can be without an orgasm, but the mind craves for it.' This seems more accurate.

afraid of the experience of falling. For this one first needs to be able to observe the three characteristics. In (Buddhagosa 1999), a classical medieval manual on insight meditation, the following is written:

> [I]mpermanence does not become apparent because it is concealed by continuity. The characteristic of nausea does not become apparent because it is concealed by the postures.
>
> [W]hen continuity *[of consciousness]* gets disrupted by discerning raise and fall *[breathing meditation]*, impermanence becomes apparent in its true nature. When the postures are exposed, ... the characteristic of nausea becomes apparent in its true nature.

Here 'impermanence' and 'nausea' refer to the first two of the three characteristics. After this has happened, (advanced) mindfulness may be applied to domesticate the fear.

One should not make the mistake of claiming that all psychoses can be cured by taking out the fundamental fear. But it certainly is worthwhile to investigate to what extent it is possible to cure or relieve them. See for example (Podvoll 2003). A personal account of a schizophrenic patient who did not get cured, but transformed her illness into something successfully manageable is (Saks 2007).

An early indication that psychoses in some cases may be cured and can lead to creative personalities is described in (Menninger et al. 1963), where the authors speak about patients becoming 'weller than well':

> Not infrequently we observe that a patient who is in a phase of recovery from what may have been a rather long illness shows continued improvement, past the point of his former 'normal' state of existence. He not only gets well, to use the vernacular; he gets as well as he was, and then continues to improve still further. He increases his productivity, he expands his life and its horizons. He develops new talents, new powers, new effectiveness. He becomes, one might say, 'weller than well.' ... Every experienced psychiatrist has seen it. ... What could it mean? It violates our conventional medical expectations, so perhaps it is often overlooked and occurs more often than we know. It may contain a clue for better prevention and better treatment. ... Transcendence does occur. And perhaps it is not an exception but a natural consequence of new insights and new concepts of treatment.

4 Interpretations

We will first show how the Cover-up Model can be interpreted within the AM. This happens 'theoretically'. But the logic of this is not the one that one finds in, say, mathematics. The style of explanation is part of a larger model as yet not fully made explicit.[5] After that we sketch how the Abhidhamma Model may be interpreted within neuroscience. This is hypothetical and needs further investigation based on empirical data.

From Cover-up to AM

The process of the Cover-up Model, consisting of the three characteristics chaos, nausea and selflessness (being beyond control) can be interpreted, using the AM, as being in a temporary lasting state in which the cetas do not fit well. This being similar to seeing an old movie of around 1920 in which there are missing picture frames making the story discontinuous. Actually this is only a weak metaphor. As the chains of cetas are forming our cognition, the discontinuity may cause impaired perception and even different forms of dissociation, e.g., desomatization, depersonalization, derealization.

The cover-up aspect of the model may be interpreted as being caused either by *cetasikas* with strong concentration effects, like those coming from greed, aversion or attachment, or by strong cognition (thinking), as being formed by coherent *vithis.* Coherent in the sense of fitting well together and forming a strong chain. The adequacy of these thoughts may be lacking. So the emotional cognitive mechanism holds us together. If we have adequate emotions and thoughts we are successful; if not we have a problem (that may be compensated by adequate coping).

The purification in the Cover-up Model is explained by the AM, by using its axiom that *cetasikas* like 'greed' may be eliminated. This is done using mindfulness, which is a cetasika giving stability to the ceta it resides in. So the method of purification exchanges the stabilizing *cetasikas*' greed, aversion or attachment by mindfulness.

From AM to neuroscience

The translation from the AM to neuroscience could go as follows. The discreteness of consciousness is not surprising, as the action potentials are discrete as well. It could be the case that action potentials are too low level to constitute consciousness. An important aspect of consciousness (the binding of the input of different senses) is caused by synchronous firing of sets of neurons, as postulated by von der Malsburg. Again the resulting mechanism is discrete. Finally, it may be that consciousness is not coded by the action potentials of single or combined neurons. Resonating waves of action potentials may be the carrier. Again there we have a fundamental discrete mechanism.

[5] When we learn elementary mechanics, the laws of Newton are not enough. In order to make deductions from them we have to use a style of reasoning that has not yet been made explicit. For example, that forces are vectors that may be decomposed along a horizontal and vertical axis.

The cetasikas that cause a state that lasts (for a while) may be implemented by neuromodulators like peptides that act through volume transmission. The AM predicts that there are three clear classes of such molecules: for attraction, repulsion, and neutral feelings (wanting to continue).

Purification through mindfulness is a harder phenomenon to be explained. It may be that some representation of our input is available at another level and is processed at cortex level without having the limbic (emotional) implications that the ordinary input has. Purification through observing nibbana as object of consciousness is the hardest phenomenon to be explained. It may have to do with the foundation[6] of consciousness, the ground for the qualia, whatever that may be. This foundation then gives stability, as it is always there.

The story can also be told in a different way. Perception and cognition of all the signals that come to us (either from outside or from the mind) have to create meaning. This is done by binding and reification, made possible by *manas*. For this reification some kind of binding neuropeptide may be necessary. Temporarily (or for longer periods) this capacity may be not functioning. Insight meditation is directed towards not being afraid of the resulting dissociation. This enables faster recovery of one's capacity to bind and reify.

5 Brouwer vs Gödel and the Common Core Thesis

Conversations with Gödel on mysticism are known from (Rucker 1983). As stated before, Gödel considered mathematics as revealing mystical experience. This is consistent with the AM view if the mystical state is identified with the *jhana* or *mental absorbtion*. In this state the object and type of consciousness are kept constant. It is said that this is possible only if the object is a 'concept' and not a sensory input. Now mathematics consists of concepts *par excellence*. The conclusion seems justified that Gödel's mysticism consisted of states of high concentration. In the AM one mentions the following aspects of these states: initial and sustained mental application, rapture, bliss and equanimity. In the more advanced 'formless absorptions' one has phenomena that Gödel undoubtedly would have liked: 'infinite space', 'infinite mind', 'nothingness' and finally 'neither perception nor non-perception'.

(Brouwer 1949C) on the other hand claimed that mathematics conceals the mystical experience. For him, descending to the foundation of consciousness is essential. This cannot be done if the mind is occupied with concepts. According to the AM the purifying consciousness is obtained by viewing the 'nibbana'. Also in the meditation tradition concepts conceal the view of nibbana, because concepts reside in a *vithi*, while nibbana within a single

[6] It is not generally believed that it is necessary to have such a foundation of consciousness, e.g., not by (Dennett 1991).

ceta. In the later Abhidharma nibbana is equated by Nagarjuna, witnessed by (Nandamalabhivamsa 2004), to pure consciousness, that is consciousness without object, which may be considered as the foundation of consciousness.

Although the state of high concentration is filled with ecstacy and bliss, its relief from suffering is only temporary. The view of nibbana provides a lasting liberation. Therefore, in the view of Nagarjuna, the foundation of consciousness is the liberating factor. This is very similar to Brouwer's description of his mystical experience.

Therefore we can replace the CCT by the Dual Core Thesis (DCT): there are two important styles of mysticism: that of bliss and ecstasy of the concentration meditation and that of permanent purification through insight meditation. Given this view it is not difficult to see it reoccurring in religious disputes, for example the incident around Hui Neng. The old fifth patriarch of Zen, Hung Jen, asked his monks to compose a poem to show their understanding of Zen, in order to choose a successor. The head monk Shenxiu wrote:

The body is a Bodhi tree,
the mind a standing mirror bright.
At all times polish it diligently,
and let no dust alight.

This requires continued work and energy. The poem of Hui Neng, the kitchen aid, was:

There is no Bodhi tree,
nor a mirror bright.
Since all is fundamentally empty,
where can the dust alight?

It was Hui Neng who got the job. Whether the story is a myth or based on reality is irrelevant. It is consistent with the DCT.

We list a few examples of opposing pairs of expressions related to mysticism, or the mystics themselves, as seen by the DCT. We briefly explain the notions. An absorption is a state of high concentration, often with a concept as object. During the liberation and a supramundane absorption the object of the *ceta* is nibbana, the foundation of consciousness. The expression 'Atman=Brahman' comes from Hindu philosophy, in which the absorption state is given a high status, as Brahman is the highest god. Actually it is not much different than the 'Mystical Union' (with God) in the Christian tradition. 'Anatta' is the selfless state of Buddhism. The mystical state of Santa Teresa has a close similarity with erotic experiences. Saint John of the Cross warned that we should not indulge in them but go further. In order to do this one has to overcome the 'dark night of the soul', probably his expression for the three characteristics.

Concentration	Insight
Absorption	Liberation & supramundane absorption
Atman=Brahman	Anatta
Shenxiu	Hui Neng
Santa Teresa	Saint John of the Cross
Gödel	Brouwer

The main advantages of the mysticism of insight meditation is that it has a lasting liberating effect and much less energy is required to obtain the state.

Brouwer's mysticism and ethical views

Let us look closer at whether it is correct that we classify the mystical experience of Brouwer under that of insight meditation. The following citations come from (Brouwer 1949C).

1. *'Consciousness ... seems to oscillate ... will-lessly and reversibly between stillness and sensation.'*
2. *'[There are] three successive phases of the exodus of consciousness from its deepest home. Of these phases the* naive *one was opened with the creation of the world sensation, the* isolated causal *one with the setting in of causal activities, and the* social *one with being involved in cooperation with other individuals.*
3. *'... security is as unattainable as unworthy.'*
4. *'Power over fellow-creatures will be avoided. ... Eastern devotion has perhaps better expressed this wisdom than any western man could have done. For instance in the ... Bhagavad-Gita ... :* "A man should not hate any living creature. ... He must be free from the delusion of I and mine." '

In 1 Brouwer comes very close to the description of baseline and cognitive consciousness in the Abhidhamma. The 'will-lessly' corresponds to the fundamental doctrine of *anatta*: selflessness. In 2 description is given of the state of 'bare-attention', known from Buddhism, and the ones that are goal-directed, in the language of Brouwer by means of *'cunning acts'*, creating *'causal sequences'* in order *'to obtain what is preferred'*. Brouwer also states later that it is not easy to return from either form of causal activity to naive being, while return to baseline is doable: *'More* [easy than the regression to the naive phase is a temporary] *refluence to the deepest home leaving aside naivety, through the free-will-perception of detachment-concentration.'* In insight meditation the *intensional concentrated observation with detachment* is the main tool towards progress. Remarkably enough the meditator arrives first at a state of naive observation, before one reaches the deepest home of consciousness. For someone who is enlightened the route via the baseline consciousness is easier indeed. Could it be thinkable that Brouwer has reached such a state of mind? He was not such an easy person to get along with and had his own desires. In the AM one speaks about several levels (four in total) of enlightenment. Only at level 3 has one rooted out desire and anger. So it is well possible that Brouwer has reached the first or second level. This is consistent with statement 3, which is a necessary view in order to reach the first level of enlightenment. In statement 4 an important ethical implication of enlightenment is stated. Brouwer also draws some consequences. The first explains the intuition needed for mathematics.

5. *'By a move of time a present sensation gives way to another present sensation in such a way that consciousness retains the former one as a past sensation.' ... 'Mathematics not bound to the world, can attain any degree of richness and clearness.'*

Science that *'reveals or playfully mathematizes naively perceptible forms and laws of nature'* is based on the naive phase of consciousness and is seen as pure.

6. *'Such science rejects human domination over nature, through its very reference.'*

If one wants to make cunning use of what one has understood, a price has to be paid.

7. *'In causal thinking and acting beauty will hardly be found.'*
8. *'Art and philosophy should not communicate with the state,* [otherwise] *they will lose their independence and degenerate.'*

And a final warning (also heard in different words in (Brouwer 1905A))

9. *'[M]ankind, possessed by the delusion of causality, will slide away in a deteriorative process of overpopulation, industrialization, serfdom, and devastation of nature, and ... when hereby first its spiritual and then its physiological conditions of life have been destroyed, it will come to an end like a colony of bacteria in the earth crust having fulfilled its task.'*

Now, in 2008, this warning starts to be heard, but it is as neglected by states as it was in 1904 and 1948.

Postscript

The Buddhist meditation tradition and its theoretical underpinning in the AM is not taken here as 'sacred'. We treat it with a critical mind. Our conclusions are based on our understanding and non-understanding and may for some part need to be readjusted. However, we feel confident about the DCT. The two classes of mystical experience come from concentration and insight meditation, respectively. Usually one does not speak about 'mystical experience' in the Western version of the tradition of insight meditation. Only for concentration meditation does one usually speak about mystical experiences, thereby pointing at mental absorptions. However, after successful training in insight meditation there is the 'fruit of the path', also called 'supramundane absorption'. This brings the yogi to the essence of consciousness and that is — at least in his description — what Brouwer experienced.

Part II

Kindred spirits

Remarks on the supposed French 'semi-' or 'pre-intuitionism'

Alain Michel

How did one come to propose, in connection with the French school of analysis at the end of the 19th century and the beginning of the 20th, represented by the names of Poincaré, Borel and Lebesgue, the respective qualifications of 'semi-intuitionism' and of 'pre-intuitionism'?

It seems that it is Brouwer who had first referred to the representatives of the French school of analysis as intuitionists *avant la lettre*, speaking of their 'pre-intuitionism'. In a lecture from 1952, he opposes the latter to what he calls the 'Old Formalist school (Dedekind, Cantor, Peano, Hilbert, Russell, Zermelo, Couturat)'. As leaders of the Pre-intuitionist school he names Poincaré, Borel, Lebesgue.[1] One finds in this text a typical reconstruction of history, due to mathematical and methodological considerations related to a later doctrinal development.[2] As for 'pre-intuitionism', it is especially due, it

[1] (Brouwer 1952B). Brouwer also occasionally quotes Baire.

[2] Brouwer examines the authors in question with respect to two fundamental issues. The first is the traditional distinction between the discrete and the continuum, the discrete being the origin of what he calls the 'separable parts of mathematics', e.g., (Brouwer 1952B, p.140). The second, proper to Brouwer, is the relation between logical language and mathematics. Having characterized the Old Formalist school by an ideal of making the differences between the languages of logic and mathematics disappear by treating the concepts hitherto considered as properly mathematical as elements of language and logic, Brouwer then affirms that the Pre-Intuitionist school rejects such an identification with regard to natural numbers, the principle of induction, and all that can be constructed from these. This source is intuitive, even observational, and does not require existence axioms. Its consistency is certain even without logical proof. The position of the Pre-Intuitionists regarding the continuum is much less clear. According to circumstances, it oscillates between, on the one hand, the acceptance of existence axioms (axiom of saturation) and thus the recovery of the Old Formalist ideal of absorbing mathematics into language and logic, and, on the other hand, the maintenance of autonomy; but they are then forced to accept the real numbers as an ad hoc system, 'ever unfinished and ever denumerable', which has the properly mathematical disadvantage of not having a measure positively differing from zero.

seems, to posterior commentators. One finds for example the expression in a note in the book that Abraham Fraenkel, Yehoshua Bar Hillel and Azriel Lévy devoted to the foundations of set theory, in the historical introduction of the chapter on the intuitionist conception of mathematics. About the work of this group of mathematicians who called themselves 'intuitionists' or 'neo-intuitionists', composed of varied and largely divergent currents, but of which a majority belongs to this century, the author affirms: 'Various names were used in the past, among them the now obsolete term of "neo-intuitionism". The current terminology will be adopted, which reserves the term "intuitionism" for the school of Brouwer. Older schools, particularly the French school, are evoked under the name of "semi-intuitionism".'[3]

These denominations were considered to be convenient and have largely remained in use. With this use comes a risk of masking their lack of legitimacy. Therefore it seems useful to me to address the question of their origin, and their well-foundedness.

1 The positioning of the question

The problem is that of the legitimacy and the conditions of use — in particular historical use — of a predicate: 'intuitionist'. As is common with a movement of thought aware of its novelty and effectuating a deliberate break with its time, promoters of intuitionism, first and foremost Brouwer, looked to history to try to find elements of justification for their own undertaking. The search for a past history, even an imperfect one, and outlines, even incomplete ones, in short, 'precursors', is a move that is all the more urgent when the undertaking in question meets opposition or obstacles — and as one knows, there was no shortage of those with intuitionism.

Besides, the problem that arises is almost always worthy of interest. It concerns here the positioning and characterization of a movement of thought prior to its birth in the strict sense. Where was, and how do we characterize, intuitionism before intuitionism?[4] Did it exist before Brouwer's first texts, and in what forms? By applying one of its principal requisites, it would be necessary, in order to assign to it a historical existence, to avail ourselves of the means of constructing it: how do we find it in history? how do we construct the history of intuitionism?

Several points related to these general questions require attention.

There is, to begin with, a purely methodological problem. It is the fact that, disregarding any consideration of its specific problems — validity of the use of the excluded middle applied to infinite collections, constructive standard for the theses of existence, etc. — , the emergence of intuitionism as a competitor of other equally legitimate standpoints contributed to making

[3] (Fraenkel et al. 1973, p.214)
[4] For the history and meaning of the term, see (Largeault 1993).

more precise and rigorous the interrogation regarding the status of procedures and of mathematical objects, or on the attention paid to doctrinal choices which, explicitly or implicitly, determine in depth the work of the mathematician. 'Intuitionism' is one of these theoretical tools which can even allow us to re-examine traditional mathematical or mathematical-philosophical works. To speak clearly, to what extent can one say that Euclid, Descartes, Cauchy were 'intuitionists' before the term was coined?

Hence the second aspect of the question, which has to do with historical methodology. It is encountered as soon as one applies to earlier works or events a category suggested in a posterior configuration, with respect to precise problems, which did not exist, or at least did not necessarily have the same sense, at these two different moments in time. One must acknowledge that the approach may arouse the suspicion of anachronism. Intuitionism as a doctrine received its sense from a singular, though typical, reaction (as its success earned later on tended to show) to some of the developments of the Cantorian theory of infinite sets (the theory of the transfinite). Naturally, there is no reason to think that the questions which worry current theorists of mathematics or of logic had been stated at that time, nor, even if they had, that they were stated in the same terms. There is no lack of examples to the contrary, of retrospective deformation that a little attention to how the authors of that time situated the problems would have been enough to weed out.

The authors who refer to French pre-intuitionism generally mention, like Brouwer, three authors: Poincaré, Borel, and Lebesgue. For the sake of brevity, we shall leave aside the case of the latter, and that of the first one, for other types of reasons. The part Poincaré played in the introduction of Cantorian set theory in France, his interventions at the time in the early discussions about the status of set-theoretic concepts and about the relations between formal logic and mathematics, his own philosophical positions, and the way they were inserted and received in the French academic philosophy of the time, are all topics that would require separate study. Admittedly, removing him altogether carries some inconveniences, since it is likely that his positions have had a direct influence on the questions discussed, and we shall evoke these positions each time it will be necessary. We shall therefore limit ourselves to the case of Borel, who is both a good intermediary, since he is a disciple of Poincaré and had discussions with Lebesgue, and a perfect representative of the French mathematical tradition that we find it important to characterize. He may, as such, supply us with a range of examples and analyses which are largely sufficient to allow us to evaluate the possibility, the significance, and the impact, of the act which consists in conferring on our authors the denomination of 'pre-' or 'semi-intuitionism'.

We should notice immediately that the aforementioned authors themselves never retained nor even discussed this manner of designating themselves. Rather, they referred to themselves as 'empirists' or as 'realists' (Poincaré preferred to evoke the name 'pragmatists'[5]), coming naturally to mobilize, in order to qualify themselves, categories which do not have a precise sense except in philosophical systems. They give to these terms a more general and vaguer sense that they never sought to specify or to clarify. Thus we must satisfy ourselves with the most immediate references. Let us say that it is a question of appealing, in the case of empiricism, to experience; and in that of realism, to reality. The term *experience* refers to that which has the features of a direct test. Indeed it refers to a kind of 'mathematical experience'. It is related to the practical activity of the mathematician. As for the expression *real* or *reality*, it qualifies a type of meaning, in relation to the objects or reasoning, which opposes it, as an area of testing, to another, which is that of pure abstraction or theory (then through the mediation of language, which inclines towards logic) — rather than a philosophically determined realism, or even a physical realism, which does not admit the existence of any thing but that which is given by sensation.[6]

Leaving aside all these complications related to denomination, the problems of interpretation remain. It seems to us that while the objects and the problems considered by our authors, and the mathematical context in general, are largely shared, the fact that Borel and Lebesgue belonged to different mathematical practices and currents of thought distinguishes them and moves them away enough from Brouwerian tradition to give up applying to them, under some form or another, the epithet of 'intuitionism'. One could even say that the differences come through all the more acutely since the problems dealt with are the same. Thus, it would certainly be of some importance to examine whether, and to what extent, intuitionism in Brouwer's sense coheres with 1) the idea that Borel and Lebesgue held of mathematics 2) their practice of mathematics, and the relation which they emphasized between the practice and the idea. In order to seize in depth these differences, I shall propose to examine, as they themselves constantly sought to do, their position regarding

[5] 'La logique de l'infini', in *Dernières pensées*, Paris, 1913; (we refer to the article published in *Scientia* in 1912 (Poincaré 1912), and not to the article of the same title which appears in the *Revue de Métaphysique et de Morale*, 1909). The pragmatists admit as objects only those that are definable in a finite number of words. Their view is akin to idealism insofar as they consider that an object exists only in thought, and therefore depends on a finite thinking subject. The realists consider things from the physical point of view. They posit the independent existence of (physical, mental, or mathematical) objects. There are two kinds of realists: the strictly physicalist realists, who are finitists, since reality in the physico-observational sense is always finite; and the Cantorians, who are realists also with respect to mathematical entities, therefore infinitists, and who may be called 'platonists'.

[6] There are of course many hesitations and waverings in the use of this terminology, as well as a tendency to oscillate between idealism and realism. See for example the study by J. Bouveresse (Bouveresse 2005), who notes curious inversions of sense in the use of terms like 'idealism' and 'realism'.

mathematical doctrines in tandem with their practice of mathematics. This relation is clearly not as strong for Brouwer himself. Many commentators asked what kind of relation the intuitionist doctrine could have had to the mathematical substance of the first topological work to be found in Brouwer. This is a matter of discussion, as it is not immediately clear that we have evidence here,[7] whereas we do for our French analysts. It is this relation that should be comprehended, and which according to us undermines the validity of the qualification as intuitionism by depriving it of its essential meaning.

2 Brouwer and the history of intuitionism

Let us recapitulate the principal features of Brouwer's position, such as it appears in the already cited text on the historical sources of intuitionism (Brouwer 1952B).

2.1 The Kantian heritage of Brouwerian epistemology

One could easily find, in the text from the lecture of 1952, three features which directly evoke Kantianism:

a) The general point of view is explicitly that of the 'mental mechanism of mathematical thought'. It is a matter of subjective methodology: one is interested in the structures of the thinking subject as it produces mathematics — which corresponds well to the Kantian roots of the problem — but of a thinking subject reduced to a mental mechanism — which is already, though common, much less faithful to the spirit of Kantianism.
b) Brouwer notes that his evolution is linked to philosophical ideas concerning the origin of mathematical certainty — a confirmation of the Kantian character of the problems — and the definition of the mathematical object.
c) Thirdly and finally, the remarkable permanence of this mental mechanism is ascribed to the belief in properties of space and time, which are themselves immutable and independent of language.

Brouwer then seeks to reconstitute the genesis of the process of mathematical knowledge. The point of departure lies with the experience, and the observation of regularities which are found in it. But an experience is not able to provide us with an absolute invariability; one is constrained to posit the latter through postulates, which inevitably take a linguistic form. From these axioms, one develops systems of properties, which give substance to what is called mathematical science.

This way of assigning an origin to mathematics explains Brouwer's adoption of a norm of judgement which he calls 'the observational standpoint'. It consists of an interpretation of the origin of mathematics, which places at this

[7] [Added in proof] See, however, (Brouwer 1921J, p.950).

origin the relation with spatio-temporal experience. Brouwer infers from this the central thesis of the independence of mathematics from logic, and from language in general. The 'observational standpoint', according to Brouwer, is that of experience, to which one simply gives a Kantian representation. It is with this idea in mind that he will consider the development of mathematics in the recent past, in order to locate the possible traces of intuitionism, and to give himself, in the same movement, the means of justifying and specifying it.

There are, in these analyses, epistemological features which unquestionably bring Brouwer's positions closer to those of the mathematicians we are here concerned with.

Initially Brouwer conceives of mathematics as a genuine form of knowledge acquisition, which makes it irreducible to a simple formal language. Here is also a recurring theme within the tradition of mathematical philosophy, from Plato and Aristotle to Descartes, Leibniz, or Kant. One claims more specifically, within the Kantian heritage of this tradition, that any knowledge comes from experience, and has as its source space-time intuition. Lastly, it will be noticed that the roots of the criticism which is addressed to the French analysts coincide with one of the most remarkable features of their mathematical production, namely convergence between on the one hand theoretical, and perhaps even more reflexive, problems of foundation, and, on the other hand, technical advances in mathematical work, caused here by the very developments of analysis — a convergence that one can interpret as a form of 'reflexivity', a typical feature of the epistemology of 20th century mathematics.

2.2 The references to Borel and Lebesgue

Brouwer distinguishes first between the case of the integer number (connected with the pure intuition of time) and that of the continuum (linked with the pure intuition of space).

According to him, as far as the integer is concerned, the French school had sustained the 'observational standpoint', which amounts to validating its empirist self-representation. Here, the source is the principle of complete induction, as formulated by Poincaré. For derived mathematical objects, an axiom of existence is not necessary, nor a demonstration of consistence. One can ignore logic.

As far as the continuum is concerned, things are different. Here, one does not have a search for the origin, which would guarantee the independence of the object from logic and language. According to Brouwer, the attitude of the French analysts kept oscillating between two attitudes: one consists in positing the continuum by a logical axiom of existence, such as the axiom of saturation; the other constrains them to mutilate the continuum. Their system of real numbers is 'always unfinished and always countable', since it is generated by a system of laws defining convergent infinite sequences of real numbers.

Hence we find two fundamental objections addressed to Borel and his colleagues. In both cases, they continue to use the laws of traditional logic. Apparently, they did not take notice of the fact that the fraction of the continuum reached in this way is precisely too restricted for analysis. A clear indication of this is mathematical: such systems are of 'measure zero', and therefore radically inapt to provide for the needs of analysis. In general, they failed to recognize fully and completely the autonomy of mathematics. In fact, they only dealt with the integer. Actually they just initiated a movement which now needs to be completed by radicalization and systematization. This will indeed be the task of Brouwer's intuitionism.

3 The example of Borel

3.1 Epistemological positions

Philosophy: 'empirist realism'

Borel postulates, as Brouwer saw it, and like him, a fundamental distinction between mathematics as a science, an objective knowledge, a way to capture reality, and language, a simple means of explaining and communicating mathematical results. This realism induces us to seek what is really or actually intended by the words 'observable realities'. The mathematician must 'take the utmost care 'not to "reason in a vacuum" '. This is the meaning of the constant contrast that Borel drew between intuition and speech. Besides, one would seek in vain a substantial significance of what he calls 'intuition'. The primary goal of the reference to the intuitive, repeated time and again, is to emphasize the opposition to the category of discourse. The intuitive is here the antidote of the purely discursive.[8]

Against the 'formalizing' approach of Weierstrass, Borel develops a realistic approach, largely inherited from Hermite, the common master of all these French analysts. Objects of mathematics are substantial beings, genuine individuals. The paradigm of representation is naturalist, it makes mathematics,

[8] One finds a very similar position in Poincaré with respect to 'nominalism'. The reference to constructions of purely symbolic systems, such as the real numbers of Dedekind (the confusion with Kronecker made by Poincaré in a first draft of chapter II of La Science et l'hypothèse is significant, since Kronecker thought that the only reality is the intuition of the integers), which are without relation to reality, without intuitive content, in particular aims to oppose verbal mathematics to real mathematics. There are only two real sources of mathematical objects, only two points of departure for their production: the integer and the geometrical continuum. They correspond, in Kantian terms, to the pure intuitions of time and space.

according to the model for biology, a science which can know and study mathematical entities in themselves, independently of their linguistic expressions, and, more generally, of procedures of construction, which express them, or allow us to capture them.[9] If this realism remains, philosophically speaking, rather cursory, this is far from being the case mathematically speaking. On the contrary, it is accompanied by a great refinement in precision, when it comes to describing the results of analytical work. Here, one thinks especially of the part played by the concept of singularity, which is important in the theory of functions of complex variables: the full information which one may have about the functions is contained not so much in its normal form as in its singularities. One of Borel's most significant methods consists in settling, so to say, right in the middle of the singularities of the functions to extract from them the amount of information necessary for the knowledge of their laws. It is a way of conceiving generalization.

Epistemology: the role of intuition

This second point is related to the first. Borel develops an intuitive approach to the mathematical object. Concerning functions for instance, if one wants to give an adequate account of the object, it is advisable to go beyond the resources of the expression, i.e., of the analytical expression. One knows the discussion, which goes back to Leibniz and Euler, but which is deeply engraved in the history of analysis, on the relationship between the analytical expression and the function. Borel expresses the requirement of knowledge of the object as a mathematical need, which is to go back, from the point of view of Weierstrass, to that of Cauchy. While restricting the point of view to that of analysis, in the general meaning of the term, i.e., composition of the function from its components (Taylor series), one deprives oneself of any means to obtain information about the objects that a more flexible approach would have authorized. Cantor has precisely given the example of a freer approach which consists in adopting initially the most general point of view, the only way not to be likely to veil, at first, unsuspected properties. There is no doubt that this generality should be restricted to the starting point. A further delimitation must guarantee the mathematical reality of the object. One shall limit oneself to objects known as 'well-defined', and thus, according to Borel, we shall make the most of the discoveries of Cantor. This is the reason why Borel gives prominence to Cauchy's definition of the function, rather than to that of Riemann. Even if the former is less general than the latter, the mathematician still gets something out of it, insofar as he is concerned with objects, functions, obtained by well-defined, and, if possible, effective, procedures from

[9] Borel devotes for example a whole passage in the introduction of his *Méthodes et problèmes de la théorie des fonctions* (Borel 1922) to developing an analogy between mathematics and biology. See also *Notice sur ses travaux scientifiques* of 1912 (Borel 1972a, pp.120–128), and the foreword and the introduction to his *Leçons sur les fonctions monogènes d'une variable complexe* (Borel 1917).

already known functions — even if it is only in an implicit way, like when a function is given as the solution of a differential equation. At the same time, since these functions are defined by series, as Weierstrass would have put it, the ideal thing to do is to approach the study of these functions by considering from a direct point of view their properties and those of their domain of definition or convergence, which are considered as sets in an abstract sense, and for the study of which one disposes of the Cantorian tools.

Concepts of method: countability, enumerability, effectivity

The sequence of integers really constitutes the nucleus of Borel's mathematical thought.

It was initially a heritage of Kant and Poincaré. The sequence of integers results from an effective act of the mind: the basic intuition of the fact that after each integer, there is another one. This is not the case with the Cantorian sequence of ordinal transfinite numbers, where one does not have a regular process of production for the sequence, since nothing can give us a clear idea that, beyond any undefined sequence of increasing functions, there is another one. Besides the sequence of integers, there is the geometrical continuum, which is given to us in nature, and to which we have access by intuition. But it is by itself insufficient to found the numerical continuum. The latter is an artificial creation, which provides us with a concept of continuum that is to some extent merely negative. Borel repeats that our only certain concept is that of the countable. The uncountable, the heart of numerical continuum, is 'hardly anything, as its denomination lets it appear, but a negative concept'. Inside the continuum (and it is what Brouwer refers to when speaking of 'a system of real numbers that is always unfinished and always countable'), there are only numbers known as 'calculable', i.e., according to Borel, defined in a precise way and without ambiguity by a finite number of words. Without playing on words, they are for Borel the only calculable numbers which are epistemologically *real*, or really existing. In the same way, the only functions which will be regarded as real, or really existing, will be functions known as 'calculable', that is, those whose value is calculable for any calculable value of the variable. The same applies to sets of points, via the concept of the characteristic function: only those sets whose characteristic functions are calculable will be regarded as realities.

The set of calculable numbers satisfies a property which is stronger than the countability: that of 'effective enumerability'. This property is defined as the possibility of indicating, by a finite number of words, a certain process which displays its members in a sequence, by ascribing without ambiguity a determined rank to each one of its members.[10] The prudence required due

[10] In today's more precise terminology, the set is 'recursively enumerable', i.e., the range of a recursive (or calculable, or algorithmic, or mechanical) function. The fundamental idea is always the requirement of a process of finite calculation.

to the flourishing paradoxes of set theory thus imply the consideration of another distinction than that between the countable and the non-countable, that between countability and effective enumerability. The first one lies within set theory, and is of a theoretical nature. The second one is essential for any reflection on set theory. It is of a practical nature. According to Borel, it should be placed at the heart of the mathematicians' treatment of the theory, once one has tried to avoid the paradoxes and taken enough precautions. These distinctions, though expressed in particularly radical terms by Borel, could be accepted by the majority of set-theorists, and constitute a kind of common program.

Moreover, Borel is perfectly aware of the fact that his point of view imposes a transformation of the traditional methods concerning definitions (we shall return to this point), postulation (for the statements, renunciation of the excluded middle), and reasoning (for sequences of statements, rejection of the axiom of choice for infinite sets).

The essential restriction relates to the objects. Borel wants to allow only 'real and normal objects'. All other more or less teratological items are more or less artificial creations, mere abstractions which do not have a mathematically clear meaning acceptable by consent among mathematicians. They must be eliminated. One will limit oneself to the 'well-defined' objects (sets and functions for the analysis), i.e., effectively enumerable ones.

One needs then to propose the basis of a program for reconstructing analysis in conformity with the adopted point of view. It is here that the specific features of the proposed treatment intervene. The choice is that of constructive procedures. A 'good definition' will be a constructive definition. A set will be regarded as well-defined if one can obtain it by addition or subtraction of intervals, except end points, given that these two operations are indefinitely iterable — but, if one decides to limit oneself to the intervals having rational ends, one performs the reasoning excluding only a countable set of points, so that the totality of the points which are left out is of measure zero.

Departing from real intervals, one obtains, as is known, what Lebesgue proposed to call in his theory of integration 'B-measurable sets' (Borel sets), which thus coincide with the well-defined sets. They constitute the field of traditional analysis, in particular that of integration. One finds here the already announced convergence between the epimathematic reflection and the work of analysis. In this specific characterization of what is required from a good definition, of restriction of mathematical reality, the 'observational standpoint', in the sense of Brouwer, finds here to some extent a mathematical and even analytical expression. We have mathematical reflexivity at work.

For the future, it is advisable to take care not to exceed the field of objects delimited in this way. In general, if one limits oneself to apply, to any family of sets already obtained by a constructive procedure, only the operations of countable union and intersection, one will be sure to obtain an admissible family of sets. Such a program was to meet some difficulties which were well analyzed by Cavaillès.[11] Reasoning which brings into play an unspecified B-measurable set actually utilizes the totality of the class II of the transfinite numbers. Borel wants to avoid regarding such a totality as being given and extends the constructive procedures which he promoted for sets and functions to the reasoning which utilizes the general properties of the B-measurable sets. This reasoning is basically inductive: 'one supposes', he says, 'that the construction has been carried out up to a certain point and has certain properties, and it is shown that these properties remain valid when a further step is made'.[12] Here we have not, Cavaillès comments, a given system, but a reality in the making — a feature retained by Brouwer in his theory of the continuum. These restrictions do not remove all the difficulties. Some of them will be emphasized by mathematicians like Lusin and Souslin, who will devote part of their work to try to clarify the nature of objects involved.[13]

4 Conclusions

Let us return to the problem which was our point of departure. Can one in the case of our authors speak of 'semi-' or 'pre-intuitionism', and, if so, in what precise sense? One should recognize that neither one of these two denominations is really satisfactory. All things considered, one should prefer, in spite of other defects, traditional philosophical terms such as 'empiricism' or 'realism'. But this is not the main point. The issue is not fundamentally that of denomination. The very use of the term, if only in the quantitatively attenuated form of the half measure ('semi') or in that of the retrospective representation ('pre'), raises more fundamental problems. Let us now return to the problems mentioned at the beginning, and which may be called problems of legitimation of historical practice. No doubt we find here what is missing in the aforementioned cases, namely proximity, not only in a chronological, but also in a theoretical sense: a homogeneity of conceptual material, a set of common problems and methods.

Let us limit ourselves to merely practical considerations: can we gain some insight by this categorization? To such a question, one would incline to give a negative answer. More important yet is to justify it positively. We shall say, to simplify, that our authors belong to a different mathematical tradition than that of Brouwerian intuitionism.

---

[11] (Cavaillès 1938, pp.12–15,19–20).

[12] (Borel 1928, p.235); quoted by Cavaillès, see footnote 11 above.

[13] See (Cavaillès 1938, pp.12–13). Projective and analytical sets enjoy remarkable properties, such as being closed under complete additivity or multiplicativity and (for the analytical sets, under certain conditions) measurable applications. They have also interesting applications, in analysis (theory of the potential), in logic, as shown by the use made of them by Tarski, Kuratowski, or Sierpinski.

This concept of mathematical tradition has its own difficulties. However, it seems essential to us in the treatment of these kind of questions.[14]

Such a tradition is initially defined in a negative manner, by the absence of doctrinal fixation. It would be difficult to say this of Brouwer, who worked at, and expressly strove to diffuse, doctrines that one may describe as 'philosophical'. Lack of interest in foundational problems is, in the case of our French authors, of course not a mere oversight. It is actually a deep feature, which reflects the way in which they apprehend mathematical activity. What Brouwer sometimes calls 'the Paris school' never constituted a school. Between our authors, there were exchanges and debates, as in the famous 'five letters on set theory', but no attempt to line up under a common banner. Nothing in any case to compare with what took place in Amsterdam or even in Göttingen. None of our authors taken in isolation ever claimed to offer a doctrine, and it would be difficult to connect them to a distinctive philosophy, the way Brouwer or Hilbert were connected to Kant.[15] Therefore, when they were summoned to do so, they rather claim their commitment to 'empiricism' or to 'pragmatism', a simple way of appealing to a standard of judgement lying outside of theoretical or doctrinal speculations. 'Je n'aime guère écrire des alephs', Borel has written,[16] in a *grand siècle* style which evokes Leibniz confessing his views on the nature of the infinitesimals. If Borel then refers to 'the more serious reasonings' for which they can be used as a guide, because of the abstract character of their concepts (transfinite numbers), for all that, he does not condemn less firmly the practice of juggling with 'symbols referring to no accessible reality', the lavishness of formal logic which looks like 'a construction without any base', these things which one calls 'infinite', 'successive powers', 'aleph', 'aleph exponent aleph', etc.[17]

As conceived by all our authors, 'reality' does not have a real meaning except in reference to mathematical work, and Cavaillès is right in stressing the importance they gave to the concept of mathematical experience.[18] One often feels, when reading them, that the only philosophy which could be of some service would be the one that managed to clear up and systematize the plain

[14] Our authors were aware of the situation, as for example Lebesgue shows in note 1 on the transfinite numbers of (Lebesgue 1928): 'this observation is not unrelated to certain ideas of Mr. Brouwer; my goal, in making it, was however just the opposite of his. I deny by no means the value of traditional logic and of reasoning by reductio ad absurdum; I only state a reminder that they should be used correctly. Fortunately, the scorning way in which Mr. Brouwer speaks about the Paris school only serves to emphasize our disagreement.' (translation ours)

[15] One could do it to a certain extent for Poincaré, who moved in the circles of the *Revue de Métaphysique et de Morale*, became acquainted with philosophers, and conversed in particular with Boutroux.

[16] See the letter to Hadamard (Borel 1928, p.158).

[17] (Borel 1928); see also his allusions to 'a logical system' (p. 174), to the 'construction of a building without any link with reality' (p.175), etc.

[18] (Cavaillès 1938, p.18).

meaning of this concept. In its absence, one can only detect infallible signs of its manifestation: the consensus of mathematicians working at their science;[19] mathematical utility or fruitfulness — such as the concept of countability for Borel, at the origin of his most important results in measure or probability theory; agreement with the ideas or images that mathematicians draw from their acquaintance with the most traditional parts of mathematics; finally, the intuition that allows discoveries — intuition which requires intervention of the subjectivity of the mathematician and shows that mathematics cannot be reduced to an objective system of deductions.[20]

Positively, this tradition would be characterized by a certain representation of mathematics as a science, from which a specific conception of its object arises.

In order to clarify this last point, let us consider, for instance, the way in which Borel represents to himself the constitution of the generality of concepts of analysis through the study of singularities. Borel very often used a method which appears very clearly from his treatment of analytical functions in his doctoral dissertation. He starts from the following observation: in the theory of the functions of complex variables, the most valuable information about the functions is contained in what are called singular points. Weierstrass had drawn attention to these singularities and to the need to take them as starting points of the formation of the generic entities of analysis. Poincaré had studied many such examples, and had shown that in theory they often made it possible to generalize functions. From the beginning, Borel means to proceed differently than Weierstrass. The point of view of the latter consists in holding as fundamental the concept of the analytical element of a function, i.e., the Taylor series of non-null radius of convergence, and to compose the function from these elements which are taken as components. According to Borel, it is nothing but an abstract and formal approach, which has its merits, especially as far as rigour is concerned, but with the serious disadvantage of imprisoning the functions within a narrow framework from which it is then difficult to escape. Borel shows that, if one takes such a series 'randomly', without particular conditions, it generally admits a convergence ring like a 'cut'. Cases where analytical prolongation is possible are exceptional, and the series represent a very particular function. He concludes from this that it is somewhat illusory to consider the Taylor series *a priori*, independently of their properly analytical origin. Here, the abstraction of the point of view is thus

[19] This is precisely what is missing in set theory, where mathematicians do not have a common representation of the objects under discussion; by contrast, Borel affirms that mathematicians have for a long time taken for granted the autonomy of the sequence of integers, independently of any metaphysical consideration, and have used it 'as if it really existed as a whole', and as if it was a perfectly clear, unambiguous concept.

[20] One sees here a strong agreement with Brouwer, limited however to a then common reflection on the value of intuition as the source of innovation, as the engine of mathematical progress.

deceptive. We would do better to give it up and adopt a direct and intuitive approach to functions, which consists in considering them as concrete objects whose diversity must be assumed — rather than seeking to dissimulate the latter artificially under a uniformity which is on the whole no more than an appearance, to settle from the start right at the singularities of functions (especially their 'essential singularities'), as if they were really existing entities, forming a 'natural species', as Hermite did not hesitate to say, and whose characteristics one must study carefully. This is how he proceeds in his study of certain series that have 'a lacunar space', by drawing from them a means of eliminating a preliminary objection which had been raised by Poincaré against the prolongation of these functions beyond one essential singular line, achieving therefore a generalization of the theory — besides an occasion to develop the concept of the measure of a set. So he will proceed, showing a remarkable unity of style, to the study of divergent series or of probabilities.

If we agree to let ourselves be inspired, for some time, by the pragmatic spirit of our authors, one will recognize that the annexation to a form of intuitionism appears rather to contribute to darkening things than enlightening them. It may conceal the originality and the specificity of the system of collective traditions and individual styles which forms the real fabric of mathematical history, and make disappear, in a factitious homogeneity, that which is undoubtedly the most valuable thing historical research can bring to us: the unfolding of a system of differences.

Acknowledgement

I thank François Clementz, my colleague at the University of Provence, and Emily Tanimura for helping me in the English translation of this text.

Poincaré: intuitionism, intuition, and convention

Gerhard Heinzmann and Philippe Nabonnand

1

Arend Heyting in the introduction of his famous volume *Mathematische Grundlagenforschung, Intuitionismus, Beweistheorie*, speaks of Poincaré's influence on 'contemporary intuitionists', *i.e.*, mathematicians who agree with the opinion that

(1) Mathematics has not only a formal but also a contentual signification.
(2) Mathematical objects are directly grasped by the thinking mind; hence mathematical knowledge is independent of experience. (Heyting 1934, p.3)[1]

In addition, according to Heyting, the second characteristic is essentially connected with ontology and admits two different interpretations:

a) Mathematical entities do not exist independently of our thinking (anti-realism).
b) Mathematical entities exist independently of us but are only knowable through construction.

Heyting attributed the first interpretation to Brouwer and called the second position semi-intuitionistic. The term was motivated by systematic considerations. One of these was the fact that the grasping of the entity might not be immediate but mediated by linguistic construction. Heyting saw the semi-intuitionistic rejection of a construction-transcendent notion of existence as approximating Kronecker's view, and he saw it even in a wider sense as the view of the 'French realists or empiricists', among whom he counted Borel, Lebesgue, and Baire, while noting that Kaufmann, Skolem and Richard held similar views. We think he is right to have omitted Poincaré from this list.

[1] '1. Mathematik hat nicht bloß formale, sondern auch inhaltliche Bedeutung. 2. Die mathematischen Gegenstände werden von dem denkenden Geist unmittelbar erfaßt; die mathematische Erkenntnis ist daher von der Erfahrung unabhängig.'

Clearly, for Poincaré, the question of mathematical entities existing independent of our mind is a question of old metaphysics about which he had nothing to say (Poincaré 1905, p.xxiv).[2]

Since Kant, mathematical intuition has been connected with construction. But constructivism could neither be identified with intuitionism (anti-realism) nor with realism. What it is that is supposed to be constructible, objects or proofs, remains a subject of debate and is linked neither with thesis (a) nor with thesis (b). Brouwer recognized mathematics as an autonomic interior *constructional* mental activity (Brouwer 1955, p.113, our emphasis). Moreover, although constructivism was a key word in Kronecker's attack against Cantor's platonism and, in a wider sense, against traditional (axiomatic) mathematics, the question of construction has a definite sense even for a Platonist. This is illustrated by Gödel's attitude towards constructible sets. Generally speaking, Gödel believed that mathematical objects can be perceived through a special cognitive ability, *i.e.*, 'intuition', in something like the same way that empirical objects are grasped through sensory perception. As Hintikka put it, the intuitionist is more interested in what we can *know* whereas the constructivist is more interested in what we can *do* (Hintikka 1996, p.237). In any case, constructivism and intuitionism are both neutral with respect to realism and anti-realism. Anti-realism doesn't imply intuitionism, and realism doesn't imply an anti-intuitionistic constructivism. In addition, historical positions don't necessarily coincide with these normative views, and their opposition is rather contrived (see (Mooij 1966, p.136)).

It was Brouwer himself who pressed the difference between (a type of) constructivism and intuitionism by comparing Poincaré's position with his own standpoint. He saw Poincaré as guilty of a 'confusion between the act of constructing mathematics and the language of mathematics' (Brouwer 1907, p.176)/(Brouwer 1975, p.96). Although, according to Brouwer, the constructional mental activity of intuitionism may have a useful linguistic expression and can be applied to an exterior world, it has 'neither in its origin nor in the essence of its method ... anything to do with language or an exterior world' (Brouwer 1955, p.113).

Poincaré was far from taking the non-linguistic and intuitive construction as the only basis for mathematics. The role of linguistic conventions and language in mathematics is a point that separates him from Brouwer. For Brouwer, 'the mistake of logistics lies in the fact that it creates nothing else than a linguistic structure, which can never be transformed into mathematics proper' (Brouwer 1907, p.176)/(Brouwer 1975, p.96); for Poincaré, mathematics proper beyond linguistic structures is not a knowable domain. The evidence of object-constructions has to be linked with their description. (By the way, this 'pragmatistic' turn of Poincaré's philosophy is the French origin

[2] According to Poincaré modern metaphysics negates all metaphysics and constitutes for this reason nevertheless a metaphysical standpoint (see letter to Camille Flammarion (Poincaré 1904)).

of analytic philosophy.) Nevertheless, Brouwer is quite right when he emphasizes that Poincaré 'reestablished on the one hand the essential difference in character between logic and mathematics, and on the other hand the autonomy of logic and of a part of mathematics' (Brouwer 1952B, p.140). In fact, according to Poincaré, the program of founding number theory on logic involves a *petitio principii* because this foundation presupposes already some number theory. It seems that even his theory of predicative definitions was taken to be independent of mathematics (cf. (Goldfarb 1988, p.64)). Expressed in other words, even though Poincaré may have agreed with Brouwer that 'mathematical language by itself can never create new mathematical systems' (Brouwer 1952B, p.141), mathematical language is nevertheless a necessary tool for mathematical creation (and not, or at least not merely, in some psychological way). We will discuss this point later.

Hence there are only some epistemological features that bring single *results* of Poincaré's position close to those of Brouwer's; however, Poincaré's thinking cannot be assimilated to Heyting's definition of semi-intuitionism.

Let us return to Heyting's characterization of intuitionists that we began with. Can one say that (1) and (2) were *influenced* by Poincaré? If so, in what sense(s)?

Concerning the first point, regarding the claim that mathematics has not only a formal but also a contentual signification, one can, following Colin McLarty, contrast three kinds of interpretation: *banal, expansive and restrictive* (McLarty 1997, p.97). The banal interpretation merely says mathematics requires something beyond formal rigor.[3] The expansive interpretation claims that the content of mathematics goes beyond any formalization. The restrictive interpretation rejects some standard mathematics as inaccessible to intuition or a constructive procedure.

Brouwer was an expansive *and* a restrictive intuitionist. The application of his mental construction principle leads to his rejecting both the universal validity of the principle of the Excluded Middle (by conceiving a restrictive alternative logic) and the continuum of classical analysis. The question is whether Poincaré was more than a banal 'intuitionist'. His oft-quoted statement [cf., for example, (Largeault 1993, p.37) or (Resnik 1996, p.459) in favor of banal intuitionism is this: 'Logic, which alone gives certainty, is the instrument of demonstration; intuition is the instrument of invention' (Poincaré 1958, p.23), (Poincaré n.d., p.130). This, however, was intended only as a summary of a discussion about the distinction between *sensible intuition* and analytic procedures. Some pages later, he underlined that *pure intuition* gives certainty too and enables us to demonstrate and to invent (Poincaré 1958, p.25).[4]

---

[3] This is expressed by Pierre Boutroux [1920, 170]: 'There is some other thing beyond the logical form [of mathematics]'.

[4] Poincaré, in considering pure intuition as a faculty of the intellect, draws on the Neo-Kantian heritage.

One may expect from the literature that, as a *philosopher*, Poincaré was also both expansive and restrictive concerning foundational questions. Concerning the restrictive character of Poincaré's philosophical position, Brouwer recalled quite rightly his predicativism and his rejection of the actual infinite that were both consequences of his pragmaticism. Poincaré, he noted, 'blames in logistics the *petitio principii* and in Cantorism the hypothesis of the actual infinite' (Brouwer 1907, p.176)/(Brouwer 1975, p.96). Clearly, Poincaré didn't reject classical logic. His view, formulated in modern terms, was that the various formal-logical theories don't express the proof-theoretical structures that constitute our understanding of mathematics, cf. (Poincaré n.d., p.149). He rejected the invariance of mathematical reasoning with regards to its contents and promoted, so to speak, a 'local' conception of mathematical reasoning according to which 'a 'gap' is no longer a logical gap but, rather, a gap in *mathematical understanding*' (Detlefsen 1992, p.360, p.366) and, as such, a gap that cannot be eliminated through formalization. 'Mathematical reasoning is thus no longer to be seen as a primarily logical relationship between *propositions*, but rather as an epistemic relationship between *judgments*' (Detlefsen 1993, p.270). In this sense Poincaré has an expansive understanding of the contentual signification of mathematics.

Let's assume that it makes sense to distinguish Poincaré as philosopher from Poincaré as mathematician. Could one also say that even the *mathematician* Poincaré was more than a banal 'intuitionist' in the sense of attributing to mathematics some contentual significance?

In order to discuss this question we need to clarify in what sense the content of mathematics goes 'beyond any formalization'.

Taking the term 'intuitionism' for the view 'that mathematics must ultimately be based on the irreducible intuition of counting, of the natural numbers as a potentially infinite sequence', Harold Edwards (Edwards 1988, p.140) proposes a common factor for proponents 'including Henri Poincaré, L.E.J. Brouwer, Hermann Weyl and Errett Bishop' (ibid.). According to this perspective, Poincaré could perhaps be called semi-intuitionist because he was intuitionist only in arithmetic, or perhaps also in set theory concerning the methods of proofs used therein. I examine only the first possibility, arithmetic. What does it mean that we have an *irreducible* intuition of counting? Here the second point of Heyting's characteristic is important: it requires an irreducible intuition of counting which, at the same time, has to be grasped by our minds in a non-mediated way and independently of non-mental experience. But unfortunately, Poincaré did not think that the sequence of natural numbers is completely independent of all experience of the outer world:

> We have the faculty of conceiving that a unit may be added to a collection of units. *Thanks to experience, we have had the occasion of*

> *exercising this faculty and are conscious of it.* (Poincaré 1905, p.24, our emphasis)[5]

In other words, the schema R for concrete concatenation of strokes I:

$$N(I)$$

$$\forall x[N(x) \rightarrow N(xI)],$$

is occasioned by experience without having to be empirical.

Now, as is well known, Poincaré considered the principle of complete induction to be a synthetic *a priori* principle because it is 'imposed upon us with such a force that we could not conceive of the contrary proposition' (Poincaré 1905, p.48).[6] Why then is it not analytical? Poincaré's concept of an analytical judgment is not the Kantian one but concerns as in Frege the means of proof: the principle is not analytic because it is 'inaccessible by analytical proof' (Poincaré 1905, p.12). One can fix an operation which, if applied to a chosen numeral n, leads to a proof of $N(n)$. But how are we to be sure that *all* numbers are covered by the range of R? And how can we secure the validity of arithmetic without applying induction? This would clearly be a vicious circle. Nor is there a uniform proof-skeleton, because the length of proof that is the number of applications of *modi ponentes* depends on n. It follows, then, that a single proof is not to be read as apprehension of a schema-instance:

> A construction only becomes interesting when it can be placed side by side with other analogous constructions for forming species of the same genus. ... The analytical process 'by construction' ... leaves us at the same level. We can only ascend by mathematical induction. (Poincaré 1905, p.16)

According to Poincaré, mathematical induction cannot be identified with an 'analytical' construction process. It is not on the same level because it expresses, so to say, an infinity of *modi ponentes*, which means that it is synthetic. Technically, we have the fact that the direct clauses of the inductive definition of numerals (that is R) do not analytically imply the final clause (or equivalently, the principle of induction).[7] The principle of induction

---

[5] Poincaré often emphasizes that experience is powerless to grasp reasoning by recurrence although it is an occasion to remind us of the capacity we have for it: 'We may say, for instance, that in any finite collection of different integers there is always one which is smaller than any other. We may readily pass from one enunciation to another ... Experiment may teach us that the rule is true for the first ten or the first hundred numbers for instance; it will not bring us to the indefinite series of numbers.' (Poincaré 1905, p.12)

[6] Poincaré's understanding of Kant on synthetic a priori judgments is debatable; see (Ben-Menahem 2001, p.476); especially note 14.

[7] Cf. (Poincaré 1906, p.303). The final clause says that the only instances are those provided by the clauses of R (cf. (Kleene 1952, pp.21–22)).

> is only the *affirmation* of the power of the mind which knows it can conceive of the indefinite repetition of the same act, when the act is once possible. *The mind has a direct intuition of this power, and experience can only be for it an occasion of using it, and thereby of becoming conscious of it.* (Poincaré 1905, p.13, our emphasis)

The awareness of the mind's power of indefinite repetition (R) is also occasioned by experience, *e.g.*, of concrete stroke concatenation, and it is the affirmation of this capacity that yields the induction principle. The intuition of the mind's power is *a priori* because the action-schema is the result of our own creation. A pure intuition can be distinguished from simple evidence by the fact that it refers to what can be *done* instead of merely to something that is. The principle of induction is synthetic because it is not generated but only *represented* by the action schema: the principle postulates a *survey* of a potentially reiterated stroke-concatenation, or a *survey* of a potentially reiterated *modus ponens.*

What is important here is that the principle is thought to be the structural component of empirical constructions: the induction principle is occasioned by experience without having to be empirical and constitutes as such the theoretical part of our knowledge. Experience is the *ratio cognoscendi* of the principle of induction. The principle of induction is a *form of understanding*, in the sense that 'we can form no conception of an experience which would be best interpreted as a violation of it' (cf. (Folina 1992, p.32)). The universal proposition of induction, as used by Poincaré, is neither used qua convention nor qua intuition in the sense of Heyting's criteria. So, Poincaré's influence on this latter, though not very deep, is nonetheless not completely disconnected.[8] In general, experience is for Poincaré by no means a sufficient source of knowledge; it plays, above all, the role of making us aware of the existence of certain structures of the mind to which we have to accommodate our experience, directly as in the case of complete induction or, by the introduction of conventions, as in geometry.

Moreover, Poincaré himself asserted that complete induction is only the simplest of all 'other similar principles, offering the same essential characteristics' (Poincaré n.d., pp.149–150). Such analogous principles are especially the awareness of our capacity to construct a continuum of any dimension, called topological intuition. This concept pre-exists, according to Poincaré, in our mind as a form of reason and the awareness of them is occasioned by experience (cf. (Poincaré 1963, pp.157,134ff.); (Poincaré 1905, p.87ff.)). We will see in the next section, that the elaboration of *conventions* in geometry is analogous to the 'genesis' of the induction principle although the *a priori* status of complete induction clearly contrasts with that of conventions.

[8] Beth's judgment seems quite striking: 'Poincaré's ideas stimulated Brouwer's, not more but not less' (Beth 1956b, p.233).

2

How does Poincaré use the term 'convention' in geometry? Prima facie, his geometrical conventionalism consists of three theses:

i) Experience does not relate to space, but to empirical bodies. Geometry deals with ideal bodies, and it can therefore be neither proved nor disproved by experience. Since the propositions of geometry cannot be analytical either, these propositions must then be conventions, neither true nor false (Poincaré 1905, p.50).
ii) The choice between conventions, and in particular between different geometries, is guided by experience (Poincaré 1905, p.50).
iii) Euclidean geometry has nothing to fear from experience, for it is the most advantageous and convenient one (Poincaré 1905, p.73).

In order to understand this threefold thesis, we will respect Poincaré's analysis of his own scientific work (Poincaré 1921, p.127) where he underlines that the first step of his reflection concerning the construction of geometrical space[9] is its psycho-physiological genesis linking sensible space to geometric space.[10] Geometrical space is obtained by choosing the language of groups to serve as the tool of reasoning about representations of muscular sensations. As a result of this genesis, Poincaré finds that certain actions, when accompanied by muscular sensations, define equivalence classes, called *displacements*, and that each set of displacement classes forms a group in the mathematical sense. Thereby the edge of Poincaré's creative virtue in arithmetic, *i.e.*, repetition, is also necessary in the psycho-physiological genesis of geometry:

> Repetition ... has given to space its essential characteristics; now, repetition supposes time; this is enough to tell that time is logically anterior to space. (Poincaré 1958, p.72)

Surely, there is a 'family resemblance' with Brouwer's time progression but only if one substitutes 'ontologically' for 'logically' (Février 1981, p.167).

Now, after having argued the idea that the genesis of space depends on the genesis of a group of transformations in his first papers about that question, Poincaré seems, in 1903 to concede to the idea of the existence of an amorphous continuum as a kind of primary frame for the genesis of geometry

[9] The essential properties of geometric space are: '1st, it is continuous; 2nd, it is infinite; 3rd, it is of three dimensions; 4th, it is homogeneous — that is to say, all its points are identical one with another, 5th, it is isotropic. Compare this now with the framework of our representations and sensations, which I may call representative space' (Poincaré 1905, p.52). In fact, 'it is neither homogeneous nor isotropic; we cannot even say that it is three-dimensional'. (Poincaré 1905, p.56)

[10] Poincaré's original conventionalism is not related to the general axiomatic method in Hilbert's sense. Influenced by Sophus Lie's theory of group transformation, he considers geometry to be nothing but the study of certain groups and subgroups.

and geometrical space. Poincaré, for some reasons of coherence, is compelled to revise parts of his original theory and to renounce his claim that there is no need to presuppose a '3-dimensional *Zahlenmannigfaltigkeit*' as the matter on which groups of transformations act. He revised his theory in two papers first published in the *Revue de métaphysique et de morale* and later in *La Valeur de la science*. In these two papers, Poincaré placed a new emphasis on topology which he presented as qualitative geometry as *prior* to quantitative geometries.

In his paper about the nature of geometry (Brouwer 1909A), after claiming the impossibility of arguing convincingly that Euclidean geometry is *a priori*, he nevertheless conceded the existence of an *a priori* form of the perception of the world of experience that he called 'the intuition of time or intuition of two-in-one' (Brouwer 1909A, p.00)(Brouwer 1975, p.116). He defined this as the mathematical intuition of continuous time, extraneous to the idea of measure as conceived by a singular subject. Because of this claim, it can be argued that Brouwer was already defending a kind of intuitionist point of view in geometry, at least in the weak understanding in the sense of Edwards or Heyting.

For Brouwer, geometry is the science of 'properties of spaces of one or several dimensions'. Geometers study and classify sets, geometrical transformations and groups of transformations in these spaces. Brouwer attached a particular status to the groups of topology (*i.e.*, the groups of continuous transformations) because of their ubiquity and because 'a figure only unfolds its internal properties while for its definition only invariants of the largest possible group are used' (Brouwer 1909A, p.00)/(Brouwer 1975, p.116).

The question is: Does Poincaré's revision of his claim and his putting *Analysis situs* at the basis of geometry make him a defender of a point of view analogous to Brouwer's? Such an idea may be reinforced by the strong link between Poincaré's definition of topological dimension and that of Brouwer (in spite of the latter's criticism of the former's argumentation). Our claim is that this analogy is totally superficial and that the geometrical conceptions of Poincaré are radically different from those of Brouwer for the same reason as in arithmetic.

The geometrical conventionalism of Poincaré is well known, but the conventional aspects of geometry are not restricted only to the choice of the most 'commodious' group of transformations. His paper, published in *The Monist* in 1898 (Poincaré 1898), was the deepest of his contributions to this question at the time. He did not seek to elude either the problems of compensation and approximation or the more general problem of the relationships between experiences and laws. Rather, he proposed that we impose certain features on Nature through certain active operations of the mind. In his view, though, such imposition continually required the use of conventions.

> When experience teaches us that a certain phenomenon does not correspond *at all* to these laws, we strike it from the list of displacements.

> When it teaches us that a certain change obeys them *only approximatively*, we consider the change, *by an artificial convention*, as the resultant of two other component changes. The first component is regarded as a displacement rigorously satisfying the laws of which I have just spoken, while the second component, which is small, is regarded as a qualitative alteration. (Poincaré 2002, p.11)

Poincaré stressed several times that one of the consequences of his theory is that there is no need for a primary notion of space; the notion of space follows from the consideration of properties of displacements (which are changes of sensations). In 1898, there was no problem for Poincaré: Geometry precedes Space.

From this point of view, Poincaré's conceptions were original compared to such predecessors as Riemann, Helmholtz, Lie or Klein, who presupposed the existence of an amorphous continuum in the development of their geometrical theories.

> [W]e must distinguish in a group the form and the matter (*material*). For Helmholtz and Lie the matter of the group existed previously to the form, and in geometry the matter is a *Zahlenmannigfaltigkeit* of three dimensions.[11] ... It is only by the introduction of the group, that they made of it a measurable magnitude, that is to say a veritable space. Again, the origin of this non-measurable continuum of three dimensions remains imperfectly explained. (Poincaré 2002, p.40)

This '3-dimensional non-measurable continuum' is the space of *analysis situs* which will be considered later by Poincaré as qualitative geometry. In 1898, Poincaré claimed that he didn't need the prerequisite of a three-dimensional continuum. Very often, the hypothesis of a general space on which the geometry is constructed comes with the idea of a general geometry or a qualitative geometry which more or less embraces all metrical geometries. We thus see why, at that time, Poincaré's philosophical thinking could not agree with the general geometry of Calinon which presupposed the existence of an amorphous space. Moreover, it is untenable to use Calinon's theory to defend a psycho-physiologic point of view of the genesis of geometry since, for Calinon, geometry has no foundation in experience. In the same manner, Poincaré rejected theories like Russell's[12] which put projective geometry at the beginning of the genesis of metrical geometry. All these theories reduce the importance of experience to the choice of the metrical geometry in which we describe our common or physical experience, while all also admit a general *a priori* geometrical framework.

In his 1898 paper, Poincaré put forward a (mathematical) explanation of the three dimensions of space. He observed that the Euclidean group, selected after many conventions, can be seen as acting on a space of three, four or five

---

[11] Poincaré adds that for him, 'on the contrary, the form exists before the matter'.

[12] See (Nabonnand 2000a).

dimensions.[13] The choice of a three-dimensional space is justified by considerations of commodity. Unfortunately, Poincaré's argument is vicious because the choice of the Euclidean group was grounded on Lie's classification of groups of transformations acting on $\mathbb{R}^3$ and admitting a fundamental invariant. In 1903, Poincaré (after his work on *analysis situs* and his first debate with Russell about the status of the axioms of geometry) returned to the question of the dimension of space in a paper titled 'L'espace et ses trois dimensions' (Space and its three dimensions) (Poincaré 1903).

This paper deals with the question of the status of the space in *analysis situs* which Poincaré presents as the common background for the Euclidean and non-Euclidean spaces. After claiming once more that Euclidean geometry could not be a 'form imposed on our sensibility', he asks if the amorphous continuum of the Analysis situs is a 'form imposed to our sensibility':

> And then comes a question: Is not this amorphous continuum that our analysis has allowed to survive a form imposed upon our sensibility? If so, we should have to enlarge the prison in which this sensibility is confined, but it would still remain a prison. (Poincaré 1958, p.40)
>
> The same questions concerning the truths of Euclidean geometry come up anew about the theorems of *analysis situs*. Are they obtainable by deductive reasoning? Are they disguised conventions? Are they experimental verities? Are they the characteristics of a form imposed either upon our sensibility or upon our understanding? (Poincaré 1958, pp.40–41)[14]

Poincaré answers these questions with the analysis of the genesis of the geometry. The general properties of the amorphous continuum are obtained with the same process as the metrical properties. In 1898, Poincaré argued that the notion of displacement was obtained as a change of the sensible (or representative) space. It is as if the displacements set up a continuous group of transformations without the group acting on the representative space:

> The complex of our sensations has without doubt furnished us with a sort of matter, but there is a striking contrast between the grossness of this matter and the subtle precision of the form of our group. (Poincaré 2002, p.41)

---

[13] For more details, see (Nabonnand 2000b).

[14] 'Et alors une question se pose : ce continuum amorphe, que notre analyse a laissé subsister, n'est-il pas une forme imposée à notre sensibilité ? Nous aurions élargi la prison dans laquelle cette sensibilité est enfermée, mais ce serait toujours une prison.' 'Les mêmes questions qui se posaient à propos des vérités de la géométrie euclidienne, se posent de nouveau à propos des théorèmes de l'Analysis Situs. Peuvent-ils être obtenus par un raisonnement déductif ? Sont-ce des conventions déguisées ? Sont-ce des vérités expérimentales ? Sont-ils les caractères d'une forme imposée soit à notre sensibilité, soit à notre entendement ?'

In addition to the experience of the composition of displacements, which makes us act as if the displacements form a group, we consider the possible repetition of displacements. When the displacement D is very small, we experience what Poincaré calls physical continuum. We can sum up this notion by the formula:

$$9D = 10D, \quad 10D = 11D, \quad 9D < 11D$$

As Poincaré claimed, this formula is intolerable for our reason and empirical data impose on us the need to create mathematical continua so as to ensure the continuity of the group of displacements. Of course, this solution imposes itself because it is in agreement with other properties of the group. If these properties were incompatible with the hypothesis of continuity, we would be forced to choose another hypothesis. For Poincaré, the group of displacements resulting from the analysis of changes of sensations and the geometrical groups of transformations entertain the same kind of relationship as do physical and mathematical continua.

> We first study its form according to the formula of the physical continuum, and since there is something repugnant to our reason in this formula we reject it and substitute for it that of the continuous group which, potentially, pre-exists in us, but which we originally know only by its form. The gross matter which is furnished to us by our sensations was but a crutch for our infirmity, and served only to force us to fix our attention upon the pure idea which we bore about in ourselves previously. (Poincaré 2002, p.41)

The choice of mathematical continuity results from convention about inferences; the geometrical space is continuous because the group of geometrical transformations is. The aim of *analysis situs* is to study properties of mathematical space which are linked with the property of continuity; the choice of the 3-dimensionality of space reflects the inferential commodiousness of considering the internal action of the group of transformations on the set of its subgroups.[15]

In 1905, Poincaré advocated a significantly different explanation of the choice of the hypothesis of the continuity of space. Poincaré began by asking what 'is meant when we say that a mathematical continuum or that a physical continuum has two or three dimensions' (Poincaré 1958, p.42). He introduced the notion of cut in the context of a physical continuum. If any two (physically) discernible elements of a physical continuum C can be linked by a path of indiscernible elements (a physically continuous path), the physical continuum C is called 'all in one piece (*d'un seul tenant*)'. A cut is an arbitrary set of elements of C. According to the chosen cut, the physical continuum C remains 'd'un seul tenant' (i.e., any two discernable elements can either be linked by a path of indiscernible elements without leaving C and without crossing the cut) or it will be divided by the cut. Poincaré pointed out that these definitions are all suggested by experience:

[15] For more details, see (Nabonnand 2000b).

> It will be noticed that all these definitions are constructed in setting out solely from this very simple fact, that two manifolds of impressions sometimes can be discriminated, sometimes can not be. (Poincaré 1958, p.43)

We say that a physical continuum is 1-dimensional if we can cut it by one or several discernable elements. A physical continuum is said to be 2-dimensional if we can cut it by one or several 1-dimensional cuts (and it cannot be cut by one or several discernable elements). And so on.[16]

This definition of a physical continuum allowed Poincaré to break the vicious circle of his former theory of the 3-dimensionality of space. For a time, Poincaré abandoned part of his early theory. Certainly, the physical continuum formed by the displacements is 6-dimensional and this result comes from experience. Poincaré claimed he could maintain his previous theory and keep inferring the 3-dimensionality of the geometrical space from the internal action of the group on its subgroup. But it would be difficult because he would still have to justify the utilization of Lie's classification of groups of transformations acting on $\mathbb{R}^3$. Poincaré didn't mention this problem. He simply argued that such a theory could not be completely satisfying because it does not respect the hierarchy of our spatial sensations:

> [A]nd when we shall have shown how the notion of this continuum [the group of displacements] can be formed and how that of space may be deduced from it, it might always be asked why space of three dimensions is much more familiar to us than this continuum of six dimensions, and consequently doubted whether it was by this detour that the notion of space was formed in the human mind. (Poincaré 1958, pp.57–58)[17]

So Poincaré added to his theory a new stage explaining why we are guided (by experience) towards a 3-dimensional physical continuum. Here, the most essential space is that which is associated with the sense of touch. Poincaré argued that a necessary condition for the setting up of a tactile space is that tactile sensations are not falsified by muscular ones which are associated with our movements. Each space which is associated with our fingers is a 3-dimensional physical continuum. Poincaré argued that, on the basis of experiences, we are led to consider them all as similar. He argued as well that we are led to assimilate the visual space to these tactile spaces.

[16] Poincaré specified that this definition is the same as the one of geometers who defined surfaces as boundaries of volumes, and lines as boundaries of surfaces and points as boundaries of lines. He added that this form of definition is 'applicable not to the mathematical continuum, but to the physical continuum, which alone is susceptible of representation' (Poincaré 1958, p.44).

[17] This kind of cognitive and genetic argument is typical of Poincaré. He refused to consider projective geometry as a type of general geometry because 'this evidently is not the manner in which our geometrical notions were formed' (Poincaré 2002, p.35).

So if Poincaré in 1905, at the beginning of his genesis of geometry, included a 3-dimensional physical continuum, he didn't do so for epistemological reasons, for reasons linked to intuition, or for mathematical reasons. Rather, he did it for reasons of coherence. He felt forced to admit a rough notion of 3-dimensional space in order to justify his utilization of Lie's classification of 6-parameter groups of transformations acting on $\mathbb{R}^3$. Similarly for admission of a significant invariant. Finally, he also wanted to explain why the 3-dimensionality of geometrical space is so immediate. He pointed out, however, that as in his previous views, experience plays an essential role at each stage of the genesis of geometry. This notwithstanding, he did not lapse into fruitless and beatific empiricism. Rather, he emphasized that 'experience does not prove to us that space has three dimensions; it only proves to us that it is convenient to attribute three to it, because the number of fillips is thus reduced to a minimum' (Poincaré 1958, p.69).

The choice of conventions is justified according to cases by considerations concerning accommodation of our experience, or by reasons of practical or mathematical usefulness. The choice of conventions can take different forms, including the definition of new objects. As Poincaré recalled, a convention is never imposed by experience. It is only the result of a capacity of our mind, and experience provides us the occasion to use this capacity. According to Poincaré, we have innate capacities to set up groups and continua. To understand the genesis of geometry is to understand how experiences provide the innate capacities of our mind the occasion to manifest themselves. So conventions are neither themselves intuitions, nor the results of intuitions. Rather, they are the result of the activation by experience of innate capacities of our minds.

So we have the same situation as for the genesis of the recurrence principle in arithmetic: experience provides our mind the occasion to use its capacities; when we do not have a choice, the role of experience comes to an end and the result is a principle of reasoning. When we have a choice, experience helps us choose and the result is a convention. It is therefore clear that Heyting's second characteristic does not apply to Poincaré. For Poincaré, mathematical theories and methods are dependent on outer experience in that it is experience which provides our minds occasion to apply its innate capacities.

3 Conclusion

We are accustomed to taking some mathematical conceptions of Poincaré and Brouwer as being rather closely related. This is perhaps especially so in the case of their criticisms of logicism and their common conviction concerning the central role of the idea of indefinite repetition. The latter serves to justify complete induction and to introduce the notions of the (mathematical) continuum or transformation groups. Both preferred the group theoretic presentation of geometry over its Hilbert's style axiomatic presentation, and both also declined to attach an *a priori* character to Euclidean geometry.

From a conceptual point of view, however, Poincaré and Brouwer are basically different in their approaches. Whereas Brouwer considered the possibility of indefinite repetition to be a sort of instantiation of the pure intuition of time (Brouwer 1913C, pp.85–86), Poincaré regarded it as the result of the excitation by experience of the mind's power 'to conceive of the indefinite repetition of the same act, when the act is once possible' (Poincaré 1905, p.41). Similarly, the notions of 'mathematical continuum' and 'group transformations' are attained thanks to the power of the mind to form continua or groups but Poincaré always took care to underline the fact that it is experience that provides the mind the occasion for such self-activation. Furthermore, he took the use of these concepts to account for experience as a type of convention.

Poincaré and Brouwer were in agreement on the idea that one should attach the greatest fundamental character to *analysis situs*. Nevertheless, Brouwer, in adopting Klein's subordination principle, justified his appraisal by the fact that the group of continuous transformations is the most general group operating over a continuum. For Poincaré, on the other hand, the special character of *analysis situs* lies in the fact that what it studies as the common part of the three Euclidean and non-Euclidean geometries are those properties that are in the relation to the continuity of the transformation group to which we relate our spatial considerations. Finally, Brouwer stipulates the preliminary existence of the continuum as the operator domain of the continuous transformations. Poincaré's approach (in 1903) is much more genetic; he argues that we acquire our experience of a three-dimensional physical continuum through our tactile sense. This experience allows us to justify the use of Lie's classification of transformation groups operating upon $\mathbb{R}^3$ to describe the process of choice in favor of the Euclidean group. The Euclidean space and its elements are constructed with regard to results obtained for the group structure.

By adopting his group theoretic conception of geometry, Poincaré expressed his dissatisfaction with a purely formal approach to the foundations of mathematics. He recognized clearly the mathematical equivalence of his own presentation to the axiomatic presentation. He nonetheless emphasized in his review of Hilbert's *Grundlagen der Geometrie* that Hilbert's preference for the axiomatic approach reflected his (*i.e.*, Hilbert's) rather too exclusive focus on the logical point of view and paid little attention to questions concerning the justification of the axioms or their psychological origins (Poincaré 2002, p.45). These considerations make it clear that Poincaré preferred his group theoretic approach for cognitive reasons despite the fact that the axiomatic presentation may satisfy all demands of rigor.[18] And in fact, it turns

[18] Poincaré specifies that all geometries (non-Archimedean, non-Pascalean...) discussed by Hilbert can be formulated as geometries associated to a group (Poincaré 2002, p.45).

out that the notion of geometrical transformation is much better adapted than the axiomatic presentation to the illustration of that part of experience which is involved in the shaping of mathematical concepts. What is most striking to us about Poincaré's philosophy, when compared to empiricism, intuitionism and formalism, is its view of the role of outer experience in the genesis of fundamental mathematical concepts. It is neither merely a prop for abstraction nor something which in the end reduces to the mere manipulation of signs. Poincaré seeks to clarify the importance of the linguistic level in mathematical thinking. Contrary to the intuitionists, the use of language is for him not only an efficient tool for memorizing or communicating but also and above all an essential part of the genesis of mathematical conceptualization. It is an integral part of the whole experiential frame on which our inner faculties are grounded.

Acknowledgement

We are grateful for comments received from Mark van Atten, Michael Detlefsen and Prosper Doh.

Some of Julius König's mathematical dreams in his *New Foundations of Logic, Arithmetic, and Set Theory*

Marcel Guillaume

1 About the author, before the book: the mathematician Julius König on set theory

Julius König, author of the *New foundations of logic, arithmetic, and set Theory* (König 1914), born in Hungary in 1849, died in 1913 before having finished writing the last pages of his book.

He was a renowned mathematician. (Guillemot 1981, pp.58–59) and (Cassinet and Guillemot 1983, pp.93–94) recall his *Introduction to the general theory of algebraic magnitudes* (König 1903). There König attests that he intends to meet Kronecker's requirements of effectivity, which in fact he did, according to a review of Jules Tannery (Tannery 1905); indeed, while he defended Cantor's set theory and contributed to it, it cannot be disputed that in other domains he followed Kronecker's pre-intuitionistic views and, as we will see below, without knowing Brouwer's first works,[1] he independently came closer to Brouwer's conceptions on some points. (Bachmann 1955, p.125) mentions, in cardinal arithmetic, what he calls 'König's formulas'. By the way, (Lusin and Sierpinski 1917, p.498) gave a new proof of 'the very remarkable theorem of König' which says that the power of the continuum cannot be the transfinite cardinal number $\aleph_\omega$; in fact, König's proof applies to every cardinal number which is the limit of an infinite series, indexed by the natural numbers, of cardinal numbers which are strictly smaller than it (Bachmann 1955, p.167). Since Paul Cohen's introduction of *forcing* in 1963, it is known that this result is the best possible: (Cohen 1964) explains how to construct, for each cardinal escaping from König's theorem, what is proved in this paper to be a model of ZFC set theory where this cardinal is the power of the continuum.

These formulas and this theorem are the solid remains of a proof, whose downfall was not the fault of its author, and whose aim was to show the impossibility of finding a well-order of the continuum; that entails the rejection

[1] On these points, see (Franchella 2000, p.47).

of the continuum hypothesis. (Guillemot 1981) and (Cassinet and Guillemot 1983, pp.93–99) recount the emotions caused among the better mathematicians of the time by König's argumentation, from his presentation before the Third International Congress of Mathematicians, and the debates which followed during and just after the Congress.

Now, the fatal error to the intended conclusion did not come from König's reasoning, but from a theorem in Felix Bernstein's thesis (Bernstein 1901) which, as nobody had until then seen, did not apply to every pair of values of two ordinal parameters, and on which the proof rested. Schoenflies, who had taken part in discussions about König's communication, relates in (Schoenflies 1922, p.100) that Hausdorff, a few days after the Congress, was in a position to clearly explain where the flaw in Bernstein's reasoning was, which he wrote straight away in (Hausdorff 1904), with a respect towards König which should be emphasized although he was also making König's proof more precise.[2] In volume 60 of the *Mathematische Annalen*, König (König 1905a) explicitly withdrew only his conclusion, and Bernstein (Bernstein 1905, p.464) recognized the limits of his result.

(Cassinet and Guillemot 1983, pp.283–287) also recount how König attempted, until the end of the year 1905, to derive a contradiction from the hypothesis that the continuum can be well-ordered. Firstly, by stating in (König 1905b) a paradox quite similar to Richard's paradox;[3] both have promptly been refuted by name in (Peano 1906, p.144). By the way, König notices that the set of finitely definable (in a given scientific language) elements of the continuum is countable; he is then led to admit not finitely definable reals and so to depart from a strict Kroneckerian observance, claiming:

> I set myself ... in conscious opposition to the assumption that to go outside 'finite laws' is not granted. The assumption used here is, to the contrary, that there are elements of the continuum which we cannot 'imagine

[2] This version of the circumstances in which the final step of König's proof was refuted is the one which emerges from the texts quoted in (Guillemot 1981) and in (Cassinet and Guillemot 1983). The version brought up in (Dauben 1979), which relies on the one of (Kowalewski 1950), and that I had briefly mentioned in my talk, is questioned by Cassinet et Guillemot, and curtly impugned by (Grattan-Guinness 2000, p.334). However, the same Cassinet et Guillemot mention a report on the Congress, by no one less than Hadamard, who says there the same thing that (Kowalewski 1950, p.202) really says: 'Zermelo ... found during that same day a demonstration of any set's capability to be well ordered' < *Zermelo ... fand sogar in jenem Tagen einen Beweis für die Wohlordnungfähigkeit einer beliebigen Menge*> (thus refuting König's main argument). Perhaps Zermelo only claimed that he had some idea towards such a proof, and may have been misunderstood; to be sure, if he really had such an idea in mind, he was slower than Hausdorff to elaborate and write it.

[3] Independently, however almost simultaneously, up to a few days, notes (van Heijenoort 1967, p.142).

to the end' and which, being 'nevertheless' free of contradiction, are then, if this indeed wholly new expression is granted, '*ideal*' elements.[4] [*My italics*].

In the same article, König begins to explain, as follows, a distinction that he makes between the concepts of *set* and *class* and that he will express differently later, in the book:

> In the formation of the concept of the continuum, the 'arbitrary' series $(a_1, a_2, \ldots, a_k, \ldots)$ is the Primal, the Original. From this one stems, through the requirement of replacing $a_1, a_2, \ldots$ by determined positive integers, a 'determined' series, an element of the continuum, which we thus also think of, if need be, as conceptually *isolated* [*My italics*] from every other element. The further requirement of imagining the *conceptual totality* of these 'well distinguished' objects, leads to the continuum.
>
> Entirely different is the matter in the case of the number *class* [*My italics*] $Z(\aleph_0)$. Its 'elements' are determined through the 'property' of being the order types of well-ordered sets of power $\aleph_0$. Of course we know such elements: $\omega, \omega + 1, \ldots$; but that property is only an abstraction, in the better case a means to distinguish between things which belong to the class and the others; but certainly not an instruction according to which *every* element of $Z(\aleph_0)$ can be formed. Here the Primal, the Original is the collective concept which I — joining Cantor's denomination — would like to designate, just owing to that, not at all as a 'set', but as a 'class'; and only from it are the elements belonging to the class *then* constructed.
>
> That the second number class $Z(\aleph_0)$ is definable as a 'completed' set of well distinguished, i.e., conceptually entirely isolated elements, cannot even, in view of the state of our set-theoretical knowledge until now, be qualified as likely.[5]

---

[4] 'Ich stelle mich ... in bewußten Gegensatz zu der Annahme, daß es nicht gestattet ist, über „endliche Gesetze" hinauszugehen. Die hier benützte Annahme ist im Gegenteil die, daß es Elemente des Kontinuums gibt, die wir nicht „zu Ende denken" können, und die „troztdem" widerspruchsfrei, also, wenn der Ausdruck in allerdings ganz neuen Sinne gestattet ist, „ideale" Elemente sind' ((König 1905b, p.158), quoted by (Cassinet and Guillemot 1983, pp.285–286).)

[5] 'Bei der Bildung des Kontinuumbegriffes ist die „beliebige" Folge $(a_1, a_2, \ldots, a_k, \ldots)$ das Primäre, Ursprüngliche. Aus dieser wird durch die Forderung, $a_1, a_2, \ldots$ durch bestimmte positive ganze Zahlen zu ersetzen, eine „bestimmte" Folge, ein Element des Kontinuums, das wir also, wenn überhaupt, auch von jedem anderen Elemente begrifflich gesondert denken. Die weitere Forderung, den *Inbegriff* dieser „wohlunterschieden" Objekte zu denken, führt dann zum Kontinuum.

Ganz anders steht die Sache bei der Zahlenklasse $Z(\aleph_0)$. Ihre „Elemente" werden durch die „Eigenschaft", Ordnungstypen wohlgeordneter Mengen von der Mächtigkeit $\aleph_0$ zu sein, bestimmt. Allerdings kennen wir solche Elemente: $\omega, \omega + 1, \ldots$; aber jene Eigenschaft ist nur eine Abstraktion, im besten Falle ein Mittel zur Unterscheidung zwischen in die Klasse gehörigen und andere Dingen; gewiß aber keine Anweisung, nach der *jedes* Element von $Z(\aleph_0)$ gebildet werden kann. Hier ist das Primäre, Ursprüngliche der Kollektivbegriff, den ich — im An-

A last attempt to disprove the continuum hypothesis in a second part (König 1906a) of the same work contains more ideas,[6] but most of them are still imprecise. Their aim is to arrive at a contradiction by appealing to the well established well-order of the ordinal numbers of Cantor's second class rather than to a hypothetical and contested well-order of the continuum; so there König is led to go a step further, admitting infinite definitions in which can be inserted a transfinite series (of any order type in Cantor's second class) of occurrences of a special sign. König did not pursue such a theorization of notations; however, he made implicit use of notations of this type at the end of his book.

2 About the history of the book

We do not know when König became interested in set theory. A glance at the list of his publications in (Guillemot 1981, pp.74–77) shows that he devotes none of them to set-theoretical problems before his intervention at the Congress of 1904 and leaves us to imagine that what led him to be interested in set theory as a researcher, is the longing to attack the first of the problems which Hilbert posed in his famous address to the Congress of 1900: what is the power of the continuum, and can it be well-ordered? Hilbert's lecture at the Congress of 1904, too, deeply marked König, as a comparison of the title of that lecture ('On the foundations of logic and arithmetic') with that of König's book ('New foundations of logic, arithmetic, and set theory') makes clear.

By his unfruitful attempts to establish that the continuum cannot be well-ordered, König was finally led to think of elaborating a doctrine of his own. At the beginning of a Note *On set theory* presented to the Paris' Academy of Science by Poincaré the 9th of July 1906, he reveals that he sets about writing such a work, while announcing he will give a proof to which he says that he attaches

schluß an Cantors Namengebung — eben deshalb gar nicht als „Menge", sondern als „Klasse" bezeichnen möchte; und erst aus diesem heraus werden *dann* der Klasse angehörende Elemente konstruiert.

Daß die zweite Zahlenklasse $Z(\aleph_0)$ als „fertige" Menge wohlunterschiedener, d.h. begrifflich durchweg gesonderte Elemente definierbar ist, kann auch nach dem bisherigen Stande unserer Mengentheoretischen Kenntnisse nicht als wahrscheinlich bezeichnet werden' (König 1905b, p.159).

[6] We do not really do justice to this aspect of this article of König's here: that would require a development of undue length with regard to the intended subject of the present work.

> a rather great importance, in view of the present discussion on the *foundations of logic, arithmetic and set theory* [*My italics*]. I intended to give it only in the exposition of the *Synthetic Logic* I hope to publish soon ... .[7]

In (König 1914, p.iii), in a footnote to the *Foreword* written by Julius König, his son Dénes confirms the change of title which occurred in the meantime. In the complement Dénes adds to this *Foreword*, he also confirms the above dating of the beginning of the work, writing that, on his book, his father

> had worked for eight years, until the last day of his life. ... Only a few pages more were lacking.[8]

He reviews next the matters his father intended to treat in the missing pages, among which his 'definitive' <*endgültig*> version of the proof of Zermelo's well-ordering theorem. He thanks, for his assistance in finalizing the manuscript, Professor Kürschák (who had himself had Dénes König as a student and who had supervised his thesis, defended in 1907; later he had also Von Neumann as a student.[9]) At the following page, Dénes writes

> I also address my sincere thanks to Prof. F. Hausdorff. By undertaking the painful and unrewarding work of proofreading with me the whole book, he has fulfilled one of the last wishes of my father.[10]

3 Questioning the methodological influence of the book

The publisher was good at doing his task of advertising: for example, the reference to König's book is inserted in the issue of the *Bulletin of the American Mathematical Society* dated November 1914, at the price of 8 marks. By its title, the book certainly aroused some interest, at least among academicians of various countries: Julius König was Secretary of the Class of mathematical and natural sciences of the Hungarian Academy of Science. To speak only of France, the catalogues localize four copies in Paris, among which two at the *École Normale Supérieure*, another in Poitiers; still another was found in Nancy.

During the months which followed the publication of the book, two fair reviews, the first by the mathematician Dmitry Mirimanoff, the second by the

[7] 'une importance assez grande, vu la discussion actuelle sur les *fondements de la logique, de l'arithmétique et de la théorie des ensembles.* Je ne voulais la donner que dans l'exposition de la Logique synthétique que j'espère publier bientôt ...' (König 1906b, p.110)

[8] 'seit acht Jahren bis zum letzten Tage seines Lebens gearbeitet hat. ... Es fehlten nur mehr einige Seiten.' (König 1914, p.v)

[9] `www-history.mcs.st-andrews.ac.uk/Mathematicians/Kurschak.html` .

[10] 'Mein aufrichtiger Dank gilt auch Prof. F. Hausdorff. Indem er die mühsame und undankbare Arbeit übernahm, eine Korrektur des ganzen Buches mitzulesen, hat er einen der letzten Wünsche meines Vaters erfüllt.' (König 1914, p.v)

philosopher of science Hugo Dingler, have been published, each with sufficient information on the treated matters to allow a potential reader to decide if he is interested in this work or not. Both apologize for not being complete; Mirimanoff:

> his book abounds in original ideas and nothing in it is useless or banal ... My analysis ... will thus inevitably be very incomplete.[11]

Dingler:

> an inextricable interlacing soon comes in place with a series of most difficult and most profound problems ... We will therefore limit ourselves to point out in broad outline how the construction is led ...[12]

Both also stress the importance of the work; Mirimanoff, who takes much heed to the logical side:

> certainly one of the deepest books among those which have been published on the principles of mathematics and logic.[13]

Dingler, who however, while giving much more detail than Mirimanoff, alternates between (sometimes big) compliments and (sometimes very) critical remarks:

> a very significant publication in the sphere of research on the foundations ...[14]

We also know that Felix Bernstein was entrusted to give a talk on König's book before the *Mathematisches Kolloquium* at Göttingen the 16th of February 1915,[15] but we know *almost nothing of what happened afterwards*; whether, and how, the book was read and perhaps studied. And this raises a historical question: *did König's book have any influence*, for better or for worse, acknowledged or denied, *on further works* or only on some aspects of their conceptions? Or did almost nobody read it, after a cursory first inspection, or following some explicit or implicit dissuasive authoritative advice? We do not have at our disposal here sufficient space to even sketch some leads on

[11] 'son livre abonde en idées originales et rien n'y est inutile ou banal ... Mon analyse ... sera donc forcément très incomplète.' (Mirimanoff 1914, p.400)

[12] 'kommt sofort ein untrennbarer Konnex mit einer Reihe schwierigster und tiefstliegender Probleme zustande ... Wir wollen uns daher darauf beschränken, anzugeben, wie der Aufbau im großen und ganzen geführt ist ...' (Dingler 1915, p.153)

[13] 'certainement l'un des livres les plus profonds qui aient été publiés sur les principes des mathématiques et de la logique.' (Mirimanoff 1914, p.402)

[14] 'eine sehr bedeutende Erscheinung auf dem Gebiete der Grundlagenforschung ...' (Dingler 1915, p.152)

[15] (Mancosu 1999, p.304) (without dealing with König at all there) presents a list of the talks posted at this seminar between 1914 and 1921.

which some clues, which would themselves need some explanation, might be extracted from various texts. And confirming or invalidating each of these hypotheses would require quite a lot of research. Hence we will content ourself with pointing out further passages, prior to the 1940s, in which some explicit appraisals of König's book appear.

The book was very succinctly reviewed in (Fraenkel 1923, pp.182–184) and yet more briefly in (Fraenkel 1929, pp.366–367), respectively the second and the third editions of Fraenkel's *Introduction to set theory.*

The following curious footnote of Von Neumann in (von Neumann 1927, p.22) calls for some comments.

> The idea of proving the freedom of contradiction through a 'valuation' stems from J. König (*New Foundation [sic]*[16] of Logic, Arithmetic and Set Theory, Leipzig 1914). But the form (inspired by Hilbert) to be used in what follows deviates from his concept of 'valuation' on a few points. The concept of 'partial valuation' rests on ideas from Hilbert.[17]

From the aforementioned comments, we will retain only that the idea of dividing up *all* formulas (of a given formal language) into *two* disjoint classes, one of which contains no contradiction but all provable formulas, is due to (Hilbert 1905) who called this tool a 'partition' <*Verteilung*>, and not to König, who re-uses it while allowing that the two disjoint classes do not form a partition, and without giving a name to it. Did Von Neumann not know this lecture of Hilbert? Or did he want to say to his German colleagues that his compatriot König brought estimable contributions in his book?

Perhaps we find a piece of the answer in footnote number 16 of (Bernays 1930), concerned with

> the investigation of Julius König ... which *closely approximates Hilbert's point of view* [*My italics*] and in which a consistency proof in the sense of proof theory is already carried out. This proof concerns only a very narrow domain of the formal realm [*des formalen Operierens*], so that *its significance is only methodological* [*My italics*].[18]

However, footnote number 1 of (Bernays 1935, p.200) is at the same time more precisely complimentary and more reserved about the work of

---

[16] The German word '*Begründung*', which Von Neumann uses here, means 'foundation-in-action'; it is a (German) synonym of the word '*Grundlagen*' of the title, whose meaning is 'foundational layers'.

[17] 'Die Idee, die Widerspruchsfreiheit der Mathematik durch eine „Wertung" zu beweisen, stammt von J. König (Neue Begründung [*sic*] der Logik, Arithmetik und Mengenlehre, Leipzig 1914). Die im folgenden zu benützende (an Hilbert anlehnende) Form weicht aber von seinem „Wertungs"-Begriffe in einigen Punkten ab. Der Begriff der „Teilwertung" beruht auf Ideen von Hilbert.

[18] From Paolo Mancosu's translation (including the original German terms in square brackets) in (Mancosu 1998, p.265).

> J. König, who in his book ... goes beyond the Heidelberg lecture, both in the more precise formulation and more thorough exposition of the methodological point of view, and in its execution. ... However, this work, which constitutes a forerunner of the later Hilbert's proof theory, had no actual influence on Hilbert.[19]

4 About the book as a pioneering testament

Today, anyone well trained in mathematics will read the book from angles which his intermediary predecessors lacked. Yet, even König's contemporaries could regret various weaknesses; some developments are *visibly unfinished*: for example, König did not see how to find 'fundamental forms' <*Grundformen*> linking the statement of the falsity of an implication (or of an equivalence) with statements of truth, falsity, or fulfilment of other conditions, on its terms. Further, remaining on this point faithful to his Kroneckerian views, he rejects any treatment of quantifications in (König 1914, p.97). And the more one knows, the more weaknesses one finds. But here and now we are not interested in these aspects.

Today's mathematics also teaches us that the book presents many views ahead of its time. Such a blend of defects and innovations is frequent in pioneering works, whose original views have frequently not been understood immediately; and when some of these ideas reappear after several years, their source may have been forgotten or not considered worthy of quoting. But the case we are considering at present is rare: we have at hand a pioneering work realized late in a life to which it is at the same time a scientific testament. Its author had become unable to defend and develop his views better, to rectify others in the light of debates with approving or critical colleagues or disciples, and to remind the silent users of the ideas of his work.

The matters of the book about which we intend to speak belong to what is nascent in the philosophical insights that König directs towards mathematics; in other words, what he dreamt of for the mathematics of the future of his time, now part of our past or present mathematics; views of which we can thus recognize, in the light of our knowledge, that he intuited as a forerunner and drafted a step ahead.[20]

---

[19] 'J. König, der in seinem Buche ... sowohl durch eine genauere Fassung und eingehendere Darlegung des methodischen Standpunktes, wie auch hinsichtlich seiner Durchführung über den Heidelberger Vortrag hinausgeht. ... Von diesem Werk, welches einen Vorläufer der späteren Hilbertsche Beweistheorie bildet, ist jedoch keine Einwirkung auf Hilbert ausgegangen.'

[20] Much information on details and on more general matters dealt with in the book can be found in (Franchella 2000).

5 König's structuralist doctrine of mathematics

Commenting upon the 'taking together' < *Zusammenfassung*> to which Cantor refers in his definition of the concept of the 'set', (König 1914, pp.38–39) writes that it can happen that

> one of the 'objects' to take together in the 'taking together' is itself generated in the 'taking together'. ... The unerring researcher's quick glance of Cantor protected him against taking in consideration ways to form sets for which a lack of attention to the mentioned circumstances leads to antinomies ... . Cantor speaks in such cases of 'inconsistent' sets, but without having gone into them in his publications.[21]

On the other hand, (Cassinet and Guillemot 1983, p.92) recall, about (Jourdain 1904), that

> Starting from the paradox of Burali-Forti ... and thus from the 'set W' consisting of all ordinal [numbers], Jourdain christens such paradoxical sets: 'inconsistent' sets ... .[22]

Thus spoke people around Cantor. At the time it was clear for everyone that 'consistency' and 'inconsistency' were not directly properties of sets, but of their *determinations* as concepts.[23] It is likely that König's thinking aimed to make this idea more precise. On the other hand, three of Hilbert's works, on the foundations of geometry, real numbers, logic and arithmetic (Hilbert 1899, 1900a, 1905), Zermelo's axiomatization of set theory (Zermelo 1908), and two of Russell's works (Russell 1903, 1908), at least,[24] gave him examples of domains of mathematical entities which were the objects of theories meant to expound and to explain, by a logically sound reasoned report, our knowledge about them, and in particular, to determine their concept and the concepts of many relations in them. *We* call such a domain a '*structure*', and say about each one of them that it is a '*model*' of the corresponding theory;

---

[21] 'eines der zusammenfassenden „Objekte" in der „Zusammenfassung" selbst erzeugt wird. ... Cantors untrüglicher Forscherblick ihn davor bewahrt hat, auch solche Mengenbildungen in Betracht zu ziehen, wo die Nichtbeachtung der erwähnten Umstände zu Antinomien ... führt. Cantor spricht in solchen Fällen von „inkonsistenten" Mengen, ohne aber in seinen Publikationen auf diese einzugehen.'

[22] 'Partant du paradoxe de Burali-Forti ... et donc de 'l'ensemble W' formé de tous les ordinaux, Jourdain baptise de tels ensembles paradoxaux : ensembles "inconsistants"' ... The addition 'numbers' is mine: contrary to (Jourdain 1904), (König 1914) distinguishes ordinals from ordinal numbers.

[23] In Frege's logic, sets even were used as denotations of concepts, with respect to an assumed absolute universal reality, an absolute against which König fought vigorously (*cf., e.g., in (König 1914): pp.78, 86–87, 124, and 136*).

[24] Of what König had the opportunity to read, we know only what he quotes (in (König 1914): pp.28, 46, 63, 69, 155, 161, 208).

for Hilbert as for us, it is the theory that is of *primary* interest; its models are secondary: knowledge of them is viewed as part of the knowledge of the theory. König's attitude is the opposite: he considered the domain as *primary*, as a mathematical object to think up (or previously thought up), to study and to know, and the theory which applies to it, as secondary, as the explanatory and logically sound reasoned report of our knowledge about it. From Husserl's views,[25] he knew the primordial cognitive role of the 'taking together'; on the other hand, he saw the set-theoretical paradoxes as urging us on to bring back to the level of the 'taking together' a broadly similar scheme of thought about what is known of it and about how the concepts coming with such an act of thinking are determined. From Kronecker's views, he retained the idea of *constructed* objects[26] which he saw as *generated* according to a predetermined *scheme* applying to predetermined *things*. *Our* 'structures' stand amongst these generated mathematical objects; to the totality of things from which the generating begins, he gave the name of '*domain of thinking*' < *Denkbereich*> ; what he calls the '*description*' < *Beschreibung*> of a domain of thinking accounts for the knowledge allowing the generating and for the stipulations needed for the conceptual determinations. Let us note that this conception, first, is not explicitly displayed in this way by König himself (even if one deprives it of the references to *our* today's notions); it depicts König's general idea as *I* see how to explain it, from the content of the many cases in which it is applied by him in the book. Second, for mathematics, it is a unifying conception, insofar as mathematics can be reduced to set theory. But the conditions to which the descriptions are in principle submitted are so restrictive that if König himself had always strictly fulfilled them, he would in general have been unable to go through with the passage to the infinite; he bypasses this obstacle by admitting *intuitions about a domain of thinking* (say D, assumed to be free of contradiction) as axioms to be added to its description in order to describe another such domain (say D'), said in (König 1914, p.141) to be *higher*, for

> the domain of thinking D, as ... a concept of collection, enters in the description of the domain of thinking D' '.[27]

[25] In (König 1914, p.73), he refers to Husserl's views on formalization, but without referring to any particular work.

[26] Yet, his notion of a construction was much less narrow than that of today's constructivists. He makes the effort to sketch his conception of the cognitive ways which lead to a progressive construction from the 'lived facts' < *Erlebnisse*> (he himself rendered this German term by the French < *faits vécus*> in (König 1906b, p.111)) to mathematical abstractions, by successive extensions of the concept of 'thing' (*cf.* in particular in (König 1914, pp.15,45), the titles of chapters II and III), for which the 'things' are incessantly 'generated' <*erzeugt*> .

[27] 'der Denkbereich D, als ... Kollektivbegriff, in die Beschreibung des Denkbereichs D' eingeht.' (König 1914, p.140) (Franchella 2000, p.54) notes this view of König's; also that it reaches the principles of contradiction and of the excluded middle.

In fact, in that place, D was what he calls the domain of thinking of pure logic, and the corresponding 'intuition about D' was the principle of contradiction, which he banished from *pure* logic — not so much however as the principle of the excluded middle that he rejected unreservedly. To the other case of the 'axiomatic domain of arithmetic' (König 1914, p.184) corresponded similarly the principle of complete induction as an intuition about it (König 1914, p.189). But all admittedly intuitive proofs of consistency given by König apply to *finitistic*[28] parts of mathematics, moreover insufficiently axiomatized; this was a great weakness in the light of Hilbert's ambitions.

When König begins to speak of 'domains of thinking' (König 1914, p.15), he says that a domain of thinking is an objectivized 'state of our consciousness' < *Zustand unseres Bewußtseins*> , 'to which certain lived facts belong, others not' <*dem gewisse Erlebnisse gehören, andere nicht*>. The state in question is explained, in fact, as being created (even if König says only, in the beginning, 'determined' <*bestimmt*>) through stipulations, which are in fact *voluntary*, about, in principle, all lived facts; for each of them, what is posited to be stipulated is either that we can have a representation of it in the state (and then it belongs to the corresponding domain) or not (and then it does not belong to this domain). König emphasizes the voluntary character of the stipulations only by the following words:

> that we cannot have a representation of it, this means that we do not want to have a representation of it, that *we want to exclude it.*[29]

If we want to speak of the 'thing' a domain of thinking is meant to 'generate', and before all to write in some form what we have to say about it, we need, as we do customarily in mathematics, a sign for that thing, even before knowing what and how that thing will be. Once this is done, one can begin to write what one will require of the thing to be generated, by establishing a more or less complicated diagram describing the generating domain of thinking which will generate the thing, or, when one knows that this diagram will be infinite, some more or less complicated procedures to generate it, the most efficient ones being the recursive ones. All these types of descriptions are present in König's book; for example, the domain of thinking of the forms[30] of pure logic is described in (König 1914, pp.99–103) by listing 'fundamental forms' which are the 'lived facts' to be incorporated initially into the domain; then, by listing inference rules, that König calls 'principles'; and finally by describing how, proceeding recursively by applying these rules, one can 'deduct', from lived facts (*i.e.*, forms) previously recognized as belonging to the domain of thinking, new lived facts (forms) belonging to the domain,

[28] I take this word, the French version of which was coined by (Poincaré 1912, p.10), from (Franchella 2000, p.47).

[29] 'daß wir es uns nicht vorstellen können, das heißt, daß wir es uns nicht vorstellen wollen, daß wir *es ausschließen wollen.*' (König 1914, p.15)

[30] 'Form' is the word König employs instead of our 'formula'.

exactly as we describe a logic today. The *structure* generated by the domain thus described can be viewed as an algebraic one; another example, however not so close to our habits today, is that of the structure generated by the domain (of thinking) of the forms of (König's) arithmetic. What a form is, is explained in (König 1914, pp.90–97) using a domain of thinking where the formations of forms occupy the place taken by formal deductions in logic. In some brief introductory explanations, König even says:

> With regard to the isolated signs, we get finite domains of thinking, and precisely, the lived facts belonging to the domain of thinking are the stipulations with which certain names are allocated to the sign in question. While we limit ourself to this domain of thinking, every other property of the sign is ruled out ... .[31]

The most simple structures are those of class and set. In his book, König explains the distinction he makes between them as follows.[32] First, a (fixed) class concept is generated by a domain of thinking whose description is reduced to a list of forms, each of which expresses that a given lived fact, represented by a sign[33] of it, answers also to the common noun of the class to be generated, itself represented by its sign; and on the lived facts given in this way (possibly implicitly as those which satisfy some suitable characteristic property),

> we say that *they are terms which fall under the class concept*, while for the things not given, use is also made of the expression that they do not fall under the class concept.[34]

Second, König dreams about an extended family of ways to form a set with fixed given elements: the manners α, β, etc. All that follows on what he calls *pure* sets is relative, separately, to each of these various manners. A pure set is generated by a domain of thinking whose description is similar to that of a domain generating a class, except, first, that the forms listed express that the named (by signs of them) lived facts are to be related as '*elements*' to the set to be generated, which is represented by its sign; and on the other hand, that these elements are given in such a way that we know how to distinguish between any two of them. Afterwards, an *ordinary* set, also

[31] 'Wir erhalten in Bezug auf die einzelnen Zeichen endliche Denkbereiche, und zwar sind die dem Denkbereiche angehörenden Erlebnisse die Festsetzungen, mit denen dem betreffenden Zeichen gewisse Namen beigelegt werden. Indem wir uns auf diesen Denkbereich beschränken, wird jede andere Eigenschaft des Zeichens ausgeschaltet ... ' (König 1914, p.90)

[32] (Franchella 2000, p.49) emphasizes more the philosophical sides of König's views on these points.

[33] *I.e.*, for König, the names of the representations in the consciousness of the lived facts themselves.

[34] 'sagen wir, daß sie *Terme sind, die unter den Klassenbegriff fallen*, während für die nicht gegebenen Dinge auch der Ausdruck gebraucht wird, daß sie nicht unter den Klassenbegriff fallen.' (König 1914, p.26)

qualified as *cantorian* by König, can be viewed as both a class and a pure set: its defining, generating domain of thinking, is described by the description of the manner under which the terms falling under the class concept are given (so that no sign of the class occurs), to which is added a form in which an indeterminate occurs and which stipulates that every such term also satisfies the relation as element to the set to be generated.

Thus a scent of Von Neumann's set theory begins to hang in the air, even if Von Neumann's and König's conceptions of the class are fundamentally different. Afterwards Bernays's 1937 set theory shows this strange feature, removed in (Gödel 1940, p.7), to resort not only to two primitive notions of collection, class and set, but also to *two* primitive notions of *membership*: the membership to a set ε, and the membership to a class η, conceived so that a class can be coextensive with a set, which then is said to *represent* this class (Bernays 1937, pp.66–67).

Historically, the first most simple elementary class of relational structures is the one of (*linearly*) ordered sets.[35] In (König 1914, p.56) it is explained that, to generate such a 'thing', one starts from the set M of its elements; to the description of the domain of thinking from which M is generated, one adds what *we* would call the description of the order's graph, so as to obtain a *new* domain of thinking $O(M)$, and that

> we say then that the set M had been ordered in $O(M)$ (*i.e.*, through the stipulations made for $O(M)$).[36]

Taking as an example[37] the case of an ordered set of three elements, extended to one of four elements by appending to the first set a new greatest element d, he explains the difference between the extended order and the initial order, which *for us* had become the order which is now *induced* on the initial set, by the absence of d and of the relation instances established by the order from the other elements to d in the description of the domain of thinking generating the first situation, with respect to the description of the domain of thinking generating the second one. Beforehand, he had stated that most generally

> Empirical relation concepts stemming from different domains of thinking are then ... also always different.[38]

Through this list of various classes of structures entering in König's book emerges the watermark of Bourbaki's structuralist programme as it is sketched

[35] By an order relation, König, like Cantor, always means a *linear* order.

[36] 'wir sagen dann, daß die Menge M in $O(M)$ (d.h. durch die für $O(M)$ getroffenen Festsetzungen) geordnet wurde.'

[37] Also displayed in (Franchella 2000, p.56).

[38] 'Aus verschiedenen Denkbereichen entstammende empirische Relationsbegriffe sind dann ... immer auch verschieden.' (König 1914, p.55)

in the first booklet of the latter on set theory (Bourbaki 1939). The background in philosophy of mathematics is the same; despite the difference of levels of mastery of various matters, the eyes are turned in the same direction. Of course, there is no question about the absence of topological structures in König's work! On the other hand, he explains a case of transfer of structure: that one which, from the set Z_k of the numerals < *Numeratoren*> from 1 to k, ordered by their natural order, numbers the successive states of what he called a '*finite process of thinking*' <*endlicher Denkprozeß*> (of length k). Here is the explanation of the transfer:

> Let x be any element of Z_k; then, one can denote by e_x the lived fact, belonging to the domain of thinking, assigned to x, and further stipulate that if and only if $x \prec y$ [holds], $e_x \prec e_y$ would also always be set. ... through this stipulation *the lived facts belonging to the domain of thinking are ordered as well.*[39]

6 König's intuitionistic views

In order to formalize the concept 'if ... then', (König 1914, p.77) introduces for it the sign $\subset$, which he calls the 'implication sign' < *Zeichen der Implikation.* To describe its content <*Inhalt*>, he explains that 'A, then B' is a name of the 'lived fact':

> *To assumption that A is a fact which cannot be refused, compels me to assume further that B is also a lived fact which cannot be refused.*[40]

But, in his book, it is without illusion that König accepts formalized axiomatisations. In his logics, the algebraic laws like, *e.g.*, the commutative law of the 'logical product' (*i.e.*, conjunction) might have been written in terms of equivalence, which he defines as usual, and used as usual by resorting to the replacement theorem; indeed, this one can be established from the axioms and rules he sets, but he never thought of such a derived rule. Hence, he added to his primitives a notion of 'isology' <*Isologie*>, problematic in his own eyes; in (König 1914, p.72) he defined it as 'non-diversity with respect to certain relationships established once and for all'.[41] It was meant to take the place of equivalence in the statement of the laws in question. Commenting on two 'fundamental forms' for isology, König writes:

[39] 'Ist ... x irgend ein Element von Z_k, so kann man dem x zugeordnete, dem Denkbereich angehörende Erlebnis mit e_x bezeichnen und ferner die Festsetzung treffen, daß dann und nur dann, wenn $x \prec y$, immer auch $e_x \prec e_y$ gesetzt werde. ... durch diese Festsetzung auch *die dem Denkbereiche angehörenden Erlebnisse geordnet sind.*' (König 1914, p.65)

[40] 'Die Annahme, daß A eine unabweisbare Tatsache ist, zwingt mich zu der weiteren Annahme daß B auch ein unabweisbares Erlebnis ist.'

[41] Following Franchella's translation in (Franchella 2000, p.51).

> What happened here, can ... with Husserl be qualified as *a formalization of the concept 'not distinct'*; we will from now on ... treat only the logical fundamental forms (Ia) and (Ib), ... and thus obtain a method which eliminates the full intuitive concept and *which uses only the signification just set in the forms* [*My italics*]. I avoid intentionally the expression 'axiomatization', for *the intuitive concept cannot be exhausted through axiomatisation* [*My italics*];[42]

and he recalls two pages farther that

> Here also, the signs $\breve{+}$ and $\breve{\times}$ arisen through formalization give *less* than the intuitive concepts 'and' and 'or'.[43]

He speaks of three families of (interpreted) *formal* logics, in fact just 'propositional calculi', one of which contains a version of the classical one. About their common formal language, he is *never* speaking of a connection <*Verknüpfung*> of negation, contrary to what he does in the cases of logical product and sum, and also of implication. On the other hand, he incorporates in this language, and handles as if they were connectors, truth and falsity *predicates*. Here, we will not use his complicated notations, but take $\mathbf{T}L$ resp. $\mathbf{F}L$ to denote (the logical form interpretable, according to him, as saying, of the form L) 'L is true resp. false'. To all forms thus constructed, he gives in (König 1914, p.92) the name of '*valence forms* (*evaluations*)' <*Valenzformen (Wertungen)*> and warns:

> A domain of thinking to which every lived fact belongs as a fact which cannot be refused or which cannot be accepted, would define the 'absolute True'. To me, such a thing seems 'impossible'.[44]
>
> Yet at the place of the 'absolute-true' may not be set the coarse empirical 'true for me'. I must be able to precisely describe the state of consciousness to which the 'true for me' refers and then I get to the truthvalue which is the subject matter of logic.
>
> *The methodological auxiliary for this description is to introduce a certain domain of thinking ..., with the further stipulation that the lived facts which belong to the domain of thinking — and only those — will be held to be facts*

[42] Was hier geschehen, kann ... mit Husserl als *Formalisierung des Begriffes „nicht verschieden"* bezeichnet werden; wir werden von nun ab ... nur die logische Grundformen (Ia) und (Ib), ... behandeln, und dadurch eine Methode gewinnen, die den vollen Anschauungsbegriff eliminiert und nur die eben in die Formen gelegte Bedeutung benutzt. Ich umgehe geflissentlich den Ausdruck „Axiomatisierung", denn der volle Anschauungsbegriff kann nicht durch Axiomatisierung erschöpft werden;' (König 1914, p.73)

[43] 'Auch hier geben die durch Formalisierung entstandenen Zeichen $\breve{+}$ und $\breve{\times}$ *weniger* als die Anschauungsbegriffe „und" und „oder". (König 1914, p.75)

[44] 'Ein Denkbereich, dem jedes Erlebnis als unabweisbare oder unannehmbare Tatsache angehört, würde das „absolut Wahre" definieren. Ein solcher scheint mir „unmöglich".' (König 1914, p.87)

> *which cannot be refused.* ... When the lived fact X belongs to the domain of thinking, then the lived fact, that the lived fact X is a fact which cannot be refused, will also belong to the domain of thinking. ...
>
> For a lived fact Z which does not belong to the domain of thinking, it can happen that the lived fact 'Z is a fact which cannot be accepted' belongs to the domain of thinking. ...
>
> *But this lived fact* [$\mathbf{F}Z$], *too, cannot belong to the domain of thinking; then nothing is precisely asserted about* Z , *neither that it is true, nor that it is false.* This lived fact ... 'is' not at all[45] ... .[46]

However, a non-negligible fragment of König's *formal* pure logic displayed in (König 1914, pp.99–102) agrees with formalizations *today* allowed by many intuitionistic trends.

The system consists of 31 'fundamental forms'[47] (*we* would call them 'axioms'), partitioned in 5 groups. The group I consists of the forms (Ia) $x \overset{\smile}{=} x$ and (Ib) $[x \overset{\smile}{=} y] \overset{\smile}{=} [y \overset{\smile}{=} x]$ already mentioned above,[48] where $\overset{\smile}{=}$ is König's sign of isology. The group II consists of the 7 forms $[x \breve{+} y] \overset{\smile}{=} [y \breve{+} x]$ and $[x \breve{\times} y] \overset{\smile}{=} [y \breve{\times} x]$; also $[x \breve{+} (y \breve{+} z) y] \overset{\smile}{=} ([x \breve{+} y) \breve{+} z]$ and $[x \breve{\times} (y \breve{\times} z)] \overset{\smile}{=} ([x \breve{\times} y) \breve{\times} z]$; then $[(x \breve{+} y) \breve{\times} z] \overset{\smile}{=} [(x \breve{\times} z) \breve{+} (y \breve{\times} z)]$; and finally $[x \breve{+} x] \overset{\smile}{=} x$ and $[x \breve{\times} x] \overset{\smile}{=} x$. The group III consists of the 13 forms $x \subset x$; $[x \breve{\times} y] \subset x$, $[x \breve{\times} y] \subset y$ and $x \subset [x \breve{+} y]$, $y \subset [x \breve{+} y]$; also $[x \subset (y \subset z)] \subset [(x \breve{\times} y) \subset z]$, $[(x \breve{\times} y) \subset z)] \subset [x \subset (y \subset z)]$; then $[(x \subset y) \breve{\times} (z \subset u)] \subset [(x \breve{\times} z) \subset (y \breve{\times} u)]$, $[(x \subset y) \breve{\times} (z \subset u)] \subset [(x \breve{+} z]) \subset (y \breve{+} u)]$ and $[x \breve{\times} (x \subset y)] \subset y$, $[(x \subset y) \breve{\times} (y \subset z)] \subset [x \subset z]$; and finally $[x \overset{\smile}{=} y)] \subset [x \subset y]$, $[x \overset{\smile}{=} y)] \subset [y \subset x]$.[49] The group IV is reduced

45 'About a thing a we say that it "exists", when it belongs to the domain ... ' < Von einem Dinge a sagen wir, es ,,existiere", wenn es dem Bereich ... angehört; ... > (Zermelo 1908, p.262)

46 'An die Stelle des ,,absolut-wahren" darf aber nicht das grobempirische ,,wahr für mich" gesetzt werden. Ich muß den Bewußtseinzustand, auf den sich das ,,wahr für mich" bezieht, genau beschreiben können und gelange dann zu dem Warheitswerte, der Gegenstand der logische Beobachtung ist.

Das methodische Hilfsmittel dieser Beschreibung ist die Einführung eines bestimmten Denkbereichs ..., mit der weiteren Festsetzung, daß die Erlebnisse, die dem Denkbereiche angehören — und nur diese — als unabweisbare Tatsache gelten sollen. ... Sollte das Erlebnis X dem Denkbereiche angehören, so soll auch das Erlebnis, daß X eine unabweisbare Tatsache ist, dem Denkbereiche angehören. ...

Für ein Erlebnis Z , das dem Denkbereiche nicht angehört, kann es vorkommen, daß das Erlebnis ,, Z ist eine unannehmbare Tatsache" dem Denkbereiche angehört. ...

Es kann *aber auch dieses Erlebnis [* $\mathbf{F}Z$ *] dem Denkbereiche nicht angehören; dann wird eben von Z nichts behauptet, weder daß es wahr, noch daß es falsch ist.* Das Erlebnis ... ,,ist" überhaupt nicht ... ' (König 1914, pp.86–87)

47 In fact, it is highly redundant.

48 The transitivity, up to isology, of the isology, is established in (König 1914, p.129) by a formal proof which is a real *tour de force.*

49 Until here, I did not follow König's order of enumeration.

to the unique form $[x \Leftrightarrow y] \stackrel{\smile}{=} [(x \subset y) \breve{\times} (y \subset x)]$, where the sign $\Leftrightarrow$ is König's (... and Bourbaki's) sign for equivalence. The group V consists of the 8 forms $\mathbf{TT}x \stackrel{\smile}{=} \mathbf{T}x$, $\mathbf{FT}x \stackrel{\smile}{=} \mathbf{F}x$, $\mathbf{TF}x \stackrel{\smile}{=} \mathbf{F}x$, $\mathbf{FF}x \stackrel{\smile}{=} \mathbf{T}x$; and $\mathbf{T}(x \breve{\times} y) \stackrel{\smile}{=} [\mathbf{T}x \breve{\times} \mathbf{T}y]$, $\mathbf{F}(x \breve{\times} y) \stackrel{\smile}{=} [\mathbf{F}x \breve{+} \mathbf{F}y]$, $\mathbf{T}(x \breve{+} y) \stackrel{\smile}{=} [\mathbf{T}x \breve{+} \mathbf{T}y]$, $\mathbf{F}(x \breve{+} y) \stackrel{\smile}{=} [\mathbf{F}x \breve{\times} \mathbf{F}y]$.[50]

The system then displays 4 '*logical principles*' <*logischen Prinzipe*> that König presents as *inference rules*. In reality, the first two are *axiom schemata*. König gives to the first one the name of '*isology principle*' <*Isologieprinzip*>; it says that for any (logical or not!) forms U, V, W, the form $[U \stackrel{\smile}{=} V] \subset [S\binom{x}{U}W \stackrel{\smile}{=} S\binom{x}{V}W]$, where $S\binom{x}{U}W$ denotes the result of the substitution of U for every occurrence of x in W, is a logical form. The second, called 'evaluation principle' <*Wertungsprinzip*> by König, says that for any form U, the forms $U \subset \mathbf{V}U$ and $\mathbf{V}U \subset U$ are logical forms.

The two genuine inference rules are the customary ones: in the so-called *modus ponens*, König sees '*the deduction, or inference, principle*' <*Deduktions- oder Schlußprinzip*>; finally, the '*substitution principle*' <*Substitutionsprinzip*> says that by substituting any form U for every occurrence of x in a *logical* form, a new logical form is obtained.

Even to outline a complete study of this system, which is very complex, would take too long here; there are traps which come from some indeterminacies in the axiomatization of the notions of isology and of valences. Nevertheless, let us give here the name of *König's positive pure logic* to the fragment of König's pure logic obtained by dropping valences from his formal language (which only leads us to discard the fundamental forms of the group V and the evaluation principle, and to restrict the allowed substitutions to forms deprived of valences). By replacing every sign of isology by a sign of equivalence in any form A which is formally provable in this logic, we obtain another thus provable form A^*. At first sight, one sees how to prove this. And when every axiom A has been replaced by A^*, it becomes obvious that all corresponding forms obtained are provable in positive intuitionistic logic.

As it is not too difficult an exercise to find formal proofs in König's positive pure logic for all of Heyting's axioms in (Heyting 1930a) which are free of occurrences of the negation sign, we can firstly hold that positive intuitionistic logic is included in König's positive pure logic. Second, from the aforementioned remarks, we may conclude that conversely, for every form A which is provable in the former logic, the form A^* is provable in positive intuitionistic logic. In this sense, *there is no essential difference between what is stated in König's positive pure logic and what is stated in positive intuitionistic logic.*

7 König's metamathematical views

Let us recall that Kleene says:

[50] The classical appearance of two of these last forms is due to the fact that König assumes implicitly that the process of recognizing that a lived fact is given, or not given, in the description of a domain of thinking, is decidable.

Metamathematics includes the description or definition of formal systems as well as the investigation of properties of formal systems. (Kleene 1952, p.62)

So, we see how and where (König 1914) is doing metamathematics, and we only need to comment on a few points. First, when he states his substitution principle, (König 1914, p.102) formalizes it under the form $[W \text{ nom. } \Lambda]$ inv. $\left[S\binom{x}{U}W \text{ nom. } \Lambda\right]$, where 'nom. Λ' means 'is a logical form' and 'inv.', which is an abbreviation for (a form of) the latin verb '*involvere*', can be read '*involves*', understood here as meaning '*entails*'. Today, to describe a formal system of logic, it is customary to leave implicit what corresponds to 'nom. Λ', and to write 'W' (or '$\vdash W$') over a sufficiently long horizontal line, and '$S\binom{x}{U}W$' under this line, in order to bring to the fore the distinction of entailment from implication; the first, in principle intuitive, metamathematical, the second, formal, mathematical. That is precisely what (König 1914, p.16) explains at the very beginning of his treatment of his 'domains of thinking' and of their descriptions; descriptions are 'metamathematical' in essence, even when, as is the case at the point where König puts his warning, there is absolutely nothing mathematical in the described domain.

All is said in the parenthesis that ends the footnote in (König 1914, p.242):

Lived fact *of* the domain of thinking and lived fact *about* the domain of thinking [*My italics*].[51]

8 König's last set-theoretical thoughts

Before departing, König had drafted some concise indications on the last creation of his mind: the notion of *ordinal*, distinct, as canonical well-order, from that of ordinal *number*, as a type of similar well-orderings. Two pages before the end of his book, we read, after some comments about his generating ideas:

Let $\bar{0}$ be the 'zero', any thing, a lived fact, the visual imprint 0 itself or whatever else. Out of this, we define, generate, a thing $\bar{1}$ through the stipulation

$$\bar{0} \mid \prec \mid' \bar{1}\,;$$

out of them anew, when $\bar{0}$, $\bar{1}$ are the things just defined, through the stipulations

$$\bar{0} \mid \prec \mid'' \bar{2}, \quad \bar{1} \mid \prec \mid'' \bar{2}$$

the thing $\bar{2}$. In the same way, a thing $\bar{3}$, through the stipulations

$$\bar{0} \mid \prec \mid''' \bar{3}, \quad \bar{1} \mid \prec \mid''' \bar{3}, \quad \bar{2} \mid \prec \mid''' \bar{3},$$

where again $\bar{0}$, $\bar{1}$, $\bar{2}$ are the already introduced 'things'.

The things $\bar{1}$, $\bar{2}$, $\bar{3}$ generated in this way will be called *ordinals*.[52]

---

[51] 'Erlebnis *des* Denkbereichs und Erlebnis *am* Denkbereich.'

[52] 'Es sei $\bar{0}$, die „Null" irgend ein Ding, ein Erlebnis, der Gesichteindruck 0 selbst oder was sonst immer. Aus diesem definieren, erzeugen wir ... ein Ding $\bar{1}$ durch die Festsetzung

This is how he describes the first steps, including the basis, of a recursive definition of the notion of ordinal. He continues by the induction step, which he calls 'application <*Anwendung*> of the $\mathfrak{o}$-process' to a set of given things entering a domain of thinking in which these given things are stipulated to come 'before' the thing generated by the domain, and notices:

> Overtly, through application of the $\mathfrak{o}$-prozess, we have 'generated', from the set which consists of the sole element $\bar{0}$, the ordinal $\bar{1}$; in the same way, from the set which consists of $\bar{0}$ and $\bar{1}$ as elements and of these only, the ordinal $\bar{2}$. In the same way, $\bar{3}$.
>
> We stipulate now:
>
> When it is 'possible' *to apply the $\mathfrak{o}$-process to the already generated set of ordinals, this process will also be thought as executed and the thing generated through the process will again be called an ordinal.*
>
> Exactly as for the numerals, we obtain here (corresponding to every numeral k) an ordinal $\bar{k}$; but, when we have formed all ordinals corresponding to the numerals, nothing prevents us from going farther and defining a thing ω so that, when k [is] an arbitrary numeral, $\bar{k}$ will come 'before' ω, where the order relation to be used for ω must however be different from each of those which was already used in generating any $\bar{k}$.[53]

Thus it is *König* who appears to be the father of the *philosophical* idea of sets well-ordered by the membership relation which in this exceptional

---

$$\bar{0} \mid \prec \mid' \bar{1};$$

aus denen wieder, wenn $\bar{0}$, $\bar{1}$ die eben definierte Dinge sind, durch die Festsetzungen

$$\bar{0} \mid \prec \mid'' \bar{2}, \quad \bar{1} \mid \prec \mid'' \bar{2}$$

das Ding $\bar{2}$. Ebenso das Ding $\bar{3}$ durch die Festsetzungen

$$\bar{0} \mid \prec \mid''' \bar{3}, \quad \bar{1} \mid \prec \mid''' \bar{3}, \quad \bar{2} \mid \prec \mid''' \bar{3},$$

wo wieder $\bar{0}$, $\bar{1}$, $\bar{2}$ die schon eingeführten „Dinge" sind.

Die so erzeugten Dinge $\bar{1}$, $\bar{2}$, $\bar{3}$ sollen *Ordinatoren* genannt werden.' (König 1914, p.257). Clearly, König is here coining the substantive *ordinal* <*Ordinator*> on the model of the substantive *numeral* <*Numerator*>.

[53] 'Wir haben offenbar durch Anwendung des $\mathfrak{o}$-Prozesses aus der Menge, die das einzige Element $\bar{0}$ enthält, den Ordinator $\bar{1}$ „erzeugt"; ebenso aus der Menge, die $\bar{0}$ und $\bar{1}$, und nur diese als Elemente enthält, den Ordinator $\bar{2}$. Ebenso $\bar{3}$.

Wenn es möglich ist, *auf die schon erzeugte Menge von Ordinatoren den $\mathfrak{o}$-Prozess anzuwenden, so soll dieser Prozess auch als ausgeführt gedacht werden und das durch den Prozeß erzeugte Ding wieder Ordinator genannt werden.*

Genau so wie bei den Numeratoren erhalten wir hier (jedem Numerator k entsprechend) einen Ordinator $\bar{k}$; wenn wir aber alle den Numeratoren entsprechenden Ordinatoren gebildet haben, hindert uns nichts, weiter zu gehen und ein Ding ω zu definieren, so daß wenn k ein beliebiger Numerator, $\bar{k}$ „vor" ω stehen soll, wo allerdings die für ω zur Verwendung gelangende Ordnungsrelation von jeder solchen verschieden sein muß, die bei der Erzeugung irgend eines $\bar{k}$ schon benutzt werde.' (König 1914, p.258)

case coincides with that of inclusion as an initial proper segment.[54] It was *v. Neumann 1923* which gave *mathematical* life and strength to this discovery, by establishing, as a part of set theory, a theory of ordinals unequalled in elegance and technical accuracy.

9 Concluding words

But 'only a few pages were lacking', in which a 'definitive' proof of Zermelo's well-ordering theorem was promised. What could indeed this intended proof have been like? The genius's guess of (von Neumann 1925), which consisted in having the universal class as image of the class of ordinals, is well known. Was this idea really that of König? We will remain without an answer for ever. That was Von Neumann's intuition.

The greatness of König's book lies in the fact that he paved the way for Von Neumann's first works and fame, but not only that. By his contributions to the elucidation of the thought processes of mathematicians in set theory, by his suggestion of a theory of structures (especially of mathematical theories), by the development he gave to Hilbert's idea of two levels of language, and by many other aspects which remain to be investigated, he strongly contributed to urging mathematicians to think about all these matters, and helped them better to conceive the unity of their science.

It is a pity that König, who was so very cautious about the negation of an implication that he thought it better to leave his work incomplete on this point, did not have knowledge of Brouwer's thesis and of his first work on negation. It is clear that König would at least have been very interested. We cannot say if he would have been able to fully agree with Brouwer's views, in particular on analysis; we can imagine that the two men would have been interested in debating their respective conceptions; but we know nothing about what could have been the conclusion of such a possible encounter. It is also likely that König could have become a precious ally for Brouwer.

So König's book deserves more searching and deepened studies, among other things on his convergences and divergences from Brouwer; in particular, his real influence on mathematical thinking during the half century after his disappearance remains to be very meticulously investigated.

Acknowledgements

I thank the organizers of the meeting for their kind invitation, as well as Emily Tanimura who with great kindness took much of her time for explaining to me how to express my views in correct English. It goes without saying that I alone assume the full responsibility for all remaining mistakes.

[54] It had yet to be realized that for this to hold until the end, it requires that one starts from the empty set, and to clarify the notion of inductive limit until its applications here.

Gödel, constructivity, impredicativity, and feasibility

Jean Fichot

1 Introduction

This paper does not pretend to be an exhaustive survey of Gödel's interpretation of intuitionism. This long and rather complicated story has already been told and analyzed by others, for instance in (Kreisel 1987b) and (Tait 2006a,b). More modestly, our first aim here is to present a different appearance of a ghost during that story, the one of impredicativity, and to show that only one case is to be taken seriously from a strict anti-realist point of view: the impredicativity of the concept of natural number. This leads to our second aim, which is to present some of the feasible versions of Gödel's Dialectica interpretation.

2 Heyting's notion of proof, constructivity and impredicativity

According to Gödel, in 1933, to provide a foundation for classical mathematics is to give axioms and rules that account for the totality of methods of proof actually used by mathematicians (Gödel 1933c). The first foundational systems, as given by Frege and Peano, were closed but also inconsistent because they admitted highly impredicative definitions like the well-known Russell set $\{x \in SET/x \notin x\} \in SET$.

So consistency became a problem. Part of the solution was to adopt other systems given by Russell and Zermelo. These systems have a strange feature: each is open-ended and may be seen as an infinity of stronger closed systems. And this feature is not accidental because it reflects an essential property of mathematics: as a totality of methods of proof, it is open-ended. They present also two problematic features: they allow non-constructive proofs and impredicative definitions of properties, sets or classes which are legitimate only if they are held not to produce their object but only to discover their existence. This justification supposes adopting a mathematical realism but Gödel, at that time, held that this philosophical standpoint could not give a convincing solution to the problem of consistency. At least, the latter should be proved by constructive and predicative methods.

Due to the failure of Hilbert 's program, intuitionism could be seen as providing the means to give a constructive justification of classical mathematics at least in the case of arithmetic. But, according to Gödel, the main defect of intuitionism is revealed by the proof of consistency of Peano arithmetic relative to Heyting arithmetic (Gödel 1933c). This may sound paradoxical but Gödel emphasized that this proof rests on a translation of the former in the fragment $\{\Rightarrow, \forall\}$ of the latter where $\neg A$ is defined by $A \Rightarrow 0 = 1$ and that in this fragment, the translation of excluded middle, given by $\neg\neg A \Rightarrow A$ and $0 = 1 \Rightarrow A$, is provable. This justification of excluded middle shows that intuitionistic logic is not constructive enough. If a non-constructive existential proposition $\exists_c x \in \mathbb{N} A(x)$ is defined by $\neg\forall x \in \mathbb{N} \neg A(x)$ one may have an intuitionistic proof of the former, but no proof of $A(n)$ for some $n \in \mathbb{N}$. There is an obvious answer to this remark: from a constructive point of view, we can distinguish this proof from a proof of $\exists x \in \mathbb{N} A(x)$ where the existential quantifier has a constructive meaning. But the distinction here rests on the adoption of Heyting's interpretation of the axioms and rules of intuitionistic logic according to which a proof of $\exists x \in \mathbb{N} A(x)$ is a proof of $A(n)$ for some $n \in \mathbb{N}$.

But there is a problem here. The proofs in Heyting's interpretation are canonical proofs and they are obviously constructive but real constructive proofs are not canonical proofs. This is due to the use of *modus ponens* that cannot be justified by means of canonical proofs. This point is well known: from canonical proofs of $B \Rightarrow \exists x \in \mathbb{N} A(x)$ and B we infer $\exists x \in \mathbb{N} A(x)$ by *modus ponens* but this proof is not a canonical proof. Brouwer faced this problem when he gave a justification of his so-called 'law of syllogism'. Suppose there is an embedding (given by) f of A in B and an embedding (given by) g of B in C; can we conclude that D is directly embedded in A? If we rephrase this law by means of a Venn diagram, with circle C contained in circle B contained in circle A, then the answer is obvious: we just have to erase B to see that C is directly contained in A. But the law of syllogism may be seen also as a generalization of *modus ponens* and now the question is not as trivial as it might seem: if f is a proof of $A \Rightarrow B$, g is a proof of $B \Rightarrow C$, how do we know that there is a direct proof of $A \Rightarrow C$ where B is not used anymore? Heyting's answer is well known: a canonical proof of $A \Rightarrow B$ should be a rule of construction that, applied to any canonical proof of A, yields a canonical proof of B. This interpretation seemed problematic to Gödel for two reasons at least.

First the notion of constructive proof cannot be replaced by that of formal proof in a specific system **S**. Gödel mentioned his translation of intuitionistic logic in a modal logic where the constant $\boldsymbol{B}$, which means intuitively 'provable', cannot be interpreted by the predicate 'provable-in-**S**' if **S** is consistent

(Gödel 1938). This justification of just one closed system, Peano arithmetic, by means of an open-ended totality of constructive proofs, is of doubtful epistemological value because, according to Heyting's interpretation, the notion of proof of an implication depends on it and, as such, seems impredicative. Gödel emphasized this point when he mentioned the '*general impredicative concept of intuitionistic proof*'. But is Heyting's concept of proof impredicative? We all know that only definitions can be impredicative, so the question should be: is Heyting's interpretation an informal definition of this concept? Here there is a distinction to be made between a definition and a characterization: a property P is a definition of a notion N if we state that something is an N if and only if it has P but P is only a characterization of N if we require only that something has P if it is an N. If Heyting's interpretation is a definition of the notion of proof, then it is impredicative because the right part of the clause refers to the totality of proofs. But if Heyting's interpretation is only a characterization, it gives us only a criterion according to which a proof can be called non-constructive. Heyting's interpretation is usually presented as an explanation of the constructive meaning of the logical constants. As such, it takes for granted the fact that mathematicians know what a proof is, but that it is not always obvious to recognize some proofs as non-constructive, at least for those who are classically minded.

Second, in 1941, Gödel was not convinced that intuitionistic logic was really constructive according to Heyting's own concept of constructive proof. And this explains Gödel's negative, and at first reading rather mysterious, remark on the notion of intuitionistically correct proof in 1941 (Gödel 1995, pp.191–192):

> [O]ne may say that it furnishes in itself a counterexample against his own admissibility, insofar as it is doubtful whether a proof utilizing this notion of a constructive proof is constructive or not.

What Gödel meant is that Heyting's interpretation cannot convince us that proofs can always be replaced by canonical proofs and, for example, that Heyting arithmetic has the so-called existence property, *i.e.*, that a proof of $\exists x \in \mathbb{N}\ A(x)$ in Heyting arithmetic yields a proof of $A(t)$ in the same theory for some computable term t. According to Kreisel, Gödel in the end of the 1930s even doubted that such a proof ensures some term t depending on the proof such that $A(t)$ (Kreisel 1987b). Kleene remembered that 'a famous logician' told him that it was impossible to give a direct interpretation of arithmetic by means of recursive functions that could allow one to prove the existence property (Kleene 1973). Kreisel adds that Kleene *coyly* referred to Gödel whose goal at that time was to find a translation D of Heyting arithmetic in a more constructive system $\mathbf{T}$, and to prove that if $\exists x \in \mathbb{N}\ A(x)$ is provable in Heyting arithmetic, then, for some computable term t, $A^D(t)$ is provable in $\mathbf{T}$.

3 Gödel's T and the Dialectica interpretation

The Dialectica intrepretation rests on an extension of finitism and a misrepresentation of intuitionism. Gödel's **T** is finitist insofar as its logical rules and axioms are those of propositional logic; induction is applied only to formulas with no bound variables; *etc.* It is an extension of finitism since it contains rules of definition of functionals for all finite type. For instance:

if a of type α and g of type $\mathbb{N}, \alpha \to \alpha$ are previously defined functionals, then f defined by $f(0) = a, f(n+1) = g(n, f(n))$ is of type $\mathbb{N} \to \alpha$

All variables have a type and each free typed variable in a formula takes its values on the class of functionals of that type defined by those rules. The Dialectica interpretation A^D of an arithmetic formula A is given by a quantifier free atomic formula A_D expressing equality between terms of type $\mathbb{N}$ built from constants and variables of functions of higher types bound by prenex existential and universal quantifications in A^D. The former may be considered as quasi-finitist, not the latter. Schematically, we have

$$A^D \equiv \overbrace{\exists \boldsymbol{f} \forall \boldsymbol{x}}^{\text{non-finitist}} \quad \overbrace{A_D(\boldsymbol{f}, \boldsymbol{x})}^{\text{quasi-finitist}}$$

where $\boldsymbol{f}, \boldsymbol{x}$ are lists of variables of finite type. The quantifiers do not belong to the language of **T** but they are used in the inductive definition of the Dialectica translation. Furthermore, this translation is a misrepresentation of intuitionism since the equivalence between a formula A and its interpretation A^D can be proved only by means of axioms that are not intuitionistically admissible on the basis of Heyting's interpretation.

However, this extension of finitism and the Dialectica interpretation give an algorithm that extracts from a proof of a formula A in Heyting arithmetic definitions of computable terms $\boldsymbol{t}$ of finite type and a proof of $A_D(\boldsymbol{t}, \boldsymbol{x})$. Since $(0 = 1)^D$ is $0 = 1$, an immediate corollary is a proof of consistency of Heyting arithmetic relative to **T**. But the more general epistemological value of the translation is to be found in the fact that each proof in Heyting arithmetic, whose constructivity is dubious, is replaced by two components: a proof in **T** that may be hold as quasi-finitist and definitions of functionals which rest on an extension of finitism. Despite the fact that this replacement was not really intuistionistically justified, Gödel had to show at least that the second component of his translation was constructively justified by giving a meaning to the formal rules of definition of functionals.

4 Computable functions of finite type and proofs

Gödel introduced the concept of computable functions of finite type that was supposed to give an interpretation of the rules of definition of functionals in **T** (Gödel 1958, 1970, 1972) and also (Tait 2006a,b). Since he wanted his concept to be more primitive than that of constructive proof, the only difficulty was to show convincingly that the former does not depend on the latter.

- The computable functions of type $\mathbb{N}$ are the natural numbers.
- f is a computable function of type $\alpha \to \beta$ if and only if it is constructively recognizable (1958), intuitionistically demonstrable (1970), constructively evident (1972) that f is an operation (1958), a well-defined mathematical procedure (1970–1972), which applied to any computable function of type α yields a computable function of type β.

Gödel's interpretation is indeed informal but the grasp of the concept of computable functions depends on the answers to two questions: First what is the meaning of 'computable operation' or 'well-defined procedure' here? Second, how are we to understand the hesitation from 1958 to 1972 between 'recognizable', 'demonstrable' and 'evident'?

From 1936 Gödel held the different definitions of mechanical computability to be absolute, independent of any formal system (Gödel 1936). Each of these definitions gives an enumeration of the algorithms that compute partial functions of finite type and it was then possible to replace the notions of operation and well-defined mathematical procedure by a mathematical definition of computation rules. But the concept of computable function requires of course total functions and there is no effective definition of the class of computable functions of type $\mathbb{N} \to \mathbb{N}$ that could produce all these functions. This means that in the right part of the clause for type $\alpha \to \beta$, the algorithm is supposed to be given with a proof or evidence that it computes a total function. This was the main problem of Gödel's interpretation: if algorithms need proofs to be recognized as computing functions, then functions are not more primitive than proofs and cannot replace them. Despite the different warnings Gödel gave his definition, for example (Gödel 1990, p.245):

> This notion is to be regarded as immediately intelligible, provided the notions 'computable function of [lower types]' are already understood.

the concept of computable functions of type $\mathbb{N} \to \mathbb{N}$, for example, is not understood 'from below' only on the basis of the understanding of the concept of natural numbers. In the 1970 version of his paper (Gödel 1970), Gödel mentioned explicitly proofs in his definition. That prompted Bernays to remark (Gödel 2003, p.281):

> On page 3 of your manuscript ... it says ... 'for which, moreover this general fact is intuitionistically demonstrable'. Here the reader could well be taken aback, since your procedure is surely intended to avoid the concept of intuitionistic proof.

At that time, Gödel had already tried to find a way to replace the concepts of proof and function in his interpretation by a single one: the concept of what he called 'reductive proofs'. This concept of proof is described in notes added to the two latter versions of his Dialectica paper. These notes are notoriously difficult: the reader sees what Gödel's goal was and the obstacles he had to overcome to reach it, but the road he took seems rather obscure. To say it roughly, it is to come back to his interpretation of his system **T**, at least the one he mentioned to Kreisel during their first meeting (Kreisel 1987a, p.106):

> [in 1955] Gödel made a point of warning me that he had not given a thought to the objects meant by his terms of finite type. The only interpretation he had in mind was formal as computation rules obtained when the equations are read from left to right.

In 1970 and 1972 Gödel's idea was to restrict his definition of 'computable function of finite type' to the ones definable in the system **T**. Those formal terms were the ones he needed in his proof of the soundness of the Dialectica interpretation. This also restricted the epistemological value of Gödel's interpretation of **T**. It could not pretend to replace the general concept of intuitionistic proof of an arithmetic formula by that of computable function. It was only the concept of formal proof in Heyting arithmetic that was replaced by the concept of function definable in **T**. Bernays noticed this point in 1973 (Gödel 2003, p.303):

> This more sharply delimited function domain [generated by the rules of **T**] stands in relation to that of all computable functions analogously as the domain of the proofs formalized in a formal system relates to that of the contentual proofs.

But this restriction or delimitation came with an interpretation of those formal definitions. This idea was already present in 1958 (Gödel 1990, p.245):

> If the notion of computable function is not to explicitly contain the notion of proof, we must see to it that it is immediately apparent from the chain of definitions that the operations can be performed, as it is the case for all functions in the system **T** specified below.

What Gödel meant here is that when we see the definition in **T** of an operation, given by a finite set of equations, we don't only recognize this definition as the one of an algorithm but as a definition of an algorithm that computes a total function of a given type. The definition is not only supposed to show how to compute the value of the function for a given argument but also to prove that the algorithm, applied to any argument, computes a value in a finite number of steps. What was a formal definition of a mechanical rule is now understood around 1970, 1972 as a contentual proof (Gödel 1990, p.275):

> A narrower concept of proof, which may be called 'reductive proof' and which, roughly speaking, is defined by the fact that, up to certain trivial supplementations, the chain of definitions of the concepts occurring in the theorem together with certain axioms about the primitive terms forms by itself a proof, i.e., an unbroken chain of immediate evidences. Note that in this context a definition is to be considered as

> a theorem stating the unicity of an object satisfying certain conditions and that, in our case, it is convenient that a statement regarding the type character of the function defined should form part of its definition.

Gödel called this 'new' system $\mathbf{T}'$ but he never gave explicitly its rules and axioms. All that can be inferred from its writings, is that in this system the rules of definition and equations are also rules of proof. What Gödel had in mind must have been very close to the actual formulation of the system $\mathbf{T}$ due to Tait, where some constants, given with their type, are used in rules of definition of typed terms and equations that allow one to compute those terms (Tait 1967). For instance

$$\boldsymbol{R} : \alpha, (\mathbb{N}, \alpha \to \alpha) \to \alpha$$

$$\left\{ \begin{array}{c} \boldsymbol{R}at\,(n+1) =_\alpha tn\boldsymbol{R}atn \\ \boldsymbol{R}at0 =_\alpha a \end{array} \right\} \qquad \frac{a : \alpha \quad t : \mathbb{N}, \alpha \to \alpha}{\boldsymbol{R}at : \mathbb{N} \to \alpha}$$

When they are read and understood as contentual proofs, formal definitions and computations of terms in this system with conclusions like $t : \alpha$ and $t =_\alpha u$ show that t computes the values of a function of type α, that t and u compute the values of the same function of type α.

So the point of departure was the proofs in intuitionistic arithmetic. Going through $\mathbf{T}$ those proofs have been broken down into two components (proofs in $\mathbf{T}$, functions defined in $\mathbf{T}$) and now, following Gödel, this second component must be seen as proofs in another system. Thanks to Howard, it is well known nowadays that the system $\mathbf{T}$, supplemented with some inessential type formation rules, can in fact express all the rules and axioms of Heyting arithmetic which can be seen as a type system when its proofs are coded by lambda terms and its formulas are notations for higher types (Howard 1980). For example

$$\frac{a : A\,(0) \qquad t : \forall y : \mathbb{N}\,(A\,(y) \Rightarrow A\,(y+1))}{\boldsymbol{R}at : \forall x : \mathbb{N}\ A\,(x)}$$

Thanks to Prawitz, it is well known that the formal rules of computation of this extension of $\mathbf{T}$ express rules of cut elimination in proofs of Heyting arithmetic (Prawitz 1965). So, this extension of Gödel's $\mathbf{T}$ takes us back to our point of departure that was the proofs in intuitionistic arithmetic. Those proofs have been broken down into two components: the quasi-finitist proofs in $\mathbf{T}$, the computable functions defined in $\mathbf{T}$. Then, following Gödel, this second component was turned into proofs in another system. And now, following Howard, Tait and Prawitz, we are back where we came from: proofs in Heyting arithmetic.

5 The impredicativity of functions and proofs in T

In 1970, Gödel noticed

> Incidentally, it should be noted that the structure of the system **T** is not quite so simple as it may seem at first sight. In particular, there exist functions of lower type which, within **T**, can only be defined in terms of functions of higher types. This is a kind of impredicativity. ... This impredicativity appears in the fact that the concept of reductive proof may itself occur in reductive proofs (just as in Heyting's logic the general concept of proof may occur in a proof). (Gödel 1970, p.14)

Since definitions in **T** are proofs in **HA**, the latter are also impredicative. But what is this kind of impredicativity exactly? And how did Gödel try to overcome it ? The first question is easy to answer. The second is ... much more difficult.

1) The impredicativity of some proofs in **T** is due to the fact that they do not have the subtype property and this is in fact already an impredicativity of some proofs in (typed) Heyting arithmetic which does not have the subformula property. For instance, even if we forget about the type of the constant $\boldsymbol{R}$ in

$$\frac{\begin{matrix}\vdots & & \vdots\\ a:\alpha & & t:\mathbb{N},\alpha\to\alpha\end{matrix}}{\boldsymbol{R}at:\mathbb{N}\to\alpha} \qquad \frac{\begin{matrix}\vdots & & \vdots\\ a:A(0) & & t:\forall y:\mathbb{N}\,(A(y)\Rightarrow A(y+1))\end{matrix}}{\boldsymbol{R}at:\forall x:\mathbb{N}\ A(x)}$$

it is clear that $\mathbb{N},\alpha\to\alpha,\forall y:\mathbb{N}\,(A(y)\Rightarrow A(y+1))$ are not subtypes or subformulas of $\mathbb{N}\to\alpha,\forall x:\mathbb{N}\ A(x)$.

2) In 1972 when Gödel mentioned explicitly this impredicativity, he seemed to content himself with the fact that at least some proofs in **T**′ are predicative, *i.e.*, have the subtype (or subformula) property and he identified them with the reductive proofs (Gödel 1990, p.276).

> [O]nly the (in comparison to Heyting's) incomparably narrower concept of reductive proof need be used in the propositions and proofs of **T**′, and that, moreover, because these proofs are uniquely determined by the theorems, quantification over 'any proof' can be avoided. Note that it is not claimed that the proofs in $T'$are reductive. This is true only in certain cases ...

One point that might seem at first reading surprising is the remark according to which this impredicativity, the fact that the concept of reductive proof may occur in reductive proofs, is related to the quantification over any proofs in Heyting's alleged definition of constructive proof. There is a distinction here that should be made between the non-reductive proofs and their justification. For instance:

	formal rule	informal justification
Heyting arithmetic	$\dfrac{A \Rightarrow B \quad A}{B}$	a proof of $A \Rightarrow B$ applied to any canonical proof of A yields a canonical proof of B
Gödel **T**	$\dfrac{t : \alpha \to \beta \quad a : \alpha}{ta : \beta}$	a function of type $\alpha \to \beta$ applied to any function of type α yields a function of type β

The use of one of these rules in a proof may result in the loss of the subformula property, making the proof non-reductive. And its informal justification, if it holds for a definition of the concept of proof, is impredicative. One possible way to get rid of this problem is to replace this justification by a mathematical theorem and to show, following Tait and Prawitz, that such steps in a proof or a definition can be avoided and to consider only the cut free proofs or the terms in normal form obtained by a mechanical procedure. Then the content of the left column is not an informal definition of the general concept of proof or computable function, but only a mathematical property of the proofs and the terms where the universal quantification is restricted to the proofs in Heyting arithmetic or to the terms typable in **T**.

mathematical property
a proof of $A \Rightarrow B$ in HA applied to any proof of A in HA yields a cut free proof of B in HA
a term of type $\alpha \to \beta$ defined in **T** applied to any term of type α defined in **T** yields a normal term of type β

Then the impredicativity of Heyting's notion of proof and the one of Gödel's interpretation is indeed avoided. But two points must be emphasized. First, this is a mathematical property and, according to (Tait 2006b), Gödel knew how to prove it in 1941. But he was of course aware that his proof cannot be formalized in Heyting arithmetic. So we need at least one proof to replace the concept of proof in Heyting arithmetic by that of normal proof. Second, some proofs in normal form do not have the subformula property. This is a general fact that does not depend on a specific formulation of the rules of Heyting arithmetic. For subsystems of Heyting arithmetic there is a relation between the subformula property and its completeness. This relation is easily shown in the case of (classical or intuitionistic) logic by the so-called tableaux method because there the completeness theorem takes the form of a disjunction: there is a model where A is false or there is a proof of A. Since the search for a model rests on the inductive definition of truth, its failure gives a normal proof of A which has the subformula property. From this it can be immediately deduced that there is a predicative proof of A if there is a proof of A. Where completeness failed, as is the case for some subsystems of Heyting arithmetic, let us say the one that corresponds to primitive recursive arithmetic, then the proofs of some formulas given in the language of this subsystem cannot have the subformula property even if they are in normal form.

An immediate example is given by the formula of primitive recursive arithmetic *CONS-PRA* which is provable in Heyting arithmetic. Due to Gödel's second incompleteness theorem, a normal proof of this formula cannot have the subformula property because it rests on the use of the induction principle for a formula of higher logical complexity than the ones of primitive recursive arithmetic. Of course, the problem is exactly the same in the case of **T**.

Now, if we identify, in Gödel's interpretation of **T**, the reductive proofs with the predicative proofs in **T**, then there are some functions defined in **T** that will not be in the interpretation. But if we identify in Gödel's interpretation the reductive proofs with the normal proofs, then there are some functions in this interpretation that are impredicatively defined. Their definitions will be given by proofs that are not uniquely determined by the theorems because the extension of the concept of proofs of formulas of primitive recursive arithmetic is an open-ended totality of proofs.

6 Impredicativity and feasability

While Gödel tried to ward off the ghost of impredicativity with his concept of reductive proof, Bernays tried to reassure him by noticing that even in finitist mathematics some impredicativity was unescapable. In 1973, he suggested that the problem, that may be not resolvable, was that a certain impredicativity remains in the concept of functionals (Gödel 2003, p.295):

> But, strictly speaking, such an impredicativity is to be found in the finitary approach ... One must still assume the assumption that if n is a constructible numeral and if furthermore, a process is intuitively described which, from given numerals, again provides a numeral, then the n-fold iteration of that process can be carried out. A corresponding assumption must also be employed for the recursive definitions of functionals.

Bernays' remark is to hold a natural number as an algorithm, *i.e.*, as a Church numeral $\boldsymbol{n} \equiv \lambda x \lambda f. f^n x$, in the definition of the functional of finite type. Then, if $a : \alpha$ and $t : \alpha \rightarrow \alpha$ are previously defined, we have $\boldsymbol{nat} = t^n a$ but then a Church numeral $\boldsymbol{n}$ is of type $\alpha, (\alpha \rightarrow \alpha) \rightarrow \alpha$ for each type and since $n : \mathbb{N}$ we must solve the equation $\mathbb{N} \equiv \alpha, (\alpha \rightarrow \alpha) \rightarrow \alpha$ for any type α. The solution is to define the type of natural numbers by $\forall \alpha\, (\alpha, (\alpha \rightarrow \alpha) \rightarrow \alpha)$ as in Girard's system **F** which is the second-order extension of **T** (Girard 1971). This type is nothing but the propositional skeleton of Dedekind's definition of the natural numbers as the elements of the smallest set that satisfies induction:

$$x \in \mathbb{N} \equiv \forall X\, [0 \in X, \forall y\, (y \in X \Rightarrow y + 1 \in X) \Rightarrow x \in X]$$

If we follow this suggestion, then we are led to the conclusion that some definitions of functionals in **T** and proofs in Heyting's arithmetic owe their impredicativity to the fact that the rules and axioms of these systems are second-order rules dressed in first-order clothing. And this is exactly the argument used by Nelson who thinks that, even formulated as a first-order schema, the induction principle is impredicative (Nelson 1986, pp.1–2).

> The reason for mistrusting the induction principle is that it involves an impredicative notion of number. It is not correct to argue that induction only involves the numbers from 0 to n; the property of n being established may be a formula with bound variables that are thought of as ranging over all numbers. That is, the induction principle assumes that the natural number system is given. A number is conceived to be an object satisfying every inductive formula; for a particular inductive formula, therefore, the bound variables are conceived to range over objects satisfying every inductive formula, including the one in question.

Nelson's argument hits the mark. The first-order induction principle is justified by the existence of the potential infinity of natural numbers but the latter is given by an impredicative definition that amounts to the former. Since proofs by induction are also definitions of functionals by recurrence, Nelson's circle and Bernays' remark, if taken seriously, should shed some doubt even on definitions by primitive recursion. As noticed by Bellantoni and Cook, if we define f by $f(0) = m, f(n+1) = g(n, f(n))$ and the set F by $F = \{y/\exists x\,(y = f(x))\}$ then the definition of F is impredicative because, in order to compute the value of f for $n+1$, we need to know that $f(n)$ is defined and this depends on the use of induction.

Usually, those who defend a predicativist point of view hold the infinite sequence of natural numbers as given, as a closed totality when they accept classical logic, as a potential infinite when they refuse the excluded middle. From an intuitionistic point of view, the concept of natural number is primitive and as such cannot be defined. But this does not settle the question, because intuitionists also hold the concept of proof as primitive. Now natural numbers when they are conceived as iterators in Girard's system **F** are proofs encoded by second-order lambda terms. Indeed, for each natural number n, the numeral $\Lambda\alpha\lambda x : \alpha\lambda f : \alpha \to \alpha.\,(f)^n\, x$ is a proof of $\forall\alpha\,(\alpha, (\alpha \to \alpha) \to \alpha)$ and one could argue that here the type expresses an impredicative property of each of its proofs. It is highly problematic in fact to draw a sharp and objective boundary between the realm of 'good' (predicative) and 'bad' (impredicative) mathematics because we do not have a criterion to distinguish between properties, or characterizations, of already given objects, and real definitions producing their objects since it depends on one's philosophical position about the existence of these objects. If we think of natural numbers as primitive objects, then an expression such as '$\mu x \in \mathbb{N}.x \in A$', with $\varnothing \neq A \subseteq \mathbb{N}$, is a property that characterizes the smallest element n of A. But if we do not think that the natural numbers are given by God or by some specific intuition, then this is a definition of n and as such an impredicative one because we have $\forall y \in A.\ n \leq y$ and $n \leq \mu x \in \mathbb{N}.x \in A$.

Now everyone agrees that some impredicativity in the definition of a natural number as the value of a function f for an argument n is harmless as long as the computation of $f(n)$ is feasible: if natural numbers are not already given, the means by which we can produce them should not be idealized. The concept of feasible function is indeed very difficult to define for different reasons. Let's mention two of them. First, this concept is related to the one of feasible number and it seems that a definition of the latter presupposes one of the former. Second, if we try to bypass this problem and just pick up a non-feasible number represented by some expression, let us say 10^{500}, and decide that this number is the limit of the feasible ones, then some functions that are not computable, according to the usual definition of the computable functions, will be feasibly computable, for instance f defined by $f(n) = 1$ if $n \leq 10^{500}$, otherwise $f(n)$ is undefined. It is generally acknowledged that a rather good definition of the concept of feasible computation is the one of function computable by an algorithm in polynomial time of the length of their input. It is well known that the adequation between this mathematical definition and the empirical concept of feasible computation must be taken with care and that the former is only an approximation of the latter, but I shall not enter into this debate here.

One solution, as given by Bellantoni and Cook, is to distinguish amongst the arguments, inputs, of a function between predicative values and impredicative ones (Bellantoni and Cook 1992). During a computation, it should not be possible to recurse on an impredicative argument, only on a feasible, predicative, one. It must be stressed that the distinction between predicative and impredicative value is not absolute but relative to the computation being performed. Formally, a function is always defined for two sorts of variables, the normal ones, for the predicative inputs, and the safe variables, for the impredicative ones; the two sorts being separated by a semicolon. Schematically we have:

$$f(\overbrace{w_1, \ldots, w_n}^{\text{normal variables}}; \overbrace{v_1, \ldots, v_m}^{\text{safe variables}})$$

Before we give a sketch of this definition, let us mention Leivant's own equivalent definition which rests explicitly on the idea of ramification and could be useful to give some meaning to the distinction between normal and safe variables (Leivant 1993, 1995). According to Leivant, we must think of the natural numbers as given in different levels. Actually only two levels are enough: $\mathbb{N}_0$ for the impredicative numbers and $\mathbb{N}_1$ for the predicative ones. It must be stressed that these two sets are isomorphic to $\mathbb{N}$ and that the levels do not reflect an absolute distinction amongst the natural numbers between feasible and unfeasible ones. The normal variables take their values on $\mathbb{N}_1$ and the safe ones on $\mathbb{N}_0$. The value of a function $f(w; v)$ applied to $n \in \mathbb{N}_1, m \in \mathbb{N}_0$, *i.e.*, $f(n; m)$, depends generally on the impredicative value m and, as such, is in $\mathbb{N}_0$. So we have $f : \mathbb{N}_1, \mathbb{N}_0 \to \mathbb{N}_0$. Keeping that idea in mind, we can give a rough presentation of Bellantoni and Cook's definition of the class **BC** of polytime functions. Lists of variables are noted with bold letters and the arguments of the functions are not unary natural numbers $S \ldots S0$ because this representation makes the length of an input grow as fast as the number itself, but are binary words on an alphabet $\{0,1\}$ with one special symbol for the empty word and two different functions for successor $S_1(w) = w1, S_0(w) = w0$. The other basic functions are predecessor, projections and a conditional. The arguments of basic functions are always safe. Safe composition of functions in **BC** is defined with one main restriction: the substitution of a term for a normal variable is allowed only if all the variables in this term are normal. Safe recursion introduces another restriction: a recursive call to the function f being defined can be made only in a safe place in the step function h in such a way that h cannot recurse on the preceding value of f.

If $g(\boldsymbol{x}; v, \boldsymbol{y})$ and $h_i(w, \boldsymbol{x}; v, \boldsymbol{y})$ for $i \in \{0,1\}$ are in **BC**, then the function $f(w, \boldsymbol{x}; \boldsymbol{y})$ defined by

$$\left\{ \begin{array}{c} f(\varepsilon, \boldsymbol{x}; \boldsymbol{y}) = g(\boldsymbol{x}; v, \boldsymbol{y}) \\ f(S_i(;w), \boldsymbol{x}; \boldsymbol{y}) = h_i(w, \boldsymbol{x}; f(w, \boldsymbol{x}; \boldsymbol{y}), \boldsymbol{y}) \end{array} \right\}$$

is in **BC**.

Bellantoni and Cook give a proof that all functions in **BC** are polytime computable. A warning must be made here: this holds only if a function is applied to arguments completely evaluated, *i.e.*, binary words built with S_i and ε. This means that the computation of an expression like $f(t_1; t_2)$ must begin by the computations of t_1 and t_2 before f can be applied to the values obtained. And they prove also that all polytime computable functions are definable in **BC**. Another warning must be made: this result means that if a function can be computed by a polytime algorithm, then a function extensionally equal, giving the same output for the same input, is in **BC**. But this does not mean that the equations defining the latter are faithful to the algorithm. In fact, the restriction imposed by safe recursion prevents some polytime algorithms, given by sets of equations, to be definable in **BC** (Hofmann 1999b). These remarks also hold for the extension of Bellantoni and Cook's definition to functionals of higher types.

7 Polytime functionals of finite type and a polytime arithmetic

When judged from a 'feasible constructive' point of view, the rules of definition for primitive recursive functions are already too permissive and have to undergo some drastic cure to become safe. But, as restrictive as it may seem, this treatment does not forbid completely definition by recurrence. If we rewrite Bellantoni and Cook's definition as a typed lambda calculus, we see that a weak form of recurrence/induction is still allowed (Hofmann 2000a).

The type of impredicatively binary words is noted **W** and the syntactic restriction on normal variables can be formulated using an operator □ obeying the rules of a modal logic in such a way that □**W** is the type of predicatively defined words.

$$\boldsymbol{R} : (\Box\mathbf{W}, \mathbf{W} \to \mathbf{W}), (\Box\mathbf{W}, \mathbf{W} \to \mathbf{W}), \mathbf{W}, \Box\mathbf{W} \to \mathbf{W}$$

$$\frac{t : \mathbf{W} \qquad \{t_i : \Box\mathbf{W}, \mathbf{W} \to \mathbf{W}\}}{\boldsymbol{R}t_0t_1t : \Box\mathbf{W} \to \mathbf{W}} i \in \{0,1\} \quad \left\{ \begin{array}{c} \boldsymbol{R}t_0t_1t\varepsilon = t \\ \boldsymbol{R}t_0t_1tS_i\,(;w) = t_iw\boldsymbol{R}t_0t_1tw \end{array} \right\}$$

where t_0, t_1, t, w are closed terms in normal form. The immediate extension of this approach to the rules of system **T** fails and one of the reasons that explain this failure is indeed very interesting. If we introduce higher-type recursion by

$$\boldsymbol{R} : (\Box\mathbf{W}, \alpha \to \alpha), (\Box\mathbf{W}, \alpha \to \alpha), \alpha, \Box\mathbf{W} \to \alpha$$

$$\frac{t : \alpha \qquad \{t_i : \Box\mathbf{W}, \alpha \to \alpha\}}{\boldsymbol{R}t_0t_1t : \Box\mathbf{W} \to \alpha} i \in \{0,1\}$$

then with the context $\Gamma = \{w : \Box\mathbf{W}, f : \mathbf{W} \to \mathbf{W}, v : \mathbf{W}\}$ we can type the term $\delta \equiv \lambda w\lambda f\lambda v.ffv$:

$$\cfrac{\cfrac{\Gamma \vdash f : \mathbf{W} \to \mathbf{W} \quad \Gamma \vdash v : \mathbf{W}}{\Gamma \vdash fv : \mathbf{W}} \qquad \Gamma \vdash f : \mathbf{W} \to \mathbf{W}}{\cfrac{\Gamma \vdash ffv : \mathbf{W}}{\vdash \lambda w\lambda f\lambda v.ffv : \Box\mathbf{W}, (\mathbf{W} \to \mathbf{W}) \to (\mathbf{W} \to \mathbf{W})}}$$

From this we get $\boldsymbol{R}\delta\delta S_0 : \Box\mathbf{W}, \mathbf{W} \to \mathbf{W}$ and this term applied to closed terms t, u of types $\Box\mathbf{W}, \mathbf{W}$ gives a closed normal term $s \equiv S_0 \ldots S_0 u$ of length $2^{|t|} + |u|$where $|t|$ and $|u|$ are the length of t, u. Obviously, the function defined by $\boldsymbol{R}\delta\delta S_0$ is not polytime and its exponential growth is due to the fact that the hypothesis $f : \mathbf{W} \to \mathbf{W}$ has been used twice in its definition.

The solution is to use an affine implication ⊸ that allows at most one use of a hypothesis in a definition by recursion unless the hypothesis is freely duplicable. In this case it must have a modal type or a ground type **W** (Hofmann 1998, 1999a, 2000b). See also (Schwichtenberg and Bellantoni 2002; Bellantoni et al. 2000). In such a system, the meaning of □ is two-fold: it is a mark for normal variables of type □**W** and it allows duplication of variables of type $\Box\alpha$ in a term of type by way of a contraction rule

$$\frac{\Gamma, x : \Box\alpha, y : \Box\alpha \vdash t : \beta}{\Gamma, z : \Box\alpha \vdash t\,[z/x, z/y] : \beta}$$

So we have the well-known decomposition due to Girard's linear logic of intuitionistic implication $\alpha \to \beta \equiv \Box\alpha \multimap \beta$ (Girard 1987). And following Hofmann, the rules for definition by recursion are given by:

$$\boldsymbol{R} : \Box (\Box \mathbf{W} \multimap (\alpha \multimap \alpha)), \Box (\Box \mathbf{W} \multimap (\alpha \multimap \alpha)), \alpha, \Box \mathbf{W} \multimap \alpha$$

$$\frac{\Gamma_0 \vdash t_0 : \Box (\Box \mathbf{W} \multimap (\alpha \multimap \alpha)) \quad \Gamma_1 \vdash t_1 : \Box (\Box \mathbf{W} \multimap (\alpha \multimap \alpha)) \quad \Gamma_2 \vdash a : \alpha}{\Gamma_0, \Gamma_1, \Gamma_2 \vdash \boldsymbol{R} t_0 t_1 a : \Box \mathbf{W} \multimap \alpha}$$

where α does not contain $\Box$ and $\Gamma_0, \Gamma_1, \Gamma_2$ are disjoint. Keeping in mind the warnings we made in **5**, the full system of typed modal affine lambda calculus is a polytime analog of Gödel's **T**, since all polytime functions can be defined in it and all the terms typable in it are polytime. The next step was to give an axiomatisation of arithmetic that could be a polytime analog of Heyting's arithmetic.

One of these systems, due to Bellantoni and Hofmann, is epistemic affine arithmetic (Bellantoni and Hofmann 2002; Schwichtenberg 2006). This arithmetic is of course as close as possible to Hofmann's system. It used the same modal operator and affine implication that allow the same decomposition of intuitionistic implication $A \Rightarrow B \equiv \Box A \multimap B$. The axioms and rules are very close to the ones of Peano arithmetic with one main difference: the axioms and rules for implication are the ones for affine implication, but a modal decoration is carefully introduced in such a way that some usual logical and arithmetical rules are reintroduced. For example, contraction $A \multimap (A \wedge A)$ is admissible only when A is $\Box B$ for some B. The induction principle can be seen as the result of the addition of some syntactic sugar to the rule of definition by recurrence in Hofmann's system when the typed terms are erased:

$$\{\Box [\forall v : \Box \mathbf{W} (A(v) \multimap A(S_i v))]\}, A(\varepsilon) \multimap \forall v : \Box \mathbf{W} A(v)$$

where A does not contain $\Box$ and $i \in \{0, 1\}$.

This arithmetic can be given a realisability interpretation by means of the typed modal affine lambda calculus and two main results follow. First, all the polytime functions definable in Bellantoni and Cook's system are provably total in epistemic affine arithmetic:

- If $f(\boldsymbol{w}; \boldsymbol{v})$ is defined by a system E of universally closed equations in **BC**, then $\Box E \multimap \forall \boldsymbol{w} : \Box \mathbf{W} \forall \boldsymbol{w} : \mathbf{W} \exists u : \mathbf{W} f(\boldsymbol{w}, \boldsymbol{v}) = u$ is provable in epistemic affine arithmetic.

Second, there is a kind of converse: all the computable functions provably total in epistemic affine arithmetic are polytime computable by a term typed in Hofmann's system.

- For each function f,
  if $\left\{ \begin{array}{c} f \text{ is defined by a system of universally closed equations } E \\ \Box E \multimap \forall \boldsymbol{w} : \Box \mathbf{W} \forall \boldsymbol{w} : \mathbf{W} \exists u : \mathbf{W} f(\boldsymbol{w}, \boldsymbol{v}) = u \text{ is provable} \end{array} \right\}$ in epistemic affine arithmetic, then f is polytime computable by a term t of type $\overbrace{\Box \mathbf{W}, \ldots, \Box \mathbf{W}}^{m}, \overbrace{\mathbf{W}, \ldots, \mathbf{W}}^{n} \multimap \mathbf{W}$.

Last, but not least, linear, or affine, implication authorizes a rather liberal use of strongly impredicative definitions. Consider the naive comprehension axioms which associate to each formula A the set $\{x/A\}$: $A\,[t/x] \vdash t \in \{x/A\}$ and $t \in \{x/A\} \vdash A\,[t/x]$. If we define the Russell set by $\boldsymbol{r} \equiv \{x/x \notin x\}$ and the Frege theorem by $A \equiv \boldsymbol{r} \in \boldsymbol{r}$, we have $\neg A \vdash A$ and $A \vdash \neg A$. The 'proof' of a contradiction in such a system is easy to give and, as noticed by Prawitz, its last step is a cut step that cannot be eliminated (Prawitz 1965).

$$\dfrac{\dfrac{\dfrac{\dfrac{x:\neg A \vdash x:A \quad y:\neg A \vdash y:\neg A}{x:\neg A, y:\neg A \vdash yx:\bot}}{z:\neg A \vdash zz:\bot}}{\vdash \lambda z.zz:\neg\neg A} \qquad \dfrac{\dfrac{\dfrac{x:A \vdash x:\neg A \quad y:A \vdash y:A}{x:A, y:A \vdash xy\bot}}{z:A \vdash zz:\bot}}{\vdash \lambda z.zz:\neg A}}{\vdash (\lambda z.zz)\,\lambda z.zz:\bot}$$

and the lambda term given by this proof has no normal form. Since, for a given system, the normalization theorem implies its consistency, this comes not as a big surprise: inconsistency implies that some proofs have no normal form. But as noticed later by Grishin, this proof depends on two steps of contraction, which duplicate the same variable z in the two subterms $\lambda z.zz$ of $(\lambda z.zz)\,\lambda z.zz$. In an affine logic, where the use of the contraction rule is restricted, this proof is forbidden and the Russell set may be defined without leading to inconsistency (Grishin 1974, 1981, 1999). Moreover, all functions provably total in such light set theory are polytime computable (Girard 1998) and also (Terui 2004).

8 Conclusion

So, our point of departure in section 5 was the alleged impredicativity of proofs in Heyting arithmetic or definitions of functional of finite type in Gödel **T**. We have seen that this impredicativity is just an unavoidable consequence of a more general phenomenon: the one of incompleteness. Bernays and Nelson's remarks have led us to the impredicativity of the induction principle and Bellantoni, Cook and Leivant have shown us how to restrict this impredicativity to reach the realm of feasible functions. Then, some extensions of this approach to functionals of higher types and proofs have revealed what was the main obstacle to overcome here: the intuitionist notion of implication is not constructive enough to be feasibly admissible and has to undergo a cure of linearity if we want to stay in this realm. And now we are back where we came from: the definition of the Russell set in sectio 2. This definition is indeed impredicative but, as such, does not lead to inconsistency; the use of classical or intuitionistic implication is another necessary ingredient in this undesirable result.

Lorenzen's operative justification of intuitionistic logic

Peter Schroeder-Heister

1 Introduction

With his *Introduction to Operative Logic and Mathematics*,[1] which first appeared in 1955, Paul Lorenzen became an exponent of an approach to the foundations of logic and mathematics, which is both formalistic and intuitionistic in spirit. *Formalistic* because its basis is the purely syntactical handling of symbols — or 'figures', as Lorenzen preferred to say — , and *intuitionistic* because the insight into the validity of admissibility statements justifies the laws of logic. It is also intuitionistic with respect to its result, as Heyting's formalism of intuitionistic logic is legitimatised this way. Along with taking formal calculi as its basis, the notion of an inductive definition becomes fundamental. Together with a theory of abstraction and the idea of transfinitely iterating inductive definitions, Lorenzen devised a novel foundation for mathematics, many aspects of which still deserve attention. When he wrote his *Operative Logic*, neither a full-fledged theory of inductive definitions nor a proof-theoretic semantics for logical constants was available. A decade later, Lorenzen's inversion principle was used and extended by Prawitz (Prawitz 1965) in his theory of natural deduction, and in the 1970s, the idea of inversion was used for a logical semantics in terms of proofs by Dummett, Martin-Löf, Prawitz and others. Another aspect which makes Lorenzen's theory interesting from a modern point of view, is that in his protologic he anticipated certain views of rule-based reasoning and free equality which much later became central to the theory of resolution and logic programming. Lorenzen's inversion principle in its general form — that is, not in its restricted application in logic — is closely related to principles of definitional reflection in logic programming (Schroeder-Heister 2007). The idea that logical introduction rules are

[1] 'Einführung in die operative Logik und Mathematik' (Lorenzen 1955), henceforth *OL*. A major earlier paper propagating the operative approach is (Lorenzen 1950). For a biography of Lorenzen see (Thiel 1996), for a bibliography of his work see (Lorenzen 1996).

but a special case of rules defining (atomic) propositions was used in a different form in Martin-Löf's theory of iterated inductive definitions (Martin-Löf 1971). Thus there are various interesting points from which we might take a closer look at *Operative Logic*.

Unfortunately, Lorenzen had already lost interest in the subject when issues such as proof-theoretic semantics and resolution-based reasoning became more popular in logic. Within the narrower realm of logic, he had already given up the operative approach in favour of dialogical logic by the end of the 1950s, perhaps motivated by discussions with Tarski in 1957–58 at the Institute for Advanced Study in Princeton (Lorenz 2001, p.257). From the early 1960s on, he broadened his approach from the philosophy of logic and mathematics to geometry, the philosophy of science, and ethics, partly in collaboration with Wilhelm Kamlah, after having taken the philosophical chair in Erlangen in 1962. He became the head of a philosophical school in Germany which, as it favoured a normative and anti-empiricist foundation of science, was opposed to the analytic philosophy of science represented in Germany by Wolfgang Stegmüller, who was strongly influenced by Carnap, and for whom the rational reconstruction of science was the primary goal. Stegmüller once even criticized the Erlangen school for developing a 'metascience of science fiction' (Stegmüller 1973, p.26). The fact that Lorenzen found philosophical allies even within hermeneutics and Habermas' critical theory of society further contributed to the split in German philosophy of science. By the end of the 1970s, when the proof-theoretic foundation of logic had become a topic within the philosophy of language, Lorenzen had begun to make completely different issues his main agenda, including the philosophy of politics. Correspondingly, he did not take notice of the rising interest in rule-based theories and proof-theoretic approaches within the realm of computer science.

In this paper we concentrate on the logical aspects of the *Operative Logic*, i.e., on Lorenzen's attempt to base intuitionistic logic on admissibility principles, leaving aside his general theory of rule-based reasoning and his foundation of mathematics. In particular, we compare his approach to the theories of Dummett and Prawitz, who consider introduction rules as the defining properties of logical constants and derive valid logical laws by using ideas closely related to Lorenzen's inversion principle. In the first section, we discuss those aspects of Lorenzen's protologic, i.e., of his general theory of calculi and admissible rules which are most relevant to the foundation of deductive logic. In the second section, we reconstruct his theory of iterated implications and meta-calculi, and extract from his remarks a validity concept for sentences alias higher-level rules. We show that Lorenzen's theory of meta-calculi can be given a rendering that makes it a sensible theory of implication, and of logical constants in general. In the third section we compare (our reconstruction of) Lorenzen's approach with proof-theoretic semantics in the tradition of Dummett and Prawitz. Although, as shown in the appendix, Lorenzen's and Prawitz's validity concepts can be translated into one another, Lorenzen's theory has certain deficiencies when looked at from an epistemological point of view. Unlike Dummett and Prawitz, Lorenzen does not pay sufficient attention to the manifestation aspect of meaning, which for constants such as conjunction and disjunction is embodied in their introduction rules, but is absent in the crucial case of implication, for which in Lorenzen's framework no structural introduction rule is defined.

2 Basic protological concepts

The first part of the *Operative Logic* is entitled 'Protologic' as it is conceptually prior to, and more general than logic. It is a theory of formal systems (calculi) and develops general principles for establishing the admissibility of inference rules in such systems. Logic in the narrower sense is a particular application of admissibility theory dealing with logical constants and with the iteration of admissibility.

2.1 Calculi

Lorenzen starts with elementary calculi (*OL*, §1), which permit one to generate words (strings of signs) over an arbitrary (finite) alphabet. The elements of the alphabet are called *atoms*, the words are called *sentences* ('Aussagen'). A calculus K is specified by giving certain initial formulas ('Anfänge') A and rules $A_1, \ldots, A_n \to A$, where an initial formula is the limiting case of a rule (for $n = 0$). In the following, we also write '$\to A$' for an initial formula. We also speak of *primitive* inference rules and (in the premiss-free case) of *axioms*, if we want to make clear that they are the rules on which inferences in K are based.[2] Formulas are words composed of atoms and variables. The variables occurring in formulas are either *eigenvariables* (*OL*, p.16) or *object variables* (*OL*, p.26ff). Eigenvariables of K can be substituted only with sentences derived in K itself, whereas object variables are to be substituted with sentences of a different calculus. Obviously, if a rule contains eigenvariables, the substitution instances of a rule are defined simultaneously with the derivations in K. For example, if

$$K_1 \begin{cases} \to | & (R_1) \\ x \to x| & (R_2) \end{cases}$$

is a calculus with the eigenvariable x, then in the derivation

$$\dfrac{\dfrac{\dfrac{}{|}\,R_1}{||}\,R_2}{|||}\,R_2$$

[2] Lorenzen avoids the term 'axiom' because of traditional connotations of something being 'evident', see *OL*, p.7.

the second line is obtained from the first one by using the substitution instance $| \to ||$ of R_2, which is only defined after the first line has been derived (and thus $|$ becomes substitutable for x). Similarly, the step from the second to the third line is based on the substitution instance $|| \to |||$ of R_3 which relies on $||$ having already been derived and is thus substitutible for x. Lorenzen needs this approach based on eigenvariables as (for him) the specification of any domain proceeds via calculi and rules, which means that one has always to begin with a calculus for which no external substitution range for object variables is given. When object variables standing for sentences derivable in a calculus K_0 are used in K, it is presupposed, of course, that each atom of K_0 is an atom of K. As a matter of terminology, we call a formula or rule *closed*, if it contains no variable, and *open* otherwise.

It is obvious that this approach is closely related to the formalist ideas of Post (Post 1921) and (Curry 1952). As formulas are just strings of atoms and variables, Lorenzen starts with an arbitrary word structure rather than the functor-argument structure common in logic. This makes his approach particularly general. The most appropriate more modern point of view would be to look at his calculi as inductive definitions (see (Aczel 1977)). Concerning the foundations of mathematics, Lorenzen's *Operative Logic* can be viewed as a programme to embed mathematics into a theory of inductive definitions.

2.2 Admissibility

The main concept on which logic is based is *admissibility* (*OL*, §2). This notion, which was coined by Lorenzen, has entered logical textbooks as a standard proof-theoretic concept. A rule R is called *admissible* in a calculus K, if its addition to the primitive rules of K — resulting in an extended calculus $K + R$ — does not enlarge the set of derivable sentences. If $\vdash_K A$ denotes the derivability of A in K, then R is admissible in K if

$$\vdash_{K+R} A \text{ implies } \vdash_K A \tag{1}$$

for every sentence A. However, as for Lorenzen admissibility is the central concept on which the notion of implication is based, he cannot give the implication (1) as its definition. Rather, admissibility is given an operative meaning by reference to the notion of an *elimination procedure* (*OL*, §3). R is admissible in K, if every application of R *can be eliminated* from every derivation in $K + R$. The implicational relation between existential statements expressed in (1) is reduced to the insight that a certain procedure reduces any given derivation in $K + R$ in such a way that the resulting derivation does no longer use R. According to Lorenzen, this is the sort of insight (evidence) on which constructive logic and mathematics is based. It goes beyond the insight that something is derivable in K, but is still something which has a 'definite' meaning. It is the admission of this sort of evidence which makes Lorenzen an intuitionist rather than a formalist. In its various variants, intuitionism has been based on the notion of a procedure or construction. The interpretation of implication in the BHK-semantics of the logical signs is based on it, as are more formalized notions like realizability. In proposing the admissibility of rules as based on elimination procedures, Lorenzen adds a new perspective to this concept.

Admissibility is to be distinguished from the derivability from assumptions. Lorenzen puts strong emphasis on this (obvious) fact, as this makes admissibility theory an enterprise, which is not merely formalistic in the sense of verifying formal derivations. For example, given the calculus

$$K_2 \begin{cases} & \rightarrow n < n| \\ m < n & \rightarrow m < n| \end{cases}$$

with m, n being object variables for lists of strokes (i.e., sentences derivable in K_1), then the rule $m < n \rightarrow m \mid < n \mid$ is admissible but not derivable (Lorenzen 1959, p.170). Here, for sentences $A_1, \ldots, A_n, A$, the derivability of a rule $A_1, \ldots, A_n \rightarrow A$ in a calculus K means that $A_1, \ldots, A_n \vdash_K A$, i.e., A is derivable from $A_1, \ldots, A_n$ as assumptions. An open rule is derivable, if all its closed substitution instances are derivable. For example, $||||<|||$ is not derivable in K_2 from the assumption $|||<||$. The standard proof-theoretic example of a rule admissible but not derivable is that of the cut rule in the first-order sequent calculus.

Lorenzen establishes five basic protological principles for the generation of admissibility statements. Due to lack of space, we only sketch the three of them which are immediately relevant to logic. The two further principles — the *induction principle* and the *equality principle* — , though highly relevant, have their main application in the foundation of mathematics. For rules $R_1, \ldots, R_n, R$ we follow Lorenzen in using the notation $R_1, \ldots, R_n \vdash_K R$ to express that R can be eliminated from derivations in K by using $R_1, \ldots, R_n$ as additional rules of inference. In particular, this means that R is admissible, if $R_1, \ldots, R_n$ are admissible.[3]

The deduction principle

Though the admissibility of a rule does not imply its derivability, the converse, which is called the *deduction principle* (*OL*, p.26), is true. If for every closed substitution instance $A'_1, \ldots, A'_n \rightarrow A'$ of a rule $A_1, \ldots, A_n \rightarrow A$ we have $A'_1, \ldots, A'_n \vdash_K A'$, then $A_1, \ldots, A_n \rightarrow A$ is an admissible rule of K. As an elimination procedure, we just have to replace every application of $A_1, \ldots, A_n \rightarrow A$ by the derivation of A' from $A'_1, \ldots, A'_n$ for the appropriate instance $A'_1, \ldots, A'_n \rightarrow A'$ of $A_1, \ldots, A_n \rightarrow A$.

[3] Lorenzen also characterizes $R_1, \ldots, R_n \vdash_K R$ as the admissibility of R in the calculus K extended with $R_1, \ldots, R_n$ as additional primitive rules (*OL*, pp.24,40). This is, however, a non-monotonic notion, and not what Lorenzen intends.

It should be noted that the deduction principle is not a formal introduction rule for an implication statement. Lorenzen does not consider Gentzen-style natural deduction rules like

$$\frac{\begin{array}{c}[A]\\ \vdots \\ B\end{array}}{A \to B}$$

where an assumption can be *discharged.* Rather, establishing the admissibility of the rule $A_1, \ldots, A_n \to A$ is a metalinguistic procedure which involves a step of reflection. It can be turned into a formal assertion of a sentence only in a so-called 'meta-calculus' (see Section 3.1).

The inversion principle

The *inversion principle* (*OL*, p.30ff) is the most relevant principle for logic. It governs the relationship between introduction and elimination rules for logical constants (in the *Operative Logic* only used for conjunction, disjunction and existential quantification, see Section 3.3). The term 'inversion principle' was later adopted by Prawitz (Prawitz 1965) in his theory of natural deduction (see Section 4). For simplicity, we just consider the propositional case without object variables. Suppose the calculus K is given by the following rules, where for each i $(1 \leq i \leq n)$, Γ_i stands for a list of sentences:

$$K \left\{ \begin{array}{l} \Gamma_1 \to A \\ \quad \vdots \\ \Gamma_n \to A \end{array} \right.$$

Then the following holds for any sentence C:

$$\text{(IP)} \quad \Gamma_1 \to C, \ \ldots \ , \Gamma_n \to C \vdash_K A \to C$$

The validity of this principle is easily seen: An application of $A \to C$ leads from A to C, where A can only be derived in K according to one of the 'introduction rules' $\Gamma_i \to A$ for A. So we have the following situation:

$$\cfrac{\cfrac{\Gamma_i}{A}\, \Gamma_i \to A}{C}\, A \to C$$

Obviously, this can be replaced with

$$\frac{\Gamma_i}{C}\, \Gamma_i \to C$$

thus eliminating the application of $A \to C$. Since $\Gamma_i \to C$ is available for every i, this elimination procedure works in any case no matter from which Γ_i the sentence A was introduced in the first place.

Reading the calculus K as an inductive definition, this principle can be regarded as an interpretation of the extremal clause sometimes used to finish an inductive definition ('nothing else is A'): Everything that follows from each defining condition of A, follows from A itself. Note, however, that this principle is not laid down as a primitive rule interpreting the inductive definition, but is proved by means of an elimination procedure. This elimination procedure, in eliminating an application of $A \to C$, removes A as a 'maximum sentence' which is first introduced by means of $\Gamma_i \to A$ and then eliminated by means of $A \to C$. In the case of $n = 1$, this principle amounts to the 'direct' inversion of a rule. Suppose

$$A_1, \ldots, A_n \to A$$

is the only primitive rule of K having A as its head. Then from

$$A_1, \ldots, A_n \to C \vdash_K A \to C$$

we obtain the admissibility of the rules:

$$\begin{array}{l} A \to A_1 \\ \quad\vdots \\ A \to A_n \end{array}$$

by choosing C in turn to be $A_1, \ldots, A_n$ and using the fact that $A_1, \ldots, A_n \to A_i$ is trivially admissible.

The inversion principle is not of much use when restricted to sentences. Even for the treatment of propositional logic, variables for sentences are necessary. Unfortunately, Lorenzen's formulation for the case with object variables is mistaken, and his own (in the second edition of *Operative Logic*) and others' (e.g., (Lorenz 1980)) attempts at repairing it failed as well. (Hermes 1959) gives a correct version which suffers, however, from certain deficiencies of the general framework, which are due to the fact that unification theory and resolution-based reasoning was not known yet. (Schroeder-Heister 2007) presents an examination of these approaches and a comparison with more modern principles, in particular with the principle of *definitional reflection* developed by Hallnäs and Schroeder-Heister (Hallnäs and Schroeder-Heister 1990, 1991) in the context of logic programming and later used as a general definitional device (Schroeder-Heister 1993, 2008a; Hallnäs 1991, 2006).

The underivability principle

The *underivability principle* (*OL*, p.36ff) says that, if the sentence A is not derivable in K, then any rule $A \to B$ is admissible in K. The reason for this result is simply that, if the premiss A of $A \to B$ is underivable in K, then $A \to B$ cannot be applied to a derivable sentence of K, so $A \to B$ can be added to the primitive rules of K without generating new derivable sentences.

Lorenzen does not consider the underivability principle as a limiting case of the inversion principle, for the case in which there is no defining rule for A available in K, but considers it a principle of its own. Obviously, the underivability principle can be used as the basis of the intuitionistic absurdity rule and the intuitionistic concept of negation, although it is not entirely clear whether this is intended by Lorenzen (see Section 3.4). In any case, this principle is needed at other places in his protologic, e.g., within his theory of inequality (*OL*, p.89ff).

3 Lorenzen's foundation of intuitionistic logic

An investigation of Lorenzen's foundation of mathematics would now study his theory of abstraction in connection with his theory of inductive definitions, in particular his idea of transfinitely iterating inductive definitions through the handling of language levels ('Sprachschichten'). Here, we discuss his justification of deductive logic based on the admissibility principles mentioned, in particular on the inversion principle, and relate it to proof-theoretic semantics in the sense of Dummett and Prawitz. For simplicity, we focus on the propositional case. Universal quantification essentially behaves like implication, and existential quantification like conjunction or disjunction.

Lorenzen divides his discussion into that of consequence logic (*OL*, §6), which contains his treatment of implication (and universal quantification), of conjunction and disjunction (and existential quantification) (§7), and of negation (§8). The separation of the treatment of implication from that of conjunction and disjunction indicates that there is a fundamental split between two sorts of connectives. He even speaks of implication (and universal quantification and negation) as 'improper' logical constants as opposed to the 'proper' ones of conjunction and disjunction (and existential quantification) (*OL*, p.172).

As a matter of terminology, we speak of 'implication' throughout when $\rightarrow$ is used as a sentence-forming operator (rather than a rule arrow), whereas Lorenzen in the *Operative Logic* does not use any specific terminology, and in his later writings propagates the term 'subjunction' (and 'adjunction' for 'disjunction'). Furthermore, we use a tree-like notation of derivations in contradistinction to Lorenzen's linear one. Finally, we only consider closed rules, i.e., rules without variables, and identify open rules with the sets of their instances, thus admitting that the set of primitive rules of a calculus is infinite. When we speak of the finite specification of calculi, this is to be understood in the sense of a specification by means of finitely many rule *schemata*.

3.1 Consequence logic and the theory of meta-calculi

Lorenzen's theory of implication is based on the idea that an implicational sentence $A \rightarrow B$ expresses the admissibility of the rule $A \rightarrow B$, so the assertion of an implication is justified if this implication, when read as a rule, is admissible. In this sense an implication expresses a meta-statement about a calculus. This has a clear meaning as long as there is no iteration of the implication sign. In order to cope with iterated implications, Lorenzen develops the idea of finitely iterated meta-calculi. In the following, we try to built a coherent theory based on his scarce remarks on the specification of meta-calculi. We take the freedom to deviate from his notation and terminology, and to extend it whenever this seems appropriate to us.

From the beginning, we have to consider the possibility that calculi are extended with inference rules for additional constants, and that such extensions are available at any meta-level. We speak of a *definitional extension* and of constants *defined* in this extension, as it is the intention of such an extension to fix their meaning. In practice, these constants will be conjunction and disjunction [and existential quantification] (see Section 3.3). A definitional extension consists of formation rules $\mathbb{F}$ and primitive inference rules $\mathbb{D}$, where $\mathbb{F}^n$ and $\mathbb{D}^n$ are the formation rules and primitive inference rules, respectively, for level n. We assume that at each level n, the formation rules $\mathbb{F}^n$ determine how to construct a level-n-sentence from atomic level-n-sentences, where in $\mathbb{F}^n$ these atomic level-n-sentences are not specified, but are just represented by variables. Starting from a basic calculus K and a definitional extension with formation rules $\mathbb{F}$, we define level-n-sentences and level-n-rules, and the sentences and rules of each meta-calculus M^nK as follows.

Level-n-sentences and level-n-rules:

(i) *Every sentence of K is a level-0-rule.*
(ii) *Every level-n-rule is an atomic level-n-sentence.*
(iii) *Every atomic level-n-sentence is a level-n-sentence.*
(iv) *Every expression generated from atomic level-n-sentences by means of the formation rules $\mathbb{F}^n$ is a level-n-sentence.*
(v) *If $A_1, \ldots, A_n, A$ are level-n-sentences, then $A_1, \ldots, A_n \to A$ is a level-$(n+1)$-rule.*

Sentences and rules of M^nK:

(i) *Every level-$(n+1)$-rule is a rule of M^nK.*
(ii) *Every level-n-sentence is a sentence of M^nK.*

We write MK for M^1K. However, in our terminology, M^0K is not just K, but the definitional extension of K (in practice: K extended with conjunction and disjunction [and existential quantification]). Although, syntactically, level-n-rules and level-n-sentences are the same, they are associated with different systems. As it transforms level-n-sentences into a level-n-sentence, a level-$(n+1)$-rule is a rule of M^nK, whereas a level-$(n+1)$-sentence is a sentence of $M^{n+1}K$. So the basic idea is that every rule over sentences of M^nK is at the same time an atomic sentence of $M^{n+1}K$. To avoid parentheses, Lorenzen uses dots to indicate precedence. For example,

$$\rightarrow A;\ B_1, B_2 \rightarrow C \mathrel{\dot{\rightarrow}} E \rightarrow F$$

is a level-2-rule (and thus a rule of MK), and at the same time an atomic level-2-sentence (and thus a sentence of M^2K).

$$A \rightarrow B_1 \wedge B_2 \mathrel{\dot{\rightarrow}} \rightarrow C \mathbin{\ddot{;}} \rightarrow E_1 \mathbin{\dot{\wedge}} D_1, D_2 \rightarrow E_2 \mathrel{\dot{\rightarrow}} F \rightarrow G \mathrel{\ddot{\rightarrow}} \rightarrow H \mathrel{\dot{\rightarrow}} I \rightarrow J$$

is a level-3-rule (and thus a rule of M^2K), and at the same time an atomic level-3-sentence (and thus a sentence of M^3K), where it is supposed that conjunctions are available (at least) at levels 0 and 1. Lorenzen also uses a notation with unequal levels such as $A \mathrel{\dot{\rightarrow}} B \rightarrow C$. This is just shorthand for $\rightarrow A \mathrel{\dot{\rightarrow}} B \rightarrow C$, and analogously for similar notations.

It remains to specify the primitive inference rules of M^nK with respect to a definitional extension $\mathbb{D}$. The primitive inference rules $\mathbb{D}^n$ available at level n govern the expressions formed by means of $\mathbb{F}^n$ and are called *definitional rules*. These rules transfer level-n-sentences into level-n-sentences, i.e., they have the form of level-$(n+1)$-rules. In practice $\mathbb{D}^n$ will contain the introduction rules for conjunction and disjunction [and existential quantification] (see Section 3.3). As the rules of $\mathbb{D}^n$ are considered definitional, they are expected to be conservative, i.e., they must not extend the class of primitive sentences which are derivable in M^nK without using $\mathbb{D}^n$.[4] In practice, this requirement is trivially fulfilled, as the introduction rules considered only generate non-atomic level-$(n+1)$-sentences, i.e., sentences containing the constant introduced.

Concerning atomic sentences, the primitive inference rules of M^0K are just the primitive rules of K. For $M^{n+1}K$, Lorenzen extends the usual notion of a finitely specified calculus in favour of a system of which it is only required that every atomic sentence derivable in $M^{n+1}K$, when read as a level-$(n+1)$-rule, be admissible in M^nK (*OL*, p.42). This means that we may add an admissible rule of M^0K (i.e., K extended with rules $\mathbb{D}^0$ for defined constants) as an axiom to MK, an admissible rule of MK as an axiom to M^2K etc. At first glance this looks like a set-theoretic closure notion of admissibility, at least for the implicational fragment without definitional extension, in which all sentences are atomic and thus have the form of rules: The rules admissible in K form the class of axioms of MK; the rules which do not properly extend this class, form the class of axioms of M^2K; and so on. However, even at higher levels, and even in the purely implicational case, Lorenzen wants to keep the idea that admissibility is tied to an operative elimination procedure. Therefore, $M^{n+1}K$ is to contain not only axioms but also proper inference rules (in addition to the definitional rules of $\mathbb{D}^{n+1}$), namely certain rules of which it has been shown beforehand that they lead from admissible level-n-rules to admissible level-n-rules. Once we have shown this for a particular rule (by means of an elimination procedure!), we may, so to speak, store this result

[4] Instead of the now common term 'conservative', Lorenzen uses the term 'relatively admissible'.

as a primitive inference rule and establish further sentences of $M^{n+1}K$ by a formal derivation. For example, the admissibility of

$$\text{(I)}\quad A_1, \ldots, A_k \rightarrow A_i \quad (1 \leq i \leq k)$$

is obvious for any calculus K, so all rules of this form can be taken as axioms of MK and therefore of any $M^{n+1}K$. Furthermore, since for Γ as $A_1, \ldots, A_k$,

$$(\Gamma \rightarrow B_1), (\Gamma \rightarrow B_m), (B_1, \ldots, B_m \rightarrow A) \vdash_K \Gamma \rightarrow A$$

holds for any K, we may use the rule

$$\text{(II)}\quad \Gamma \rightarrow B_1;\ \ldots\ ; \Gamma \rightarrow B_m; B_1, \ldots, B_m \rightarrow A \ \dot{\rightarrow}\ \Gamma \rightarrow A$$

as a primitive inference rule of MK and therefore of any $M^{n+1}K$. Rules like (I) and (II), which are available for any calculus K, are called *universally admissible* (*OL*, pp.42–44). Now we can, for example, establish the admissibility of the level-2-rule

$$A \rightarrow B_1; B_1, B_2 \rightarrow B \ \dot{\rightarrow}\ A, B_2 \rightarrow B$$

by formally deriving $A, B_2 \rightarrow B$ from $A \rightarrow B_1$ and $B_1, B_2 \rightarrow B$ in MK using (I) and (II) (*OL*, p.43).

Hence, even if the meta-calculi are not fully formalized by specifying finite sets of axioms and rules, certain formal rules are available even in the implicational fragment, which allow for formal derivations. Meta-calculi are *open systems* in the sense that inference principles once justified may be used as formal rules.

So far, as formal rules of the meta-calculi, we have discussed definitional rules $\mathbb{D}^n$ and universally admissible inference principles like (I) or (II). Further rules result from the admissibility principles, especially from the inversion principle. For example, if, for sentences A, B and C, K contains the rules

$$\begin{cases} A \rightarrow C \\ B \rightarrow C \end{cases}$$

then the inversion principle says that

$$A \rightarrow D, B \rightarrow D \vdash_K C \rightarrow D$$

which justifies using the rule

$$A \rightarrow D; B \rightarrow D \ \dot{\rightarrow}\ C \rightarrow D$$

as a primitive rule of MK (and at the same time as an axiom of M^2K). In general, each admissibility principle discussed in Section 2.2 generates formal inferences in appropriate meta-calculi.

Viewed in this way, in a meta-calculus a variety of formal rules is available. Therefore, even though it is an open system, the protological admissibility principles are applicable to a meta-calculus directly. For example, the *deduction principle* holds without restriction: If in M^nK we can formally prove A from $A_1, \ldots, A_k$ (for level-n-rules alias M^nK-sentences $A_1, \ldots, A_k, A$), we have established the sentence $A_1, \ldots, A_k \rightarrow A$ as an axiom of $M^{n+1}K$. Even the *inversion principle*, though in general dependent on a fixed set of rules and therefore non-monotonic, can be used at the meta-level in certain cases, namely if we know that, in the open system, there cannot be further introduction rules for a sentence C beyond those of a certain form. This can be achieved by definition, e.g., when C is a conjunction or disjunction, so that the elimination inferences inverting the introduction rules for conjunction and disjunction are available at any level.

Furthermore, as a universal principle to generate formal level-$(n+1)$-rules from admissible level-n-rules, we may use that $\rightarrow A_1, \ldots, \rightarrow A_k \dot{\rightarrow} \rightarrow A$ is a derivable rule of $M^{n+1}K$ if $A_1, \ldots, A_k \rightarrow A$ is admissible in M^nK. This results from

$$\frac{\rightarrow A_1, \ldots, \rightarrow A_k \quad A_1, \ldots, A_k \rightarrow A}{\rightarrow A},$$

which is an application of the universally admissible rule (II), by removing the level-1-axiom (= admissible rule of K) $A_1, \ldots, A_k \rightarrow A$.[5] We call this the *principle of lifting*. In particular, since $\rightarrow A$ is admissible in M^nK iff A is derivable in M^nK, $\rightarrow A$ is derivable in $M^{n+1}K$ iff A is derivable in M^nK. When we use the downwards direction of this law, we speak of *downlifting*.

Thus Lorenzen's much criticized doctrine of meta-calculi is perfectly sensible when a meta-calculus $M^{n+1}K$ is viewed as an open system, which not only collects admissible level-n-rules as axioms, but in which general admissibility principles can be used as formal rules of inference. Thus derivations in the open system M^nK with respect to a definitional extension $\mathbb{D}$ proceed by the following primitive inference rules.

Primitive rules of M^nK:

(i) Every rule admissible in M^nK can be chosen as an axiom of $M^{n+1}K$.
(ii) Every rule that leads from rules admissible in M^nK to rules admissible in M^nK can be chosen as a primitive inference rule of $M^{n+1}K$.
(iii) Every rule of $\mathbb{D}^n$ is a primitive rule of M^nK.

If clause (iii) is lacking, i.e., in the purely implicational case, then in every $M^{n+1}K$, all admissible rules of $M^{n+1}K$ may be considered to be derivable. However, as we have an open system, it is not intended that these inference rules are available as primitive rules. As primitive rules one would rather choose those rules, of which it has already been demonstrated that they lead

[5] This follows by using (II). Lorenzen only remarks that this rule is admissible in $M^{n+1}K$ (Lorenzen 1955, 44).

from admissible rules to admissible rules of M^nK. Derivability and admissibility are distinct anyway at the ground level (i.e., in K), and also when a definitional extension is present. For example, when the definitional extension contains introduction rules for conjunction, the elimination rules for conjunction are available as primitive rules only at the next higher level (see Section 3.3).

Besides (I) and (II), Lorenzen considers the universally admissible rules of importation and exportation to be fundamental (*OL*, p.46).

(III)

$$A_1;\ldots;A_m \dot{\rightarrow} A_{m+1},\ldots,A_n \rightarrow A \ \ddot{\rightarrow}\ A_1;\ldots;A_{m-1} \dot{\rightarrow} A_m,\ldots,A_n \rightarrow A$$

$$A_1;\ldots;A_{m-1} \dot{\rightarrow} A_m,\ldots,A_n \rightarrow A \ \ddot{\rightarrow}\ A_1;\ldots;A_m \dot{\rightarrow} A_{m+1},\ldots,A_n \rightarrow A$$

Based on our discussion, the following notions of a valid sentence and a valid consequence can be distilled out of Lorenzen's theory (Lorenzen himself does not use a concept of validity):

Lorenzen's concept of validity:

(i) *A level-n-sentence A is valid with respect to K and $\mathbb{D}$, if it is derivable in M^nK.*
(ii) *A level-n-sentence A is a valid consequence of level-n-sentences $A_1,\ldots,A_n$ with respect to K and $\mathbb{D}$, if $A_1,\ldots,A_n \rightarrow A$ is a valid level-$(n+1)$-sentence, i.e., a sentence derivable in $M^{n+1}K$ and therefore a rule admissible in M^nK.*
(iii) *A level-$(n+1)$-sentence is valid with respect to $\mathbb{D}$, if for every K it is valid with respect to K and $\mathbb{D}$, i.e., if A is universally admissible with respect to $\mathbb{D}$ as a level-$(n+1)$-rule.*
(iv) *A level-$(n+1)$-sentence is logically valid, if it is valid with respect to $\mathbb{D}$, where $\mathbb{D}$ contains exactly the introduction rules for the logical operators (see Section 3.3).*
(v) *A level-$(n+1)$-sentence is a valid sentence of positive implicational logic, if it is valid with respect to $\mathbb{D} = \emptyset$.*

This is not a formalistic definition of validity, as it is not reducible to derivability in a fully formalized system, but relates to open meta-calculi. It is not an exclusively semantical definition either, as it is based on fully formalized derivations on the ground level and rule-based definitional extensions at all levels. This position in between a formalistic way of thinking and an exclusively semantical approach puts Lorenzen's theory into a close relationship to proof-theoretic semantics in the Dummett-Prawitz-style (see Section 4).

Lorenzen finally arrives at the positive implicational calculus governed by the rules I–III, which is the intuitionistically accepted part of implicational logic. However, he cannot prove that there are no valid sentences beyond those derivable in positive implicational logic. He proves some sort of closure property of the system which is, however, different from semantic completeness.[6] So Lorenzen is faced with a completeness conjecture he cannot prove, quite similar to the one later made by Prawitz (Prawitz 1973, p.246). It should also be noted that Lorenzen does not fully distinguish between implication as a connective and consequence as a relation between sentences. Consequence is an admissibility statement which at the same time can be asserted as a formal sentence. This is due to the fact that by means of a reflection step, any admissibility statement can be formally expressed at the next higher level as a sentence. This reflection step is not an inference step within a single system, which, as we shall argue, is a fundamental weakness of Lorenzen's approach (see Section 4.2).

3.2 Digression on Lorenzen and natural deduction

Since according to Lorenzen, establishing an implication always means a jump to a higher level, we are not allowed to pass *from* A to B in K whenever $A \to B$ is admissible. For example, if K is extended with $A, B \to A \wedge B$ as an introduction rule for conjunction, the elimination $A \wedge B \to A$ is admissible in this extension; however, as a primitive inference rule, we are only entitled to use it at the meta-level. This means that

$$\frac{A \wedge B}{A}$$

would not be a valid step at the ground level. However,

$$\frac{\to A \wedge B}{\to A}$$

would be a correct step at the meta-level due to lifting. This shows that there are derivable rules of the form $\to A_1; \ldots; \to A_k \dot{\to} \to A$ at the meta-level, whose object-level counterparts $A_1, \ldots, A_k \to A$ are only admissible. The meta-level complements the object level by making admissible rules available (in 'lifted' shape) as formal derivation rules.

The fact that for every rule $A_1, \ldots, A_k \to A$ derivable (and therefore admissible) in K the rule $\to A_1; \ldots; \to A_k \dot{\to} \to A$ can be assumed as a primitive rule of MK, shows that any derivation $\mathcal{D}$ of A from $A_1, \ldots, A_k$ in K can be represented as a derivation of $\to A$ from $\to A_1, \ldots, \to A_k$ in MK. This means that even reasoning in natural deduction can be formally reconstructed within Lorenzen's framework, even though originally no formal inference rules are available that discharge assumptions. The introduction rule for implication can be made understandable in the meta-calculus by means of

[6] Although Lorenzen claims that this property is near to completeness (*OL*, p.49).

$$\frac{\begin{array}{c}[\to A]\\ \vdots\\ \to B\end{array}}{A \to B} \tag{2}$$

Since $A \to B$ is admissible in K whenever $\to B$ is derivable from $\to A$ in MK, we can safely add (2) to MK and create a natural-deduction-style $\to$-introduction rule.

Using (2), and denoting by $\overset{m}{\to} A$ the m-fold iteration of the arrow, we can even associate with any natural-deduction derivation of A from $A_1, \ldots, A_k$ a derivation of $\overset{m}{\to} A$ from $\overset{m_1}{\to} A_1, \ldots, \overset{m_k}{\to} A_k$ in $M^n K$ for certain $m_1, \ldots, m_k, m, n$, where the iterations of the arrow not only equalize the (possibly different) levels of $A_1, \ldots, A_k, A$, but where the numbers $m_1, \ldots, m_k, m$ also depend on the levels of the sentences internally reached within the natural-deduction derivation.[7]

Therefore, it would be wrong to claim that Lorenzen's way of reasoning is entirely foreign to the idea of natural deduction. The basic difference is that in Lorenzen $A \to B$ means that the derivability of B follows from that of A (i.e., $\to B$ from $\to A$), not that B itself follows from A itself. This means that the semantical content of $A \to B$ is properly kept. Even according to Dummett and Prawitz, the semantical meaning of an open derivation of B from A consists in that a (valid) closed derivation of B can be generated given a (valid) closed derivation of A.

3.3 Conjunction and disjunction

Lorenzen's treatment of conjunction and disjunction (and existential quantification) proceeds by adding introduction rules for these connectives to a calculus K and showing that the corresponding elimination inferences are admissible. Assuming that $\wedge$ and $\vee$ do not occur as atoms of K, then given the additional rules

$$\begin{cases} A, B \to A \wedge B \\ \quad A \to A \vee B \\ \quad B \to A \vee B \end{cases}$$

the rules

$$A \wedge B \to A$$
$$A \wedge B \to B$$

and

[7] In fact, we might consider a system which switches between the different levels. In that case the arrow would have to be treated as a kind of modal operator. We would always be allowed to pass from $\to A$ to A, but from A to $\to A$ only if all assumptions are prefixed with $\to$.

$$A \to C; B \to C \;\dot{\to}\; A \vee B \to C$$

are admissible. They are both straightforward applications of the inversion principle. For $\vee$ this is obvious, for $\wedge$ this becomes obvious if we infer $\wedge$-elimination from

$$A, B \to C \;\dot{\to}\; A \wedge B \to C$$

using the universally admissible rules $A, B \to A$ and $A, B \to B$, which are instances of (I). Combining these rules with the laws of consequence logic, the system of positive intuitionistic propositional logic is obtained.

It should be noted that these rules are available at any level. For example, if A, B, C, D are sentences of K, we can infer $A \to B \mathbin{\dot{\wedge}} C \to D$ from $A \to B$ and $C \to D$ in MK, etc. This means in particular, as the conjunction sign is a fresh symbol at every level, that by means of inversion we obtain the elimination rule at this level. This is an example of an application of the inversion principle within an open (i.e., not fully formalized) system.

Besides conjunction and disjunction rules intrinsic to every level, there are lifted translations of the introduction and elimination rules of the lower levels available. In MK, from $\to A$ and $\to B$ we can infer both $\to A \mathbin{\dot{\wedge}} \to B$ (by conjunction introduction in MK) and $\to A \wedge B$ (by conjunction introduction in K, lifted to MK). Conversely, from $\to A \wedge B$ we can infer both $\to A$ and $\to B$ (by lifting $\wedge$-elimination, which is admissible in M^0K, to MK) and thus $\to A \mathbin{\dot{\wedge}} \to B$. From $\to A \mathbin{\dot{\wedge}} \to B$ we do not obtain $\to A \wedge B$, as $\dot{\wedge}$-elimination is only admissible in MK but not derivable.

Lorenzen was the first to formulate an inversion principle in sufficient detail and to apply it to infer elimination rules from introduction rules. This is one of the lasting achievements of his *Operative Logic*.

3.4 Negation

As one would expect, negation is introduced via absurdity. Define a *falsum* sentence $\bot$[8] as a sentence such that $\bot \to A$ is admissible for every sentence A of K. Then the negation $\neg A$ of A is defined as $A \to \bot$. This is unambiguous as for any two *falsum* sentences $\bot$, $\bot'$, the rules $\bot \to \bot'$ and $\bot' \to \bot$ are admissible. Together with the laws of implication and the admissibility of *ex falso quodlibet*, we obtain intuitionistic propositional logic in the usual way.

Since a *falsum* sentence is by definition a sentence for which *ex falso quodlibet* holds, the underivability principle does not formally enter the theory of negation. In fact, Lorenzen presupposes that a *falsum* sentence is always available in the calculus considered, or that otherwise we could find a surrogate *ad hoc*.[9] However, if we have to construct a *falsum* sentence on the basis of the

[8] Lorenzen uses the notation $\bigwedge$ (1st ed. of *OL*) or $\curlywedge$ (2nd ed.), but does not give it a name.

[9] By the latter he means the conjunction of the propositions under consideration (see *OL*, pp.63,75).

given calculus K, we would enlarge K by a fresh constant $\bot$ without an introduction rule. When applied to the empty set of introduction rules for $\bot$, the underivability principle establishes $\bot$ as a *falsum* sentence. By presupposing a *falsum* sentence to be given in the underlying calculus K, Lorenzen actually argues in favour of minimal logic with *ex falso quodlibet* being a property of K and not of its extension with logical constants.[10] This is even more surprizing as Lorenzen uses the underivability principle anyway, namely in his theory of equality (OL, p.89ff). As an explanation we can only speculate that for Lorenzen a symbol which does not occur in any rule does not have an operative meaning.

3.5 Classical logic

To interpret classical logic within his framework, Lorenzen uses an adaptation of Gödel's double-negation translation devised by E. Wette (OL, pp.80–84). He shows that, when starting from a basic calculus K_0, all of whose sentences A are *stable* in that the double-negation law $\neg\neg A \rightarrow A$ is valid, i.e., $A \rightarrow \bot \overset{.}{\rightarrow} \bot \overset{..}{\rightarrow} A$ is admissible in M^2K, then the following holds. Suppose K is an extension of K_0 such that the heads of all primitive rules of K, which are not at the same time primitive rules of K_0, do not belong to K_0. (This trivially implies that K is a conservative extension of K_0.) Then every sentence of K_0 derived in K by means of classical principles (e.g., by assuming the double-negation law $\neg\neg B \rightarrow B$ or the *tertium non datur* $B \vee \neg B$ for arbitrary sentences B of K) is already derivable in K_0, i.e., the classical extension of K is a conservative extension of K_0.

4 Lorenzen's approach and Dummett-Prawitz-style proof-theoretic semantics

Lorenzen's idea to justify certain inference rules by admissibility proofs, and in particular his inversion principle as a central tool, is closely related to Dummett's and Prawitz's approaches towards proof-theoretic semantics. Technically, these approaches are based on proof-theoretic ideas and results originally developed by Prawitz (Prawitz 1965) in the context of natural deduction and later extended and used by Tait (Tait 1967), Martin-Löf (Martin-Löf 1971), Girard (Girard 1971), and Prawitz (Prawitz 1971) in the context of logic and type theories, in particular in proofs of strong normalization. Philosophically, they received much underpinning within Dummett's 'verificationist' theory

[10] A careful reading of §8 of *OL* supports this view. There Lorenzen speaks of the 'complete' reduction of negation to implication (p.75). On the other hand, in §5 he speaks of the underivability principle as the justification of *ex falso quodlibet* (p.37). So there is some tension between different views.

of meaning.[11] They can be traced back to Gentzen's remark that the introduction inferences in natural deduction may be viewed as definitions, and the elimination inferences as a sort of consequences thereof (Gentzen 1934, p.189). Prawitz (Prawitz 1965) used Lorenzen's term 'inversion principle' to describe the relationship between introduction and elimination rules in general, not only for the cases of conjunction and disjunction [and existential quantification], in which Lorenzen used the inversion principle. In the following we compare Prawitz's notion of validity with Lorenzen's notions of admissibility and derivability in meta-calculi.[12] We do not include a discussion of Martin-Löf's approach, which is in the same spirit, since due to his distinction between proofs as acts and proofs as objects he adds an additional layer of understanding which is beyond the scope of our paper.[13]

4.1 Prawitz's definition of validity

In the following we give a simplified version of Prawitz's definition of validity for positive propositional logic (i.e., logic based on implication, conjunction and disjunction). Prawitz defined the validity of derivations with respect to atomic systems. We may identify such an atomic system with a calculus in Lorenzen's sense, although we are aware that Prawitz from the very beginning uses a term-formula structure rather than an arbitrary word structure. This difference is not really essential when it comes to logic and logical constants. Here and in the following we speak of 'derivations' throughout, even though these 'derivations' are intended to carry a meaning and to be epistemologically significant. The term 'proof' will only be used in the context of metalinguistic considerations. As the concept of a *valid derivation* presupposes a neutral idea of *derivation* which includes invalid ones, by a derivation we understand a tree-like structure composed of arbitrary inference steps which, in the context of Prawitz-validity, may discharge assumptions. The introduction rules for the logical connectives are considered to be valid by definition, i.e., introduction rules are 'self-justifying' in Dummett's terminology (Dummett 1991a, p.251). Furthermore, it is assumed that certain reduction procedures are available which reduce derivations to other derivations of the same end formula (from

[11] (Dummett 1991b) is a convenient reference as it contains the essence of Dummett's logical semantics in monograph form. Dummett's basic papers on the philosophical basis of intuitionistic logic date back to the 1970s (see (Schroeder-Heister 2005), which contains an extensive bibliography).

[12] For a detailed discussion of Prawitz's notion of validity and the notion of proof-theoretic semantics in general see (Schroeder-Heister 2006) and (Kahle and Schroeder-Heister 2006). As original papers by Prawitz, see, e.g., (Prawitz 1973, 1974, 1978, 1985, 2006).

[13] It might be mentioned that Per-Martin Löf's definition of *computability* is closely related to Prawitz's definition discussed below. However, although it was conceived earlier, Prawitz's definition is not merely an adaptation of it. For a discussion of this issue see (Schroeder-Heister 2006).

the same or fewer open assumptions). In a more elaborated form these procedures form a parameter of the definition of validity. Here we assume for simplicity that they are implicitly fixed. Typically, the standard reductions used in normalization proofs are assumed to be given, but other reductions can be thought of as well. A derivation using an introduction rule in the last step is called *canonical.*

Prawitz's definition of validity

(i) *Every closed derivation in the atomic system is valid.*
(ii) *A closed canonical derivation is valid, if its immediate subderivations are valid.*
(iii) *A closed non-canonical derivation is valid, if it reduces to a valid canonical derivation.*
(iv) *An open derivation* $\begin{matrix} A_1 \ldots A_n \\ \mathcal{D} \\ A \end{matrix}$ *is valid, if for every list of closed valid derivations* $\begin{matrix} \mathcal{D}_i \\ A_i \end{matrix}$ $(1 \leq i \leq n)$*, the derivation* $\begin{matrix} \mathcal{D}_1 \quad \mathcal{D}_n \\ A_1 \ldots A_n \\ \mathcal{D} \\ A \end{matrix}$ *is valid.*

The following three basic features of this definition deserve to be pointed out.

(i) The definition applies to derivations, not just to rules. The validity of a rule $\dfrac{A_1 \ \ldots \ A_n}{A}$ results as a limiting case, if that rule is considered a one-step derivation with the open assumptions $A_1, \ldots, A_n$ and the end formula A. This rule is valid, if for all valid closed derivations $\begin{matrix} \mathcal{D}_i \\ A_i \end{matrix}$ $(1 \leq i \leq n)$, the closed derivation $\begin{matrix} \mathcal{D}_1 \quad \mathcal{D}_n \\ A_1 \ldots A_n \\ A \end{matrix}$ is valid. So the validity of a rule is explained in terms of the validity of derivations, rather than the other way round, according to which a derivation would be valid if it consists of applications of valid rules. In that whole derivations and not rules are the primary focus, Prawitz's definition of validity is *global* rather than *local.*

(ii) Closed derivations are semantically prior to open ones. A valid open derivation of A from $A_1, \ldots, A_n$ is semantically interpreted as something that yields a valid closed derivation of A when supplemented with valid closed derivations of $A_1, \ldots, A_n$. This means that open assumptions in natural deduction derivations are interpreted as placeholders for closed derivations. If we call closed derivations *categorical* and open derivations *hypothetical,* we may speak of the priority of the categorical over the hypothetical.

(iii) The definition of validity rests on the distinction between canonical and non-canonical derivations. Canonical derivations are valid if they end with an introduction step applied to valid premiss derivations, whereas non-canonical derivations are valid, if they *reduce* to valid canonical derivations. For this *canonical vs. non-canonical* distinction it is crucial that a natural-deduction model of derivation is chosen, as only there is a genuine introduction rule for implication available.

Comparing Prawitz's approach with Lorenzen's, there are both similarities and differences.

Ad (i) Lorenzen's approach is definitely not global. The validity predicate assigned to him in Section 3.1 means the validity of a rule in the sense of its admissibility — or the derivability of this rule read as a sentence in a meta-calculus — rather than the validity of a derivation as a whole. When we described meta-calculi as open systems, we nevertheless understood them as something in which to proceed from axioms according to certain rules. So validity as derivability in a meta-calculus ultimately requires the justification of certain inference rules. However, this difference is not so big as it might appear at first glance. If we consider a rule a one-step derivation, then the Prawitz-validity of this derivation means essentially the same as admissibility: We have to demonstrate that it does not extend the class of valid closed derivations, i.e., that it is admissible as a rule with respect to this class.[14] Conversely, we could extend Lorenzen's definition of the admissibility of a rule to that of the admissibility of a derivation by considering any derivation $\begin{array}{c} A_1, \ldots, A_n \\ \mathcal{D} \\ A \end{array}$ of A from $A_1, \ldots, A_n$ to be admissible, if the rule $A_1, \ldots, A_n \to A$ is admissible. In practice, when applying Prawitz's definition of validity, the relevant reduction cases are the standard elimination inferences, which are one-step derivations, and most of the more complicated derivations would, even in Prawitz conception, be handled as composed of single steps which are justified separately. That a Prawitz-valid derivation establishes a Lorenzen-valid sentence or rule, and vice versa, is shown in the appendix.

Ad (ii) In Lorenzen, too, closed derivations are primary as compared to open derivations. That $A \to B$ is admissible in K means that any application of $A \to B$ can be eliminated from a closed derivation in $K + \{A \to B\}$. Otherwise the crucial distinction between admissible and derivable inference rules would break down. In this respect both Lorenzen and Prawitz unequivocally belong to the intuitionistic tradition of defining a construction in the categorical sense first and

[14] This is made clear in (Moriconi and Tesconi 2008). See also (Schroeder-Heister 2008b).

then define a hypothetical construction by the transmission of the categorical concept. This is, incidentally, also the position of classical semantics, in which the categorical concept of truth (in a structure) is defined first, and the hypothetical concept of consequence is defined by reference to the transmission of truth. An alternative position, which takes the concept of consequence as primary, would go beyond both Lorenzen's and Prawitz's conception of validity, and also beyond the classical view of truth as the basis of consequence. Such ideas are developed in (Hallnäs 1991, 2006; Schroeder-Heister 1993, 2004, 2008a; Schroeder-Heister and Contu 2005).

Ad (iii) As the *canonical vs. non-canonical* distinction constitutes the fundamental difference between the two approaches, this point is dealt with in a separate section.

4.2 Implication and the *canonical vs. non-canonical* distinction

The distinction between canonical and non-canonical derivations, which figures prominently in Dummett-Prawitz-style semantics, has an analogue in Lorenzen's conception only in the cases of conjunction and disjunction [and existential quantification], but not in the case of implication [and negation and universal quantification]. For conjunction and disjunction proper introduction rules are given, and corresponding elimination inferences are established by applying the inversion principle. The split between conjunction/disjunction and implication is for Lorenzen one between proper and improper connectives and constitutes a difference between proper calculi and improper meta-calculi. For Lorenzen, establishing an implication $A \to B$ means either a reflection step resulting from establishing the admissibility of the rule $A \to B$, or by inferring $A \to B$ in the meta-calculus using inference principles available there (such as the universally admissible rules like (I)–(III), or the inversion principle IP), which are themselves based on such reflection steps. Semantically, Prawitz's definition of validity is in the same spirit, as the validity of an open derivation is defined in a way corresponding to an admissibility statement. However, for Prawitz-validity it is crucial that for implication there is a *formal* canonical introduction step available in analogy to conjunction and disjunction, namely the formal step from an open derivation $\begin{matrix} A \\ \mathcal{D} \\ B \end{matrix}$ to the assertion of $A \to B$. This is a *syntactical* step within an extension of the basic system, which is possible in a natural-deduction style framework permitting the discharging of assumptions. The syntactic application of the $\to$-introduction rule is different from any semantic validity (or admissibility) claim. From the standpoint of Prawitz's conception, Lorenzen's introduction of implication as an assertion in the meta-calculus passes directly to the semantical level. From the standpoint of Lorenzen's conception, Prawitz adds an intermediate syntactical step when defining the validity of a derivation of an implication.

There are at least three points Prawitz can claim in his favour.

(i) *Uniformity.* Implication is treated on par with conjunction and disjunction. Correspondingly, the inversion principle can be applied to generate the elimination inferences for all logical constants, whereas in Lorenzen it applies only to conjunction and disjunction [and existential quantification], which are the distinguished constants with explicit introduction rules.
Commentary. Uniformity is not a value in itself. Even according to Prawitz, the distinguished character of implication remains present in the way open derivations are handled. The natural-deduction-style introduction rule for $\rightarrow$ forces him to combine the validity of open derivations with that of closed derivations in a joint inductive definition, and to deal with the introduction rules at the same level as the rules justified as valid. Lorenzen's idea to interpret implication by a jump to the meta-calculus separates these levels. For him the rule $A \wedge B \rightarrow A$, as a rule justified by the inversion principle, would reside as an axiom only in the meta-calculus, unlike the rule $A, B \rightarrow A \wedge B$ which 'defines' conjunction and is therefore already available in the basic definitional extension of K.

(ii) *Direct vs. indirect derivations.* A basic feature of Dummett-Prawitz-style proof-theoretic semantics is the distinction between direct and indirect closed derivations. A direct derivation arrives at its conclusion by means of a 'self-justifying' inference step and is thus called canonical, whereas an indirect derivation is one that reduces (= can be transferred) to such a form and is called non-canonical. The distinction between canonical and non-canonical derivations corresponds to the one between direct and indirect knowledge or evidence. This crucial epistemological distinction is lacking in Lorenzen in the case of implication.
Commentary. For implication a distinction between direct and indirect derivations can even be drawn within Lorenzen's conception in a natural way. An implication sentence $A \rightarrow B$ can be asserted (i) as an axiom of MK based on a demonstration of the admissibility in K of the rule $A \rightarrow B$, or (ii) by means of a formal derivation in MK using axioms and rules already shown to be valid. (i) is a direct derivation of $A \rightarrow B$ as it relies on a demonstration of admissibility, whereas (ii) is an indirect derivation as it relies on other admissibility demonstrations carried out beforehand, e.g., on proofs that certain rules are universally admissible. An indirect derivation according to (ii) has the feature that it reduces to a direct one according to (i) by spelling out the admissibility proofs relied on, thus arriving at an admissibility proof of $A \rightarrow B$. So the postulate that there be a *direct vs. indirect* distinction does not speak unequivocally in favour of Prawitz.

(iii) *Behaviourally significant knowledge of meaning.* Proof-theoretic semantics in the Dummett-Prawitz tradition is an epistemological approach to semantics, according to which we must distinguish between mere knowledge of meaning and inferences correctly drawn on the basis of this knowledge. This knowledge of meaning must manifest itself in observable behaviour. This manifestation is the correct application of introduction inferences. It presupposes that introduction inferences are solely structurally and thus syntactically specified, and therefore decidable patterns. This condition is not met in Lorenzen, as even the 'direct' establishment of an implication such as $A \wedge B \rightarrow A$ is not a syntactical step but rests on the insight into a (elimination) procedure.

Commentary. This is the decisive argument against Lorenzen's treatment of implication. As Lorenzen requires even of a most direct derivation of an implication sentence that an admissibility proof be carried out, there is no behaviourally significant way of showing that a reasoner knows the meaning of implication. The proper epistemological approach demands that a certain part of the semantically correct usage be behaviourally significant. Syntactically specified introduction rules satisfy exactly this requirement. In a Gentzen-style introduction rule for implication the conclusion $A \rightarrow B$ captures a structural feature of the deduction, namely that there is a derivation of B from A, still independent of whether this hypothetical derivation itself is valid, quite analogously to the fact that, e.g., the introduction rule for a conjunction $A \wedge B$ captures the structural feature that there is a pair of derivations, one of A and one of B, still independent of the validity of the derivations of A and B themselves. Even though in Dummett-Prawitz-semantics the validity of a hypothetical derivation is defined by reference to a transformation procedure, i.e., in a way closely related to admissibility, the canonical derivation of the corresponding implication is based on a separate syntactical step from a hypothetical derivation to a derivation of an implication. For example, though the validity of the one-step derivation

$$\frac{A \wedge B}{A}$$

rests on a transformation, establishing the implication $A \wedge B \rightarrow A$ rests on an additional canonical introduction step:

$$\cfrac{\cfrac{[A \wedge B]^{(1)}}{A}}{A \rightarrow B}\,(1)$$

Dummett-Prawitz-semantics is based on a subtle analysis of the epistemological content of a derivation which goes beyond the standard intuitionistic way embodied in BHK-semantics and in Lorenzen-style semantics. Like BHK-semantics, Lorenzen's semantics of implication is an attempt to give an epistemological semantics of implication without paying sufficient respect to the way knowledge of meaning manifests itself. Although Lorenzen puts much emphasis on the operative handling of syntactic figures, the manifestation-free intuitionistic epistemology is not overcome by Lorenzen.

4.3 Lorenzen's way out: Dialogue games

Lorenzen's turn away from operative logic to dialogical logic and game-theoretical semantics can be seen as a way out of these deficiencies. As dialogue semantics distinguishes between matters of games and matters of strategies, we can assign the knowledge of meaning to the game level and the validity of consequence statements to the strategy level. Someone who masters the syntactically specified game rules for the logical connectives would be considered to know their meaning, whereas the availability of a strategy, which needs the handling of a constructive procedure (thus resembling admissibility), would be responsible for the logical facts which turn out to hold. There are some indications in Lorenzen's texts supporting this view.[15]

If this is correct, then both Dummett-Prawitz-style proof-theoretic semantics and Lorenzen-style dialogical semantics would be ways out of the shortcomings of the operative approach (and of the intuitionistic BHK-approach as well). Both Dummett-Prawitz semantics and dialogical semantics would be epistemological approaches doing justice to the requirement of a behavioural manifestation of meaning.

5 Appendix: Prawitz-validity vs. Lorenzen-validity

First we define a translation which assigns to each sentence and rule of positive propositional logic a level-n-sentence in Lorenzen's sense, which essentially means equalizing unequal implication levels. Let $\stackrel{m}{\to}$ denote the m-fold iteration of the arrow, and let $\ell(A) := n$, if A is a Lorenzen-style level-n-sentence (i.e., $\ell(A)$ is the level of A).

Translation of sentences of positive propositional logic into Lorenzen-style sentences:

$A' := A$, *if* A *is atomic.*
$(A \,\Box\, B)' := \stackrel{m}{\to} A' \,\Box \stackrel{n}{\to} B'$,
where $m = max(0, \ell(B') - \ell(A')), n = max(0, \ell(A') - \ell(B'))$, *if* $\Box$ *is* $\wedge$ *or* $\vee$.
$(A_1, \ldots, A_k \to A_{k+1})' := \stackrel{m_1}{\to} A'_1, \ldots, \stackrel{m_k}{\to} A'_k \to \stackrel{m_{k+1}}{\to} A'_{k+1}$
where $m_i = max(0, max(\ell(A'_1), \ldots, \ell(A'_{k+1})) - \ell(A'_i))$.

Conversely, to each level-n-sentence in Lorenzen's sense we assign a sentence of positive propositional logic, which essentially means replacing the comma by conjunction and removing rules without premises:

[15] See (Lorenzen 1961) and (Lorenz 2001).

Translation of Lorenzen-style sentences into sentences of positive propositional logic:

$A^* := A$, *if* A *is atomic.*

$(A \,\square\, B)^* := A^* \,\square\, B^*$, *if* $\square$ *is* $\wedge$ *or* $\vee$.
$(A_1, \dots, A_k \rightarrow A)^* := A_1^* \wedge \ \dots \ \wedge A_k^* \ \rightarrow \ A^*$.
$(\rightarrow A)^* := A^*$.

We assume that the notions of validity are relativized to a fixed atomic calculus K and that the definitional extension $\mathbb{D}$ considered for Lorenzen-validity contains the introduction rules for conjunction and disjunction at every level, and nothing else. Then we can show the following:

Theorem. For all sentences of positive propositional logic $A_1, \dots, A_k, A$ *the following holds:*

(i) There is an open Prawitz-valid derivation $\begin{matrix} A_1, \dots, A_k \\ \mathcal{D} \\ A \end{matrix}$ *iff* $(A_1, \dots, A_k \rightarrow A)'$ *is Lorenzen-valid.*

(ii) There is a closed Prawitz-valid derivation $\begin{matrix} \mathcal{D} \\ A \end{matrix}$ *iff* A' *is Lorenzen-valid.*

Proof. We first remark that the second assertion can be viewed as a special case of the first one for empty $k = 0$, since $\rightarrow A$ is a Lorenzen-valid level-$(n+1)$-sentence [i.e., a rule admissible in M^nK] iff A is a Lorenzen-valid level-n-sentence [i.e., a sentence derivable in M^nK]. We prove (i) simultaneously by induction on the complexity of $A_1, \dots, A_k, A$. For a closed derivation of an atomic sentence, which is a derivation in K, there is nothing to show.

Suppose A is a conjunction $A_1 \wedge A_2$, and a closed Prawitz-valid derivation of A is given. Then according to the definition of Prawitz-validity, this derivation reduces to one of A using $\wedge$-introduction in the last step. We apply the induction hypothesis to its premiss derivations and obtain Lorenzen-style derivations of A_1' and A_2' in appropriate meta-calculi. If they are of unequal levels, we use lifting to obtain derivations of $\overset{m_1}{\rightarrow} A_1'$ and $\overset{m_2}{\rightarrow} A_2'$ in the same meta-calculus, from which by $\wedge$-introduction in this meta-calculus we obtain A'. Conversely, if $A_1' \wedge A_2'$ is Lorenzen-valid, then there is a Lorenzen-style derivation in the appropriate meta-calculus using the $\wedge$-introduction rule in the last step. From its premiss derivations we obtain derivations of A_1' and A_2' (using downlifting if necessary). Application of the induction hypothesis gives us closed Prawitz-valid derivations of A_1 and A_2, and thus a closed Prawitz-valid derivation of A by means of $\wedge$-introduction. — If A is a disjunction, we argue analogously.

Suppose A is an implication $A_1 \rightarrow A_2$, and a closed derivation of A is given. Then this derivation is — or reduces to — a derivation using $\rightarrow$-introduction in the last step, with a Prawitz-valid derivation $\begin{matrix} A_1 \\ \mathcal{D} \\ A_2 \end{matrix}$ as its premiss. By induction hypothesis we know that $(A_1 \to A_2)'$ is Lorenzen-valid. Conversely, suppose that $(A_1 \to A_2)'$ is a Lorenzen-valid level-$(n+1)$-sentence, i.e., that it is derivable in $M^{n+1}K$. We know that $(A_1 \to A_2)'$ has the form $\overset{m_1}{\to} A_1' \to \overset{m_2}{\to} A_2'$ for appropriate m_1 and m_2, i.e., that it is an atomic level-$(n+1)$-sentence, i.e., that it has the form of a level-$(n+1)$-rule. Therefore, its derivability in $M^{n+1}K$ implies that it is (as a rule) admissible in M^nK, i.e., we can transform every derivation (in M^nK) of $\overset{m_1}{\to} A_1'$ into one of $\overset{m_2}{\to} A_2'$. Since by induction hypothesis we know that, if there is a closed valid Prawitz-derivation $\begin{matrix} \mathcal{D}' \\ A_1 \end{matrix}$, then A_1' is Lorenzen-valid, i.e., derivable in the appropriate metacalculus, we also know (by lifting) that, if there is a closed valid Prawitz-derivation $\begin{matrix} \mathcal{D}' \\ A_1 \end{matrix}$, $\overset{m_1}{\to} A_1'$ is derivable in M^nK. Transforming this derivation into one of $\overset{m_2}{\to} A_2'$, and observing that this yields a derivation of A_2' in the appropriate meta-calculus (using downlifting, if necessary), by induction hypothesis we obtain a closed Prawitz-valid derivation of A_2. This yields a closed Prawitz-valid derivation of the form

$$\dfrac{\dfrac{[A_1]}{A_2}}{A_1 \to A_2}$$

where the one-step derivation from A_1 to A_2 is justified by the procedure just described.

Now suppose there is an open Prawitz-valid derivation

$$\begin{matrix} A_1, \ldots, A_k \\ \mathcal{D} \\ A_{k+1} \end{matrix}$$

To show that $(A_1, \ldots, A_k \to A_{k+1})'$ is Lorenzen-valid, we suppose that $(A_1, \ldots, A_k \to A_{k+1})'$ is a level-$(n+1)$-sentence and that derivations in M^nK of $\overset{m_1}{\to} A_1', \ldots, \overset{m_k}{\to} A_k'$ are given. By downlifting (if necessary), we know that there are derivations of $A_1', \ldots, A_k'$ in appropriate meta-calculi, which means that $A_1', \ldots, A_k'$ are Lorenzen-valid. By induction hypothesis there are closed Prawitz-valid derivations $\begin{matrix} \mathcal{D}_1 \\ A_1 \end{matrix}, \ldots, \begin{matrix} \mathcal{D}_k \\ A_k \end{matrix}$. Combining them with the open Prawitz-valid derivation which is given, yields a closed Prawitz-valid derivation of A_{k+1}, from which by induction hypothesis we conclude that A_{k+1}' is Lorenzen-valid. By applying lifting (if necessary), we obtain that $\overset{m_{k+1}}{\to} A_{k+1}'$ is Lorenzen-valid, i.e., derivable in M^nK. Conversely, suppose that $(A_1, \ldots, A_k \to A_{k+1})'$ is Lorenzen-valid. Then we know that, for some n, $(A_1, \ldots, A_k \to A_{k+1})'$ is an an atomic level-$(n+1)$-sentence of the form $\overset{m_1}{\to} A_1', \ldots, \overset{m_k}{\to} A_k' \to \overset{m_{k+1}}{\to} A_{k+1}'$. Since it is derivable in $M^{n+1}K$, it is (as a rule) admissible in $M^n K$, i.e., we can transform every list of derivations in $M^n K$ of the level-n-sentences $\stackrel{m_1}{\rightarrow} A'_1, \ldots, \stackrel{m_k}{\rightarrow} A'_k$ into a derivation in $M^n K$ of $\stackrel{m_{k+1}}{\rightarrow} A'_{k+1}$. By applying the induction hypothesis (and lifting, if necessary), we obtain a procedure which transforms every list of closed Prawitz-valid derivations $\genfrac{}{}{0pt}{}{\mathcal{D}_1}{A_1}, \ldots, \genfrac{}{}{0pt}{}{\mathcal{D}_k}{A_k}$ into a closed Prawitz-valid derivation of A_{k+1}. Thus the open one-step derivation $\dfrac{A_1 \ \ldots \ A_k}{A_{k+1}}$ is Prawitz-valid.

Theorem. A level-n-sentence A is Lorenzen-valid iff for the sentence A^* there is a closed Prawitz-valid derivation.

Proof. This follows from the previous result by observing that for any level-n-sentence A, we have that $(A^*)'$ is the same as A modulo lifting, i.e., modulo replacing subsentences B with a lifted form $\stackrel{k}{\rightarrow} B$.

Part III

Mathematical perspectives

The Hilbert-Brouwer controversy resolved?

Per Martin-Löf

Since this lecture is called the Beth lecture, it is natural to ask oneself: what connection is there between Beth and intuitionism? and, to my mind, the most obvious answer is: the Beth models for intuitionistic propositional and predicate logic. I would like to remind you of a certain picture that we all use when thinking in terms of possible worlds, namely the picture in which we are always at a certain stage, considering a number of possible future alternatives, and then one of those alternatives is realized and we get to a new stage, at which we are faced with some other alternatives, one of those alternatives materializes, and then the whole process continues in the same way. That means that, at each moment, we are at the end of a certain path, which I will draw by a thick line, symbolizing that which has already materialized, and we are considering all possible future alternatives, one of those alternatives materializes, as a result of which the thick path gets extended, and then the pattern repeats itself:

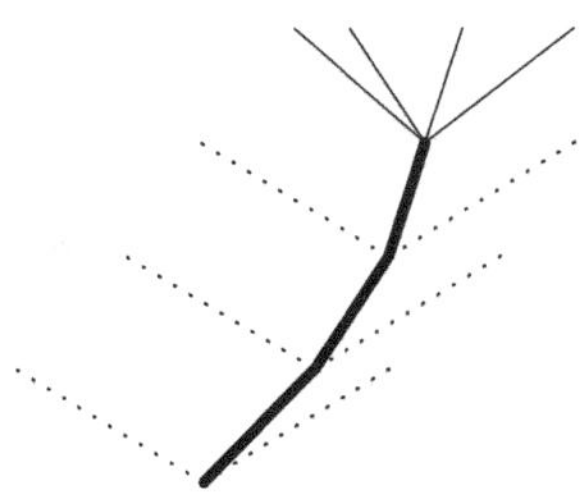

We all think in terms of this picture: it underlies possible world semantics, and it is natural to ask: where does it really come from? I mean: who had this picture for the first time? Most of us have probably met it in connection with Kripke semantics (Kripke 1965), but actually Kripke semantics was predated by Beth semantics by a couple of years: Beth's papers were published in 1956 and 1959 (Beth 1956a, 1959), and Beth in turn drew on Brouwer here, because he took the nodes, or stages, to form a spread in Brouwer's sense

(Brouwer 1918B). So the picture that we standardly draw in possible world semantics is actually the picture of a spread, and maybe this is the most important, or at least most widespread, use that the notion of spread has been put to. Moreover, the thick path, which gets extended as soon as one of the future alternatives with which we are faced materializes, is precisely a choice sequence in Brouwer's sense, of which only a finite initial segment has been determined at each stage of its development. So there is this connection between Beth and intuitionism, which makes it not quite inappropriate that this lecture is called the Beth lecture, although this is a meeting celebrating the centenary of Brouwer's intuitionism.

Now the other thing one may wonder about is: why is this meeting held in France and not in the Netherlands as one would expect? And, again, this is not so strange as one might think, and several of the previous speakers have already addressed this question. It has to do with the fact that, as far as mathematics is concerned, Brouwer's intellectual lineage, which he referred to in many places, derives from Poincaré and proceeds via Borel to himself, and anyone who is familiar with Brouwer's work knows that this is not just a rationalization but an adequate account of how he was formed intellectually. There is a touching statement of this that he himself made in 1949, the last time, I suppose, that he was in Paris on a professional visit. Thanks to Van Dalen, this has now become available to anybody who wants to read it, in the second volume of his Brouwer biography, and so I will quote this passage here:

> Addressing you here, I also feel an emotion of recollection. It is now exactly forty years ago, it was in December 1909 and January 1910, that I inhaled for the first time the scientific air of Paris. And I recall vividly and with a deep gratitude the encouraging benevolence, with which I, a young beginner, was received by the grand old men of that time, whose names had been tied to the grandiose evolution of the mathematical sciences which was taking place, and to whom, through their published courses and their monographs my generation owes the greater part of its knowledge and a considerable part of its inspiration. It was the five classical memoirs on analysis situs of Poincaré and the reflections on dimensionality of Poincaré that were to open the perspectives in which my thoughts on topology developed themselves, and it was the studies of Poincaré and Borel, in particular the manner in which the latter had introduced the notion of measure, that made me glimpse the direction in which I had to seek the primordial origin of mathematical thought. It is a very special sensation to find of these great old men of the past, after a so considerable lapse, some who survive full of vigour (van Dalen 2005, pp.848–849, translation modified).[1]

[1] I am grateful to Dirk van Dalen for providing me with the French original of this passage of Brouwer's Paris address.

And I imagine that those who survived full of vigour in front of him were Borel, in the first place, and presumably also Hadamard, who was older than Borel but nevertheless full of vigour in 1949. So this was Brouwer's final tribute to his French ancestors, Poincaré and Borel.

Another thing that has been alluded to by several of the previous speakers is the fact that Brouwer did not introduce the terms intuitionism and formalism until 1911 in his review of a book on elementary mathematics by Mannoury (Brouwer 1911A). There the terms appear for the first time, but what he had in mind then with intuitionism was the French school, Poincaré and Borel in the first place, and when, in his inaugural lecture of 1912, he wanted to refer to his own viewpoint, he called that neo-intuitionism (Brouwer 1912A). In fact, it was not until the twenties that he took the shrewd step of calling his own conception intuitionism tout court, qualifying his predecessors instead as either the pre-intuitionists or the old-intuitionists, by Heyting also called the semi-intuitionists. So this much about the appropriateness of holding this meeting in France.

Now to the proper subject of my talk, the Hilbert-Brouwer controversy. What that controversy was about I take it that we all know, but I would like to recall the crucial steps in this controversy, which started exactly in 1900 and stretched over a period of almost thirty years, until 1928. That year I take to be the natural stopping point as far as the involvement of Hilbert and Brouwer themselves is concerned. In 1900, it was in the *Mathematical Problems* paper (Hilbert 1900b) that Hilbert stated for the first time what we now call the second Hilbert problem, the problem of the consistency of the arithmetical axioms, as he called it, which means that really this story begins with the previous paper of his *Über den Zahlbegriff* (Hilbert 1900a), published in the same year, in which he introduced for the first time the modern axiom system for the real numbers. Now it is in his detailed statement of the second problem that he conceives of the idea that one should prove the consistency of the axioms, not in the natural way by constructing a model, as one had always done in geometry, but by a study of the proof figures, viewed as purely combinatorial sign configurations. So this is the birthplace of Hilbert's program, and then he gave a somewhat more detailed sketch of it in his Heidelberg lecture of 1904, *Über die Grundlagen der Logik und der Arithmetik* (Hilbert 1905).

This was all that was available to Brouwer when he wrote his thesis in 1907, which we are here to commemorate (Brouwer 1907), and it is remarkable that he was able to give such a penetrating criticism of Hilbert's program at this early stage, when one remembers that the only items that were available to him were the Heidelberg lecture of 1904 and the *Mathematical Problems* paper from 1900. And the most important point for the future development, and also for my talk, was no doubt his criticism already there of Hilbert's solvability axiom, or solvability conviction, if you prefer, the conviction that every mathematical problem can be solved positively or negatively. The importance he attached to this point is clear from the fact that he put it last in his series of theses, saying outright that this conviction of Hilbert's is unfounded, which is to say that he thought that there was no evidence at all for this conviction. The following year, 1908, this was turned into the criticism of the law of excluded middle by simply identifying the law of excluded middle, in its constructive interpretation, with Hilbert's solvability axiom. Then, in 1912, he gave his inaugural lecture *Intuitionism and Formalism* (Brouwer 1912A): so now the terms intuitionism and formalism have come in, but it did not contain so much new actually: it was only an inaugural lecture after all.

A new phase in this controversy began in 1921 with the publication of Weyl's paper *Über die neue Grundlagenkrise der Mathematik* (Weyl 1921), and that is what really fired it and made it so bitter. (And it seems clear that it had to do with the fact that Weyl was after all Hilbert's doctoral student: he took his doctor's degree with Hilbert, and I do not know, but presumably Hilbert thought of him as the best of his doctoral students over the years.) So Weyl had been converted to intuitionism and wrote this paper, which is very rhetorical, and where had he learned this kind of rhetoric? From Hilbert, no doubt. Hilbert was a master in writing rhetorical papers: if you think of his *Axiomatisches Denken* (Hilbert 1918), for instance, it is a very beautiful piece of rhetoric, and now he suddenly had this quite rhetorical paper directed against him, and that fired him quite clearly, mathematically as well as rhetorically, and his response was to give lectures at various universities, all of them on essentially the same topic. The most well-known is the one that he gave in Hamburg in 1921: so that was immediately after the Weyl paper, and that was printed in 1922, called *Neubegründung der Mathematik* (Hilbert 1922). There we find several of the memorable quotations that we have all read, and that were his answer to Weyl's rhetoric. After all, he wanted to make clear that he was not only the mathematical master but also the rhetorical master, and I do not know what one should say about this rhetoric that came into the purely scientific debate as an additional element: it has its negative sides, we probably all agree, but it is not entirely negative, producing as it did some of the more quotable sentences that we have in this area, and if we are not now letting us be seduced by this rhetoric, the quotations can with profit be used on various occasions to bring the purely scientific issues out in sharper relief. The following year, there appeared a quite similar paper of Hilbert's, *Die logischen Grundlagen der Mathematik* (Hilbert 1923), and then, in 1925, *Über das Unendliche* (Hilbert 1926), again with a lot of rhetoric in it and, as we now know, also mathematical mistakes. The final Hilbert paper that I want to refer to, containing as it does some of Hilbert's most outrageous sentences about Brouwer and intuitionism, is *Grundlagen der Mathematik* (Hilbert 1928), with whose publication we are already in 1928.

Now Brouwer, who is always said to have been such a difficult person, was actually relatively quiet, pursuing his mathematics, during these years. His basic papers on intuitionism, his reconstruction papers, date after all from the years 1918–1927, and I have only been able to locate one paper, albeit a very well-known one, *Über die Bedeutung des Satzes vom ausgeschlossenen Dritten in der Mathematik, insbesondere in der Funktionentheorie* (Brouwer 1924N), which contributes to this rhetorical fight in any clear way. It is that paper which contains the often quoted half of a sentence:

> eine durch keinen Widerspruch zu hemmende unrichtige Theorie ist darum nicht weniger unrichtig, so wie eine durch kein reprimierendes Gericht zu hemmende verbrecherische Politik darum nicht weniger verbrecherisch ist (Brouwer 1924N, p.3).[2]

Of course, that was quite in parity with Hilbert as far as rhetoric is concerned, but it seems to be the single place where he took to this kind of language, with the exception of his final contribution to the controversy with Hilbert, which was his 1928 paper *Intuitionistische Betrachtungen über den Formalismus* (Brouwer 1928A2), a paper which I take it that we have all studied in one way or the other. So this is the controversy as it has come down to us through their own writings, and Brouwer himself characterized the situation, already towards the end of his inaugural lecture of 1912 in the following way:

> So far my exposition of the fundamental issue, which divides the mathematical world. There are eminent scholars on both sides and the chance of reaching an agreement within a finite period is practically excluded (Brouwer 1913C, p.96).

which means that he thought that it was almost excluded that this controversy would not go on forever. Now my own attitude towards this is that, when there is very serious disagreement, to the extent that controversy is a better word, and you cannot make up your mind, what is the natural attitude? Well, keep quiet and just wait and see what happens: maybe eventually new information will accumulate that will make us look at the controversy in a different way. And, indeed, this is to my mind what has happened: although it did not take an endless time, it took a very long time, and I would say that it is only around ten years ago, in the nineties that is to say, that we have come to get a proper understanding of the scientific content of this controversy, an understanding which makes it not unreasonable to speak of a settlement of the controversy, so that it finally ceases to be a problem for us.

So now I will tell two different stories that have eventually provided the information that makes the Hilbert-Brouwer controversy look different now from what it looked like in 1928, and the first of these is the double-negation interpretation. Naturally, the mathematical community was at the time very much bothered by this controversy, and especially those who were young enough to have their formative years in the twenties, they eagerly wanted to learn: what was it all about, and could something be done about it? One of them was Kolmogorov, and one can see in his paper *On the law of the excluded middle*

[2] an incorrect theory, even if it cannot be inhibited by any contradiction that would refute it, is none the less incorrect, just as a criminal policy is none the less criminal even if it cannot be inhibited by any court that would curb it (van Heijenoort 1967, p.336).

from 1925 (Kolmogorov 1925) a very strong wish to reach some kind of objective verdict in this controversy, but not only in that paper: there is also a much less known paper that was published in a popular scientific journal in 1929, *Contemporary debates on the nature of mathematics* (Kolmogorov 1929), which testifies to the same strong wish. And, if you read the Résumé which you find at the end of his 1925 paper, it is clear that he thought actually at the time that he had been able to settle this controversy. And how did he settle it? Well, he settled it by forgetting about Brouwer's distaste for language and logic: he sat down with the Hilbert axioms for propositional and predicate logic, removing those that did not look all right on a constructive interpretation and retaining the other ones, except for the crucial law ex falso quodlibet, which it remained for Heyting to add to the system in 1930, and then he simply interpreted all of classical propositional and predicate logic by means of constructive propositional and predicate logic, utilizing what we now call the double-negation interpretation. And, indeed, that does give a constructive interpretation of classical logic, and he thought that, having achieved that, the problem of giving a constructive interpretation of classical mathematics in its entirety was essentially solved. This was the mistake, because mathematics cannot be built on propositional and predicate logic alone: there has to be some amount of set theory also, and two questions then arise, first: what should be the laws of this set theory? and, second: does this set theory with classical logic, which is to say classical mathematics, allow the double-negation interpretation to go through? These questions were not addressed in Kolmogorov's 1925 paper, so he was overoptimistic at the time.

Then the double-negation interpretation was independently rediscovered, as we all know, by Gödel and Gentzen in 1933 (Gödel 1933a), but, more importantly, they extended it from pure logic to first-order arithmetic, and that immediately gave a constructive interpretation, and therefore a constructive consistency proof, of first-order arithmetic, but the question remained: could it be extended beyond first-order to second- or higher-order arithmetic? On this point, a crucial role was played by Spector's last paper, published posthumously in 1962 (Spector 1962), since he died the year before. Spector's paper contained a consistency proof of analysis by an extension of principles formulated in current intuitionistic mathematics, and the extension was by the principle of so-called bar recursion of higher type. The paper gave new hope that maybe one could after all give a constructive interpretation of all of classical second-order arithmetic, which would yield a positive solution to Hilbert's second problem. Now Spector's principle of bar recursion of higher type was a quite complicated scheme of axioms, and so the question immediately arose: do we at all have an interpretation that validates them? The natural interpretation turns out to be by means of the so-called continuous functionals of finite type, and if you start thinking about what abstractions you need in order to prove that these axioms are validated under the continuous functional interpretation, the conclusion is: we need all of classical second-order arithmetic, strengthened by the axiom of dependent choices, which means that no reduction whatsoever has been obtained.[3]

This led to Kreisel's proposal, in 1968, of accepting instead intuitionistic second-order arithmetic, which he showed to have the very good property that you immediately interpret classical second-order arithmetic in it by means of the old double-negation interpretation (Kreisel 1968). So, if intuitionistic second-order arithmetic is constructively acceptable, then we do indeed have a solution to the second Hilbert problem, because Hilbert's second problem was at an early stage associated precisely with classical second-order arithmetic. So optimism at this time.

Then this was further extended by Myhill in 1971, with a correction in 1974, from second-order logic to full simple type theory (Myhill 1971, 1974). So now we had intuitionistic simple type theory over arithmetic at the bottom, and again the double-negation interpretation extended: classical higher-order arithmetic is easily translatable into the corresponding intuitionistic theory by means of the double-negation interpretation, so if this intuitionistic theory is comprehensible, or acceptable, then we have indeed, among other things, a consistency proof for a full-scale system of classical mathematics. And then Friedman (Friedman 1973) carried it even a bit further, to Zermelo-Fraenkel set theory, by modifying the axioms of Zermelo-Fraenkel set theory so that they became formally intuitionistic, by which I mean that the logic is intuitionistic, and he proved that classical Zermelo-Fraenkel set theory is interpretable in this intuitionistic, or formally intuitionistic, version of Zermelo-Fraenkel, thereby bringing this line of development to its natural conclusion. It looked very promising: maybe we just had to accept these impredicative intuitionistic systems as comprehensible.

This is what I did myself during a year and a half, from 1970 to 1971, as a result of Girard's extension of the proof of normalization to second-order logic by means of his so-called candidates of reducibility in 1970 (Girard 1971). That had been a long-standing open problem, to prove normalization for second-order logic, and, indeed, if you allow yourself to use second-order abstractions on the metalevel, you could do that, as Girard showed. I was naturally impressed by this and leaned towards simply accepting those principles: at least we had learned how to use them to prove this particular combinatorial result on normalization. But this optimism did not last very long in my case, because at the same time approximately, in 1969, the Curry-Howard correspondence, or isomorphism, between propositions and sets had been discovered (Howard 1980). I will use the term set, but you should understand it in the sense of individual domain, or quantificational domain: in first-order predicate logic, you always quantify over a domain, and Curry-Howard is a correspondence, or isomorphism, between propositions and sets in that sense. The Curry-Howard correspondence was to me from the beginning the natural

[3] I am grateful to Thierry Coquand for criticism which has made me let the text differ at this point from the talk as I gave it at the conference.

completion of the Brouwer-Heyting-Kolmogorov interpretation of intuitionistic first-order predicate logic, just drawing the full consequences, so to say, of the Brouwer-Heyting-Kolmogorov interpretation. Now the paradox that Girard discovered in 1971 (Girard 1972), in the first version of constructive type theory, then called intuitionistic type theory, showed that Curry-Howard is incompatible with impredicativity: if you combine the Curry-Howard identification of propositions and sets with impredicativity, that is, the possibility of quantifying over propositions, over classes, over relations, and so on, you end up with a contradiction. So one of them has to go: either Curry-Howard has to go or there is some problem with impredicativity, with which there had been problems from the very beginning: when the notion itself was introduced by Russell in 1906 (Russell 1906), it was precisely because it was a problematic notion. So it seems fair to say that Girard's paradox put an end to the optimism about impredicative intuitionism in the 1960s, and maybe the form in which it convinced most people was the form that was given to it in 1986, which is after the calculus of constructions had come into being, independently and at the same time by Coquand in France and by Hooke and Howe as well as Mitchell in the United States, who proved that, if you combine impredicative second-order existential quantification with the strong rules of existential elimination, which are clearly valid under the Brouwer-Heyting-Kolmogorov interpretation, you already arrive at an inconsistency (Coquand 1986; Hook and Howe 1986; Mitchell and Harper 1988). So, hopeful as impredicative intuitionism had seemed for some time, during the approximate decade from Spector's paper to Girard's paradox, it nevertheless ended in disillusion.

My second story is about ordinal proof theory, more particularly, about the proof-theoretic scale, which is the result of a long development in German proof theory, originally aiming at fulfilling Hilbert's program, which is to say producing finally the dreamt of constructive consistency proof of classical analysis. The development of this particular branch of Hilbert's proof theory begins as we all know with Gentzen's proof of the consistency of first-order arithmetic by means of purely combinatorial reasoning extended by transfinite induction up to ε_0, an ordinal about whose constructive accessibility nobody has any doubt. Now this ordinal measures such a natural level of abstractions that there are many systems which lie on this level, not only classical but also intuitionistic first-order arithmetic, Gödel's theory T, Heyting arithmetic extended to all finite types, HA^ω, as it is usually called, possibly even with the constructive axiom of choice that can be formulated in that system, and also constructive type theory without any well-orderings and universes is a theory on this level. The year in which it was first reached, by Gentzen, is 1936 (Gentzen 1936).

Then, after the Second World War, this line of investigations was continued by Lorenzen and Schütte (Schütte 1951) in the fifties. The next natural stopping point is arithmetical analysis, whose proof-theoretic ordinal was determined to be $\varepsilon_{\varepsilon_0}$, and, if you begin iterating the process any finite number of times, you will get up to the first critical epsilon number. This is the ordinal of full ramified analysis, which is to say arithmetic with ramified second-order quantification of all finite levels. Then, beyond that, there is an ordinal of some significance, which is usually denoted by $\phi_{\varepsilon_0}(0)$. It is the so-called Veblen notation that is used, and this ordinal was determined by Friedman in the late sixties (Friedman 1968) to be the ordinal of classical analysis, which is to say second-order arithmetic, with no comprehension axiom but instead with the so-called Σ^1_1 axiom of choice, which is to say the axiom of choice limited to formulas of Σ^1_1 form. And, again, there is a corresponding version of type theory: you just need to add to constructive type theory the axioms for one universe, and you get a system of precisely this proof-theoretical strength.

Then we get to the very well-known ordinal Γ_0, which was identified by Schütte and Feferman in 1963–64 as the ordinal of full predicative analysis, that is, predicative analysis with not only finite but also transfinite ramification levels (Feferman 1964; Schütte 1965). And, again, this corresponds to a natural system of constructive type theory, namely constructive type theory with not just one universe but a whole sequence of successive universes $U_0, U_1, \ldots$ indexed by the sequence of natural numbers.

Then, beyond this, we reach a very significant level of abstractions: in terms of ordinals, it is the Howard ordinal $\phi_{\varepsilon_{\Omega+1}}(0)$, whose notation involves the so-called Bachmann extension of the Veblen hierarchy. This is a very significant level, because it is the level where, in addition to ordinary mathematical induction and recursion, you have transfinite induction and recursion, so it is precisely the kind of abstractions that Brouwer learned from Borel: if you remember the quotation that I gave from his lecture at the Sorbonne in 1949, he said that he was particularly indebted to Borel's definition of measure, and how did that go? Well, it was by transfinite recursion on the build-up of the Borel set, and that is a transfinite recursion on countable well-founded trees, precisely the kind of reasoning that he himself came to use in connection with the bar theorem. So this is roughly the proof-theoretic strength of the methods used by Brouwer himself in his writings, since he did not go to any higher number classes, and the theory which has this strength is usually referred to as ID_1. The Howard ordinal is also the ordinal of Kripke-Platek set theory, and it is the ordinal of Aczel's constructive version of Zermelo-Fraenkel set theory, which, although it has this comparatively limited proof-theoretic strength, is nevertheless such that, when you add the law of excluded middle to it, you get full classical Zermelo-Fraenkel set theory.

Then, if you start iterating the inductive definitions, so as to have not only the second number class but also the third and the higher number classes, finitely high, you get to an ordinal which I can only write down in a notation such as $\theta(\Omega_\omega)$, where θ is technically called a collapsing function in this kind of proof theory. The only thing you need to know about it is that it is a function that takes an ordinal of any size and collapses it into a countable ordinal, and that, for the particular large ordinals, like Ω_ω, that we are going to collapse, the countable ordinals obtained by the collapsing procedure will be given by notation systems which are defined in a purely combinatorial manner. The ordinal $\theta(\Omega_\omega)$ measures the proof-theoretic strength of the theory of finitely iterated generalized inductive definitions, and it is also the ordinal determined by Takeuti in 1967 for classical analysis with the Π^1_1 comprehension axiom and the schema of mathematical induction limited to Π^1_1 formulas (Takeuti 1967). Viewed type-theoretically, it is the ordinal of constructive type theory, without universes, but now with the general well-ordering operation W. On the other hand, if you take instead classical analysis, still with the Π^1_1 comprehension axiom, but now without the restriction on the induction scheme, you get to a slightly larger ordinal here, which happens to be $\theta(\Omega_\omega \cdot \varepsilon_0)$, but this is a rather unnatural ordinal, as you see, much less natural than $\theta(\Omega_\omega)$. On the other hand, if you not only add ordinary mathematical induction for formulas of arbitrary complexity but strengthen that from ordinary induction to bar induction, then you get even further here to an ordinal which is the collapse of the first epsilon number beyond Ω_ω, $\theta(\varepsilon_{\Omega_\omega+1})$, again a relatively technical and unnatural ordinal in comparison with $\theta(\Omega_\omega)$.

The next really important step further was taken in 1982 by Pohlers and Jaeger (Pohlers and Jäger 1983), who determined the proof-theoretic strength of classical analysis with the Δ^1_2 comprehension axiom plus the scheme of bar induction for formulas of arbitrary complexity to be $\theta(\varepsilon_{I+1})$. Now what makes this particularly interesting is that, for the first time, a large cardinal appears in the ordinal notation: it is the collapse of the first epsilon number beyond the first inaccessible, which is what I denote by I here. At the earlier stages, we have had not so wildly big cardinals, or initial ordinals, coming in, but now, remarkably enough, an inaccessible cardinal makes its appearance in the ordinal notation, and this in a piece of constructive proof theory. The ordinal $\theta(\varepsilon_{I+1})$ is also the proof-theoretic ordinal of Aczel's constructive set theory, when it is strengthened by the axiom that he calls the regular extension axiom. To reach this level with constructive type theory, you need to have both the well-ordering operation and a universe which is closed under the well-ordering operation. As shown by Setzer in his thesis of 1993 (Setzer 1993), the resulting system of type theory has the proof-theoretic ordinal $\theta(\Omega_{I+\omega})$, an ordinal which is even slightly bigger than $\theta(\varepsilon_{I+1})$.

We are now already in the nineties, when one has ventured to extend these results even further, and the only way in which one has succeeded in doing that has been by adding new large cardinal axioms, although they have been given constructive meaning. There was first the step taken by Rathjen in 1990–91 of replacing the inaccessible that we have already seen by a Mahlo cardinal instead (Rathjen 1990), thereby reaching $\theta(\varepsilon_{M+1})$, and proving this to be the ordinal of Kripke-Platek set theory, when strengthened by an axiom which forces the universe to be Mahlo (Rathjen 1991). Analogously, in 1996, Setzer added the axioms for a Mahlo universe to type theory and determined the proof-theoretic ordinal of the resulting system of type theory to be $\theta(\Omega_{M+\omega})$ (Setzer 1996). So this is the ordinal strength of constructive type theory with the well-ordering operation, as before, but now adding axioms for, not the usual kind of universe, but a so-called Mahlo universe. Actually, one yet further step has been taken, both by Rathjen in terms of Kripke-Platek set theory and by Setzer in terms of constructive type theory, replacing the Mahlo cardinal by a weakly compact cardinal instead, but I have to leave this last step out.

The crucial point now is that what we see in front of us at this stage is some kind of abyss, or chasm, which we do not seem able to pass, and that the classical type- as well as set-theoretic systems, second- and higher-order arithmetic, Zermelo's set theory and Zermelo-Fraenkel set theory, they are all on the other side of this abyss, and it is not only when you go to full second-order arithmetic but already when you go to the next form of the comprehension axiom, which is to say Π^1_2 comprehension, that you find yourself on the other side of the abyss. On this level of proof-theoretic strengh, progress has been possible to make, again by Rathjen, only by beginning to postulate certain large cardinals with the right properties, so that you get notation systems which are big enough, but this is a complete change in the methodology of constructive proof theory, which brings it rather into contact with current set theory: you simply begin to postulate instead of building things up by means of evident principles, and, as it seems, that is the only way that is available if we want to proceed into this region, that is, to reach something like classical second-order arithmetic, which was Hilbert's original problem, or the corresponding intuitionistic theory, Heyting arithmetic of second-order.

So this is the picture that has emerged as the result of persistent work by the German proof-theoretic school during a period of more than seventy years. The original aim was to obtain a constructive consistency proof for classical analysis, which early on came to be identified with full second-order arithmetic, but we have now so much information that we know that this is out of our reach, and why? Well, if this is to be a constructive consistency proof, it will have to use constructively acceptable principles, and we know by now what are the strongest constructively acceptable principles that are available to us at the moment. It is of course never fixed at any absolutely precise level: as soon as you fix it, you can go beyond it by some kind of reflection principle, but basically we have at present exhausted the principles for which we can claim evidence, and this is a completely new situation. If you go back to the late sixties, for instance, when Takeuti did his fundamental work, it was still wide open how far you could go in this kind of proof theory. So we know by now that we cannot pass this abyss unless we are able to think up some brand-new strong constructive principles, and there is no sign whatsoever of that at the moment.

This means in particular that we have to take a different attitude to the systems on the other side of the abyss, like second-order arithmetic, and above all to the law of excluded middle. You see, some of the systems on this side of the abyss, namely Aczel's system CZF of constructive set theory on the one hand and constructive type theory on the other hand, are such that, when you add the law of excluded middle to them, you get full classical mathematics, which we know that we cannot interpret constructively unless a miracle intervenes. This is what I propose to call the second failure of the Hilbert program. The first failure of the Hilbert program was the one which was discovered by Gödel and of which we are now all aware, but that gave rise to the revised, or modified, Hilbert program, whose characteristic is that we no longer allow merely combinatorial methods in the consistency proof but arbitrarily strong constructive methods. But even this revised, or modified, Hilbert program has come to an end in the nineties, or has failed in the nineties, so it is the second failure of the original Hilbert program, which I cannot interpret in any other way than that we have to give up the dream of being able to establish the consistency of classical mathematics by constructive means.

We find ourselves in a new situation: it is a new picture that has emerged here, if you compare it with the optimistic picture that we had in the sixties, and to my mind the natural conclusion is that we have to look at the law of excluded middle in a different way, the way set theorists look at large cardinal and determinacy axioms rather. So we have constructive mathematics, and one non-evident principle that we can add to it is the law of excluded middle. Then some kind of explosion takes place: we pass from a fully understood system, which is on this side of the abyss, to full classical mathematics, which is on the other side of the abyss, and it is in the nature of things that we cannot be certain of the consistency of the classical system that we arrive at, because we transcend the principles that have evidence. On the other hand, we have inductive evidence of the consistency of the classical system in the sense that the classical set-theoretical principles have been investigated with respect to their consequences during such a long time and in such detail without there being any sign of an inconsistency arising from them. But, since we do go beyond what is intuitively evident when we adjoin the law of excluded middle, we are not certain that it cannot give rise to a contradiction in the way we are on the previous level, which is to say before the abyss. We are rather in the situation that Woodin has characterized by the sentence, or half of a sentence:

> just as those who study large cardinals must admit the possibility that the notions are not consistent (Woodin 1998, p.330).

So the law of excluded middle has to be ranged together with the large cardinal and determinacy axioms rather than with the axioms of which we have a constructive, or contentual, understanding. This is in line with the fact that, if we reformulate the Brouwer-Heyting-Kolmogorov interpretation of intuitionistic propositional and predicate logic in terms of games and winning strategies, as suggested by Ranta in his paper *Propositions as games as types* (Ranta 1988), the law of excluded middle appears as the simplest and most general axiom of determinacy, which says that every game is determinate.

Now some final words on the Hilbert-Brouwer controversy, so as to justify the choice of the title of my talk: The Hilbert-Brouwer controversy resolved? To my mind, it is indeed a kind of resolution that we have reached as a result of the new information that has accumulated during the period from the sixties to the nineties, and that I have reported on. So let us look at one side first, the Hilbert side.

We know what was Hilbert's foundational program: to formalize mathematics, allowing classical logic in the formalization, and then justifying it proof-theoretically, on the metalevel, by means of a consistency proof, which was first required to be purely combinatorial, like the consistency statement itself, and then liberalized to be merely constructive, allowing of any means as long as they carry conviction. This is precisely what has failed, not only in the first, but also in the second, more liberal, case. That means that the two-storeyed building of mathematics that Hilbert had in mind, that simply has not worked out: it is not to be avoided that we have to go through mathematics as it exists and disentangle the constructive parts of it from the uses of the law of excluded middle and Zermelo's axiom of choice. This is precisely what Brouwer started and what is still going on: at the present time, actually, a lot has been achieved in this direction. So this much about the Hilbert side.

Then, if you turn to Brouwer's views, the formulations that come first to my mind at least are:

> mathematics is a mental construction (Brouwer 1947, p.477),

or:

> mathematical objects are mental constructions,

on the one hand, and:

> mathematics is a languageless activity of the mind
> (Brouwer 1952B, p.141),

on the other. Now these statements are very understandable psychologically if one remembers that he was opposed to Hilbertian formalism: he was opposed to the idea that the mathematical objects are mere sign configurations, combinatorial sign configurations, as Hilbert thought, the classical formulation being:

> am Anfang ist das Zeichen (Hilbert 1922, p.163),[4]

alluding to both Goethe's Faust and the Gospel of St. John. And Brouwer's way of contradicting this as much as possible was to say that, on the contrary, mathematical objects have nothing at all to do with signs, or with language: they are mental constructions, or thought constructions, but that seems to me to be a very difficult position to maintain nowadays. I think we must admit that Brouwer's views on language were of a rather old-fashioned nature: it is clear that he thought of language in the traditional way as the expression of thought, which is to say that the thoughts had to be there first, before possibly being expressed in language, and that view of language has after all been replaced, in the course of the last century, by the view of language as a

[4] in the beginning is the sign

means of communication. And it is not for nothing that this shift has taken place: if you take Brouwer's view that the mathematical objects are mental constructions, or thought constructions, then there arises immediately the problem: how do we get to know them? and not much self-reflection is needed to see how we do get to know them. After all, there is no mathematical zoo to which we can be taken in order to have them displayed to us without any linguistic mediation: we go to listen to some lectures or to read some books or articles, which means that we are dependent on language and symbols from the very start. So this view of Brouwer's is hardly credible any longer because of the developments in the philosophy of language during the last century, as Dummett has emphasized for the first time in his paper *The philosophical basis of intuitionistic logic* (Dummett 1975). But there is a kind of mediating position, namely that the mathematical objects are not just purely formal sign configurations: they are meaningful sign configurations, and that is what gives them properties which do not come from their combinatorial nature, but come from the meaning with which they are endowed, as beautifully stated by Gödel in the beginning of his *Dialectica* paper (Gödel 1958). It is this view of the mathematical objects, mediating as it is between Hilbert and Brouwer, which has turned out to be the most credible one to my mind: at least it is the one which underlies all of my own work from 1974 until the present day. So, even if there is a question mark in my title, a cautious question mark, it seems to me that the progress that has been made during the thirty-year period from the sixties to the nineties is such as to make it not unreasonable to speak of a resolution of the Hilbert-Brouwer controversy, part of this resolution being that we have had to change our attitude towards the law of excluded middle in a quite drastic way.

Acknowledgement

I am greatly indebted to Mark van Atten, not only for recording the talk as I gave it at the conference on tape, but also for subsequently providing me with the technical equipment necessary for its transcription.

Proof theory and Martin-Löf Type Theory

Anton Setzer

1 Introduction

It is the goal of ordinal theoretic proof theory to reduce the consistency of theories for formalising mathematical proofs to the well-foundedness of ordinal notation systems. In order to obtain a satisfactory solution to the consistency problem, this reduction needs to be supplemented by a second step, namely by proofs of the well-foundedness of the ordinal notation systems in 'safe' theories, for which one has an argument that everything shown in these theories is valid. Because of Gödel's incompleteness theorem, a mathematical correctness proof can prove only relative correctness, but never provide an absolute proof of correctness. In order to obtain an argument which provides absolute trust, the only possibility is to have a philosophical correctness argument. Although there are other theories that could serve as such safe theories, the theories for which such arguments have best been worked out at present are extensions of Martin-Löf type theory (*MLTT*).

In this article we will give an overview of the techniques used in this program of reducing the consistency of mathematical theories to MLTT. We will start by giving an introduction into the origins and some basic techniques of ordinal theoretic proof theory. This will motivate the notion of the proof theoretic strength. We will then indicate how to use MLTT in order to obtain a complete consistency proof, which is necessarily based on a philosophical argument. Then we will go through some key steps on the proof theoretic scale: we consider the following variants of MLTT: MLTT with natural numbers; the additional extension by the W-type; the additional extension by one universe; MLTT with inductive-recursive definitions; the red and black Mahlo universes. We will finally look briefly at the hyper-Mahlo, hyper-α-Mahlo, autonomous Mahlo, and Π_3-reflecting universes.

Notations. When introducing new notions, we write them in italic, if they occur as text, and underline them, if they occur as mathematical formulae.

We write terms in functional style, e.g., in the form $\mathrm{C}\ a_0 \cdots a_n$ rather than in mathematical style $\mathrm{C}(a_0, \ldots, a_n)$. In running text we put, if necessary, brackets around it, i.e., we write $(\mathrm{C}\ a_0 \cdots a_n)$ instead of $\mathrm{C}\ a_0 \cdots a_n$.

Upper bounds for the proof theoretic strength are determined by modelling MLTT in extensions of Kripke-Platek set theory. When developing such models for proof theoretic reasons, we will write $[\![A]\!]$ for the interpretation of set A. Although we will use as our main model of MLTT the *PER* (*Partial Equivalence Relation*) model, where $[\![A]\!]$ is a set of pairs of terms, namely those which are equal elements of A, we usually do for simplicity as if $[\![A]\!]$ were just a set of terms and identify therefore $[\![A]\!]$ with its domain. (The PER model was originally introduced by G. Rosolini in order to deal with the impredicativity of Girard's system F. It is closely related to Martin-Löf's meaning explanations, and is well suited for modelling MLTT, even though MLTT is predicative.)

We write $a[x := b]$ for the result of substituting of all free occurrences of the variable x in a by b. We write $a[x]$ for an expression a possibly depending on a variable x. Once we have used $a[x]$, $a[b]$ denotes $a[x := b]$, where we assume that any clashes between bound and free variables are resolved by applying α-conversion (all α-equivalent expressions are identified).

2 The Rôle of Type Theory in a Proof Theoretic Program

Proof theory is a discipline of mathematical logic, which was founded by David Hilbert (Hilbert 1926, 1928) at the beginning of the 20th century. At that time various axiom systems for carrying out mathematical proofs had been developed, of which some had turned out to be inconsistent. In order to guarantee that the axiom systems which were actually used, don't contain any inconsistencies, Hilbert proposed to prove the consistency of mathematical axiom systems (Hilbert 1900b). He observed that, if one shows the consistency of a theory for formalising mathematics in the same or an even stronger theory, one has not achieved anything: if the original theory is inconsistent, it proves everything, even its own consistency. So in order to achieve something, one has to do more: namely to show the consistency using methods that are considered to be safe. According to Hilbert, finitary methods were safe (Hilbert 1926). By finitary methods he considered finitary calculations which can be carried out on a piece of paper.

There are two main approaches for carrying out consistency proofs. One is to introduce a model of the system in question in the meta-theory. However, it seems to be implausible to assume that in this way one can prove the consistency of a theory by using finitary methods, since such methods do not allow the use of sets. Hilbert realized this and suggested therefore that one should instead analyse proofs and in this way show directly that it is not possible to derive a contradiction in the formal system in question. He called the mathematical discipline in which such investigations are carried out, proof theory (Hilbert 1926, 1928).

Gödel's second incompleteness theorem and the failure of Hilbert's original programme. In 1931 Gödel (Gödel 1931) showed in his second incompleteness theorem that Hilbert's original programme cannot be carried out — assuming minimal conditions on a theory T, he could show that a consistent theory T does not prove its own consistency. Since finitary methods should be formalisable in any reasonable theory T, it follows that the consistency of theories fulfilling those minimal conditions cannot be shown by finitary means. Most natural theories, except for extremely weak ones, fulfil the premise of the last sentence — Hilbert's original programme had failed.

Gentzen's proof of the consistency of Peano Arithmetic. In 1936 Gerhard Gentzen (Gentzen 1936, 1974) showed the consistency of *Peano Arithmetic* (PA) in PRA (primitive recursive arithmetic, as defined later) extended by quantifier free transfinite induction up to ϵ_0. This was the birth of ordinal theoretic proof theory. His methods have been improved since. The shortest and most elegant presentation is due to Buchholz (e.g., in (Buchholz 1997)), which is as follows:

One develops PA in a one-sided calculus, the *Tait calculus* (Tait 1968), in which we derive finite sets of closed formulae Γ, where the intended meaning of Γ is essentially the disjunction of the formulae in Γ. A statement $A_1, \ldots, A_n \vdash B_1, \ldots, B_k$ is translated into the one-sided sequent form as $\neg A_1, \ldots, \neg A_n, B_1, \ldots, B_k$ (this elegant approach works only for classical logic).

We develop an infinitary theory PA*, in which we will interpret proofs from PA. PA* uses the Tait calculus, and formulae in PA* are closed formulae from PA. Each formula corresponds to a finitary or infinitary disjunction or conjunction, and we write $A \simeq \&_{i\in I} A_i$ for formula A being associated with the conjunction of A_i for $i \in I$, similarly $A \simeq \bigvee_{i\in I} A_i$ for the association of a disjunction. For a true prime formula A we have $A \simeq \&_{i\in\emptyset} A_i$ (there is no A_i); for a false prime formula we have $A \simeq \bigvee_{i\in\emptyset} A_i$; we have $A_0 \mathbin{\&} A_1 \simeq \&_{i\in\{0,1\}} A_i$; $A_0 \vee A_1 \simeq \bigvee_{i\in\{0,1\}} A_i$; $\forall x.A(x) \simeq \&_{i\in\mathbb{N}} A(i)$; $\exists x.A(x) \simeq \bigvee_{i\in\mathbb{N}} A(i)$. By the DeMorgan rules we can consider negation as a defined operation (assuming we have for each atomic formula its negation as an atomic formula as well). Note that the formulae used are finitary expressions, which are associated with possibly infinitary conjunctions and disjunctions. When writing $\&_{i\in I} A_i$ or $\bigvee_{i\in I} A_i$, we mean in the following any formula A such that $A \simeq \&_{i\in I} A_i$ or $A \simeq \bigvee_{i\in I} A_i$, respectively.

PA* has three rules:

$$\frac{\Gamma, A_i \quad (\text{all } i \in I)}{\Gamma, \&_{i\in I} A_i}\ (\& -\text{intro}) \qquad \frac{\Gamma, A_i \quad (\text{some } i \in I)}{\Gamma, \bigvee_{i\in I} A_i}\ (\vee - \text{intro})$$

$$\frac{\Gamma, A \qquad \Gamma, \neg A}{\Gamma}\ (\text{Cut})$$

The formula A in the rule (Cut) is called the *cut formula* of this rule. Derivations in PA* are well-founded derivations constructed from those rules, where *well-foundedness* means that there is no infinite descending sequence in the proof tree.

One can see that PA* allows one to interpret the logical rules of classical logic. The assumption rule (in standard sequent calculus written as $A \vdash A$) reads $\neg A, A$ and is provable in this calculus by induction on the rank of A (where the *rank* is the number of logical connectives in A — note that the formulae are closed formulae from PA, which are finite expressions). For instance, the step from $\neg A(n), A(n)$ for all $n \in \mathbb{N}$ to $\neg(\forall x.A(x)), \forall x.A(x)$, which is $\exists x.\neg A(x), \forall x.A(x)$, reads

$$\cfrac{\cdots \quad \cfrac{\neg A(n), A(n)}{\exists x.\neg A(x), A(n)} \quad \cdots \quad (n \in \mathbb{N})}{\exists x.\neg A(x), \forall x.A(x)}$$

By $\vee$-introduction we obtain immediately $A \vee \neg A$ as well.

Modus Ponens, which derives from $A \to B$, which is $\neg A \vee B$, and A the formula B, is interpreted as follows (we use here the fact that PA* is closed under *weakening*, i.e., from a proof of Γ, we can obtain a proof of Γ, Δ; this is used here to weaken the proof of A to a proof of A, B; note that $\neg(\neg A \vee B) = A \mathbin{\&} \neg B$):

$$\cfrac{\neg A \vee B \qquad \cfrac{A \qquad \neg B, B}{A \mathbin{\&} \neg B, B}}{B}\ (\mathrm{Cut})$$

Similarly, arithmetical and equality axioms can be interpreted. True atomic formulae correspond to the empty conjunction, which is provable by the rule (& –intro) with no premises in this instance; this allows one then by the rules for & and $\vee$ to prove all true propositional closed formulae. By the infinitary introduction rule for $\forall$ we can prove all true closed formulae of the form $\forall x.A(x)$, where $A(x)$ is quantifier free. This allows one as well to prove the equality axioms (where the congruence axiom ($s = t \to A(s) \to A(t)$) is restricted to quantifier free formulae; this is sufficient to prove congruence for all formulae). Now we can interpret proofs in PA as infinitary proofs in PA*.

The only principle we haven't interpreted in PA* yet is the induction rule. So assume PA $\vdash A(0)$ and PA $\vdash \forall x.A(x) \to A(x+1)$, from which the induction rule derives PA $\vdash \forall x.A(x)$. By induction hypothesis we have PA$^* \vdash A(0)$ and PA$^* \vdash \forall x.A(x) \to A(x+1)$. Then we obtain in PA* proofs of $A(n)$ for all $n \in \mathbb{N}$. One can easily show $(\forall x.A(x) \to A(x+1)) \to (A(n) \to A(n+1))$ for all n, therefore we obtain $A(n) \to A(n+1)$. Now using iterated Modus Ponens, we can, using PA$^* \vdash A(0)$, prove PA$^* \vdash A(n)$ for all $n \in \mathbb{N}$. By &-introduction we obtain therefore PA$^* \vdash \forall x.A(x)$. Note that this proof has infinite height, since, in order to prove the premise $A(n)$, at least n applications of Modus Ponens were used.

The *cut rank* of a proof in PA* is the supremum of the rank of cut formulae occurring in it. One can see that the cut rank of a proof in PA*, which is the translation of a proof in PA, is finite.

If we omit the cut rule, PA* is consistent, since falsity is interpreted as the empty disjunction, and there is no rule for deriving the empty disjunction. Therefore, in order to prove the consistency of PA, it suffices to show that from each proof of falsity with finite cut rank we obtain a cut free proof of falsity. In fact, the cut-elimination result holds for arbitrary proofs of finite cut rank (the theorem can be extended to infinite cut rank as well) in PA*: all proofs can be transformed into cut free proofs of the same set of formulae.

The height of well-founded trees and ordinal notation systems. The proof of cut-elimination is carried out by induction over the (well-founded) derivations. Since we have used infinitary proofs, the height of a proof can no longer be measured by a natural number. Proofs are well-founded, and the *height of well-founded trees* can be measured by ordinals. In set theory, *ordinals* form a class Ord, which is *well-ordered* (which means linearly ordered and well-founded), so there are no infinitely descending sequences. For any other well-ordered set $(A, \prec)$ we can find an ordinal α such that $(\{\beta \in \mathrm{Ord} \mid \beta < \alpha\}, <)$ (which is equal to α) is isomorphic to $(A, \prec)$. In this situation α is called the *order type* of $(A, \prec)$. Using this fact one can derive that the height of any well-founded tree, which is a set, can be measured by a unique ordinal α.

Ordinals from set theory are needed only for heuristic reasons — in ordinal theoretic proof theory one works with ordinal notation systems instead. An *ordinal notation system* is a primitive recursive subset OT of $\mathbb{N}$ together with a primitive-recursive binary relation $\prec$ such that $(\mathrm{OT}, \prec)$ is well-ordered. Note that elements of OT are therefore just finitary objects. Any $a \in \mathrm{OT}$ denotes the ordinal α, where α is the order type of $\{b \in \mathrm{OT} \mid b \prec a\}$ — this relates ordinal notations to set-theoretic ordinals.

Gentzen was able to replace the induction over proof trees, which is needed in the proof of the cut-elimination theorem for PA*, by transfinite induction over an ordinal notation system, the order type of which is denoted by ϵ_0. All infinitary derivations can be replaced by finitary objects (Gentzen worked directly in a finitary system), and Buchholz has developed in the article (Buchholz 1997) mentioned before a very simple and concise technique for achieving this.

All other arguments can be carried out in *primitive recursive arithmetic PRA* (which is PA extended by symbols for primitive recursive functions and their defining axioms, and induction restricted to quantifier free formulae), which is usually considered as a formalisation of Hilbert's finitary methods. (See Tait's article (Tait 1981) for an argument supporting this thesis.) Furthermore, transfinite induction over ϵ_0 can be restricted to quantifier free formulae. Gentzen proved the consistency of PA by using PRA extended by quantifier free transfinite induction up to ϵ_0. It is easy to convince oneself directly that transfinite induction up to ϵ_0 is a valid principle, and in this sense Gentzen obtained a consistency proof of PA.

Proof theoretic strength. Since PRA can be interpreted in PA, PA extended by transfinite induction up to ϵ_0 proves the consistency of PA. Therefore, by Gödel's incompleteness theorem, PA does not prove transfinite induction up to ϵ_0 (unless it is inconsistent). One can show that for any $\alpha < \epsilon_0$ PA proves transfinite induction up to α. We define the *proof theoretic strength* $|\mathrm{T}|$ of a theory T as the supremum of all α such that T proves transfinite induction up to α (more precisely one needs to restrict oneself to natural ordinal notation systems, as they are called; up to now there exists no formal definition of natural ordinal notation systems in general, but those which have been used — see (Pohlers 1996) for a short overview over the techniques used for developing them — are all considered as examples of natural ordinal notation systems). Then we obtain $|\mathrm{PA}| = \epsilon_0$.

The need for a constructive foundation of proof theory. Gentzen's approach has later been extended to systems of increasing strength and the most powerful result was the analysis of second-order arithmetic restricted to the comprehension axiom for Π^1_2-formulae by M. Rathjen (Rathjen 1995b,a, 2005c,b) and, independently, by T. Arai (Arai 1996a,b,c, 1997a,b,c). In fact, Rathjen has carried out an analysis up to the strength of $(\Pi^1_2 - \mathrm{CA}) + (\mathrm{BI})$, but at present the fully published versions reach $(\Delta^1_2 - \mathrm{CA}) + (\mathrm{BI}) + (\Pi^1_2 - \mathrm{CA})^-$. Here $\underline{(\Pi^1_2 - \mathrm{CA})}$ is the comprehension axiom for Π^1_2-formulae (other comprehension axioms are denoted similarly), $\underline{(\Pi^1_2 - \mathrm{CA})^-}$ is the parameter free Π^1_2-comprehension axiom, and $\underline{\mathrm{BI}}$ stands for Bar induction. T. Arai has carried out an analysis up to the strength of $(\Sigma^1_3 - \mathrm{DC}) + (\mathrm{BI})$ ($\underline{\mathrm{DC}}$ stands for the axiom of dependent choice), but most of his work beyond Kripke-Platek set theory plus Π_3-reflection exists at present only in the form of (carefully worked out) preprints.

With increasing proof theoretic strength the ordinal notation systems used become increasingly complicated, and it is no longer possible to get a direct insight into their well-foundedness. (There is an approach (Setzer 1999, 2000a) by the author called ordinal systems, to obtain directly intuitive arguments for as strong as possible ordinal notation systems.) What can be done instead is to prove the well-foundedness in another theory, a theory $\mathrm{T}_{\mathrm{Good}}$, for which we have a more direct insight into that all its theorems are valid. If PRA can be interpreted in those theories, which is expected, we obtain that $\mathrm{T}_{\mathrm{Good}}$ proves that T is consistent. This way we obtain therefore a consistency proof of T, assuming we believe that everything proved in $\mathrm{T}_{\mathrm{Good}}$ is valid.

The most successful (but not only) theories providing such a direct insight into their validity are constructive theories, and at present the best developed theories for this purpose are extensions of *Martin-Löf Type Theory* (*MLTT*). The argument that everything proved in MLTT is valid is given by *meaning explanations.* (See Martin-Löf's article (Martin-Löf 1996) for meaning explanations for the logical connectives and his book (Martin-Löf 1984) for meaning explanations for MLTT.) Meaning explanations rely substantially on the philosophy of language, which goes beyond the author's expertise, and therefore no meaning explanations are given in this article. We note that at present no meaning explanations have been given for theories of strength the black Mahlo universe and beyond.

One should note that there is another main approach to providing theories which could serve as $\mathrm{T}_{\mathrm{Good}}$, namely Feferman's theories of explicit mathematics (Feferman 1975). However, their philosophical foundations are not as well developed, and most research on these theories has been focused on classical and therefore non-constructive variants.

Therefore, MLTT can be used as a substitute for Hilbert's finitary methods. In order to prove the consistency of strong theories for carrying out mathematical proofs, we need therefore proof theoretically strong extensions of MLTT.

For this program we only need to prove lower bounds for the proof theoretic strength of MLTT. However, in order to determine the limits of a certain extension, it is important to know as well an upper bound. Therefore, a constructive underpinning of proof theory is achieved by developing proof-theoretically strong extensions of MLTT, for which we have an insight into their correctness, and by determining their precise proof theoretic strength.

Note that this is as well of relevance for a full constructivist, who is only interested in proofs in constructive theories. A constructivist should be interested in proving as many theorems as possible, and therefore has an interest in obtaining an as strong as possible constructive theory. It is as well important to compare this strength with non-constructive theories, in order to determine the limits of what can be proved in the constructive theory.

3 From Natural Numbers to Universes

The basic framework of MLTT. For historic reasons, the lowest type level in MLTT is called '$\underline{\mathrm{Set}}$'. Outside MLTT this type level will usually be denoted by 'Type', whereas in MLTT '$\underline{\mathrm{Type}}$' denotes a type level on top of Set. We will not make use of 'Type' until we reach the theory of inductive-recursive definitions. The use of Type outside MLTT explains why we will, as usual in type theory, refer later to Π-types, Σ-types etc., even though they are elements of Set and should therefore be called Π-sets, Σ-sets, etc.

In MLTT one has *non-dependent judgements* $\underline{A : \mathrm{Set}}$ (A is a set), $\underline{A = B : \mathrm{Set}}$ (A and B are equal sets), $\underline{a : A}$ (a is an element of set A), and $\underline{a = b : A}$ (a and b are equal elements of set A). Furthermore, we have *dependent judgements* $\underline{x_1 : A_1, \ldots, x_n : A_n \Rightarrow \theta}$, where θ has the form of a non-dependent judgement, but might depend on $x_1, \ldots, x_n$. Here A_i might depend on $x_1, \ldots, x_{i-1}$. We usually write capital Greek letters $\underline{\Gamma}, \underline{\Delta}$ for contexts of the form $x_1 : A_1, \ldots, x_n : A_n$.

When introducing sets by declaring what their elements are, we will in the following mean in fact canonical elements of these sets. *Canonical elements* are those which are in head normal form, i.e., they start with a constructor. Arbitrary elements of such a set are those which reduce to canonical elements.

The basic theory of MLTT. *Basic MLTT* (note that Martin-Löf usually adds $\mathbb{N}$, the W-type, and universes to his type theory) consists of the rules for the following sets, which we call the *basic set constructions*:

- the *finite sets* $\underline{\mathrm{N}_n}$ having n elements $\underline{\mathrm{A}^n_0, \ldots, \mathrm{A}^n_{n-1}}$;
- the Σ*-type* $\underline{\Sigma x : A.B[x]}$, the elements of which are pairs $\underline{(\mathrm{p}\ a\ b)}$, where $a : A$ and $b : B[a]$;
- the Π*-type* $\underline{\Pi x : A.B[x]}$ (dependent function type) with elements $\underline{\lambda y.t[y]}$ such that $y : A \Rightarrow t[y] : B[y]$; we define $\underline{A \to B} := \Pi x : A.B$ for some fresh variable x, and see that the non-dependent function type is an instance of the Π-type;
- the *disjoint union* $\underline{A + B}$ of sets $A + B$ with elements $\underline{(\mathrm{inl}\ a)}$ for $a : A$ and $\underline{(\mathrm{inr}\ b)}$ for $b : B$;
- the *identity type* $\underline{(a =_A a')}$ for $A : \mathrm{Set}$, $a : A$, $a' : A$, with element $\underline{\mathrm{refl}\ a}$: $a =_A a$ for $a : A$; $(a =_A a')$ stands for the proposition that a and a' are equal elements of A.

MLTT with natural numbers. The weakest type theory we consider is MLTT extended by the rules for the set of natural numbers $\mathbb{N}$, and a microscopic universe $\underline{\mathrm{Atom}}$. Atom is needed, since without it one can show that one cannot prove Peano's fourth axiom, namely $\neg(0 =_{\mathbb{N}} 1)$. In order to define Atom, we first give better names to N_2 and its elements by defining $\underline{\mathrm{Bool}} := \mathrm{N}_2$, and calling its two elements $\underline{\mathrm{tt}}$ and $\underline{\mathrm{ff}}$. Then the rules for Atom express that, depending on b : Bool, we have Atom b : Set, with equality rules Atom ff $= \mathrm{N}_0$, Atom tt $= \mathrm{N}_1$. Note that Atom converts a Boolean value into the corresponding atomic formula (therefore the name Atom; we have learned this name and its usage for converting Boolean values into propositions from H. Schwichtenberg): (Atom b) is inhabited (i.e., provable) if and only if b is true. The introduction rules for $\mathbb{N}$ express that it contains $\underline{0}$, and that, whenever $n : \mathbb{N}$, then $\underline{\mathrm{S}\ n} : \mathbb{N}$. The elimination rule for $\mathbb{N}$ corresponds to the principle of induction (or primitive recursion into an arbitrary dependent type). Let the resulting type theory be called $\underline{\mathrm{MLTT} + \mathrm{Atom} + \mathbb{N}}$.

It is easy to show that all formulae provable in *Heyting Arithmetic* $\underline{\mathrm{HA}}$ (which is PA with classical logic replaced by intuitionistic logic; we have $|\mathrm{PA}| = |\mathrm{HA}|$) are provable in MLTT + Atom + $\mathbb{N}$ as well, and from this fact one can derive $|\mathrm{MLTT} + \mathrm{Atom} + \mathbb{N}| \geq |\mathrm{HA}|$. Furthermore, each judgement $\Gamma \Rightarrow \theta$ provable in MLTT + Atom + $\mathbb{N}$ can be interpreted as a closed provable formula in HA, from which one can derive that $|\mathrm{MLTT} + \mathrm{Atom} + \mathbb{N}| \leq |\mathrm{HA}|$. Therefore we have $|\mathrm{MLTT} + \mathrm{Atom} + \mathbb{N}| = |\mathrm{HA}| = |\mathrm{PA}| = \epsilon_0$. Note that we have therefore reduced PA to an extension of MLTT and therefore given a constructive justification of PA.

If one looks at this type theory from a consistency point of view, one observes that one has to understand first the basic setup of type theory. Once one has done this, the only problematic construction which remains to be understood is the set of natural numbers $\mathbb{N}$. In order to understand this set, one needs to understand the meaning of a least set closed under finitary introduction rules. This is rather unproblematic, since each element of $\mathbb{N}$ can be introduced by only finitely many applications of the introduction rules. Note that there is an analogy with time: if we start with 0 and in regular time intervals apply the successor function to this element, we will reach each natural number after a finite amount of time.

The W-type. The next step to increase the strength of MLTT is the addition of the *W-type*. Assume A : Set, $x : A \Rightarrow B[x]$: Set. Then the formation rule for the W-type expresses that $\mathrm{W}x : A.B[x]$: Set. The elements of $\mathrm{W}x : A.B[x]$ are labelled well-founded trees, with labels in A, where nodes with label $a : A$ have branching degree $B[a]$. So whenever we have $a : A$ and $f : B[a] \to \mathrm{W}x : A.B[x]$, we can introduce a new element $(\mathrm{sup}\ a\ f) : \mathrm{W}x : A.B[x]$. Here $(\mathrm{sup}\ a\ f)$ is a tree with root $(\mathrm{sup}\ a\ f)$, which is labelled with a, has therefore branching degree $B[a]$, and subtrees $(f\ b)$ for $b : B[a]$. Let $\mathrm{MLTT} + \mathrm{Atom} + \mathbb{N} + \mathrm{W}$ be the extension of the previous theory by the rules for the W-type.

We introduce a standard model of this type theory. The most difficult construction is to interpret the W-type. We define $[\![\, \mathrm{W}x : A.B \,]\!] := [\![\, \mathrm{W} \,]\!]([\![\, A \,]\!], \lambda a \in [\![\, A \,]\!].[\![\, B \,]\!]_{x \mapsto a})$. Here $[\![\, \mathrm{W} \,]\!]$ takes a set A of terms and a function F, mapping A to a set of terms, and returns the least set X of terms closed under reductions, such that, if $a \in A$ and $f \in F(a)\ [\![\to]\!]\ X$, then $(\mathrm{sup}\ a\ f) \in X$. Here $f \in F(a)\ [\![\to]\!]\ X$ means that for $x \in F(a)$, $f\ x \in X$. *Closure under reduction* means that, if b is a term that reduces to $c \in X$, then $b \in X$. More precisely the above has to be reformulated in terms of PERs rather than sets of terms. If we unfold the operators $[\![\, \mathrm{W} \,]\!]$ and $[\![\to]\!]$ in order to obtain a direct definition of $[\![\, \mathrm{W}x : A.B \,]\!]$, we see that $[\![\, \mathrm{W}x : A.B \,]\!]$ can be interpreted as a set defined by a strictly positive inductive definition.

Using this idea one can interpret all statements provable in $\mathrm{MLTT}+\mathrm{Atom}+\mathbb{N}+\mathrm{W}$ in the theory of finitely iterated inductive definitions $\mathrm{ID}_{<\omega}$, and therefore obtain $|\mathrm{MLTT} + \mathrm{Atom} + \mathbb{N} + \mathrm{W}| \leq |\mathrm{ID}_{<\omega}|$. Here $\mathrm{ID}_{<\omega}$ extends PA by adding predicate symbols indexed by strictly positive formulae and axioms stating that the sets given by those predicates are the least fixed points given by the operators corresponding to those formulae. (See (Buchholz et al. 1981) for a monograph on the proof theory of iterated inductive definitions.)

It is known (Buchholz et al. 1981) that the classical theory $\mathrm{ID}_{<\omega}$ has the same strength as the theory $\mathrm{ID}^{\mathrm{int},\mathcal{O}}_{<\omega}$, which is the restriction of $\mathrm{ID}_{<\omega}$ to intuitionistic logic, and where we take as inductive definitions only the predicates corresponding to the nth constructive number classes. An easy way to introduce those number classes O_i is to refer to their representation in type theory. There we define $\mathrm{O}_0 := \mathrm{N}_0$ (the empty set), $\mathrm{O}_1 := \mathrm{N}_1$ (the singleton set), $\mathrm{O}_2 := \mathbb{N}$, and for $n \geq 3$ $\mathrm{O}_n := \mathrm{W}x : \mathrm{N}_n.B[x]$, where $B[\mathrm{A}^n_i] = \mathrm{O}_i$. (More precisely we define $B[x]$ using the identity type in such a way that $B[\mathrm{A}^n_i]$ is isomorphic to O_i.) Then the nth constructive number class can be represented as O_{n+1}. There is a straightforward encoding of O_i as a set O'_i

of natural numbers, and O'_3 turns out to be Kleene's O. O'_{n+2} can be modelled by an n-times nested inductive definition. $\mathrm{ID}^{\mathrm{int},\mathcal{O}}_{<\omega}$ is the theory having as inductive definitions exactly those defining O'_{n+2} for $n \in \mathbb{N}$. It is easy to model $\mathrm{ID}^{\mathrm{int},\mathcal{O}}_{<\omega}$ in MLTT + Atom + $\mathbb{N}$ + W, and we therefore obtain $|\mathrm{ID}_{<\omega}| = |\mathrm{ID}^{\mathrm{int},\mathcal{O}}_{<\omega}| \leq |\mathrm{MLTT} + \mathrm{Atom} + \mathbb{N} + \mathrm{W}| \leq |\mathrm{ID}_{<\omega}|$. It is known as well that $|(\Pi^1_1 - \mathrm{CA})_0| = |\mathrm{ID}_{<\omega}|$. Here $(\Pi^1_1 - \mathrm{CA})_0$ is the second-order theory of Π^1_1-comprehension with induction formulated as an axiom, i.e., induction can only be applied to sets definable by comprehension. Since all proof theoretic equivalences can be shown in HA (although this is not done explicitly, it is well-known in the proof theoretic community that in principle it can be done) and therefore as well in MLTT + Atom + $\mathbb{N}$ + W, it follows that MLTT + Atom + $\mathbb{N}$ + W proves the consistency of approximations of $\mathrm{ID}_{<\omega}$, $\mathrm{ID}^{\mathrm{int},\mathcal{O}}_{<\omega}$, $(\Pi^1_1 - \mathrm{CA})_0$, and therefore we obtain a constructive consistency proof for those theories. By the results of reverse Mathematics (see for instance the monographs (Simpson 1999, 2005)) it is known that $(\Pi^1_1 - \mathrm{CA})_0$ allows one to prove almost all theorems of ordinary mathematics, so already at this stage most of mathematics can be secured constructively.

Relationship to Kripke-Platek set theory. For the understanding of the proof theory of further extensions of type theory, it is important to understand the relationship to extensions of Kripke-Platek set theory (Platek 1966; Barwise 1975; Jäger 1986) and admissible ordinals. One approach to introducing admissible ordinals (we follow here (Barwise 1975, p.3)) is to define an idealised computer, which performs computations involving less than α steps. The functions F from ordinals to ordinals computed by such a computer are called *α-recursive.* An ordinal α is *admissible*, if for every α-recursive function F we have that α is closed under F, i.e., $F \restriction \alpha : \alpha \to \alpha$. Barwise (Barwise 1977) has shown that for any admissible ordinal there exists a structure $\mathcal{M}$ with a definable pairing function such that α is the limit of all closure ordinals of first-order positive inductive definitions in $\mathcal{M}$, and that for any such $\mathcal{M}$ this limit is admissible.

Here a *positive inductive definition* defines the least fixed point of an operator Γ mapping sets of natural numbers to sets of natural numbers, where $\Gamma(X) := \{x \mid \varphi(x, X)\}$ with X occurring only positively in the formula $\varphi(x, X)$. By the positivity of X in φ we have that Γ is *monotone*, i.e., $X \subseteq Y$ implies $\Gamma(X) \subseteq \Gamma(Y)$. In general an *operator* is a function mapping sets of natural numbers to sets of natural numbers. Then we can define by recursion on set-theoretic ordinals α the sets $\Gamma^\alpha \subseteq \mathbb{N}$ by $\Gamma^\alpha := \Gamma(\Gamma^{<\alpha}) \cup \Gamma^{<\alpha}$, where $\Gamma^{<\alpha} = \bigcup_{\beta<\alpha} \Gamma^\beta$. (In case of Γ being monotone, as it is the case here, we see easily that $\Gamma^\alpha = \Gamma(\Gamma^{<\alpha})$). We can easily see that $\Gamma^\alpha \subseteq \Gamma^\beta$ for $\alpha < \beta$. By a straightforward cardinality argument there exists a minimal α such that $\Gamma(\Gamma^{<\alpha}) = \Gamma^{<\alpha}$. This α is called the *closure ordinal* of the operator Γ, and of any corresponding inductive definition. In case of Γ being monotone we obtain that $\Gamma^\alpha = \Gamma^{<\alpha}$ is the least fixed point of Γ.

Kripke-Platek set theory (KP) is a weak axiomatisation of set theory. It is designed in such a way that in the constructible hierarchy we have: L_α is a model of KP if and only if α is admissible. (L_α is the αth constructible set as introduced by Gödel (Gödel 1940) — $\mathrm{L} := \bigcup_{\alpha\in\mathrm{Ord}} \mathrm{L}_\alpha$ forms a model of ZFC + GHC). One can easily show that the height of O_{2+i} is the ith admissible ordinal τ_i, sometimes denoted by ω_i^{ck}. In general we have that, if $[\![A]\!] \in \mathrm{L}_\alpha$, and if for $a \in [\![A]\!]$ we have $[\![B]\!]_{x\mapsto a} \in \mathrm{L}_\alpha$, then $[\![\mathrm{W}x : A.B]\!] \in \mathrm{L}_{\alpha^+}$. Using this idea one can develop a model of any approximation of MLTT+Atom+$\mathbb{N}$+W in the theory $\mathrm{KPl^r}$, and obtain therefore $|\mathrm{MLTT}+\mathrm{Atom}+\mathbb{N}+\mathrm{W}| \leq |\mathrm{KPl^r}|$. Here $\mathrm{KPl^r}$ is a variant of KP, which essentially formulates the existence of finitely many admissible ordinals, so if $\tau_{<\omega} := \sup_{n<\omega} \tau_n$, then $\mathrm{L}_{\tau_{<\omega}}$ is the standard model of $\mathrm{KPl^r}$ (the superscript r refers to the fact that induction over $\mathbb{N}$ and over sets is restricted to Δ_0-formulae.) Using a proof theoretic analysis of $\mathrm{KPl^r}$ and a well-ordering proof in MLTT + Atom + $\mathbb{N}$ + W for the ordinal notation system used in this analysis, one can show as well $|\mathrm{KPl^r}| \leq |\mathrm{MLTT}+\mathrm{Atom}+\mathbb{N}+\mathrm{W}|$, therefore $|\mathrm{MLTT}+\mathrm{Atom}+\mathbb{N}+\mathrm{W}| = |\mathrm{KPl^r}| = |\mathrm{ID}_{<\omega}|$. Especially we have therefore obtained a constructive consistency proof of $\mathrm{KPl^r}$.

If one looks at MLTT+Atom+$\mathbb{N}$+W from a consistency point of view, we observe that one needs to understand, in addition to the constructions studied before, the W-type. For this we need to understand $\mathrm{W}x : A.B$ as the least set closed under the introduction rule, which introduces elements of the form (sup a f). This is much more problematic than understanding $\mathbb{N}$ as discussed before, since the elements can no longer be introduced in finitely many steps. Note as well that the analogy with time, as we had it for the natural numbers, is broken — one needs to understand the analogy of some kind of transfinite time, which goes beyond our direct experience.

Universes. The next set construction added to type theory, which substantially increases its strength, is a universe. A *universe* is a family of sets. This is given by a set U : Set of codes for sets, together with a decoding function T, such that $u : \mathrm{U} \Rightarrow \mathrm{T}\ u : \mathrm{Set}$. Then (T u) is the set denoted by the code u.

The introduction rules express that U is closed under the previous set constructions N_n, Σ, Π, $+$, $=$, $\mathbb{N}$, W, which we call the *standard set constructions.* This is expressed in case of closure under $\mathbb{N}$ as follows: we have a code $\widehat{\mathbb{N}} : \mathrm{U}$ with equality rule $\mathrm{T}\ \widehat{\mathbb{N}} = \mathbb{N} : \mathrm{Set}$. In case of closure under the W-type we have: if $a : \mathrm{U}$, $b : \mathrm{T}\ a \to \mathrm{U}$, then $\widehat{\mathrm{W}}\ a\ b : \mathrm{U}$, and $\mathrm{T}\ (\widehat{\mathrm{W}}\ a\ b) = \mathrm{W}x : \mathrm{T}\ a.\mathrm{T}\ (b\ x) : \mathrm{Set}$.

In the original formulation by Martin-Löf there are no elimination rules for universes. One can add such rules (see, e.g., (Palmgren 1998)), but they do not add any strength to the theory. In fact, it turns out that these rules are not very useful. If one, for instance, added another universe $\mathrm{U}' : \mathrm{Set}$ together with $u : \mathrm{U}' \Rightarrow \mathrm{T}'\ u : \mathrm{Set}$ to the type theory, such that U' has the same closure properties as U (with constructors $\widehat{\mathbb{N}}'$, $\widehat{\mathrm{W}}'$ etc), then the elimination rules don't seem to allow one to define a function $f : \mathrm{U} \to \mathrm{U}'$ such that $f\ \widehat{\mathbb{N}} = \widehat{\mathbb{N}}'$, $f\ (\widehat{\mathrm{W}}\ a\ b) = \widehat{\mathrm{W}}'\ (f\ a)\ (\lambda x.f\ (b\ x))$ unless one extends type theory substantially. (It is believed that the definition of f is not possible, but this hasn't been shown yet. One extension which allows one to define

f is the addition of an extensional equality both on sets and on Set itself, and of large elimination for U.) Let $\mathrm{MLTT} + \mathbb{N} + \mathrm{W} + \mathrm{U}$ be the extension of $\mathrm{MLTT} + \mathrm{Atom} + \mathbb{N} + \mathrm{W}$ by the rules for a universe closed under the W-type, and by omitting the rules for Atom (Atom is definable using the universe).

The universe as introduced above is the one originally introduced by Martin-Löf (Martin-Löf 1984), which we call the *standard universe.* (Martin-Löf considered in his book (Martin-Löf 1984) finitely iterated universes as well). One can consider universes with different closure properties. In its most general form we obtain the concept of an inductive-recursive definition, as discussed in the next section.

Because of their closure under the W-type, standard universes form some kind of inductive definitions, which is closed under the step to the next inductive definition. More precisely, universes correspond to *recursively inaccessible ordinals.* Recursively inaccessible ordinals are admissible ordinals I, which are closed under the step to the next admissible ordinal, i.e., if $\alpha < \mathrm{I}$, then $\alpha^+ < \mathrm{I}$, where α^+ is the next admissible ordinal above α.

Richter has shown ((Richter 1971), see as well (Aczel 1977, Prop. 3.5.1), (Jäger 2001)) that the first recursively inaccessible is the limit of the closure ordinals of $[\Pi^0_1, \Pi^0_0]$-*non-monotone inductive definitions.* A $[\Pi^0_1, \Pi^0_0]$-non-monotone operator is an operator Γ such that $\Gamma(X) = \{x \in \mathbb{N} \mid x \in \Gamma_0(X) \vee (\Gamma_0(X) \subseteq X \;\&\; x \in \Gamma_1(X))\}$, where Γ_0 is given by a Π^0_1-formula and Γ_1 by a Π^0_0-formula. So iteration of Γ means that we iterate Γ_0, until we have reached a fixed point. Then we apply Γ_1 to it. Then we continue applying Γ_0, until we reach the next fixed point, etc. This characterisation of the first recursively inaccessible ordinal as the limit of such closure ordinals makes precise the statement that the first recursively inaccessible corresponds to an inductive definition closed under the step to the next inductive definition.

We will not make use of this characterisation of the first recursively inaccessible ordinal, but illustrate, in which sense a recursively inaccessible ordinal occurs naturally when constructing a model of $\mathrm{MLTT} + \mathbb{N} + \mathrm{W} + \mathrm{U}$: We define by recursion on α sets of terms U^α and functions T^α, mapping elements of U^α to sets of terms, such that $(\mathrm{U}^\alpha, \mathrm{T}^\alpha)_{\alpha \in \mathrm{Ord}}$ is an *increasing sequence of family of sets.* This means that, if $\alpha < \beta$, then $\mathrm{U}^\alpha \subseteq \mathrm{U}^\beta$ and $\mathrm{T}^\beta \restriction \mathrm{U}^\alpha = \mathrm{T}^\alpha$. Let $\mathrm{U}^{<\alpha} = \bigcup_{\beta<\alpha} \mathrm{U}^\beta$, similarly for $\mathrm{T}^{<\alpha}$. Then U^α is defined by closing $(\mathrm{U}^{<\alpha}, \mathrm{T}^{<\alpha})$ under the one-step application of the closure operators for the universe and under reductions. So we have if a reduces to $b \in \mathrm{U}^{<\alpha}$, then $a \in \mathrm{U}^\alpha$ and $\mathrm{T}^\alpha(a) = \mathrm{T}^{<\alpha}(b)$. $\widehat{\mathbb{N}} \in \mathrm{U}^\alpha$ and $\mathrm{T}^\alpha(\widehat{\mathbb{N}}) = [\![\mathbb{N}]\!]$. If $a \in \mathrm{U}^{<\alpha}$ and $b : \mathrm{T}^{<\alpha}(a) \,[\![\to]\!]\, \mathrm{U}^{<\alpha}$, then $\widehat{\mathrm{W}}\ a\ b \in \mathrm{U}^\alpha$ and $\mathrm{T}^\alpha(\widehat{\mathrm{W}}\ a\ b) = [\![\mathrm{W}]\!](\mathrm{T}^{<\alpha}(a), \lambda y \in \mathrm{T}^{<\alpha}(a).\mathrm{T}^{<\alpha}(b\ y))$. Then we have that $\mathrm{U}^\alpha, \mathrm{T}^\alpha \in \mathrm{L}_{\tau_{1+\alpha}}$. The reason why we need to refer to $\tau_{1+\alpha}$ is that $\mathrm{T}^\alpha(\widehat{\mathrm{W}}\ a\ b)$ is obtained by iterating the operator for inductively defining the W-type up to the next admissible ordinal, so in each step of the construction of the universe, we have to go to the next admissible ordinal.

We can define $\llbracket \mathrm{U} \rrbracket = \mathrm{U}^{<\mathrm{I}}$, $\llbracket \mathrm{T}\ a \rrbracket = \mathrm{T}^{<\mathrm{I}}(a)$, where I is the first recursively inaccessible ordinal. The main case in showing that $\llbracket \mathrm{U} \rrbracket$ is closed under the introduction rule for the universe is to prove that if $a \in \llbracket \mathrm{U} \rrbracket$ and $b \in \mathrm{T}^{<\mathrm{I}}(a) \llbracket \rightarrow \rrbracket \llbracket \mathrm{U} \rrbracket$, then $\widehat{\mathrm{W}}\ a\ b \in \llbracket \mathrm{U} \rrbracket$. This can be shown using the fact that $\mathrm{T}^{<\mathrm{I}}(a) \in \mathrm{L}_\gamma$ for some $\gamma < \mathrm{I}$ (using the fact that I is closed under $\lambda\alpha.\tau_\alpha$) and the fact that I is admissible, from which one can deduce that $a \in \mathrm{U}^{<\alpha}$ and $b \in \mathrm{T}^{<\alpha}(a) \llbracket \rightarrow \rrbracket \mathrm{U}^{<\alpha}$ for some $\alpha < \mathrm{I}$, so $\widehat{\mathrm{W}}\ a\ b \in \mathrm{U}^\alpha \subseteq \llbracket \mathrm{U} \rrbracket$.

Using this construction (see (Setzer 1993, 2008) for details) the author has given a model of $\mathrm{MLTT} + \mathbb{N} + \mathrm{W} + \mathrm{U}$ in the theory KPI^+. Here KPI^+ is a variant of KP, which formalises the existence of one recursively inaccessible ordinal and finitely many admissible ordinals above it. We therefore obtain that $|\mathrm{MLTT}+\mathbb{N}+\mathrm{W}+\mathrm{U}| \leq |\mathrm{KPI}^+|$. The author has shown using a technically complex well-ordering proof (Setzer 1993, 1998a,b) that this bound is sharp, so we have $|\mathrm{MLTT} + \mathbb{N} + \mathrm{W} + \mathrm{U}| = |\mathrm{KPI}^+| = \psi_{\Omega_1}(\Omega_{\mathrm{I}+\omega})$. Here $\Omega_\alpha = \aleph_\alpha$, and ψ_κ is the collapsing function, which collapses ordinals to ordinals $< \kappa$. ψ_κ is the main function used in impredicative ordinal notation systems and has a technically difficult definition. I is the first inaccessible cardinal. With some effort cardinals can be replaced by their recursive analogues (admissibles and limits of admissibles; see (Schlüter, n.d.)). We note here that a variant of this theory has been analysed (Griffor and Rathjen 1994) by E. Griffor and M. Rathjen. Note that the above gives a constructive justification of KPI^+, and, since the theories $(\Delta^1_2 - \mathrm{CA}) + (\mathrm{BI})$ (second-order analysis with comprehension restricted to Δ^1_2-formulae and the addition of bar induction) and KPI (KP plus inaccessibility of the universe) are proof theoretically slightly weaker, those two theories are constructively justified as well.

Many more types have been added to MLTT, but the W-type and extended universes are the main ingredients of standard extensions of MLTT, which add proof theoretic strength. E. Palmgren has added superuniverses (Palmgren 1991) (Rathjen (Rathjen 2000, 2001) has analysed the variants one obtains by omitting the W-type, which are very weak) and higher type universes to type theory (Palmgren 1998), which go substantially beyond the theory above, but are instances of indexed inductive-recursive definitions as discussed in the next section.

From a consistency point of view it seems that once one has accepted the W-type, it is easy to accept as well a universe. A universe is, although technically slightly more complicated than the W-type, nothing but an inductive definition, in which we define, whenever we have introduced a new element, recursively T applied to this element, and make use of T applied to previous elements.

4 Inductive-Recursive Definitions and the Mahlo Universe

The logical framework. The theory of inductive-recursive definitions is best introduced using the *logical framework (LF)*. There we have two type levels, namely Set and Type. We have Set : Type, and if A : Set then A : Type.

Set and Type are closed under the LF *dependent function type* $\underline{(x : A) \to B[x]}$. So if A : Type, and $x : A \Rightarrow B[x]$: Type then $\underline{(x : A) \to B[x]}$: Type, similarly with Type replaced by Set. $(x : A) \to B[x]$ has essentially the same rules as $\Pi x : A.B[x]$, and in addition the η-rule. Elements introduced by the introduction rule are denoted by $\underline{(y)b[y]}$ where $y : A \Rightarrow b[y] : B[y]$. We will not use the Π-type in this article in the presence of the LF, and redefine, when using the LF, $\underline{\lambda y.b[y]} := (y)b[y]$, which is a more readable notation, and $\underline{A \to B} := (x : A) \to B$ for some fresh variable x.

In order to easily formulate the theory of inductive recursive definitions, one closes Set and Type as well under the *dependent sum type* $\underline{(x : A) \times B[x]}$ with the same formation rule as $(x : A) \to B[x]$. $(x : A) \times B[x]$ has essentially the same introduction, elimination and equality rules as $\Sigma x : A.B[x]$, and in addition the η-rule. Elements introduced by the introduction rule are denoted by $\underline{\vdash a, b} : (x : A) \times B[x]$ for $a : A$ and $b : B[a]$, and the elimination rule is given as projections: for $a : (x : A) \times B[x]$ we have $\underline{\pi_0\ a} : A$ and $\underline{\pi_1\ a} : B[\pi_0\ a]$.

We add as well $\underline{\{*\}}$: Set, which has essentially the same rules as N_1 and in addition the η-rule.

The LF allows one to introduce type constructors in a more convenient way: For instance, in case of a universe U, T we have T : U $\to$ Set (for which we need that U $\to$ Set : Type), rather than having to write x : U $\Rightarrow$ T x : Set with an extra rule that if $u = u'$: U then T u = T u' : Set.

When modelling type theory with the LF, one has, in order to make sense of types such as Set $\to$ Set, to introduce an interpretation $[\![\,\text{Set}\,]\!]$ of Set, which will usually be a collection of terms (or more precisely a PER). This means that in the model $[\![\,\text{Set}\,]\!]$ is a closed object. Without the LF, there is no need to define $[\![\,\text{Set}\,]\!]$, all that is needed is to make sure that, if we derive A : Set, then $[\![\,A\,]\!]$ is defined and fulfils certain correctness conditions. Therefore, the LF adds complications to the model construction and to the meaning explanations, and therefore we usually avoid the LF. We note that the LF usually doesn't add any strength. In order to reflect this in the model, one can interpret elements of Set as sets, and elements of Type as classes in set theory.

Inductive recursive definitions (IRD) were introduced by P. Dybjer (Dybjer 2000) as a concept, which generalises inductive definitions and universes. IRD allow one to define U : Set and T : U $\to D$ for an arbitrary type D, using strictly positive constructors C. In case of $D = \{*\}$, the function T is trivial (by the η-rule it is $\lambda x.*$), and we obtain the special case of strictly-positive inductive definitions. In case of D = Set we obtain universes.

The idea behind an *IRD* is that we define U inductively. Whenever we introduce an element of U, we recursively define T. Therefore, when referring in the inductive definition of U to elements of U previously introduced, we can make use of T applied to them.

The constructors of U are supposed to be *strictly positive* in U, which means the following: The constructors have the form $\mathrm{C} : (x_1 : A_1, \ldots, x_n : A_n) \to \mathrm{U}$, where the set A_i refers either to sets introduced before one started to define U, T, or $A_i = (y_1 : B_1, \ldots, y_l : B_l) \to \mathrm{U}$, where B_j were introduced before introducing U, T. If A_i doesn't refer to U, then x_i is called a *non-inductive argument*, otherwise it is called an *inductive argument*.

What is crucial is the dependency of A_i on $x_1, \ldots, x_{i-1}$. A_i can directly depend on a non-inductive argument x_j for $j < i$. If x_j is an inductive argument $x_j : A_i = (y_1 : B_1, \ldots, y_l : B_l) \to \mathrm{U}$, we cannot make use of x_j directly: we are still at the stage of forming the set U, and have therefore not defined yet how to introduce other sets from an element in U. However, we can make use of the simultaneously recursively defined function T applied to it, i.e., on $\mathrm{T}\ (x_j\ b_1 \cdots b_l)$, which has been introduced before adding $(\mathrm{C}\ x_1 \cdots x_n)$ to U.

In order to define T recursively, one has to define $\mathrm{T}\ (\mathrm{C}\ a_1 \cdots a_n)$, and this can be defined in an arbitrary way (by using sets introduced before U, T are introduced), but the dependency on the arguments a_i is the same as the dependency of later arguments on previous arguments.

If one wants to make the above precise, one sees that before we can introduce an IRD we are required to carry out some derivations beforehand. For instance, in order to accept the constructor $\mathrm{C} : (x : A, y : B[x] \to \mathrm{U}, z : E[x, y]) \to \mathrm{U}$, $\mathrm{T}\ (\mathrm{C}\ x\ y\ z) = F[x, y, z]$, one needs first to derive the following judgements:

- $A : \mathrm{Set}$;
- $x : A \Rightarrow B[x] : \mathrm{Set}$;
- $E[x, y] = E'[x, \mathrm{T} \circ y]$ for some E' such that $x : A, y' : B[x] \to D \Rightarrow E'[x, y'] : \mathrm{Set}$;
- $F[x, y, z] = F'[x, \mathrm{T} \circ y, z]$, such that $x : A, y' : B[x] \to D, z : E'[x, y'] \Rightarrow F'[x, y', z] : \mathrm{Set}$.

Therefore, a precise formalisation of a theory of IRDs needs to interleave derivations with the introduction of new IRDs. In order to define such a theory using finitely many rules only, the author has together with Peter Dybjer (Dybjer and Setzer 1999, 2003) introduced a data type OP_D of IRDs $\mathrm{U} : \mathrm{Set}$, $\mathrm{T} : \mathrm{U} \to D$. Elements of OP_D are codes for IRDs. If $\gamma : \mathrm{OP}_D$, then one has rules for defining $\mathrm{U}_\gamma : \mathrm{Set}$ inductively while defining $\mathrm{T}_\gamma : \mathrm{U}_\gamma \to D$ recursively. Note that OP_D together with $\lambda\gamma.\mathrm{U}_\gamma$ and $\lambda\gamma.\mathrm{T}_\gamma$ forms a generalised universe (which is a true type and no longer inductive-recursive), the elements of which are inductive-recursively defined sets.

In his original paper (Dybjer 2000), P. Dybjer introduced in fact a slight generalisation of IRD, called *indexed inductive-recursive definitions* (*IIRDs*). The generalisation allows one to define inductively simultaneously several sets

U i : Set for $i : I$, where I : Set is an index set introduced before, while simultaneously recursively defining T i : U $i \to D[i]$. Here $i : I \Rightarrow D[i]$: Type. A closed formalisation of IIRDs was formulated by the author and P. Dybjer in (Dybjer and Setzer 2001, 2006).

Apart from the LF we need only rules for introducing IIRDs, all standard set constructions are instances of IIRDs. The theory of IIRDs allows one to define practically all standard extensions of MLTT considered in the literature including Palmgren's super universes (Palmgren 1991) and higher type universes (Palmgren 1998).

At present, we have introduced only a full set-theoretic model of the theories of IRD and IIRD, which doesn't provide any realistic proof theoretic bound. However, it seems not too complicated to transform this model into a model in the theory KPM^+. Here $\underline{\mathrm{KPM}^+}$ is the variant of KP, which formulates the existence of a recursively Mahlo ordinal and finitely many admissibles above it. A *recursively Mahlo ordinal* is an admissible ordinal M such that, whenever we have for a Δ_0-formula φ that $\forall x \in \mathrm{L_M}.\exists y \in \mathrm{L_M}.\varphi(x,y)$ holds, then there exists an admissible ordinal $\kappa < \mathrm{M}$ such that $\forall x \in \mathrm{L}_\kappa.\exists y \in \mathrm{L}_\kappa.\varphi(x,y)$. The model construction for the theory of IRD and IIRD is rather complicated, we will sketch a model for the Mahlo universe below, which is a theory slightly stronger than the theory of IIRD, and which captures the essence of IIRD.

If such a model is established, we obtain $|\mathrm{IRD}| \leq |\mathrm{IIRD}| \leq |\mathrm{KPM}^+| = \psi_{\Omega_1}(\Omega_{\mathrm{M}+\omega})$, where $\underline{\mathrm{M}}$ is a Mahlo cardinal. A lower bound can be obtained by showing that IRD and IIRD allow to interpret the *red Mahlo universe*, which is a variant of the ordinary Mahlo universe, sometimes called the *black Mahlo universe.* We will first develop the black Mahlo universe, then the red Mahlo universe, and we will illustrate what is known about their strengths. We will see that $\psi_{\Omega_1}(\epsilon_{\mathrm{M}+1}) \leq |\mathrm{IIRD}| \leq |\mathrm{IRD} + \mathrm{ext}|$, where $\psi_{\Omega_1}(\epsilon_{\mathrm{M}+1})$ is slightly smaller than the upper bound $\psi_{\Omega_1}(\Omega_{\mathrm{M}+\omega})$ and $\underline{\mathrm{ext}}$ are the rules of extensionality. The precise strengths of IIRD and IRD are unknown at present.
The (black) Mahlo universe is a universe which makes use of non-positive constructors. It was developed by the author (Setzer 2000b) in order to capture the notion of a recursively Mahlo ordinal. We will in the following develop the formulation of the black Mahlo universe as motivated by the definition of a recursively Mahlo ordinal: Remember that a recursively Mahlo ordinal is an admissible ordinal M such that whenever we have for a Δ_0-formula φ that $\psi := \forall x \in \mathrm{L_M}.\exists y \in \mathrm{L_M}.\varphi(x,y)$ holds, then there exists an admissible ordinal $\kappa < \mathrm{M}$ such that $\psi' := \forall x \in \mathrm{L}_\kappa.\exists y \in \mathrm{L}_\kappa.\varphi(x,y)$ holds as well. If we replace 'admissible' by 'recursively inaccessible', we obtain an equivalent definition. Recursively inaccessible ordinals correspond in type theory to universes. The existence of a recursively inaccessible M is translated into the existence of a universe $\underline{\mathrm{V}}$: Set, $\underline{\mathrm{T_V}} : \mathrm{V} \to$ Set, which is closed under the standard set constructions. $\mathrm{L_M}$ are sets definable up to level M, which correspond to elements of Fam(V) for the corresponding universe. Here $\underline{\mathrm{Fam(V)}} := (x : \mathrm{V}) \times (\mathrm{T_V}\ x \to \mathrm{V})$ is the set of families of sets in V, where $\vdash a, b : \mathrm{Fam(V)}$ is the family which us-

ing pseudo set-theoretic notation might be denoted by $\{\mathrm{T_V}\ (b\ x) \mid x : \mathrm{T_V}\ a\}$. The formula ψ can be represented in type theory as the assumption of some $f : \mathrm{Fam(V)} \to \mathrm{Fam(V)}$. The existence of κ is translated into the existence of a subuniverse $\underline{\mathrm{U}_f} : \mathrm{Set}$, $\underline{\widehat{\mathrm{T}}_f} : \mathrm{U}_f \to \mathrm{V}$. Let $\underline{\mathrm{T}_f} := \mathrm{T_V} \circ \widehat{\mathrm{T}}_f$. That κ is recursively inaccessible translates into $\widehat{\mathrm{U}}_f, \widehat{\mathrm{T}}_f$ being closed under the standard set constructions. Let $\underline{\mathrm{Fam}(\mathrm{U}_f)} := (x : \mathrm{U}_f) \times (\mathrm{T}_f\ x \to \mathrm{U}_f)$. That L_κ is closed under ψ' is represented as U_f having constructors, which reflect f: Assume $a : \mathrm{Fam}(\mathrm{U}_f)$, and lift it to an element $a_\mathrm{V} : \mathrm{Fam(V)}$. Then the constructors introduce the two components of an element $b : \mathrm{Fam}(\mathrm{U}_f)$, such that its lifting to Fam(V) is equal to $(f\ a_\mathrm{V})$. Finally, we need to model that $\kappa \in \mathrm{M}$, which is expressed as the existence of $\underline{\widehat{\mathrm{U}}_f} : \mathrm{V}$ such that $\mathrm{T_V}\ \widehat{\mathrm{U}}_f = \mathrm{U}_f$.

The precise formulation of the Mahlo universe avoids the LF. For this we uncurry f and split it into two functions. So assume $f_0 : (x : \mathrm{V}, y : \mathrm{T_V}\ x \to \mathrm{V}) \to \mathrm{V}$ and $f_1 : (x : \mathrm{V}, y : \mathrm{T_V}\ x \to \mathrm{V}, z : \mathrm{T_V}(f_0\ x\ y)) \to \mathrm{V}$, and let $\mathbf{f} := f_0, f_1$. (So, from f above we would obtain $f_0 = \lambda x.\lambda y.\pi_0\ (f\ \vdash x, y)$ and $f_1 = \lambda x.\lambda y.\lambda z.(\pi_1\ (f\ \vdash x, y))\ z$.) In the following, when we say that $\mathbf{f}$ is a *function from families of sets inside a universe to itself*, we mean that $\mathbf{f} = f_0, f_1$ for the two uncurried components of such a function. Assuming such $\mathbf{f}$, there exists $\underline{\mathrm{U}_\mathbf{f}} : \mathrm{Set}$ and $\underline{\widehat{\mathrm{T}}_\mathbf{f}} : \mathrm{U}_\mathbf{f} \to \mathrm{V}$, such that with $x : \mathrm{U}_\mathbf{f} \Rightarrow \mathrm{T}_\mathbf{f}\ x := \mathrm{T_V}\ (\widehat{\mathrm{T}}_\mathbf{f}\ x) : \mathrm{Set}$ we have:

- $\mathrm{U}_\mathbf{f}, \widehat{\mathrm{T}}_\mathbf{f}$ is closed under the standard set constructions. Closure under $\mathbb{N}$ means that we have $\underline{\widehat{\mathbb{N}}_\mathbf{f}} : \mathrm{U}_\mathbf{f}$, $\widehat{\mathrm{T}}_\mathbf{f}\ \widehat{\mathbb{N}}_\mathbf{f} = \widehat{\mathbb{N}}_\mathrm{V} : \mathrm{V}$, where $\mathrm{T_V}\ \widehat{\mathbb{N}}_\mathrm{V} = \mathbb{N}$. Closure under W means that we have if $a : \mathrm{U}_\mathbf{f}$ and $b : \mathrm{T}_\mathbf{f}\ a \to \mathrm{U}_\mathbf{f}$, then $\underline{\widehat{\mathrm{W}}_\mathbf{f}\ a\ b} : \mathrm{U}_\mathbf{f}$ and $\widehat{\mathrm{T}}_\mathbf{f}\ (\widehat{\mathrm{W}}_\mathbf{f}\ a\ b) = \widehat{\mathrm{W}}_\mathrm{V}\ (\widehat{\mathrm{T}}_\mathbf{f}\ a)\ (\widehat{\mathrm{T}}_\mathbf{f} \circ b)$ where $\mathrm{T_V}\ (\widehat{\mathrm{W}}_\mathrm{V}\ c\ d) = \mathrm{W}x : (\mathrm{T_V}\ c).\mathrm{T_V}\ (d\ x)$.
- $\mathrm{U}_\mathbf{f}$, $\widehat{\mathrm{T}}_\mathbf{f}$ is closed under f_0, f_1. So we have, if $a : \mathrm{U}_\mathbf{f}$ and $b : \mathrm{T}_\mathbf{f}\ a \to \mathrm{U}_\mathbf{f}$, and $a_\mathrm{V} := \widehat{\mathrm{T}}_\mathbf{f}\ a : \mathrm{V}$, $b_\mathrm{V} := \widehat{\mathrm{T}}_\mathbf{f} \circ b : \mathrm{T_V}\ a_\mathrm{V} \to \mathrm{V}$ are the corresponding elements in V, then $\underline{\widehat{\mathrm{f}}_{\mathbf{f},0}\ a\ b} : \mathrm{U}_\mathbf{f}$ and $\widehat{\mathrm{T}}_\mathbf{f}\ (\widehat{\mathrm{f}}_{\mathbf{f},0}\ a\ b) = f_0\ a_\mathrm{V}\ b_\mathrm{V}$. If, in addition, $c : \mathrm{T}_\mathbf{f}\ (f_0\ a_\mathrm{V}\ b_\mathrm{V})$, then $\underline{\widehat{\mathrm{f}}_{\mathbf{f},1}\ a\ b\ c} : \mathrm{U}_\mathbf{f}$ and $\widehat{\mathrm{T}}_\mathbf{f}\ (\widehat{\mathrm{f}}_{\mathbf{f},1}\ a\ b\ c) = f_1\ a_\mathrm{V}\ b_\mathrm{V}\ c$.
- We have a constructor $\underline{\widehat{\mathrm{U}}_\mathbf{f}} : \mathrm{V}$ such that $\mathrm{T_V}\ \widehat{\mathrm{U}}_\mathbf{f} = \mathrm{U}_\mathbf{f}$.

A model of the Mahlo universe can be constructed in KPM^+ as follows (some of the ideas of this model are due to U. Berger (private communication)): For simplicity one treats $\mathrm{U}_\mathbf{f}$ as a subset of V, and therefore treats $\widehat{\mathrm{T}}_\mathbf{f}$ as the identity function, identifies $\widehat{\mathbb{N}}_\mathbf{f}$ with $\widehat{\mathbb{N}}_\mathrm{V}$, $\widehat{\mathrm{W}}_\mathbf{f}$ with $\widehat{\mathrm{W}}_\mathrm{V}$, etc., $\widehat{\mathrm{f}}_{f_0,f_1,0}$ with f_0, and $\widehat{\mathrm{f}}_{f_0,f_1,1}$ with f_1. Furthermore, we omit the subscript V of $\mathrm{T_V}$, $\widehat{\mathrm{W}}_\mathrm{V}$ etc. As for the universe one defines by recursion on α sets $\underline{\mathrm{V}^\alpha}$ and functions $\underline{\mathrm{T}^\alpha}$ with domain V^α by closing them under the same operations as before, such that $(\mathrm{V}^\alpha, \mathrm{T}^\alpha)_{\alpha\in\mathrm{Ord}}$ is an increasing sequence of families of sets.

In addition one adds closure under $\widehat{\mathrm{U}}_\mathbf{f}$ as follows: One defines for every pair of terms $\mathbf{f} = f_0, f_1$ a set $\underline{\mathrm{U}^\alpha_\mathbf{f}} \subseteq \mathrm{V}$. $\mathrm{U}^\alpha_\mathbf{f}$ is the least subset of V^α which is closed under the standard set constructions and under $\mathbf{f}$, provided the results are in

$\mathrm{V}^{<\alpha}$. So if $a \in \underline{\mathrm{U}^{<\alpha}_{\mathbf{f}}}$ $(:= \bigcup_{\beta<\alpha} \mathrm{U}^{\beta}_{\mathbf{f}})$, $b \in \mathrm{T}^{<\alpha}(a)\ [\![\rightarrow]\!]\ \mathrm{U}^{<\alpha}_{\mathbf{f}}$, and $\widehat{\mathrm{W}}\ a\ b \in \mathrm{V}^{<\alpha}$, then $\widehat{\mathrm{W}}\ a\ b \in \mathrm{U}^{\alpha}_{\mathbf{f}}$, similarly for the other standard set constructions. If for the same a, b we have $f_0\ a\ b \in \mathrm{V}^{<\alpha}$, then $f_0\ a\ b \in \mathrm{U}^{\alpha}_{\mathbf{f}}$. If furthermore $c \in \mathrm{T}^{<\alpha}(f_0\ a\ b)$ and $f_1\ a\ b\ c \in \mathrm{V}^{<\alpha}$, then $f_1\ a\ b\ c \in \mathrm{U}^{\alpha}_{\mathbf{f}}$.

By $\mathrm{U}^{\alpha}_{\mathbf{f}}$ being *closed* we mean that the condition of $\mathrm{V}^{<\alpha}$ in the definition of $\mathrm{U}^{\alpha}_{\mathbf{f}}$ is always fulfilled: we have under the above conditions for a and b that $\widehat{\mathrm{W}}\ a\ b \in \mathrm{V}^{<\alpha}$, similarly for the other standard set constructions; furthermore we have for the same a and b that $f_0\ a\ b \in \mathrm{V}^{<\alpha}$ holds, and for the c as above that $f_1\ a\ b\ c \in \mathrm{V}^{<\alpha}$ holds. Note that if $\mathrm{U}^{\alpha}_{\mathbf{f}}$ is closed, then, because of the fact that if $\alpha < \beta$ then $\mathrm{V}^{\alpha} \subseteq \mathrm{V}^{\beta}$ and $\mathrm{T}^{\beta} \restriction \mathrm{V}^{\alpha} = \mathrm{T}^{\alpha}$ we have that $\mathrm{U}^{\alpha}_{\mathbf{f}} = \mathrm{U}^{\beta}_{\mathbf{f}}$ for all $\beta > \alpha$, i.e., $\mathrm{U}^{\beta}_{\mathbf{f}}$ doesn't change any more. If $\mathrm{U}^{\alpha}_{\mathbf{f}}$ is closed, then $\mathrm{U}^{\alpha}_{\mathbf{f}}$ is a complete subuniverse of V closed under $\mathbf{f}$, and we add $\widehat{\mathrm{U}}_{\mathbf{f}}$ to V^{α}, with $\mathrm{T}^{\alpha}(\widehat{\mathrm{U}}_{\mathbf{f}}) = \mathrm{U}^{\alpha}_{\mathbf{f}}$.

Finally, one defines for a recursively Mahlo ordinal M the interpretation of V as $\underline{[\![\mathrm{V}]\!]} := \mathrm{V}^{\mathrm{M}}$, $\underline{[\![\mathrm{T}_{\mathrm{V}}\ a]\!]} := \mathrm{T}^{\mathrm{M}}(a)$, $\underline{[\![\widehat{\mathrm{U}}_{\mathbf{f}}]\!]} := \mathrm{U}^{\mathrm{M}}_{\mathbf{f}}$. That $[\![\mathrm{V}]\!]$ is closed under the introduction rule for $\widehat{\mathrm{U}}_{\mathbf{f}}$ is shown as follows: Assume f_0, f_1 is a function from elements of $[\![\mathrm{V}]\!]$ to itself. This property can be, using the fact that M is admissible, expressed as a formula $\forall\alpha < \mathrm{M}.\exists\beta < \mathrm{M}.\varphi(\alpha,\beta)$, which means that $\mathrm{U}^{\mathrm{M}}_{\mathbf{f}}$ is closed (we use the fact that if α is inaccessible, then V^{α} is closed under the standard set constructions). Then this property holds as well with M replaced by some inaccessible $\kappa < \mathrm{M}$. But then V^{κ} and therefore as well $\mathrm{U}^{\kappa}_{\mathbf{f}}$ are closed under the standard set constructions, and by $\forall\alpha < \kappa.\exists\beta < \kappa.\varphi(\alpha,\beta)$ it follows that $\mathrm{U}^{\kappa}_{\mathbf{f}}$ is closed under f_0, f_1. Therefore, $\mathrm{U}^{\kappa}_{\mathbf{f}}$ is closed and we obtain $\widehat{\mathrm{U}}_{\mathbf{f}} \in \mathrm{V}^{\kappa} \subseteq [\![\mathrm{V}]\!]$.

It is easy to verify that all other rules (including those for $\mathrm{U}_{\mathbf{f}}$, $\widehat{\mathrm{T}}_{\mathbf{f}}$) hold as well. Note that this model construction is predicative in nature: While defining approximations of V, we were never referring to V as a whole, but only to elements of the approximation. Especially the definition of $\mathrm{U}^{\alpha}_{\mathbf{f}}$ only refers to elements of $\mathrm{V}^{<\alpha}$, i.e., elements defined before. Therefore, the model construction reveals that the Mahlo universe is predicative in nature, and gives an insight into its consistency.

One should note however that $[\![\mathrm{V}]\!]$ contains more elements than those justified by the introduction rules for the Mahlo universe: When adding $\widehat{\mathrm{U}}_{\mathbf{f}}$ to V^{α} we have only guaranteed that there exists a subuniverse of V^{α} closed under $\mathbf{f}$. This doesn't guarantee that $\mathbf{f}$ forms a function from families of $[\![\mathrm{V}]\!]$ to itself, only that it is one on families of $[\![\mathrm{U}_{\mathbf{f}}]\!]$.

It would be very satisfactory if one could define a variant of the Mahlo universe that would be in accordance with its standard model, so that one could see immediately that it is predicative in nature, and that one could develop meaning explanations based on this idea. The problem is that the model refers to the collection of all terms (when referring to arbitrary f_0, f_1), and we don't have access to the collection of all terms in MLTT.

Let $\underline{\mathrm{MLTT} + \mathbb{N} + \mathrm{W} + \mathrm{Mahlo}}$ be the type theory formulating the Mahlo universe closed under the W-type. Using the model (Setzer 1996, 2008) the author has shown that $|\text{MLTT} + \mathbb{N} + \text{W} + \text{Mahlo}| \leq |\text{KPM}^+| = \psi_{\Omega_1}(\Omega_{\text{M}+\omega})$, and, using a well-ordering proof (Setzer 2000b), he has shown as well that $\psi_{\Omega_1}(\Omega_{\text{M}+\omega}) \leq |\text{MLTT} + \mathbb{N} + \text{W} + \text{Mahlo}|$, therefore the bound is sharp. Here M is a Mahlo cardinal. If one accepts the consistency of the Mahlo universe one gets therefore a constructive justification of the rather strong theories KPM^+ and KPM (where KPM is the extension of KP by the fact that the set-theoretic universe has the Mahlo property).

The red Mahlo Universe. When looking at IIRDs, we see that the type Set has the same closure properties as the Mahlo universe: Assume $\mathbf{f} = f_0, f_1$ are the two components of a function $((A : \text{Set}) \times (A \to \text{Set})) \to ((A : \text{Set}) \times (A \to \text{Set}))$. Then the rules of induction-recursion show that there exists a universe $\text{U}_{\mathbf{f}} : \text{Set}$, $\text{T}_{\mathbf{f}} : \text{U}_{\mathbf{f}} \to \text{Set}$, such that $\text{U}_{\mathbf{f}}$ is closed under the standard set constructions and under $\mathbf{f}$.

We can now define a theory of the red Mahlo universe as follows: It contains the rules for the standard set constructions, the rules LF, and the rules RedMahlo expressing that for every $\mathbf{f}$ as above there exists a universe $\text{U}_{\mathbf{f}}$, $\text{T}_{\mathbf{f}}$ closed under $\mathbf{f}$ and the standard set constructions. Furthermore, we add *large elimination* (i.e., elimination into a type, which might depend on the argument, rather than into a set) for the W-type and for $\mathbb{N}$. We call the resulting theory $\text{MLTT} + \mathbb{N} + \text{W} + \text{LF} + \text{RedMahlo}$. The well-ordering proof for the black Mahlo universe can easily be adapted in order to show $|\text{MLTT} + \mathbb{N} + \text{W} + \text{LF} + \text{RedMahlo}| \geq \psi_{\Omega_1}(\epsilon_{\text{M}+\omega}) = |\text{KPM}|$ (see the proof by the author in (Dybjer and Setzer 2003)). It should not be too difficult to define a model of $\text{MLTT} + \mathbb{N} + \text{W} + \text{LF} + \text{RedMahlo}$ by interpreting sets in type theory as PERs which are sets, types as PERs which are classes, and $\text{U}_{\mathbf{f}}$ as the least set closed under $\mathbf{f}$ and the standard set constructions (assuming $\mathbf{f}$ is in KPM a term representing a function from families of sets to families of sets). This would show that $|\text{MLTT} + \mathbb{N} + \text{W} + \text{LF} + \text{RedMahlo}| \leq |\text{KPM}|$, and therefore that this bound is sharp.

The red Mahlo universe is a subtheory of the theory of IIRD, which in turn can at least in the presence of an extensional equality be simulated by IRD (unpublished result by P. Dybjer and the author). Assuming large elimination, the red Mahlo universe is as well directly a subtheory of IRD. That the theory of the red Mahlo universe can be interpreted in IIRD even without large elimination can be seen as follows: $\text{MLTT} + \mathbb{N} + \text{W}$, and U_f are instances of IIRD. Furthermore, large elimination can be simulated by using IIRDs. If one wants to define $f : (y : (\text{W}x : A.B[x])) \to D[y]$ where $y : (\text{W}x : A.B[x]) \Rightarrow D[y] : \text{Type}$, which is defined induction on $\text{W}x : A.B[x]$, then one defines first inductive recursively $\text{U}' : (\text{W}x : A.B) \to \text{Set}$ together with $\text{T}' : (y : \text{W}x : A.B) \to \text{U}'\ y \to D[y]$. If the desired equality rule is $f\ (\text{sup}\ a\ b) = g\ a\ b\ (f \circ b)$, where $g : (a : A, b : B[a] \to \text{W}x : A.B, (y : B[a]) \to D[b\ y]) \to D[(\text{sup}\ a\ b)]$ then we take as constructor of U' the function $\text{C} : (a : A, b : B[a] \to \text{W}x : A.B, (y : B[a]) \to \text{U}'[b\ y]) \to \text{U}'\ (\text{sup}\ a\ b)$ with $\text{T}'\ (\text{sup}\ a\ b)\ (\text{C}\ a\ b\ c) = g\ a\ b\ (\lambda y.\text{T}'\ (b\ y)\ (c\ y))$. Now we can define

$f' : (y : \mathrm{W}x : A.B) \to \mathrm{U}'\ y$ by $f'\ (\sup\ a\ b) = \mathrm{C}\ a\ b\ (\lambda y.f'\ (b\ y))$, define $f = \lambda y.\mathrm{T}'\ y\ (f'\ y)$, and verify that f fulfils the equation $f\ (\sup\ a\ b) = g\ a\ b\ (f \circ b)$.

From this it follows that $|\mathrm{KPM}| \leq |\mathrm{MLTT} + \mathbb{N} + \mathrm{W} + \mathrm{LF} + \mathrm{RedMahlo}| \leq |\mathrm{IIRD}| \leq |\mathrm{IRD} + \mathrm{ext}|$ and $|\mathrm{MLTT} + \mathbb{N} + \mathrm{W} + \mathrm{LF} + \mathrm{RedMahlo}| \leq |\mathrm{IRD} + \mathrm{Largeelim}|$, where $\underline{\mathrm{ext}}$ are the rules of extensionality, and $\underline{\mathrm{Largeelim}}$ are the rules of large elimination. This doesn't provide a sharp bound for the theories of IRD and IIRD, we only obtain that their strength (assuming extensionality or large elimination in case of IRD) is in the interval $[|\mathrm{KPM}|, |\mathrm{KPM}^+|]$. Since the black Mahlo universe has strength KPM^+, we can say that it captures the essence of IIRD (for instance, assuming the model of IIRD is developed in KPM^+, it shows the well-foundedness of a sequence of ordinals with supremum the upper bound of the proof theoretic strength of IIRD and therefore the consistency of approximations of IIRD).

5 The Π_3-Reflecting Universe

The Hyper-Mahlo Universe. The first step towards the Π_3-reflecting universe is the hyper-Mahlo Universe. The *hyper-Mahlo Universe* is a universe $\underline{\mathrm{U}_2}, \underline{\mathrm{T}_2}$ such that for every function $\mathbf{f} = f_0, f_1$ from families of sets in U_2 to itself, there exists a subuniverse $\underline{\mathrm{U}_{\mathbf{f},1}}$: Set together with $\underline{\widehat{\mathrm{T}}_{\mathbf{f},1}} : \mathrm{U}_{\mathbf{f},1} \to \mathrm{U}_2$, which is Mahlo, closed under $\mathbf{f}$ (all universes in this section will be closed under the standard set constructions as well), and represented in $\mathrm{U}_{\mathbf{f},1}$ as a code $\underline{\widehat{\mathrm{U}}_{\mathbf{f},2}}$.

Let $\underline{\mathrm{T}_{\mathbf{f},1}}\ a = \mathrm{T}_2\ (\widehat{\mathrm{T}}_{\mathbf{f},1}\ x)$. That $\mathrm{U}_{\mathbf{f},1}$ is a Mahlo universe is expressed by the rule that for every function $\mathbf{g}$ from families of sets in $\mathrm{U}_{\mathbf{f},1}$ to itself there exists a subuniverse $\underline{\mathrm{U}_{\mathbf{f},\mathbf{g},0}}$ together with $\underline{\widehat{\mathrm{T}}_{\mathbf{f},\mathbf{g},0}} : \mathrm{U}_{\mathbf{f},\mathbf{g},0} \to \mathrm{U}_{\mathbf{f},1}$, which is closed under $\mathbf{g}$ and represented as a code $\underline{\widehat{\mathrm{U}}_{\mathbf{f},\mathbf{g},0}}$ in $\mathrm{U}_{\mathbf{f},1}$. Note that this means that we need to have an element representing $\mathrm{U}_{\mathbf{f},\mathbf{g},0}$ in U_2 as well, in order to define $\widehat{\mathrm{T}}_{\mathbf{f},1}\ \widehat{\mathrm{U}}_{\mathbf{f},\mathbf{g},0}$, which means that there needs to be an additional constructor for U_2. There are two ways for achieving this: One possibility would be to add a constructor which introduces directly a code for $\widehat{\mathrm{T}}_{\mathbf{f},1}\ \widehat{\mathrm{U}}_{\mathbf{f},\mathbf{g},0}$ in U_2. When moving to hyper^n-Mahlo universes and beyond it becomes technically rather complicated to make sure that all elements created by a subuniverse are passed on to all universes containing it. The other more simple possibility is to consider $\widehat{\mathrm{T}}_{\mathbf{f},1}$ as a constructor of elements of U_2 rather than a recursively defined function. The disadvantage of this is that we get doubling of elements, e.g., $\widehat{\mathbb{N}}_2$ and $\widehat{\mathrm{T}}_{\mathbf{f},1}\ \widehat{\mathbb{N}}_{\mathbf{f},1}$ are now two different codes for the natural numbers in U_2. This doesn't cause any problems from a proof theoretic point of view, and reduces the technicalities in the following. So in the following, all subuniverses are given as '*inductive subuniverses*', which means that $\widehat{\mathrm{T}}$ is a constructor.

Although a lot of work needs to be carried out in order to analyse this theory, it seems very plausible that this theory has the strength of

$\underline{\mathrm{KP - Hyper - M^+}}$, which is the extension of KP by the existence of a recursively hyper-Mahlo ordinal and finitely many admissible ordinals above it. Here a *recursively hyper-recursively Mahlo ordinal* is an ordinal, which fulfils the same condition as a recursively Mahlo ordinal M, except that the $\kappa < \mathrm{M}$ in the definition of a recursively Mahlo ordinal, which was claimed to exist, can be chosen to be recursively Mahlo rather than simply admissible.

The hyper-n- and hyper-α-Mahlo universes. The next step is to define a *hyper-n-Mahlo universe* which is done by simply iterating the step towards the hyper-Mahlo universe further. Similarly, one can define, assuming a fixed type of ordinals Ord, *hyper-α-Mahlo universes* for $\alpha : \mathrm{Ord}$, which correspond to recursively hyper-α-Mahlo ordinals. Here the definition of *recursively hyper-α-Mahlo ordinals* is similar to that of recursively hyper-Mahlo, except that the κ can for any $\beta < \alpha$ be chosen to be hyper-β-Mahlo.

The autonomous Mahlo universe. (See as well the article (Setzer 2005) by the author.) A *recursively autonomous Mahlo ordinal* is an ordinal κ, which is recursively hyper-κ-Mahlo. If we translate recursively inaccessible ordinals as universes, and the ordinals of a universe $\mathrm{V}, \mathrm{T_V}$ as $\mathrm{W}x : \mathrm{V}.\mathrm{T_V}\ x$, we arrive at the following definition of an autonomous Mahlo universe: It is a universe $\underline{\mathrm{V}}, \underline{\mathrm{T_V}}$, which is hyper-$d$-Mahlo for every $d : (\mathrm{W}x : \mathrm{V}.\mathrm{T_V}\ x)$. Since $(\mathrm{W}x : \mathrm{V}.\mathrm{T_V}\ x)$ can only be defined once the definition of V is complete, we replace this set by the set $\underline{\mathrm{Deg}}$ of *Mahlo degrees*, which is the least set introduced by introduction rule $\underline{\mathrm{sup}'} : (x : \mathrm{V}, y : \mathrm{T_V}\ x \to \mathrm{Deg}) \to \mathrm{Deg}$. Deg is isomorphic to $\mathrm{W}x : \mathrm{V}.\mathrm{T_V}\ x$, but can be defined simultaneously with V and the other constructions defined in the following. We note that elements of Deg depend only locally on V: For every element of $d : \mathrm{Deg}$ we can define an $a : \mathrm{V}$ and $b : \mathrm{T_V}\ x \to \mathrm{V}$, such that all elements of V used in Deg are of the form $(b\ x)$ for some $x : \mathrm{T_V}\ x$.

The autonomous Mahlo universe $\mathrm{V}, \mathrm{T_V}$ is hyper-d-Mahlo for every $d : \mathrm{Deg}$. If we consider $\mathrm{V}, \mathrm{T_V}$ as being inductively defined by the introduction rules, we see that, as new elements are added to V, new elements of Deg, i.e., new Mahlo degrees become available, which result in the possibility of using the Mahlo reflection rule for V into new Mahlo degrees. The existence of the hyper-Mahlo universe claims that this process reaches a fixed point.

The precise formulation of the autonomous Mahlo universe is as follows: We first define depending on $d : \mathrm{Deg}$ a set $\underline{\mathrm{Univ}_d}$ of codes for universes of Mahlo degree d, together with for $u : \mathrm{Univ}_d$ the universes $\underline{\mathrm{U}_{d,u}} : \mathrm{Set}$, $\underline{\mathrm{T}_{d,u}} : \mathrm{U}_{d,u} \to \mathrm{Set}$ given by u. Let $d = \mathrm{sup}'\ a\ b$. That $\mathrm{U}_{d,u}, \mathrm{T}_{d,u}$ has Mahlo degree d means that for every function $\mathbf{f}$ from families of sets in $\mathrm{U}_{d,u}$ to itself and for every $c : \mathrm{T_V}\ a$, i.e., for every subdegree $d' := b\ c$ of d, there exists an element $\underline{\mathrm{u}_{0,d,u,\mathbf{f},c}} : \mathrm{Univ}_{d'}$. Let temporarily $u' := \mathrm{u}_{0,d,u,\mathbf{f},c}$. Then there exists a constructor $\underline{\widehat{\mathrm{T}}_{0,d,u,\mathbf{f},c}} : \mathrm{U}_{d',u'} \to \mathrm{U}_{d,u}$ preserving the denoted sets, and $\mathrm{U}_{d',u'}$ is closed under $\mathbf{f}$. Furthermore, $\mathrm{U}_{d',u'}$ is represented in $\mathrm{U}_{d,u}$, i.e., there exists $\underline{\widehat{\mathrm{U}}_{0,d,u,\mathbf{f},c}} : \mathrm{U}_{d,u}$ such that $\mathrm{T}_{d,u}\ \widehat{\mathrm{U}}_{0,d,u,\mathbf{f},c} = \mathrm{U}_{d',u'}$.

That V, T_V is hyper-d-Mahlo for every $d : \mathrm{Deg}$ means that for every such d and function $\mathbf{f}$ from families of sets in V to itself there exists an element $\underline{u_{1,d,\mathbf{f}}} : \mathrm{Univ}_d$. Let temporarily $u' := u_{1,d,\mathbf{f}}$. Then there exists a constructor $\underline{\widehat{T}_{1,d,\mathbf{f}}} : U_{d,u'} \to V$ with $T_V\ (\widehat{T}_{1,d,\mathbf{f}}\ c) = T_{d,u'}\ c$. $U_{d,u'}$ is closed under $\mathbf{f}$. Furthermore, there exists $\underline{\widehat{U}_{1,d,\mathbf{f}}} : V$ such that $T_V\ \widehat{U}_{1,d,\mathbf{f}} = U_{1,d,\mathbf{f}}$.

The Π_3-reflecting Universe. The subuniverses of the autonomous Mahlo universe have *static Mahlo degrees* $d : \mathrm{Deg}$. However, we cannot assign directly to the autonomous Mahlo universe a static Mahlo degree itself — the subdegrees depend on families of sets in V, T_V (remember that for each $d : \mathrm{Deg}$ we could determine an $a : \mathrm{Fam}(V)$ such that d refers only to elements of V in a). In the Π_3-reflecting universe this is generalised by having *dynamic Mahlo degrees.* The new introduction rule for *(dynamic) Mahlo degrees* $\underline{\mathrm{Deg}}$ is that if $f_0 : (a : V, b : T_V\ a \to V) \to V$, $f_1 : (a : V, b : T_V\ a \to V, c : T_V\ (f_0\ a\ b)) \to \mathrm{Deg}$ are the two components of a function from families of sets in V to families of elements of Deg indexed by V, then there exists $\underline{\mathrm{sup}'}\ f_0\ f_1 : \mathrm{Deg}$. $\underline{\mathrm{Univ}_d}$, $\underline{U_{d,u}}$, $\underline{T_{d,u}}$ are formed as before, but we have as well a constructor $\underline{\widehat{T}_{2,d,u}} : U_{d,u} \to V$ such that $T_V\ (\widehat{T}_{2,d,u}\ a) = T_{d,u}\ a$. Assume $d = \mathrm{sup}'\ f_0\ f_1$, $a : U_{d,u}$, $b : T_{d,u}\ a \to U_{d,u}$, i.e., $\vdash a, b : \mathrm{Fam}(U_{d,u})$. Let $a_V := \widehat{T}_{2,d,u}\ a : V$, $b_V := \widehat{T}_{2,d,u} \circ b : T_V\ a \to V$. Then $\widehat{f}_{0,d,u}\ a\ b : U_{d,u}$ and $T_{d,u}\ (\widehat{f}_{0,d,u}\ a\ b) = T_V\ (f_0\ a_V\ b_V)$, so the first component of d is reflected in $U_{d,u}$. Furthermore, assume $c : T_V\ (f_0\ a_V\ b_V)$. Then $d' := f_1\ a_V\ b_V\ c$ is a subdegree of d relative to $U_{d,u}$ as given by $\vdash a, b : \mathrm{Fam}(U_{d,u})$. Assume $\mathbf{g}$ is a function from families of sets in $U_{d,u}$ to itself. Then there exists a subuniverse of $U_{d,u}$ of Mahlo degree d' closed under $\mathbf{g}$ and represented in $U_{d,u}$. Furthermore, assume $d : \mathrm{Deg}$, $\mathbf{f}$ a function from families of V to itself. Then there exists a subuniverse of V which is an element of Univ_d, represented in V and closed under $\mathbf{f}$. This concludes the definition of the Π_3-reflecting universe.

In (Setzer 2005) the author gives a model of the Π_3-reflecting universe in the theory $\mathrm{KP}\Pi_3^+$, which is defined following a similar methodology as the model of the Mahlo universe, although it is technically much more complicated. Here $\underline{\mathrm{KP}\Pi_3^+}$ is the theory which formulates the existence of a Π_3-reflecting ordinal α, and finitely many admissibles above it. That α is Π_3-*reflecting* means that for every Π_3-formula ψ relative to L_α, where $\psi = \forall x \in L_\alpha.\exists y \in L_\alpha.\forall z \in L_\alpha.\varphi(x, y, z)$, there exists a $u \in L_\alpha$ which is transitive, not empty, and which reflects ψ, which means $\forall x \in u.\exists y \in u.\forall z \in u.\varphi(x, y, z)$. The model of the Π_3-reflecting universe shows that the Π_3-reflecting universe has at most the strength of $|\mathrm{KP}\Pi_3^+|$. We have a sketch of a well-ordering proof which shows that the Π_3-reflecting universe has the strength of $\mathrm{KP}\Pi_3^+$, therefore this bound is sharp. Therefore we obtain, assuming one accepts the consistency of the Π_3-reflecting universe, a constructive justification of $\mathrm{KP}\Pi_3^+$.

6 Future Work

We have seen in the course of this article that there are many theorems about the precise strength of certain theories which still need to be worked out in detail. Especially, the precise strength of the theories of IRD and IIRD still needs to be determined. The next steps towards stronger theories seem to be to develop Π_n-reflecting, Π_α-reflecting, and Π^1_1-reflecting universes. The author has some sketches of type theories of that strength, but he doesn't know yet whether they have the strength of KP extended by corresponding recursively large admissibles. A big step would be to reach the strength of $(\Pi^1_2 - \mathrm{CA}) + (\mathrm{BI})$. M. Rathjen has argued in (Rathjen 2005a) that there is a limit for the proof theoretic strength of extensions of MLTT, which is due to the fact that sets in type theory can be interpreted as non-monotone inductive definitions. This is an interesting point of view, but one might see a paradigm shift happening in type theory, which allows MLTT to go beyond this boundary, see the author's review (Setzer 2006) of Rathjen's article. It remains to be seen whether MLTT can move beyond those boundaries, but we are optimistic that, if a mathematical theory can be analysed proof theoretically, then it will be possible to develop an extension of MLTT, which reaches its strength.

7 Conclusion

We have seen how to develop extensions of MLTT of increasing strength. Whereas for the first theories a direct insight into their consistency was easy, this became increasingly more difficult when moving to stronger theories. This is by Gödel's incompleteness theorem unavoidable. If we formulate any consistency argument mathematically precisely, it becomes a proof in a theory, the strength of which is stronger than the theory in question. Therefore, the stronger the theory, the stronger the theory needs to be in which such a consistency argument is formalised. There are of course technical problems, which add to the complexity of a consistency argument without requiring any strength, and it is a mathematical task to develop theories so that the mathematical technicalities of the consistency argument are as simple as possible (see (Feferman 2000) for an essay on the significance of this area which is called reductive proof theory). But once one has rid theories of any unnecessary mathematical ballast, the real problem of getting an insight is condensed to having to use principles which have a certain proof theoretic strength. All we can do is to prove mathematically that certain theories we want to use are equivalent or reducible to certain other reference theories, for which we have an as easy consistency argument as possible. But the strength of the principles used in this consistency argument will necessarily increase and our trust in theories weakens as we proceed along the proof theoretic scale.

Some remarks on linear logic*

Mitsuhiro Okada

1 Introductory remarks

Linear logic and its variants, often called substructural logics, have shed light on various new aspects of traditional logic. In this note we shall discuss some of them.

For example, linear logic distinguishes between stable and unstable propositions. This distinction provides a different view of the traditional logic. The distinction between stable and unstable propositions is very simple, but it is one of the most important distinctions which linear logic offers us, although the distinction had not been recognized until the late twentieth century in the 2300-year history of formal logic. The distinction appeared for the first time with the discovery of linear logic (by Jean-Yves Girard) in the late 1980s. There have been, of course, various preliminary advances and valuable contributions towards the formalization of this important logical distinction; various pioneering works of the so-called 'substructural' logics succeeded in obtaining the structure of the unstable propositions part of logic, by deleting the stable propositions part. Some of such pioneering 'pre-linear' logical works include, for example, the Lambek calculus and related logics (or, more properly speaking, type theory) of categorial grammars, relevant logics, logic without contraction rules, and others. The traditional view of the logical connectives

* A related version of this paper was published as (Okada 2004). A different earlier version may be found in (Benoist et al. 2006), although the earlier standpoint of the author is different from that in this paper. For a technical introduction to linear logic, J.-Y. Girard's (Girard 1987) and his other writings are always most suggestive and informative, although some philosophical standpoints are not the same as those of the present paper. A minimal technical introduction for non-experts may be found in (Okada 1999). In particular, the reader can see some concrete examples of the use of the non-modal state description of linear logic on the state-description level, which is a lower level than the truth-falsehood level of the traditional logics.

will have to be changed when we become conscious of the use of the structural inference rules from the linear logical or substructure-logical standpoint. As is well known, and as will be explained shortly, traditional logic is decomposed into the modality-free part and a single modal operator. Traditional logic is represented naturally as the modality-free logic wrapped by the modal operator; more precisely speaking, a traditional logical 'proposition' (or the state-of-affairs of the proposition, either in the sense of Husserl or of Wittgenstein's *Tractatus*) can be considered as a modally-wrapped state-description. In other words, a non-modal logically articulated state description can become a logical proposition or a stable logical 'state-of-affairs'; as for the former (a non-modal logical state description), it makes no sense to talk about 'truth' or 'falsehood', while, as for the latter (a modally-wrapped one), it makes sense to talk about its truth value. We call this modal operator the 'stabilizer' operator (and we use the box notation, such as $\Box A$, in this paper, following the tradition of the modal logic community although we usually use the notation $!A$ in the linear logic and process calculus communities).

Now for the moment we consider the modality-free part. Since the linear logical 'stability' modality is nothing other than the logical explicit representation of the structural rules (especially, the contraction rule, which had usually been only implicit in the 2300 years' history of logic until Gentzen). The modality-free part means the part obtained by the critical structural rules of logic. Each of the main logical connectives in traditional logic is now separated into two different logical connectives; for example, the traditional logical conjunction 'and' is divided into two kinds of 'and', i.e., the superpositional 'and', usually denoted as $\otimes$, (or often called 'tensor product') and the selective 'and', usually denoted as &, (often called 'with'). The superpositional 'and' means 'and' in the sense of superpositional state; for example, 'A and B' means the state that a state A and a state B are superposed. On the other hand, the selective 'and' means 'and' in the sense of selection; for example, 'A and B' means the selective state that either a state A or a state B *may* be selected.

a) These distinctions made over various connectives lead us to a global distinction between the superpositional structure of logic and the selective structure of logic.

For example, one could also introduce a distinction between the superpositional 'or' and the selective 'or' by referring to the superposition/selection distinction. The superpositional 'or' represents 'or' in the sense of superposed states, usually denoted as $\invamp$; 'A or B' means the state that a state A and a state B are superposed in such a disjunctive sense that whenever an orthogonal state of A is realized, the superposed state 'A or B' converges into a state B, while whenever an orthogonal state of B is realized, the superposed state 'A or B' converges into a state A, and unless any phenomenon orthogonal (or contradictory in a broad sense) to A or B appears, the 'or'-superposition

'A or B' does not determine one out of the two. On the other hand, the selective state 'A or B' (usually denoted as $\oplus$) means that state A or state B has already been selected by the environment independently of the subject's choice (or, one could also say, by your interlocutor or by your opponent, if you are situated in a dialectic environment). (Note that the selective 'and' is also concerned with selection, but it is, by contrast, the subject's selection; 'A and B' means that the subject *may* choose either A or B as he or she wishes.) The distinction between superpositional and selective connectives can be extended to all logical connectives and logical constants in principle.[1]

One could take a slight look at a rich expressive power by, e.g., an example given in section 2 of (Okada 1999). In terms of the above state-based

[1] The following table shows some correspondences between the logical connectives of linear logic and the traditional logical connectives/rules.

Linear Logic	Traditional Logics
$\otimes$ (superpositional 'and')	$\wedge$ (' and')
& (selective 'and')	
$\parr$ (superpositional 'or')	$\vee$ ('or')
$\oplus$ (selective 'or')	
$\neg A$ (linear negation)	$\neg A$ (traditional negation)
$\multimap$ (linear implication)	$\rightarrow$ (traditional implication)
$\Box A$ ('necessity' modality)	structural rules
$\Diamond A$ ('possibility' modality)	(weakening and contraction)
1 ('one')	$\top$ ('true')
$\top$ ('top')	
$\bot$ ('bottom')	$\bot$ ('false')
0 ('zero')	

As can be seen in the above example, the traditional logical connective $\wedge$ ('and') splits into two different connectives, $\otimes$, &, in linear logic, and the traditional logical connective $\vee$ ('or') into two different connectives, $\parr$, $\oplus$. $\otimes$ and $\parr$ belong to the superpositionals family, and & and $\oplus$ belong to the selectives family. The linear logical implication $A \multimap B$ can be defined in terms of $\neg A \parr B$ (or equivalently $\neg(A \otimes \neg B)$) from the point of view of the duality definition, hence belongs to the superpositional family. The traditional structural inference rules (which are explicit in Gentzen's Sequent Calculus formulation of traditional logics) are now represented by the modality connectives $\Box$, $\Diamond$. (Cf. (Okada 1999)) Girard uses the symbol ! for the necessity-operator, while we use $\Box$ in this chapter since this symbol is more familiar in the context of philosophical discussions. Girard always chooses the symbolism carefully, and this is also the case with the linear logical symbols. There are important suggestive meanings on the chosen symbols. So, we do not intend to propose the change of symbols used in linear logic, here. We just change some in this chapter, in order to fit our symbols to make our philosophical discussion more natural; in particular, we often use the usual implication symbol $\rightarrow$ instead of $\multimap$, and $\neg A$ instead of linear negation $A^{\bot}$.

meaning of the linear logical connectives, one may specify a course menu at a restaurant as follows (This kind of example was first introduced by Y. Lafont in his lectures, as far as the author knows. Here, the example is modified and extended by the author.):

<u>Today's course menu</u>

((Vegetable Soup $\oplus$ Consommé Soup) & Salad) $\otimes$ (Red Wine $⅋$ White Wine) $\otimes$ (Fish & Meat) $\otimes$ $\Box$(Coffee) $\otimes$ (Two euros $\multimap$ (Cake & Ice Cream))

The above means according to the basic state-based linear logical representations as follows: you as a customer get either Soup or Salad as an entrée (by your selection), and when you have chosen Soup, either Vegetable Soup or Consommé Soup will be served, (which cannot be chosen by you as the customer; it is a selection by your environment, such as the chef in the restaurant). A glass of red wine and a glass of white wine are served on the table for you and your guest. You can get one glass as soon as your guest has decided between the two. You as a customer also get a main dish, for which you can choose either Fish or Meat as you select (a selection by yourself). You also get Coffee, which can be refilled as many times as you wish (including 0 times, which means that you could skip Coffee). You also get a dessert if you pay extra two euros, for which you can choose either Cake or Ice Cream. We now remark about the basic distinction between the superpositionals and the selectives.

b) The superpositional logical structure and the selective logical structure are completely dual. Usually linear logicians and substructural logicians use the superpositional implication and the superpositional negation as the basis of logical syntactic rules. However, if one takes the selective implication/negation as the basis for describing the logical syntactic rules, the selective rules become exactly the same form as the original superpositional rules and vice-versa. This means there is no distinction between the two sets of the logical connectives, and it is a free choice to make which set as the superpositional or the selective. So, our interpretation on the 'superpositional' world view and the 'selective' world view is also dual; the distinction between 'superpositional' and 'selective' is itself dual and is a switchable interpretation.

As for b) above, more precisely speaking, one could switch the selective and the superpositional by replacing the interpretation of the sequent $A_1, \ldots, A_n \vdash B_1, \ldots, B_m$. In the usual representation of the linear logic syntax, a sequent $A_1, \ldots, A_n \vdash B_1, \ldots, B_m$ is interpreted by the superpositional sequent in the sense that $\vdash$ means the superpositional implication, namely $A \vdash B$ means $A \multimap B$, which is identical with $(\neg A) ⅋ B$. If we consider the selective $\vdash$ by interpreting $\multimap$ as a selective connective, i.e., by $(\neg A) \oplus B$, and reformulate the sequent calculus, the form of inference rules of & gets exactly the same as the original form of the superpositional inference rules of $\otimes$ with the original superpositional $\vdash$. It is also the case for all other connectives; for example, the form of inference rules of $\otimes$ based on the selective sequent $\vdash$ becomes exactly the same form as the original selective & based on the superpositional sequent $\vdash$; the form of inference rules of $\oplus$ based on the selective sequent $\vdash$ becomes exactly the same form as the original superpositional $\parr$ based on the superpositional $\vdash$.

c) This basic distinction (between the superpositional sub-structure of the world and the selective logical sub-structure of the world) will be convergent into one when we consider the stable states by using the stability modal operator (linear logical modal operator), which is usually representable in the form of the Girard Principle (which we shall explain in the next section).

On the other hand, there has been a tradition of logic which extends a logical inference system to a modal logical inference system by adding modal expressions and modal propositions with the use of modal operators since Aristotle (who augmented the classical syllogistic inferences with modalities) and, more recently, in the early twentieth century, since C.I. Lewis (who augmented modern symbolic logic by adding modalities). The resulting logic with modalities has become a modal logical inference system where the traditional classical logical inference system is a subsystem of the modal system. Ignoring the modal operators in the modal logic inference rules usually yields the non-modal traditional logical inference rules as special cases of the modal system. (In this sense, a traditional normal modal system is a conservative extension of the non-modal traditional logic.[2])

Linear logic, discovered and presented by a French proof theorist, Jean-Yves Girard, as is well known, is traditional logic as a whole; the only difference is that the modal (or, what we call stable) part and the non-modal part are divided *inside* traditional logic itself; in other words, linear logic provides a fine-grained structure of traditional logic where traditional logic itself is expressed as a modal extension of the non-modal or basic part of logic.

[2] This fact of conservation can be easily shown by deleting the modality symbols from any modal proof as the resulting proof structure is still a well-formed non-modal proof, for the usual normal modal systems.

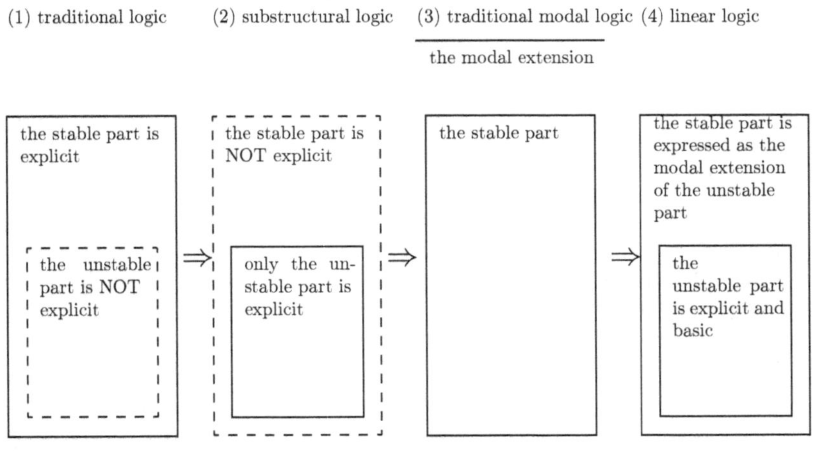

It is to be noted that different modal operator axioms of modal logic give different forms of traditional logic; for example, the most commonly used modal operators, S5 and S4 of Lewis, result in classical logic and intuitionistic logic, respectively. (See (Okada 2004) for a non-formal explanation of this.) There is also a possibility of setting up different forms of traditional logic by using different modal operators; for example, some modal systems give feasible versions of traditional logic, which are often called light logics or light linear logics, where light logics provide modal inference rules which are computationally feasible or realizable. For example, the modal rules for system T give Girard's elementary logic, one of the most basic light logics, which characterizes the class of elementary functions in the Grzegorczyk Hierarchy of computational complexity. The modalities of linear logic play a most important role in understanding logic on various levels. The usual S4 and S5 modalities of linear logic provide the structures of logical stability to form the stable propositions part of linear logic. The S4 linear logical modal propositions behave as the traditional intuitionistic logic, while the S5 linear logical modal propositions behave as the traditional classical logic. Note that the usual setting of linear logic with the S4 modality satisfies the De Morgan duality principles which have been considered a characteristic of classical logic, not of intuitionistic logic; the linear logical representation of intuitionistic logic gives a different view, namely, that the De Morgan principles and intuitionistic logic may be in harmony with each other. A further investigation showed us, in our opinion, how the De Morgan dualities of linear logic could be understood from a more basic fragment of linear logic. Although it has been well known in the linear logic community for a long time that the following three logical systems are essentially the same:

(a) the traditional intuitionistic logic (of Heyting),
(b) the (classical) linear logic, and
(c) the intuitionistic version of linear logic,

the Minimal Logic (in Prawitz' sense) fragment of linear logic no longer has the same logical strength as the above three and is in fact a strict subsystem. We tentatively call this 'Minimal Logic'[3] version of linear logic 'Simple Logic' (Okada and Terui 2003). Simple Logic has various good logical properties; for example, the state-based complete semantics for Simple Logic is much more natural than the one for full linear logic; note that it is well known that the usual complete semantics for linear logic requires a certain pseudo-topological notion to interpret the basic linear logical connectives; $A \otimes B$ in linear logic (A 'and' B in the superpositional sense) is typically interpreted as the 'neighborhood' states to entail A and B, while $A \otimes B$ in Simple Logic simply and exactly means the superposition of the states to entail A and the states to entail B. This is also the case with the interpretation of modality; for a complete semantics of linear logic, the interpretation of modality requires the pseudo-topological notion 'neighborhood' and one says that, roughly speaking, $\Box A$ (linear logical necessity of A) is interpreted as the neighborhood of the states which can *always* realize state A (or, equivalently, which realizes state A as many occasions as needed); on the other hand, $\Box A$ of complete semantics for Simple Logic simply and exactly stands for the states which can *always* realize A. (Note that non-modal A stands for the states which can realize state A, but not necessarily repeatedly more than once.) Another nice property which Simple Logic has, in relation with the full linear logic, is that it explains the need for the pseudo-topological notion of 'neighborhood' in the full linear logical (complete) semantics, on one hand, and the appearance of the basic De Morgan dualities of the full linear logical syntax on the other hand. Both semantic 'neighborhoodness' (or, sometimes called 'closure') and syntactic 'dualities' come with the Gödel-Kolmogorov double-negation interpretation of the full linear logical connectives in terms of the Simple Logic's logical connectives.[4]

The stability operator (linear logical modal operator) *inside* the traditional logic gives a connection to the notion of truth as the (eternal or non-spatio-temporal) state. In other words, well-formed propositions in the (non-modal) linear logic represent changeable states or the procedural descriptions on the state change level, while the stability-modal operator operating on linear logical expressions represent a stable (unchangeable) state, or non-spatio-temporal 'state of affairs'. In this way, a stable (modal) linear logical proposition could be understood as a representation of (the traditional notion of) 'proposition', and an unchangeable (stable) state as (the traditional notion of) 'state of affairs' (for example, in the sense of Husserl), and its denotation (or value) could be understood as a truth value (in the sense of Frege's reference of a proposition). When a logical proposition represents a (changeable) state, a logical implication (such as 'A implies B') can be naturally interpreted as

[3] Minimal in the sense of (Prawitz 1965), for the traditional logics. We found the Minimal Logic correspondence of linear logic and named it 'Simple Logic'.

[4] See the next section for further discussions on this.

a state-change, namely 'a state A changes into a state B'. In fact, the basic implication on the linear logical level can be viewed as 'may-change' of states, which is often in harmony with the most widely accepted resource-sensitive interpretation of the linear logical (and substructure-logical) implication: if the state is A and the linear logical implication 'A implies B' is *applicable*, then it means that state A *may be* replaced by state B on the consumption of the original state A (obviously in the sense that state A is a pre-condition for the possibility of changing into state B and that A disappears when B is realized).[5]

Obviously, there could be various other interpretations of the basic representations of linear logical propositions. In fact, we think the most basic understanding of the linear logical implication $A \multimap B$ is the interpretation by means of the linear logically internal equivalent expression $(\neg A) \parr B$ ('(not A) or B', where $\neg$ ('not') is the linear logical orthogonal and 'or' is the superpositional disjunction $\parr$ (called 'par' in the linear logic research community). This means that the basic implication of linear logic can be understood as a form of the classical 'material' implication although it does not make sense to talk about truth and falsehood (hence, it does not make sense that 'A implies B' of linear logic is 'A is false or B is true'.); $(\neg A) \parr B$ (equivalently $A \multimap B$ in linear logic and in 'any' variants of linear logic) means 'not A' and 'B' are in the 'disjunctive superpositional' state. This disjunctive superposition of 'not A' and 'B' means that the state 'not A' (or, more precisely speaking, 'an orthogonal state with the state A') and the state 'B' are superposed and this superposed state is converged into one of the two superposed states when the other of the two states appears to be (linear logically) contradictory; for example, $A \multimap B$ (A implies B linear logically), or equivalently, $(\neg A) \parr B$ ('the orthogonal of A' and 'B' are disjunctly superposed) is the case that, whenever a state A explicitly appears the superposed state of '$\neg A$' and 'B' becomes convergent into state 'B' because the possible state $\neg A$ becomes contradictory (and whenever state '$\neg B$' appears, the superposed state of '$\neg A$' and 'B' becomes convergent into state '$\neg A$' because the possible state 'B' becomes contradictory.) This more basic interpretation of $A \multimap B$ tells us how the appearance of a state A implies the change of A to B by $A \multimap B$. Hence, state $A \multimap B$ can be understood as a waiting state, waiting for A in order to realize B 'and' waiting for $\neg B$ in order to realize $\neg A$, where 'and' is the selective 'and'. (See the next section on this.) The usual 'resource consumption' interpretation of the linear logical implication (which had been widely employed for a naive first lesson in a linear logic introductory course, interpreting '$A \multimap B$' as 'when a resource A is given, B is generated at the expense of A') is viewed as one of special cases of employing the more basic state-based interpretation of $(\neg A) \parr B$. Note that in linear logic, $A \multimap B$ is officially nothing but an abbreviation (i.e., a

[5] Moreover, the state 'A implies B' itself is consumed when state A and the state 'A implies B' are used to realize state B, unless 'A implies B' is wrapped by the stabilizer modality.

shorthand) of $(\neg A) \parr B$. Both interpretations for the linear logical (and some substructure-logical) expressions with the notion of states and of consumptions are, nevertheless, among the most natural interpretations widely used in the literature, even if those issues are often implicit. However, we think the state-based interpretation is more basic than the resource-consumption-based interpretation, as we explained above. There are also other interesting interpretations of the logical operators and logical proofs, such as ones using functional analysis/operator algebras, matrix operators on vector space and Hilbert spaces, and other more sophisticated mathematical concepts and techniques (see recent works by Jean-Yves Girard and his group). In this paper, our emphasis and purpose are put on the state-change interpretation of linear logical expressions (in particular, implication) so that we can explain how the traditional notions of proposition, of (stable, or non-spatio-temporal) states of affairs, of truth and falsehood, can be understood by means of the modal operators of linear logic. The explanation for the emergence of the stable state of affairs or stable propositions could, in our opinion, be seen as the emergence of the denotations, objectivities or truth-makers, or objectification, from the linear logical viewpoint. The traditional twentieth century's logic was often committed to a dualism; syntax-denotational semantics, apophantic judgement form-objectivity, and other dualisms, which often view the denotations, references, objects, or objectivity side to be separated from the linguistic logical syntactic side of the two sides.[6]

Our view of the linear logical formation of stable propositions and stable states (of affairs) offers a different view that the logical formations do not have to depend on any states of affairs at the beginning, but do explain the states of affairs (or stable states in our sense) in a monist manner.

2 Remarks on some linear logical connectives

In this section, we shall explain why 'classical linear logic is a representation of intuitionistic logic'. In order to connect linear logic to intuitionistic logic, the role of modal operator in linear logic turns out to be of a special philosophical importance. The claim that 'classical linear logic is a representation of intuitionistic logic' has some important consequences. Since linear logic has the basic logical dualities such as all the De Morgan laws, the excluded middle, the double negation-elimination law etc., the claim implies the compatibility between intuitionistic logic and the classical dualities on a logically basic level. At first glance, this might seem contradictory with the famous controversy on the De Morgan dualities (and the excluded middle) vs. the intuitionist standpoint, in the history of philosophy of mathematics. But, in what follows, we shall explain why this new situation is not contradictory.

[6] This tendency might be seen more explicitly in the dualism of truth-bearer and truth-maker, where a truth-bearer such as a proposition depends on truth-maker such as a state of affairs.

Now we shall investigate the notion of modality in linear logic more closely from a philosophical point of view.[7] The original linear logic of Girard is based on the so-called S4-style necessity-operator (and its De Morgan dual: the possibility-operator). The notion of modality in the sense of S4 is most widely used and well known to philosophical logicians. Note that $\Box A$ ('A necessarily holds') in the linear logical sense is not the same as in the traditional modal logical sense, where the underlying logic is full classical logic. The traditional modal system S4 is given by adding several new inference rules for the use of modalities to the underlying classical logic. As we pointed out, the modality-free part of linear logic is logically much weaker than classical logic. The basic inference rules of linear logic are obtained from the traditional (classical) logic by removing the contraction rules and the weakening rules (typically in the framework of Gentzen's Sequent Calculi). Then, the inference rules for the modal operator $\Box$ ('necessity') consist of the usual S4-modal rules, the additional 'explicit' contraction rules and the weakening rules. Essentially, the contraction rule means that the necessary proposition can be 're-used' or is 're-constructible' repeatedly as needed at any time (namely $\Box A \rightarrow \Box A \otimes \Box A$ in symbols). The weakening rule means that a (linear logical) contradiction produces any possibility proposition (namely $\bot \rightarrow \Diamond A$ or equivalently $\Box A \rightarrow \mathbf{1}$ in symbols). These rules tell us that modal propositions can ignore state-change, resource-consumption, and concurrency.[8]

S5-necessity (Leibniz-necessity) can also be considered in the same way as S4-necessity with the S5 rules of the traditional modal logic (by C.I. Lewis).[9]

We first note that with the appearance of the linear logical S4-necessity modal operator, the distinction between the two kinds of 'and' (conjunction), $\otimes$ and $\&$, collapses into the traditional unique 'and'. It is also the case with other connectives such as two kinds of 'or' (disjunction),[10] as well as two kinds of 'true',[11] and two kinds of 'false'.[12] This integration is seen in the

[7] There have been many important and pioneering works and results in the study of substructural logic. The discussion on contraction rules-free logics (Ono and Komori 1985) is one of the pioneering studies. There are also a substantial number of studies regarding the weakening rule-free logic in the context of 'relevant logic', as is well known. There was also an important pre-history of linear logic from the mathematical viewpoint for an application to formal linguistics, namely from the categorial grammar or Lambek calculus (Lambek 1958). However, those pioneering works did not include modality as their main topic, while linear logic does.

[8] In other words, multiple $\Box A$s are constructed concurrently, and nothing is constructed from $\Box A$. (Note that $\mathbf{1}$ is constructed without any resource in linear logic, hence the above weakening rule means the possibility of producing nothing at the expense of $\Box A$).

[9] See (Okada 2004) for the precise forms of linear logical S4-modality and S5-modality.

[10] $⅋$ and $\oplus$

[11] $\mathbf{1}$ and $\top$

[12] $\bot$ and $\mathbf{0}$

following most important and fundamental equivalence relation, which we call the 'Girard Principle' of linear logic: 'always A and always B superpositionally (at the same time)' if and only if 'always one can select either A or B'. In terms of our state-based interpretation, it reads: the conjunctive superposed state of the state 'always A' and the state 'always B' is equivalent to 'always the conjunctive selective state of A or B'. In symbols,

Girard Principle:

$$\Box A \otimes \Box B \Leftrightarrow \Box(A \mathbin{\&} B)$$

In this paper, the usual distinction between additive and multiplicative connectives in the literature of linear logic is explained in terms of the distinction between selections and superpositions. In fact, the multiplicatives could be deepened further by viewing them in terms of 'superpositions' on the state-description level. We now explain this a little further. The multiplicative, which we rename as superpositional, 'A and B' ($A \otimes B$ in symbols) may be understood as 'a state A and a state B are superposed in the sense of and', namely, the state $A \otimes B$ gets the properties of state A and state B at once, in our state-based understanding, while state $A \parr B$ ('A or B' in the multiplicative/superpositional sense) may be understood as 'A and B are superposed in the sense of or', which means that the superposed state $A \parr B$ converges into the single state A whenever the orthogonal state of state B (not B) is realized, and it converges into the single state B whenever the orthogonal state of state A (not A) is realized; on the other hand, unless the orthogonal of B or the orthogonal of A happens, the superposed state $A \parr B$ cannot tell which. Let us return to a restaurant table to describe a little bit further the linear logical Today's Menu (of restaurant 'Chez Lafont' maybe). On the restaurant table, there is a glass of red wine and a glass of white wine.

$$\text{Red Wine} \otimes \text{White Wine}$$

may be interpreted as the 'and'-superposition, which means that you can take both red wine and white wine, while in the Menu presented in section 1 describes

$$\text{Red Wine} \parr \text{White Wine},$$

which can be interpreted as the 'or'-superposition, which means that there are both red wine and white wine on your table, but that the two wine glasses are shared with, say, your partner or your guest at the table; when the situation becomes 'orthogonal of red wine' which could be 'it has turned out that your partner has taken the red wine glass on the table', it has also turned out that the red wine glass is not yours, then it means that the white wine glass is yours. This situation could be linear logically written as:

$$((\text{not Red Wine}) \otimes (\text{Red Wine} \parr \text{White Wine})) \multimap (\text{White Wine})$$

Note that this can be identified with a conditional syllogism, linear Modus Ponens. Namely, 'Red Wine $\parr$ White Wine' could behave like '(not Red Wine)

$\multimap$ (White Wine)'. If the partner has taken the white wine glass on the restaurant table, it has turned out that the white wine glass is not yours, then it means that the red wine glass is yours immediately, namely:

$$((\text{not White Wine}) \otimes (\text{Red Wine} \parr \text{White Wine})) \multimap (\text{Red Wine}),$$

where 'Red Wine $\parr$ White Wine' could behave like '(not White Wine) $\multimap$ (Red Wine)'. In fact, linear logically all the following are equivalent:

$$\begin{gathered}(\text{White Wine}) \parr (\text{Red Wine})\\ (\text{not White Wine}) \multimap (\text{Red Wine})\\ (\text{not Red Wine}) \multimap (\text{White Wine})\\ ((\text{not White Wine}) \multimap (\text{Red Wine})) \,\&\, ((\text{not Red Wine}) \multimap (\text{White Wine}))\end{gathered}$$

Namely, or-superposition $A \parr B$ could be naturally interpreted that 'if an evidence of the orthogonal of A is given, the superposed state $A \parr B$ converges into B, "and" (&) if an evidence of the orthogonal of B is given, the superposed state $A \parr B$ converges into A'. The usual linear logical implication $C \multimap D$ can be understood as (not C)$\parr$ D in our or-superposition interpretation of the multiplicative/superpositional 'or'. Then as we have seen by the example, $A \multimap B$ gets the 'usual' resource-consumption interpretation, via the more basic $\parr$ (multiplicative 'or'), which is in harmony with the basic De Morgan duality (and the basic 'material implication' interpretation of the logical implication that 'A implies B' means 'not A or B'.)

3 Remarks on a linear logical view of intuitionistic logic and of the classical De Morgan dualities

The above-mentioned Girard Principle $\Box A \otimes \Box B \leftrightarrow \Box(A \,\&\, B)$ suggests two ways of interpreting the traditional intuitionistic conjunction, $\Box A \otimes \Box B$ and $\Box(A \,\&\, B)$. A more systematic interpretation of the traditional intuitionistic logical connectives by means of the basic logical connectives of linear logic is to insert the necessity $\Box$ into all linear logical syntactic articulations and define the intuitionistic $A \wedge B$ as $\Box(\Box A \otimes \Box B)$, which is equivalent to $\Box(\Box A \& \Box B)$, as a simple corollary of the Girard Principle. Each traditional proposition (used in the traditional intuitionistic logic) can be interpreted as a specific modal proposition of linear logic in the way in which every appearance of an atomic subproposition and of a logical connective is wrapped by the 'necessity'. In this manner, a proposition, say A, of the intuitionistic logic is translated as a 'stable' proposition A^I of linear logic, where every subformula B of A is also translated as a stable proposition B^I.[13] Note that a stable proposition

[13] More formally speaking, A^I is defined as follows; $P^I \equiv \Box P$ (for an atomic proposition P), $(A \wedge B)^I \equiv \Box(A^I \,\&\, B^I) \equiv \Box(A^I \otimes B^I)$, $(A \vee B)^I \equiv \Box(A^I \oplus B^I)$, $(A \to B)^I \equiv \Box(A^I \multimap B^I) \equiv \Box(\neg A^I \parr B^I)$ (The last equivalence holds because $A \multimap B$ is defined as $\neg A \parr B$ in linear logic.), $(\neg A)^I \equiv \Box(A^I \multimap \bot) \equiv \Box(\neg A^I)$, $(\forall x A(x))^I \equiv \Box \forall x (A(x)^I)$, $(\exists x A(x))^I \equiv \Box \exists x (A(x)^I)$.

A^I is equivalent to $\Box A^I$.[14] Moreover, we do not need to add the outermost $\Box$ for $(A \wedge B)^I$ and $(A \vee B)^I$ if we take $A^I \otimes B^I$ and $A^I \oplus B^I$, respectively, since $\Box(A^I \otimes B^I) \equiv A^I \otimes B^I$ and $\Box(A^I \oplus B^I) \equiv A^I \oplus B^I$. The intuitionistic logical reasoning (in the sense of Heyting's intuitionistic logic) can be perfectly expressed as the reasoning about 'stable-propositions' of classical linear logic for the intuitionistic logical language composed of $\rightarrow, \vee, \wedge, (\forall, \exists)$. If we include $\perp$ and $\neg$ in A, we need to presume the weakening rule in linear logic, namely we presume the so-called affine version of linear logic in the sense that (linear) contradiction $\perp$ can produce any conclusion A (namely, $\perp \rightarrow A$ for any A), which is in fact a problematic rule from the constructivist view.[15]

Fact: *A proposition* A *of intuitionistic logic is provable in intuitionistic logic if and only if the corresponding stable proposition* A^I *of linear logic is provable in linear logic*

An intuitionistic logical proof of a proposition, say A, essentially corresponds to a linear logical proof of the corresponding stable proposition A^I, and vice versa.[16]

The above observation tells us that linear logic accommodates both the logical duality laws and intuitionistic logical reasoning, while it forbids non-constructive reasoning on the selectives. Hence, linear logic is not only considered a logic with dualities which has intuitionistic logic as a part (as the stable propositions part), but *is* a natural representation of intuitionistic logic. From the linear logical viewpoint, the traditional duality laws (of the traditional classical logic) do not represent the fundamental logical duality. This is

---

[14] In fact, the above stability interpretation can be simplified in various ways. For example, $P^I \equiv \Box P, (A \wedge B)^I \equiv (A^I \otimes B^I), (A \vee B)^I \equiv A^I \oplus B^I, (A \rightarrow B)^I \equiv \Box(A^I \rightarrow B^I) \equiv \Box(\neg A^I ⅋ B^I), (\forall x A(x))^I \equiv \Box\forall x(A(x)^I), (\exists x A(x))^I \equiv \exists x(A(x)^I), (\neg A)^I \equiv \Box(\neg A^I)$. In particular, the simplification of the interpretation for $(A \wedge B)^I$, $(A \vee B)^I$ and $(\exists x A(x))^I$ is possible since $\Box(A^* \otimes B^*) \equiv A^* \otimes B^*$, $\Box(A^* \oplus B^*) \equiv A^* \oplus B^*$ and $\Box\exists x A^*(x) \equiv \exists x A^*(x)$ in linear logic.

[15] In the framework of original linear logic, the stability propositions consist of the so-called 'minimal logic' reasoning, rather than the intuitionistic logic reasoning. Here, the 'minimal logic' is obtained from intuitionistic logic by deleting the absurdity inference rule in Gentzen's Natural Deduction system, which says 'any proposition is implied by a contradiction'. (Cf. (Troelstra and Schwichtenberg 2000) and (Prawitz 1965) for the minimal logic.) Although Heyting's formulation of intuitionistic logic contains the absurdity rule, intuitionistic logic without the absurdity rule seems the essential part of intuitionistic logic since the constructibility of any mathematical object from $\perp$ does not play any important role in the Brouwerian constructivity interpretation of intuitionistic mathematics. One could use **0** (the second negative constant) to define the intuitionistic negation in the original linear logic (as used in (Girard 1987)), but it violates the De Morgan dualities.

[16] The reverse direction can be shown modulo cut-elimination.

because, when the linear logical modalities are explicit, the negation of 'necessarily A' does not mean that necessity of the negation of 'necessarily A'. For example, in linear logic, $\neg\neg A$ (the double linear-negations of A) is the same as A. But it is not the case for $\neg\neg A$ (the traditional double-negations of A), since $\neg\neg A$ means $\Box\neg\Box\neg A$, which is equivalent to the proposition that the possibility of A is necessary, $\Box\Diamond A$[17], which is different from A itself in general. This shows why the duality laws and constructive logical reasoning are compatible; this is because we take more fundamental logical dualities than those of the traditional classical logics, which are actually not logical dualities unless one imposes the additional (stronger) reasoning on the modalities (which are S5-modalities[18]).

With a naturally set S5-modality (Leibnizian modality), instead of S4-modality (Girard's original modality), linear logic becomes the (deep structure of) traditional *classical logic*. Namely, classical (standard) logical reasoning could be expressed as reasoning about 'stable-propositions', say A^C of classical linear logic, in the same way as A^I. In fact, with the presence of the S5-rule, A^C can be simplified by deleting all modalities except for atoms.[19]

Fact *A proposition A (of classical logic) is provable in classical logic if and only if a proposition A^C (of linear logic) is provable in classical linear logic with the S5-modality*

Kolmogorov, Gödel and Gentzen showed that the traditional classical logic can be interpreted inside intuitionistic logic by putting the double negation $\neg\neg A$ at each subexpression of a proposition. Since, as was seen, intuitionistic logic itself can be constructed inside linear logic by means of the linear logical S4-modality, one could use the double-negation interpretation of Kolmogorov-Gödel-Gentzen on top of the intuitionistic logic construction of the previous section in order to obtain classical logic from linear logic. Then, one could obtain an alternative classical interpretation based on the S4-modality indirectly by combining it with the double-negation interpretation.) In fact, the linear logical possibility $\Diamond$ serves as a double-negation interpretation (since $\Diamond$ is equivalent to $\neg\Box\neg$, by the De Morgan duality, and the use of $\Diamond$ has

---

[17] Here we use the linear logical duality laws $\neg\Box\neg A \Leftrightarrow \Diamond A$

[18] We discussed the S5-modality in section 3 above.

[19] $P^C \equiv \Box P$ (for an atomic proposition P), $(A \wedge B)^C \equiv \Box(A^C \otimes B^C) \equiv \Box(A^C \,\&\, B^C)$, $(A \vee B)^C \equiv \Box(A^C \oplus B^C) \equiv \Box(A^C \parr B^C)$, $(\neg A)^C \equiv \Box(\neg(A^C))$, $(\forall x A(x))^C \equiv \Box\forall x(A(x)^C)$, $(\exists x A(x))^C \equiv \Box\exists x(A(x)^C)$, $\bot^C \equiv \bot(\equiv \Box\bot)$. The classical (material) implication is defined as usual; $(A \rightarrow B)^C \equiv (\neg A \vee B)^C$ in terms of De Morgan dualities. However, when all atoms P are interpreted as $P^C \equiv \Box P$, all other modalities can be deleted. Namely, $A^C \wedge B^C \equiv A^C \otimes B^C \equiv A^C \,\&\, B^C, A^C \vee B^C \equiv A^C \parr B^C \equiv A^C \oplus B^C, \neg A^C \equiv \neg A^C$.

the effect of the double-negation interpretation. Namely, the classical stable proposition A^D is defined by accompanying atoms, $\neg, ⅋$ with $\Diamond$.[20] Then,

Fact *A proposition A of classical logic is provable in classical logic if and only if the corresponding stable proposition A^D of linear logic is provable in linear logic*

The above classical interpretation works for $\bot$, too, without assuming the affine version.[21] The classical interpretation A^D (based on the possibility-modality) looks simple, if we take the same notion of the boundless consumability interpretation of $\Box$ as the basis.

In fact, it is well known and very obvious from the proof-theoretic view that the S4-modality is much more in harmony with the rest (i.e., the non-modal) part of linear logic (see (Okada 2004), for example) for the harmony in terms of the proof-theoretic fundamental property (the cut-eliminability) (Okada 2004)). From this proof-theoretic point of view, we claim that the S4-modality-based (standard) linear logic is an appropriate representation of full linear logic, as was claimed by Girard, and the S4-modality-based linear logic could be essentially identified with the traditional intuitionistic logic, as many linear logicians claimed including Girard, where the linear logical representation of the traditional intuitionistic logic has the following characteristics:

(A) Full linear logic is a classical system in the sense that the De Morgan dualities (and the excluded middle and elimination of the double negation laws, which are associated closely with the De Morgan duality laws) hold. And at the same time:
(B) Full linear logic has a natural correspondence to the traditional intuitionistic logic (more precisely, the standard linear logical propositions sufficiently wrapped by the linear logical S4-modality correspond to the traditional intuitionistic logical propositions). This fact is true not only for the provable formulas level (hence, the provability is essentially equivalent between the S4-modally wrapped linear logical propositions and the non-modal traditional intuitionistic propositions) but also for the *proofs* level (hence, the non-modal intuitionistic logical proofs are identified with the linear logical proofs with the linear logical S4-modality).
(C) While the traditional intuitionistic logic is naturally representable in linear logic, as is stated in (B), the traditional classical logic is also representable with some modification with the essential use of the (linear logical) negation, as we sketched above.

---

[20] A^D is defined as; $P^D \equiv \Diamond P, (A \vee B)^D \equiv \Diamond(A^D ⅋ B^D) \equiv A^D ⅋ B^D (\equiv \Diamond(A^D \oplus B^D)), (\neg A)^D \equiv \Diamond \neg A^D$.

[21] $(\bot)^D$ can be defined as $?\bot$. On the other hand, $(\top)^D$ is defined as, say, $(A \vee \neg A)^D$. Here, the affine version simplifies this definition of $(\top)^D$ to $?\top$.

We shall repeat a few comments on these (A),(B),(C). First, the coexistence of the two characteristics (A) and (B) might not be seen to be in harmony with each other, as one of the most major controversies in the most advanced philosophy of mathematics in the early 20th century was the controversy between the standpoint with the De Morgan duality and its denial (for example, the Hilbert School wanted to save the classical duality among some others, such as the use of the notions of infinity, infinite sets, etc., by showing *consistency* of the classical use of them, while Brouwer, Weyl and some other intuitionists or constructivists wanted to restrict the use of the classical dualities based on their anti-classical standpoints). The claim of linear logic is that linear logic is a natural logical representation of intuitionistic logic *with* the classical (De Morgan) dualities. Although some readers who are familiar with the historical context of philosophy of mathematics might be confused at first glance about (A) and (B), it is not so surprising. There is no specific magic here for this claim that linear logic is intuitionistic logic with the classical dualities: We remind the reader that linear logic has the S4-style modality to express (explicitly) the structural inference rules (such as the contraction rule, which had been implicit for the first 2300 years' history of logic). We also remind the reader that Gödel showed essentially that the classical logical propositions sufficiently wrapped with the S4-modality[22] are identified with the intuitionistic logical propositions. As we remarked in Section 1 of this paper, Gödel's S4-modal characterization of intuitionistic logic was made in the framework of the traditional S4-modal system, which is already an extended system of classical logic, while intuitionistic logic is a strict subsystem of classical logic. This means that the Gödel-style original characterization of intuitionistic logic was made by the use of a much larger logical framework than intuitionistic logic itself. Although the whole machinery is based on the traditional S4-modal system, which accommodates the De Morgan dualities (including the excluded middle), the traditional S4-modal system itself, in our opinion, cannot be considered an appropriate representation of intuitionistic logic *with* the De Morgan dualities because of its excessively big logical machinery to characterize intuitionistic logic. Linear logic can be, in our opinion, considered an appropriate representation of intuitionistic logic *with* De Morgan dualities because the (non-modal) logical structure is a much weaker (than intuitionistic logic) but very core logical structure with the De Morgan dualities, and the linear logical S4-modal extension of the core part corresponds to intuitionistic logic by means of Gödel's idea (but with a more appropriate setting to characterize intuitionistic logic *with* the De Morgan dualities). Our remark (C) above is to be understood as a simple application of the Gödel-Kolmogorov double-negation interpretation in addition to the linear logical representation of the S4-based intuitionistic logic.

---

[22] Gödel used a variant of S4-modality in his (short) paper, but ever since it has been well known that Gödel's claim works with the S4-modality.

Now we could ask ourselves from where the basic De Morgan dualities of linear logic comes. As for this question, we could make the following remark (D):

(D) From the traditional intuitionist view, the classical De Morgan dualities are usually explained in terms of the Gödel-Kolmogorov double-negation interpretation of the intuitionistic logical propositions.

This intuitionist view offers 'a' coherent understanding of the classical De Morgan dualities from the constructivist standpoint.

Now we ask ourselves how to explain the appearance of the De Morgan dualities in linear logic which is, from our view, essentially equivalent to intuitionistic logic (we repeat our view that 'linear logic *is* intuitionistic logic *with* the De Morgan dualities'). An obvious first try to answer to this might use the same strategy that the traditional intuitionistic logicians used, and say that the De Morgan dualities of linear logic could be explained by means of the Gödel-Kolmogorov double-negation interpretation of the 'intuitionistic' linear logical propositions. In fact, one could set up the 'intuitionistic' version of linear logic in the most natural way and could employ exactly the same Gödel-Kolmogorov procedure to construct the De Morgan dualities on the 'intuitionistic' linear logic. However, one could see quickly that although the whole procedure and machinery work perfectly well in all the technical senses, the contentual logical result would be just empty. This is because it is well known that 'intuitionistic' linear logic is essentially logically equivalent to (the full classical) linear logic itself.[23] In fact, since intuitionistic linear logic is already almost classical linear logic, one could not expect intuitionistic linear logic to play such a major role of explaining classical linear logical (De Morgan) dualities by means of the Gödel-Kolmogorov double-negation (on the basis of intuitionistic linear logic). Now, our real question is: which subsystem of linear logic is the proper subsystem which can be employed for the Gödel-Kolmogorov double-negation to yield (the classical full) linear logic (hence, to explain the De Morgan dualities of linear logic)? An answer that we offer for this is the system of Simple Logic, which we mentioned in Section 1.

Simple Logic is a strict subsystem of intuitionistic linear logic (which is, as well known, almost (classical) linear logic). Simple Logic contains only

[23] This essentially (almost) equivalence relation between intuitionistic linear logic and (classical) linear logic has been known since Danos-Joinet-Schellinx's proof theoretic results. For a more recent purely semantical result, see (Kanovich et al. 2006). In any case, the fact that intuitionistic linear logic is (almost) the same as (classical) linear logic is perfectly in harmony with our claim that (classical) linear logic is a natural representation of (the traditional non-linear) intuitionistic logic *with* the De Morgan dualities, since if intuitionistic linear logic were much more restricted (more constructivist) logic than the full (classical) linear logic it would be impossible for us to claim that (classical) linear logic is a representation of intuitionistic logic.

the intuitionistic rules of $\otimes$ (superpositional 'and'), $\oplus$ (selective 'or'), the intuitionistic S4-style (linear logical) modality $\Box$ (or, ! in the standard linear logical notation), and a very restricted intuitionistic linear implication (which we shall explain below) as well as a few constants.[24] For simplicity, we omit the quantifiers, but we could include the intuitionistic $\exists$ when we wish the quantified logic version of Simple Logic. (Note that the intuitionistic $\exists$ is, of course, a selective connective, as the usual linear logical quantifiers are always selective-operators.) We presume a list of atomic formulas (say, $P_1, P_2, P_3, \ldots$) in the language of Simple Logic. We could identify one of the atomic formulas as the linear logical constant $\perp$ (Bottom), as we usually do for any setting of (any variants of) usual linear logic: Interpreting $\perp$ means, in any setting of linear logic, setting what would be the contradictory situations (the contradictory states in our state-based interpretation) in a semantic model, in terms of the linear logical version of formal semantics, for example. One could denote the first atom P_1 as $\perp$, without loss of generality. The implication of Simple Logic is defined as such a restricted intuitionistic linear implication $A \multimap B$ that the form of B is restricted to the atoms (including $\perp$, which is P_1). As usual, $A \multimap \perp$ can be abbreviated as $\neg A$ (or, $A^{\perp}$ in the standard notation in the linear logic research community). By this means, the notion of orthogonal (the linear logical negation) is introduced, although we did not contain $\neg$ (negation) at the beginning of the Simple Logic setting. Although Simple Logic looks weak (significantly weaker than intuitionistic linear logic), one could easily see that one could employ the Gödel-Kolmogorov double-negation theory on Simple Logic, and as the theory predicts easily, wrapping by the Gödel-Kolmogorov double-negations on the Simple Logic formulas yields all classical linear logical propositions. Especially, the missing primitive logical connectives in the beginning of the Simple Logic setting, such as $\parr$ (the superpositional 'or') and & (the selective 'and') can be naturally defined by the double-negation interpretation A^* in the following way; $B \parr C$ is defined as $\neg(\neg B^* \otimes \neg C^*)$, $B \,\&\, C$ is defined as $\neg(\neg B^* \oplus \neg C^*)$. The linear logical $\otimes$, $\oplus$ and $\Box$ also need to be revised from the original $\otimes$ and $\oplus$ of Simple Logic. For readability, we tentatively write $\otimes'$, $\oplus'$, and $\Box'$ for the (full classical) linear logical connectives, and $\otimes$, $\oplus$, and $\Box$ for those of Simple Logic. $B \otimes' C$ is defined as $\neg\neg(B^* \otimes C^*)$, $B \oplus' C$ is defined as $\neg\neg(B^* \oplus C^*)$, $\Box' B$ is defined as $\neg\neg(\Box B^*)$. The linear logical constants can be obtained from the Simple Logic's $\perp$ (P_1) and **1** in the same way.

From the same viewpoint that the intuitionist understands classical logic and the De Morgan dualities and the excluded middle by means of the Gödel-Kolmogorov double-negation wrappings, one could understand the full classical linear logic (which is essentially the traditional intuitionistic logic in our view) and the fundamental De Morgan dualities and the (superpositional) excluded middle. The classical (full) linear logic with the classical dualities is exactly represented as the Gödel-Kolmogorov way of identifying the classical

[24] Essentially, only $\top$ is enough.

system on the basis of the intuitionistic negation, which can be expressed in Simple Logic (where $\neg A$ is an abbreviation of $A \multimap \bot$, which is an abbreviation of $A \multimap P_1$, where P_1 is an arbitrarily prefixed atomic formula (atom)). It is noted that the 'neighborhood' notion (often called 'closure' in the sense of a certain pseudo-topology in the literature (including Girard's famous introductions to linear logic,[25] Sambin's series of pioneering works on the domain, and work by the present author and his group)) which appears in the standard formal semantics (and its algebraic semantics dialects) is nothing but Gödel-Kolmogorov double-negation on Simple Logic. As one could easily expect from this, the complete formal semantics of Simple Logic does not need any notion of the pseudo-topological notion of 'closure' or 'neighborhood' to interpret any expressions of Simple Logic, and the 'closure' or 'neighborhood' which appears in the complete formal semantics (which we often call 'phase semantics' after Girard's naming) is exactly the semantic correspondence to the Gödel-Kolmogorov double-negation in the Simple Logic syntax.[26]

---

[25] For example, (Girard 1995).

[26] For this 'exact' correspondence between the formal semantics (phase semantics) of the classical full linear logic and the Gödel-Kolmogorov double-negation syntax of Simple Logic, one needs to restrict the Simple Logic's modal expressions in the form of $\Box(A \multimap B)$, where A may contain any nested modalities of the same form while B is an atom. Note that the restriction on B is the basic restriction to define Simple Logic. We also note that this additional restriction of the form of modal expressions does not effect the Gödel-Kolmogorov double negation nor the logical strength of Simple Logic, for the following reasons:

(1) The full linear logical modal expression $\Box' A^*$ is an abbreviation of the double-negation of the Simple Logic modal expression, namely, $\Box' A^*$ is $\neg\neg\Box A^*$ where A^* is of the form of $\neg B^*$ for some B because of the double-negation syntactic interpretation, which is an abbreviation of the form $(B^* \multimap P_1)$. Hence, $\Box' A^*$ is an abbreviation of $(((\Box(B^* \multimap P_1)) \multimap P_1) \multimap P_1)$, which satisfies the additional restriction on Simple Logic.

(2) The version of Simple Logic with the additional restriction on the form of modal expressions is a subsystem of the original Simple Logic without the additional restriction. However, it is a *conservative* subsystem of the original Simple Logic. Hence, the logical provability strengths of the two systems are exactly the same. This conservation theorem can be shown via the syntactic cut-elimination theorem of Simple Logic as well as via purely semantic arguments through completeness and soundness. One could also prove the cut-elimination theorem purely semantically in the same way as the author's well-known proof-theoretic trick on phase semantics. The completeness of the restricted version of Simple Logic holds with respect to the 'simple' (or sometimes called 'naive') phase semantics without the closure notion. Usually the interpretation A^* of A is defined as the set of state-models which may realize A in the simple (or naive) phase semantics, while the usual (non-naive) phase semantics requires the closure (the double-negation), hence the interpretation A^* of A is defined as the closure of the set of state-models which may realize A in the usual (non-naive) phase semantics. For the complete phase semantics of the restricted version of Simple Logic, any possible interpretation A^* of any expression A should contain

Conclusion

We have shown that the logical duality principles and the intuitionistic logical reasoning are compatible in linear logic, where linear logic itself is of constructivist nature. This was shown by separating the non-constructive notion of selective connectives and the constructively admissible notion of superpositive connectives, which are mixed in the traditional logical connectives. We also emphasized the state-based understanding of the linear logical descriptions. We explain the basic distinction between the 'superpositional' (or, often called 'multiplicative') and the 'selective' (or, often called 'additive') connectives. We explained, for example, how the linear logical implication can be understood by means of the notion of superpositional state. We also explained how the state-descriptions could become the states-of-affairs which make sense to talk about truth or falsehood; the switch was explained in terms of Gödel's S4-interpretation inside linear logic. We also explained how one could understand the origin of the fundamental linear logical classical dualities (of the De Morgan laws and the excluded middle law) by means of a Minimal Logic version (which we call Simple Logic) of linear logic. We also tried to explain that linear logical analysis is useful for traditional questions on logic, and that it opens up a new view on various old issues.

Acknowledgement

The author would like to express his thanks to Profs. Jocelyn Benoist, Jean-Yves Girard, Michel Bourdeau, Mark van Atten, the participants of the

any combination of the stable sub-states. Note that A^* is a set of state-models. Each state-model is a (possibly infinitely many) multi-set (in theoretical computer science terminology) or a commutative monoid (in abstract algebra terminology) of primitive sub-states (or, in the resource-consumption interpretation, instead of in the state-based interpretation, each model is a (possibly infinitely many) multi-set of primitive tokens (or, more precisely, of primitive resource tokens)). The sub-multi-set composed of all the stable primitive sub-states (or of all the stable primitive tokens) is presented as (sub-)'set', instead of just a sub-multi-set. This presence of 'set' from the multi-set environment in the (denotational, in Russell's sense) formal semantics exactly corresponds to the content of a contraction rule in the modality. Note that set $\{1, 1, 2, 2, 2\}$ is identical with set $\{1, 2\}$ or set $\{1, 1, 1, 1, 2, 2\}$ in the sense of set (where the contraction rule is exactly here for the set-theoretic objectification or 'denotation'), while these three are different in the sense of multi-set.

The complete semantics of the full Simple Logic without the additional restriction on the use of modality requires a little additional requirement on the simple (naive) phase models; a certain divisibility requirement, which has no obvious correspondence to the standard classical phase semantics of the full linear logic. However, one can easily see that this modification does not change the provability strength, namely the validity of the expressions permitted in the restricted version of Simple Logic, which results in the conservation theorem.

Cerisy Conference *1907–2007: One Hundred Years of Intuitionism*, and Mr. Shigeyuki Nakayama for useful discussions and comments during the preparation stage of this paper.

Two applications of dynamic constructivism: Brouwer's continuity principle and choice sequences in formal topology

Giovanni Sambin

Tristo è quel discepolo che non avanza il suo maestro.

Leonardo da Vinci[1]

1 Beyond Brouwer: the dynamic process

We are here to celebrate a master, namely L.E.J. Brouwer, and one hundred years from the birth of his intuitionism. I believe that the best way to keep Brouwer alive is to try and go beyond Brouwer himself. That is why it seems proper to start with this interesting quote from Leonardo da Vinci. To go beyond Brouwer means to learn his lesson, but also try to fix his mistakes and to soften (in ourselves) some hardness in his personality. And in this way to keep him alive. To confine Brouwer into the prison of what he has materially written, would mean to kill him (again).

I feel consonant with Brouwer's view on almost everything, and actually I feel him as very congenial to my way of thinking, including, as you will see, topics like choice sequences and continuity principles, which have a very controversial status and are often overlooked. As you do with a person you like, I would like to take away his defects, and that is why I address the difficulties in his thought.

The main contradiction in Brouwer, or his main mistake, is his dismissal of the importance of communication, from which his fame of being a solipsist, which means that the others do not exist, on one hand, and on the other hand his whole span of life dedicated to the fight for intuitionistic mathematics. A real solipsist would have kept his thoughts for himself, while he made a great effort to *write*, which means that the others do exist after all, at least those who are supposed to read.

[1] 'Miserable is that disciple who does not go beyond his master.'

The solution in my opinion is simply to accept with no hesitation a matter of fact, namely that others do indeed exist. And hence that one must keep communication in the picture, not only in practice, also in a theoretical sense.

Communication is an essential ingredient of our knowledge. We have no scientific evidence supporting the existence of absolute truth, from above us; this is a commonly accepted concept, and it gives cheap solutions, but it is not easy at all to define it properly, without vicious circles or acts of faith. Human truth can only be constructed from below, bit by bit, very humbly. And it is obtained through communication, which means that human objectivity is the same as intersubjective agreement. This is reached by confronting one's views of reality with the views of other people.

It is convenient if there is agreement between what one believes, and what other people believe, that is teachers, books, fellows,... But this is not obtained for free, once for all. There will be some views which one will accept from others, because one feels they are convincing. And some others which don't convince at all, even after trying hard to understand them. Then one will try to understand the questions better in one's way, and later try to convince somebody else about them, and so on... It is a dynamic process.

The same can be seen between pupil and master. A pupil must follow his master, but not too passively, up to considering his teachings as completed and static; this, as I said before, would be like killing him. But, on the other hand, a pupil is also not supposed to kill the master with acts, for instance by going against all that he says, or some other openly negative attitude.

So it is a little bit of fighting for our own opinions, a little bit of adjusting to other people's opinions. It is convenient that there is an agreement between our views and those of others, but without sacrifying the way of thinking of either.

My general claim is that all abstract concepts are produced by, or even can be identified with, a dynamic process. All words, abstract concepts in particular, are mostly a social construction. And this applies also to concepts in mathematics.[2]

The dynamic process in mathematics takes place between intuition and formalization, that is, one could say, between what we can understand with our mind and what we can express and communicate, or formalize mathematically in the usual way. This is very similar to the interplay between expression and content, or even between an abstract noun and the individuals it refers to.

In my opinion, to grasp the meaning of an expression can be defined as being able to understand the interplay between that expression and its content, that is, between a noun or concept, and the things which we want to fall under it. So in this sense meaning itself is *defined* as a dynamic process, and hence it needs two different levels of abstraction: the word expressing a concept is an abstraction from the things which fall under it.

---

[2] I have argued for this claim most openly in (Sambin 2001). There are also some related previous papers, unfortunately still in Italian.

Also, to express things to oneself one has to communicate. To see this, just recall what you do when you want to keep the memory of something you have grasped: you write it down, and you expect to be able to read it the next day or in a week's time. At least, this happens to me, my memory is definitely not infinite and not extendible, and so I have to write a lot; but if I don't write properly, I will forget.

However, a good communication within oneself, that is a good correspondence between different levels of understanding and involvement, is not only a matter of convenience, it really improves the quality of understanding and of awareness.

Some kind of detachment is necessary, and it is similar to self-detachment for meditation as in Henk Barendregt's talk. Instead of touching thoughts all the time, as it is when you keep them in your mind, you try to step aside and look at them as objectified, as something distinct from you. And this is what you do when you write something for publication, or when you express a piece of mathematics in a formal system.

What Brouwer apparently did not appreciate in full terms is that, even if one really behaves as a solipsist and speaks only with oneself, to reach full awareness of one's thoughts it is useful to express them (in the sense of its etymology *ex-primere*), for instance by writing, even if just for oneself. In fact, this starts the interplay between two different levels of thought, the content, or intuition (from *in-tueri*), which is a time-less and word-less activity, and its linguistic expression (in particular, if one reads oneself after a week or so ...). And again dynamics is what gives life.

In this talk, we will see specific examples of the fact that one can improve awareness by considering also communication, hence language, hence formal systems. We will see that one obtains a much better understanding of matters. As we all know, in contemporary study of foundations and logic, the use of formal systems is the standard practice. And we can call ourselves constructivists and intuitionists, even if Brouwer never used a formal system in the proper sense of the word.

So one can certainly say that communication is the same as, or similar to, objectification. And we can also agree with Brouwer who said, as recalled by Dirk van Dalen in his talk, that this objectification, this passionless expression, is like killing something. But this is true in my opinion only if the process linking content with expression is artificially blocked. If this happens, certainly language becomes a dead thing. In my opinion, this is what happens with formalism; indeed, that is why we are here to celebrate Brouwer and not Hilbert.

2 Logic before mathematics

As an example of the relevance of the dynamic process, I will show that if one takes dynamics seriously, one can obtain a justification of logic, which is simple and in my opinion quite convincing. The claim is that every logical constant is obtained from the interplay between language and metalanguage, more precisely as the solution of the dynamics between language and metalanguage.

Let's look at one example. It is better to consider the connective of disjunction, rather than conjunction, because in this way one can see that the explanation is not trivial. One starts from a definitional equation,

$$\Gamma, A \vee B \vdash \Delta \quad \text{if and only if} \quad \Gamma, A \vdash \Delta \text{ and } \Gamma, B \vdash \Delta$$

which says that the connective of disjunction $\vee$, that is a sign which links two propositions to form a new one, is the reflection of a link *and* between judgements, so a metalinguistic link.

My claim, which I am going to prove in a few lines, is that all you want to know about disjunction $\vee$ is contained implicitly in that equation. We say that a certain connective obeys the principle of reflection, or is defined by following the principle of reflection, if we can obtain its definition by solving an equation like this.

The equation is solved very easily. First, one direction is just the rule of formation, which is written here with the horizontal bar:

$\vee$-formation:

$$\frac{\Gamma, A \vdash \Delta \quad \Gamma, B \vdash \Delta}{\Gamma, A \vee B \vdash \Delta}$$

It is called formation because it says under which conditions one is entitled to assert the new proposition.

The other direction is what I call implicit reflection:

$$\frac{\Gamma, A \vee B \vdash \Delta}{\Gamma, A \vdash \Delta} \qquad \frac{\Gamma, A \vee B \vdash \Delta}{\Gamma, B \vdash \Delta}$$

We look only at the case for A, since the other case, for B, is completely symmetric. By what I call 'trivialize', namely forcing the premiss to be an identity, that is choosing Δ to be $A \vee B$ and Γ to be $\emptyset$, one obtains that

$$A \vdash A \vee B \qquad B \vdash A \vee B$$

holds, and this I call the $\vee$-axiom. Then by composition of $\Gamma \vdash A$ with $A \vdash A \vee B$, that is by performing a cut, one obtains the rule

$\vee$-reflection:

$$\frac{\Gamma \vdash A}{\Gamma \vdash A \vee B} \qquad \frac{\Gamma \vdash B}{\Gamma \vdash A \vee B}$$

To check that this final rule is a good one, one has to show that it is equivalent to what we started from. This is easily seen, because if one takes Γ to be A, then from $\vee$-reflection one obtains as a consequence the axiom $A \vdash A \vee B$, and then by a simple cut, or composition, from this axiom one has implicit reflection.

So the good rules are those named $\vee$-formation and $\vee$-reflection. Why are they good with respect to the other ones? Because they introduce the connective by an inductive definition. While if we stop at implicit reflection, it is a circular definition, because it requires the disjunction to be known to be able to deduce something. Technically, in implicit reflection the formula $A \vee B$ is in the premiss, while in $\vee$-reflection it is in the conclusion of the rule.

Simple as it is, this in my opinion is a very convincing definition of what the logical constant $\vee$ is. We have shown in all details, but I don't repeat it here of course, that exactly the same holds for all connectives (in (Sambin et al. 2000)) and also the quantifiers (in (Maietti and Sambin 2005)). And moreover the solution is found by following in *all* cases exactly the same scheme as illustrated above. Namely, always formation and implicit reflection as the two directions of a definitional equation, axiom obtained always by trivializing and then reflection by composing; and conversely first by trivializing one gets the axioms back, and finally implicit reflection by trivializing. So certainly one has the feeling that there is a deep structure here.

There is a very delicate point, however, to which one should pay some careful attention. When one writes the equation, one necessarily is assuming $A \vee B$ to be a proposition. And this applies also to its solution: in the step 'trivialize' from formation to axioms, for instance, we said that one must put Δ equal to $A \vee B$. But a proposition $A \vee B$, which satisfies the definitional equation, is precisely what we want to define. There seems to be a circularity.

The circularity is only apparent, as I try to explain in all details by way of the following scheme:

1. First assume that $A \vee B$ is a proposition. This is just a hypothesis, for a moment.
2. Assume that $A \vee B$ satisfies the definitional equation written above.
3. Then, solve the equation, as we just did. That is, obtain the inference rules as a consequence.
4. Now, forget about the assumption, and simply *define* $A \vee B$ by induction through the two rules of formation and reflection. This is an inductive definition, and hence fully predicative.
5. Finally, *prove* that the two assumptions are reasonable: $A \vee B$ is a proposition because it is properly defined, given the propositions A and B, and the definitional equation actually holds, because it is equivalent to the two rules, as we just showed.

As one can see, there is no circularity. The scheme is exactly the same as when one solves an equation in mathematics. One does not know the value of the unknown x, but still one assumes it satisfies some condition, usually given by an equation. Then, solving the equation, one finds what the value of x must be, and then finally one shows that the equation is satisfied by that value.

The means by which one can solve the definitional equations are some very simple properties, which one should better specify in advance, and these are the properties of *and* and of *yields*. They are the standard properties of assertions: identity, composition (or cut), associativity of *yields*, and the fact that from the composite assertion, with *and* between two assertions, one can obtain both the first and the second, and conversely.

If one is still puzzled by the question of circularity, there is a nice example from everyday life which should be convincing. Imagine we want to hang a painting, and we want it to be exactly in a specific position. Then we assume the task is solved: that is, we hold the painting in our hands exactly where we want it in the end to be, we check where the nail must be to obtain that effect, we make a little mark with a pencil, we take away the painting, we bang the nail in the wall over the little sign, and then finally we hang the painting. Done! The problem is solved, with no circularity. The reason is simply that by assuming the problem to be solved, we obtain some information, which we use to actually obtain the solution.

One must be a bit careful here: this does not mean that every equation is solvable! We can solve a definitional equation for all logical constants, but for instance second-order definitions of logical constants are still impredicative, until one finds how to solve them. In other words, we are not assuming that whatever implicit definition one writes can be solved. That would really mean assuming that some entity can be created by pure linguistic expressions.

The equation is an implicit definition, or a wish which one would like to achieve. It is solved, and the wish fulfilled, by showing that one has an explicit inductive definition of a proposition which satisfies it. In the end the definition of disjunction, as well as of any other logical constant, is the usual one, and it is given by the rules which describe its use, namely, the two rules of sequent calculus $\vee$-formation and $\vee$-reflection. The example of $\vee$ was chosen on purpose to show that the solution is not just pure playing with symbols. By writing the same equation, but with proposition A and B on the right of *yields*, it is easy to confuse the outcoming connective & with the link *and* between assertions, that is, to confuse the level of propositions (object language) with that of assertions (meta-language).

3 The dynamics between the real and the ideal in mathematics

It is interesting to realize that a very similar dynamics is present also in mathematics, between real and ideal aspects. This is very clear in formal topology. By real/ideal I mean here more or less what Hilbert meant, but more generally just two aspects of mathematics, finitary/infinitary, effective/non-effective. The presence of both aspects in topology is a matter of fact, which was known long ago. I believe the first careful expression is in Martin-Löf's thesis (Martin-Löf 1970) in 1970. The leading idea there is that if you start from open subsets, for instance intervals of real numbers rather than points, then you can develop a really constructive approach to topology since the structure of open subsets can be given effectively. Points are the limit, so to say, of all their approximations given by open subsets. So open subsets are real, while points are ideal entities.

Another example is in type theory, or more generally in a predicative foundation, where sets can be given a very effective interpretation, since one may think of the notion of set as including rules to generate it. Then it will be different from a collection, like that of all subsets of a set, which cannot be generated.

I am going to show that in formal topology there is a dynamic interplay between real and ideal, that is between observables and points, which is similar to the one I just described between metalanguage and object language in logic. I thus follow the same scheme as before, in which the same numbers stand for corresponding content. Contrary to the previous case, at the end we will see that it cannot be solved. But still following the same scheme is interesting because it gives a lot of information.

1. Assume we have points. We don't actually have them yet, but assume we do. And moreover assume they are real objects, in the sense that they form a set X, which can be generated.

2. Assume that points are well linked with a set S of observables, in the sense that the extensions of observables form a base for a topological space. The best way to express this is to introduce a relation $\Vdash$ between points and observables, and $x \Vdash a$ says that a certain point x of X is in the neighbourhood with name $a \in S$, or satisfies the observable a. One can easily find out which are the conditions on the relation $\Vdash$ which allow one to say that we obtain a topological space on the set of points. However, it is not relevant here because the aim is only to illustrate the methodological scheme.

3. Given that the link in 2 corresponds to the definitional equation in the case of logic, what is its solution? We want to solve the link, that is, we want to introduce a structure which is defined through that link, but which lives on one side only, namely on the formal side, that of observables. So the solution is to find out which structure is induced on the set S by the set X which is supposed to exist. One can see that every relation, as $\Vdash$ is, induces on S two natural relations between elements and subsets, which are defined in terms of points of X, namely a cover $a \lhd U$, which is defined by: all points which are in a are also in some $b \in U$, and a positivity $a \ltimes U$, which is defined by: there exists a point x in a all of whose neighbourhoods are in U. So we get a structure $\mathcal{S} = (S, \lhd, \ltimes)$.

4. Now, as we did before in the case of logic, we forget about the equation we want to solve, and about how we obtained that structure, and just *take* the properties of $\lhd$ and $\ltimes$ as an autonomous axiomatic definition. We are absolutely free to do that. We can give any axiomatic definition we wish, as long as it is helpful. And in this case, not only it is very helpful, but it is obtained by a conceptually clear method.

So we call a basic topology every structure $\mathcal{S} = (S, \lhd, \ltimes)$ satisfying those properties (also when there are no X and $\Vdash$ linked with S). We call it a formal topology if it also satisfies a condition on $\lhd$, called convergence, which is induced on S when the neighbourhoods form a base for a topology on X.

5. Then, can we prove 1 and 2 above? We can do it only impredicatively, which in a sense means begging the question. Why is it so? If we have a formal topology $\mathcal{S}$, we can define what a formal point is, namely we can characterize properties that subsets of observables must have for us to be able to conceive the fact that they are converging to something. This is not difficult: we start from the case in which we *have* points, we look at the subset $\{a \in S : x \Vdash a\}$ which is induced by a point x on the set of observables, and by describing its properties (with respect to the structure of $\mathcal{S}$, that is $\lhd$ and $\ltimes$) we will obtain the definition of formal point. It is easy to find out that, for any subset $\alpha \subseteq S$, the conditions we must require on α to call it a formal point are that it is inhabited, convergent (that is, if $a \epsilon \alpha$ and $b \epsilon \alpha$, then there exists $c \epsilon \alpha$ which is covered both by a and b), splits covers (that is, if $a \epsilon \alpha$ and $a \lhd U$, then there exists $b \epsilon U$ with $b \epsilon \alpha$) and enters positivity (if $a \epsilon \alpha$ and $\alpha \subseteq U$, then $a \ltimes U$). Then we define a formal space to be simply the collection $\mathcal{P}t(\mathcal{S})$ of all formal points. But, if we insist on predicativity, this is not a set! In fact, it is formed by subsets of a set, which is not a set.

So we cannot prove 1 and 2, but we can learn a lot from this analysis. In fact we learn when we can do it, what we can do otherwise, how far we are from having a set of points, etc. This is what I like about formal topology. It keeps the distinction between what we really have effectively, and that is the structure on observables, and what is only an ideal definition, as those of formal point and of formal space. It is very interesting to realize that, to be able to keep this distinction, one *must* adopt a predicative foundation, such as type theory.

Another benefit that we get by following this methodology and the hypothetical link with a set X (as in 2) is the discovery of the positivity relation $\ltimes$ between observables, besides the cover $\lhd$. This is an important outcome of the dynamics between X and S. It makes one realize that actually every topological notion is accompanied by its dual. The notion of open, which is defined by a formula with logical structure $\forall\exists$, is accompanied by that of closed, which constructively cannot be defined by complementation, but by a formula having dual form $\exists\forall$. Positivity is dual to cover in the same sense. This idea can be carried out and extended to a treatment of all notions of topology. The result is a theory which I have called the Basic Picture and on which a whole book (Sambin, forthcoming a) is going to appear.

4 Minimalist foundation

The relation between master and pupil, as I said before, is always dynamic. And even if it remains clear who is the master and who is the pupil, often the pupil gives his active contribution, at least with questions, discussions, ... One could say that this is part of the role of a pupil. When can we say that the pupil goes beyond and produces something really *new* with respect to the master? When he comes out with something which is against what the master believes, or on a field that the master had not considered. That's a very concrete way to see that it is his own contribution. Unfortunately, some pupils reverse the criterion, and believe that just by going against the master they can show that they are contributing with something original. The result is that they try to attack all the time, often with no value.

That is why, before saying something different from what Per Martin-Löf says, I went on reflecting on it for years. When I wrote the notes of Per's book (Martin-Löf 1984) around 1982, I tried to convince myself of all that is contained there, but there was one single idea which I did not agree with, and that is the identification of propositions with sets, which means in the end the identification of logic with set theory. I was impressed by its elegance, and I could certainly appreciate this, but I was unable to convince myself that it was really correct to assume it.

In my opinion logic must become before mathematics, must be independent of it, and it must be possible to understand logic, as for example a lawyer does, also without having an education in mathematics. But this difference with respect to the master came out extremely slowly, and it took over twenty years for me to reach a precise formulation.

A posteriori, there is a sign showing that this opinion was deeply rooted in my mind, and that is the fact that in all my papers on formal topology, starting from (Sambin 1987), I never made use of the axiom of choice, which is the most relevant consequence of the propositions-as-sets interpretation. Which means that in my mind I did not assume it. I became fully aware of this perhaps only ten years later, and even later, with the help of Milly Maietti, I could see some more compelling, technical problems which my approach could solve. Only recently have we been able to find a clear formulation (Maietti and Sambin 2005).

So let us see which are the problems to be solved, and how a dynamic attitude can solve them. We must recall two precise propositions; for a simple proof, we refer to (Maietti and Sambin 2005). We denote by AC the axiom of choice, which follows from the propositions-as-sets interpretation, by CT the internal Church Thesis, that is the assumption that every function is recursive, and by EXT the principle of extensionality, in particular extensionality of functions $\forall x(fx = gx) \rightarrow f = g$, which says that if two functions f and g produce equal outputs on every argument, then they are equal. Each of these three assumptions seems quite reasonable, when considered in itself; but if we put them together they clash, in the sense that they produce a plain contradiction:

$$AC + CT + EXT \vdash 0 = 1$$

I take it as understood that together with these assumptions one is assuming also a theory just strong enough to put them into action.

I was not fully aware of the relevance of this result until relatively recently, and I have the impression that somehow it has not been considered with due attention. In fact, it seems to me that it can be interpreted as saying that the computational view of mathematics, which should be consistent with AC and CT, is in fact incompatible with the usual extensional mathematical practice, in which EXT is given for granted (it is often even part of the definition of function).

Another problem is perhaps more well known, and this is the fact that the powerset axiom PSA, which says that if X is a set then also all the subsets of X form a set, and the axiom of choice are constructively incompatible, in the sense that they jointly entail the classical law of excluded middle LEM (saying that $A \vee \neg A$ is true for every proposition A):

$$PSA + AC \vdash LEM$$

This can be proved either over set theory, or over type theory, or even in categories (where actually it was first proven, and it follows from Diaconescu's theorem). So it is a very stable result.

I believe that we have to deal with these two facts, and face openly the problems that they point to. My claim is that one can solve them, and obtain a better picture in general, by looking at matters dynamically. The dynamics, which in the case of logic takes place between language and metalanguage, here takes place between *two* levels of abstraction that one must keep distinct (as first proposed in (Maietti and Sambin 2005)). One is what we call minimal type theory. This is an intensional type theory, which roughly speaking is obtained from Martin-Löf's type theory, as in (Nordström et al. 1990), by avoiding the propositions-as-sets interpretation, that is, the equality between propositions and sets. We realized more recently that it is better to provide it also with a treatment of collections (Maietti, forthcoming). The other is an extensional theory, which is obtained from the intensional theory by abstraction, which in practice means that sets and collections can be equipped with an extensional notion of equality, or in other terms are closed under quotients.

The distinction of two levels solves the first problem, since in this way no single theory is required to be consistent with AC, CT and EXT together. It is the intensional theory at the ground level which is compatible with AC and CT; by the first result, we know that it just *cannot* be extensional. While of course the extensional theory satisfies extensionality, and hence is consistent with EXT.

The dynamic interaction between these two levels is guaranteed by the requirement that they must be linked with each other in a suitable way. So not only the extensional theory must be obtained by abstraction, which means that some information is disregarded or 'forgotten', but also each of its statements must admit an implementation in the ground intensional theory, that is to say, it must be possible to always restore the forgotten information (for some comments on how this is possible, see (Sambin, forthcoming b)). In other words, one can abstract and obtain extensional tools only if one knows how one can implement them back to the ground level. The purpose of this requirement is to obtain a foundational theory over which mathematics can be developed in the usual way (that is, extensionally), but still knowing that one does not lose any computational content.

A typical extensional tool is the notion of truth of a proposition; it is obtained from the intensional theory (in which proof-terms are present) by considering all possible proofs of that proposition to be equal. The notion of subset is obtained in a similar way. One defines subsets to be the same as propositional functions U (over a given set S), but declares two subsets to be equal if they have the same elements. We say that $a \in S$ is an element of U when $U(a)$ is true, independently of the proof one has for it.

We can see that with the minimalist foundation, that is by adopting minimal type theory and by keeping it distinct from the extensional theory, one can solve also the second problem, that of constructive incompatibility between AC and PSA. Indeed, one can see that the minimalist foundation is compatible with all usual foundational theories. The trick is that one can obtain them by expanding either the intensional *or* the extensional level with some further assumptions. Martin-Löf's type theory is obtained simply by putting back the identification of propositions with sets. So in particular one has consistency with AC. Moreover, just to recall that you can have other intensional theories, if you also allow impredicative aspects, you obtain a theory equivalent to that of the language Coq.

But if you pass to the extensional level, as shown by Maietti in (Maietti 2005), by adding PSA you obtain topos theory, or better the theory of the internal language of a topos. And of course adding also AC as we have seen is enough to have excluded middle, and hence essentially axiomatic set theory ZFC.

In the next section, it will be essential to recall that at the minimal level we don't have logic given through the propositions-as-sets interpretation. As a consequence, we don't have the axiom of choice AC, and we even do not have the so-called axiom of unique choice $AC!$, which says that any total and single-valued relation contains a function (whose graph then coincides with the relation).

5 Choice sequences

In several talks, mostly in the talk by Carl Posy, there has been a discussion about the notion of choice sequence itself. Assuming that a mathematical object must be independent of experience, even if one still has to make this clear, the main question is: are choice sequences mathematical objects? Are they independent of experience, or experiments, or any contingent reality? My answer should be clear by the end of this section.

First we start from the main idea, which was very clear to Per Martin-Löf and myself when we first worked on formal topology and which is present in the first paper (Sambin 1987), namely that a choice sequence should be the same as a formal point over a formal topology. Let us see a bit of the formal aspects, just to have at least an approximate idea of the precise definitions. We have a tree, whose elements are called nodes; each node a is connected to a subset $C(a)$ of all its immediate successors.

The notion of choice sequence is visually clear: it is any path starting from the root of the tree, and going up indefinitely. This is not a sufficiently clear mathematical definition; we would like to look at it as a formal point over a certain formal topology, a cover in particular, defined on nodes of that tree.

The formal cover on the tree is the most natural one, namely we define a node a to be covered by the subset $C(a)$ of all its immediate successors. Then we must close under certain conditions to obtain the standard characterization of covers, but these are technical details which at the moment we can leave out. It is enough to recall that in this way for a subset U of nodes, the proposition $a \lhd U$ essentially says the same thing as Brouwer's notion that U is an inductive bar of a. Let $\mathcal{S}$ be the resulting formal topology.

Now we can define a choice sequence to be any subset of nodes α which is a formal point with respect to the cover. If $a \epsilon \alpha$, we say that α passes through a, and write $\alpha \Vdash a$. The fact that α is a formal point means that it is inhabited and splits covers, that is: if α passes through the node a, that is $\alpha \Vdash a$, and $a \lhd U$, then there exists $b \epsilon U$ such that α passes through b, that is $\alpha \Vdash b$. Moreover, α being convergent means that it proceeds without bifurcations.

This looks fine, but already in the 1980s we were aware of a problem. Namely, that if we assume the axiom of choice, and actually the axiom of unique choice is enough, then we can prove that there is a single function f telling for each node a such that $\alpha \Vdash a$ that α will proceed by going through $f(a)$. In other terms α is lawlike:

$$AC! \vdash \alpha : \mathcal{P}t(\mathcal{S}) \rightarrow \alpha \text{ is lawlike}$$

The proof (taken from (Maietti and Sambin, in preparation)) is simple enough to be given here in some detail. Given the subset α, we define a relation R between nodes by putting:

$$R(a, b) \equiv \alpha \Vdash a \rightarrow \alpha \Vdash b \;\&\; b \,\epsilon\, C(a).$$

Now assume that $\alpha \Vdash a$ for an arbitrary a. Recalling that $a \lhd C(a)$ and that a formal point splits covers, we obtain that there exists $b \epsilon C(a)$ such that $\alpha \Vdash b$. That is, $\forall a \exists b R(a, b)$, which says that R is total. It is easy to see that R is single-valued because α is convergent. So by $AC!$ there exists a function f whose graph coincides with R, and hence α is lawlike.

One must be very careful here, because in many foundations $AC!$ is just the definition of what a function is. But if we want to keep the link with implementation, we must keep the notion of function different from the notion of total and single-valued (extensional) relation. It is one thing to have the instructions to compute a function, and another just to know that a relation is total and single-valued: we know that, for any node we reach, there is only one successor, but we have no instructions on how to find it! To see better that they are different, imagine we start with a function, in the sense of some instructions, fixed in advance, which tells us how to go on with the sequence. And then imagine that we pass to its graph. One can think of it as the union of all finite approximations of the function, that is the finite sequences of nodes through which the function is passing. But then it is clear that, if we pass from f to its graph and keep only that, since it is just the union of an infinite family of finite sequences, we lose the instructions completely, and there is no way to recover them. While the axiom of choice, also in its very weak form $AC!$, would say that we can. That is why I believe it is reasonable to have a foundation in which we don't assume $AC!$, to be able to distinguish between two situations which seem to have a different informational content.

There is a second fact which we must consider, and that is the following result by Kleene (see (Troelstra and van Dalen 1988, p.221)):

$$AC! + CT + BI \vdash 0 = 1$$

By using CT, one has that every lawlike sequence is given by a recursive function, and then the claim is proved in the same style as Kleene's clever counterexamples. Again, one must be careful with $AC!$, because sometimes this result is stated without explicitly mentioning it, and it remains hidden in the formulation of CT. So it is better to make it explicit.

Here BI stands for Bar Induction. And what is it? In the language of formal topology, following the main idea that choice sequence equals formal point, its statement is extremely clear. It becomes simply:

$$\forall \alpha (\alpha \Vdash a \rightarrow \alpha \between U) \rightarrow a \lhd U$$

(where $\alpha \between U$ is an abbreviation for $(\exists b \in U) \alpha \Vdash b$). Here we must assume that $\forall \alpha (\alpha \Vdash a \rightarrow \alpha \between U)$ is a proposition, even if it contains a quantification over formal points, that is subsets. If one moreover admits impredicative definitions, that this proposition defines a relation, it is easy to see that it is a cover. So, impredicatively, BI can be interpreted simply as: the cover defined by a quantification over formal points coincides with the cover defined inductively.

Since the main reason for introducing the notion of choice sequence is that it allows one to state BI, we cannot do without BI. But it is a fact that the three assumptions $AC!$, CT and BI together are simply contradictory! The solution is that one must take away $AC!$, which means to adopt the minimalist foundation. Then the first idea is still right and one can simply define choice sequences to be the same as formal points, with no contradictions.

Of course, to show that one is not falling into a system totally deprived of effectivity, one must show that the two assumptions BI and CT are compatible. This has been done in (Maietti and Sambin, in preparation).

So, regarding the question whether choice sequences are mathematical objects, the answer is: yes, they are mathematical objects, because they are the formal points in a certain formal topology (and nobody doubts that these two definitions are mathematical). But one must be very careful which foundation one chooses. So there is still some subjective choice here, namely, the choice of foundation: we are free to do that according to taste. But it is objective that if one works with the minimalist foundation that we suggest, then the identification of choice sequences with formal points works. Otherwise, due to the presence of $AC!$ in all other known foundations, this is not possible, because by $AC!$ all formal points become lawlike sequences.

Finally, it is very interesting to note that in his book (Martin-Löf 1970) Martin-Löf already had a notion of point. In the case of trees, it is exactly what now we call a formal point according to the general definition, which we gave together 15 years later, except for the further condition that in (Martin-Löf 1970) a point is recursively enumerable. So some kind of effectivity was implicit in the very definition, while here we just take subsets. And that is also one of the reasons why the identification of choice sequences with formal points is meaningful, since otherwise it would give only lawlike sequences.

6 Continuity principles

Formal topology also provides a good framework to deal with another controversial issue introduced by Brouwer, that is the result saying that every total function defined on real numbers is continuous. One can *prove* Brouwer's continuity theorem in the following form. Suppose that functions between formal spaces are proper ideal entities. That is, suppose that you can obtain them *only* starting from observables by means of some idealization. Then every function between formal spaces is continuous.

So, let $\mathcal{P}t(\mathcal{S})$ and $\mathcal{P}t(\mathcal{T})$ be two formal spaces, that is the collections of formal points over two formal topologies $\mathcal{S}$ and $\mathcal{T}$. Since $\mathcal{S}$ and $\mathcal{T}$ are the effective structures from which the formal spaces $\mathcal{P}t(\mathcal{S})$ and $\mathcal{P}t(\mathcal{T})$ are obtained via the idealization of formal points, it is natural to assume that every function $f : \mathcal{P}t(\mathcal{S}) \to \mathcal{P}t(\mathcal{T})$ is similarly defined by idealization from some effective entity connecting $\mathcal{S}$ and $\mathcal{T}$, that is a relation s between the sets of observables S and T.

$$\begin{array}{ccc} \mathcal{P}t(\mathcal{S}) & & \mathcal{S} \\ \downarrow{\scriptstyle f} & & \downarrow{\scriptstyle s} \\ \mathcal{P}t(\mathcal{T}) & & \mathcal{T} \end{array}$$

One can see that this means that f and s must be linked by

$$\forall\alpha(\alpha \Vdash a \to f\alpha \Vdash b) \text{ if and only if } a\,s\,b$$

for every $a \in S$ and $b \in T$. To understand this, imagine that f is given 'by miracle', and try to approximate it by means of a relation between observables. What one can say is that a will be related to b when every $\alpha : \mathcal{P}t(\mathcal{S})$

approximated by a is sent by f to a point $f\alpha : \mathcal{P}t(\mathcal{T})$ which is approximated by b. Of course, one actually starts from the relation s, which can be given fully effectively, and then defines f in such a way that the above equivalence is satisfied. Then one can see that it is compulsory to define $f\alpha$ to be $s\alpha \equiv \{b \in T : a\, s\, b \;\&\; \alpha \Vdash a \text{ for some } a\}$, that is the existential image of the subset α under the relation s.

There is a very natural definition of what it means for a relation like s above to be continuous: one just has to require that it respects the structure of formal topology (see (Sambin, forthcoming a) for details). Then one can prove, at least at this point with usual mathematical rigour, that:

1. the assignment $\alpha \mapsto s\alpha$ is a function iff s is a continuous relation.

More precisely, the direction from right to left is always true, while the other direction is true under the assumption that $\mathcal{S}$ and $\mathcal{T}$ are bi-spatial (that is, the propositions $\forall\alpha(\alpha \Vdash a \to \alpha \between U)$ and $\exists\alpha(\alpha \Vdash a \;\&\; \alpha \subseteq U)$ are equivalent to $a \lhd U$ and $a \ltimes U$, respectively).

One can also easily check that:

2. if $s : \mathcal{S} \to \mathcal{T}$ is a continuous relation, then the induced function $f : \mathcal{P}t(\mathcal{S}) \to \mathcal{P}t(\mathcal{T})$, defined by $f : \alpha \mapsto s\alpha$, is a continuous function.

So one can prove the continuity theorem in the following sense. Assume that a function f between formal spaces $\mathcal{P}t(\mathcal{S})$ and $\mathcal{P}t(\mathcal{T})$ can be defined constructively only as an assignment $\alpha \mapsto s\alpha$ for some relation s between S and T. Then by 1, s must be a continuous relation, and by 2, f is continuous. This argument rests on the assumption that f is defined starting from a relation s. But it seems to me that this is the only way to define constructively a function between formal spaces (I am so convinced, that I am ready to bet that, whenever one can produce an f constructively, I will be able to find the s by which it is induced).

In conclusion, I am convinced that when the notion of functions between any formal spaces, in particular the formal space of real numbers, is thought through constructively, it becomes evident that every such function is automatically continuous, with no use of debatable assumptions.

Acknowledgement

I thank Milly Maietti for discussions on content and the editors for suggestions improving exposition.

A reverse look at Brouwer's Fan Theorem

Douglas Bridges

Classical reverse mathematics is a research programme, started over thirty years ago by Harvey Friedman (Friedman 1975), in which the aim is to place the theorems of large parts (perhaps all?) of mathematics into a number of equivalence classes, the theorems in any given class being equivalent to some natural set-existence principle. Currently, as presented in the compendious text by Simpson (Simpson 1999), five such equivalence classes are used, whose representing set-existence principles can be placed in increasing order of logical complexity. For example, it is known that both the Bolzano-Weierstrass theorem on the extraction of convergent subsequences and the Ascoli-Arzelà theorem on compact subsets of function spaces are equivalent to the second principle in that hierarchy (Simpson 1984, Theorem 4.2).

Around the turn of this century, Veldman (Veldman 2005), in the Netherlands, and Ishihara (Ishihara 2005, 2006), in Japan, independently initiated a programme of *constructive* reverse mathematics, **CRM**, which uses intuitionistic, rather than classical, logic. (Priority in the modern era of CRM actually lies with Julian and Richman, whose paper (Julian and Richman 1984) was, for nearly 20 years, an isolated singularity of the subject.) The framework of our less formal approach to CRM is Bishop-style constructive mathematics (**BISH**) — that is, mathematics with intuitionistic logic and some appropriate set-theory such as those found in (Aczel and Rathjen 2001; Friedman 1977; Myhill 1975). There are several models of **BISH**, the most significant of which seem to be:

- **INT**: Brouwer's intuitionistic mathematics, which, freeing ourselves from the shackles of philosophy, we regard as **BISH** plus Brouwer's continuity principle and Fan theorem (see (Beeson 1985; Bridges and Richman 1987; Dummett 2000b; Troelstra and van Dalen 1988));
- **RUSS**: Markov's recursive constructive mathematics, which we regard as **BISH** plus the Church-Markov-Turing thesis and, where needed, Markov's

principle[1] (see (Beeson 1985; Bridges and Richman 1987; Kushner 1985)); and

- **CLASS**: Classical (traditional) mathematics, which we regard as **BISH** plus the law of excluded middle.

The first two of these models of **BISH** are constructive in an expanded sense: they use only intuitionistic logic. The aim of the CRM program is to classify the theorems in the three models according to which principles we must, and need only, add to **BISH** in order to prove them.

We should make it clear at this stage that we are referring here to *informal* CRM, in which we take for granted the principles of function- and set-construction as described in the first chapters of (Bishop 1967; Bishop and Bridges 1985; Bridges and Richman 1987; Bridges and Vîţă 2006), and we work in the informal, though rigorous, style of the practising analyst, algebraist, topologist, ... Consider, for example, the following result of CRM:

Theorem 1. *The following are equivalent (over* ***BISH****):*

(i) *For every real number x, either $x = 0$ or $|x| > 0$.*
(ii) ***LPO****: For every binary sequence $(a_n)_{n\geqslant 1}$, either $a_n = 0$ for all n or else there exists N such that $a_N = 1$ (the* ***limited principle of omniscience****).*

The informal rigorous proof is simple. To prove that (i) $\Rightarrow$ (ii), let $(a_n)_{n\geqslant 1}$ be any binary sequence and apply (i) to the real number $\sum_{n=1}^{\infty} 2^{-n}a_n$; to prove that (ii) $\Rightarrow$ (i), let x be any real number and apply (ii) to a sequence $(a_n)_{n\geqslant 1}$ constructed so that

$$
\begin{aligned}
a_n &= 0 \Rightarrow \forall_{k\leqslant n}\left(|x| < 2^{-k}\right), \\
a_n &= 1 \Rightarrow \exists_{k\leqslant n}\left(|x| > 2^{-k-1}\right).
\end{aligned}
$$

Not only is the style of this proof informal, but also we have taken for granted such function-construction principles as the principle of countable choice, which is used to construct the sequence $(a_n)_{n\geqslant 1}$ in the proof that (ii) $\Rightarrow$ (i).

In practice, CRM splits naturally into two parts. In the first of these, we consider a theorem T of **INT** or **RUSS**, and try to find some principle, valid in that model and other than T itself, whose addition to **BISH** is necessary and sufficient for a constructive proof of T. In the second part of CRM we deal with a theorem T of **CLASS** that we suspect is non-constructive, and we try to prove that T is equivalent, over **BISH**, to one of a number of known

[1] Markov's principle, viewed with suspicion by most constructivists and with horror by some, says that for any binary sequence, if it is impossible for all the terms to be 0, then there exists a term equal to 1.

essentially non-constructive principles,[2] such as **LPO** (stated in Theorem 1) or the following:

- **LLPO**: For every binary sequence $(a_n)_{n\geqslant 1}$ with at most one term equal to 1, either $a_n = 0$ for all even n, or else $a_n = 0$ for all odd n (the **lesser limited principle of omniscience**);
- **LEM**: For every syntactically correct mathematical statement P, either P is provable or $\neg P$ is provable (the **law of excluded middle**).

Theorem 1 is one example of this aspect of CRM. Here are some others.

Theorem 2. *The following statements are equivalent (over **BISH**) to **LLPO**:*

(i) *For every real number x, either $x \geqslant 0$ or $x \leqslant 0$.*
(ii) *For each continuous function $f : [0,1] \to \mathbb{R}$, if $f(0)f(1) < 0$, then there exists $\xi \in (0,1)$ such that $f(\xi) = 0$ (the **intermediate value theorem**).*

We use the word **inhabited** of a set S to signify that we can construct an element of S; this is a constructively stronger notion than the impossibility of emptiness. A subset S of a set X is said to be **detachable** from X if for each $x \in X$, either $x \in S$ of $x \notin S$.

Theorem 3. *The following are equivalent (over **BISH**) to **LEM**:*

(i) *Every inhabited set of real numbers that is bounded above has a supremum.*
(ii) *Every subset of $\{0, 1\}$ is detachable.*

We omit the proofs of Theorems 2 and 3 (more information on which can be found in the various references on general constructive mathematics that we have provided), save for the amusing argument showing that statement (i) of Theorem 3 implies **LEM**. Given a syntactically correct mathematical statement P, consider the set

$$S \equiv \{0\} \cup \{x \in \mathbb{R} : x = 1 \wedge (P \vee \neg P)\}.$$

(Note, again, our acceptance of the informal set-construction rules in **BISH**.) This set contains 0 and is bounded above by 1. Suppose that it has a supremum σ. Then either $\sigma > 0$ or $\sigma < 1$ (a decision that we can make, using carefully chosen rational approximations to σ). The case $\sigma < 1$ is ruled out, since it implies that $1 \notin S$ and therefore $\neg (P \vee \neg P)$, which is equivalent to the absurdity $\neg P \wedge \neg\neg P$. Thus we must have $\sigma > 0$. It follows from this and the definition of *supremum* that there exists a point s of S with $s > 0$; whence

[2] For our present purposes, we consider a classically valid principle to be essentially non-constructive if it is provably false in both **INT** (unadorned by extras such as Kripke's schema or the theory of the creating subject — see (Dummett 2000b)) and **RUSS**.

$s = 1$ and $P \vee \neg P$ holds. Note that the same set S can be used to prove that statement (ii) of Theorem 3 implies **LEM**.

Incidentally, Theorems 1–3 embody Brouwerian counterexamples, a genre that goes back to Brouwer's early work (Brouwer 1921A1). So it is not unreasonable to trace the origins of CRM back to Brouwer, many decades before the appearance of (Julian and Richman 1984).

1 Equivalents of Fan Theorems

In the rest of this paper we concentrate on theorems of the type

$$\mathbf{BISH} \vdash \mathbf{FT}_? \Leftrightarrow T,$$

where $\mathbf{FT}_?$ is some form of Brouwer's fan theorem, and T is a theorem of **INT**. To do so, we need to introduce some terminology and notation.

The set 2^* of all strings (finite sequences) in $\{0,1\}$ is called the **complete binary fan**. Let $\alpha \equiv (\alpha_1, \alpha_2, \ldots)$ be a finite or infinite binary sequence. Then for each applicable $n \in \mathbb{N}$,

$$\overline{\alpha}n \equiv (\alpha_1, \ldots, \alpha_n)$$

is called a **restriction** of α. The **concatenation** of α with another string $\beta \equiv (\beta_1, \ldots, \beta_k)$ is

$$\alpha\frown\beta \equiv (\alpha_1, \ldots, \alpha_n, \beta_1, \ldots, \beta_k).$$

For $b \in \{0,1\}$ we write $\alpha\frown b$ rather than $\alpha\frown(b)$. By a *c*-**subset** of 2^* we mean a set $B \subset 2^*$ such that

$$B = \{u \in 2^* : \forall_{v \in 2^*} (u\frown v \in D)\} \tag{1}$$

for some detachable subset D of 2^*. Every detachable subset of 2^* is a *c*-subset. On the other hand, by a Π_1^0-**subset** of 2^* we mean a set $B \subset 2^*$ with the following property: there exists a detachable subset S of $2^* \times \mathbb{N}$ such that

$$\forall_{u \in 2^*} \forall_{n \in \mathbb{N}} ((u, n) \in S \Rightarrow (u\frown 0, n) \in S \wedge (u\frown 1, n) \in S)$$

and

$$B = \{u \in 2^* : \forall_{n \in \mathbb{N}} ((u, n) \in S)\}.$$

Every *c*-subset B of 2^* is a Π_1^0-subset: simply take $S \equiv D \times \mathbb{N}$, where D is a detachable subset of 2^* such that (1) holds.

By a **path** in 2^* we mean a finite or infinite binary sequence. Let B be a subset of 2^*. We say that

- a path α is **blocked** by B if some restriction of α is in B; that

- B is a **bar** for 2^* if each infinite path of 2^* is blocked by B; and that
- B is a **uniform bar** for 2^* if there exists a positive integer n such that each finite path of length n is blocked by B.

Every uniform bar is a bar. If ? denotes a property of subsets of 2^*, then Brouwer's **fan theorem for ?-bars, FT$_?$**, tells us that every bar with the property ? is a uniform bar. We are particularly interested in the **fan theorem for detachable bars**,

FT$_D$ *Every detachable bar of the complete binary fan is uniform;*

the **fan theorem for** c**-bars** (that is, bars that are c-subsets),

FT$_c$ *Every c-bar of the complete binary fan is uniform;*

the **fan theorem for** Π_1^0-bars (that is, bars that are Π_1^0-subsets),

FT$_{\Pi_1^0}$ *Every Π_1^0-bar of the complete binary fan is uniform;*

and the **full fan theorem:**

FT *Every bar of the complete binary fan is uniform.*

Note that, relative to **BISH**,

$$\mathbf{FT} \Rightarrow \mathbf{FT}_{\Pi_1^0} \Rightarrow \mathbf{FT}_c \Rightarrow \mathbf{FT}_D,$$

but it is not known whether any of these implications can be reversed.

Typically, we want to prove that **FT**$_?$ is equivalent, over **BISH**, to the proposition that, for every set S of an appropriate sort, some pointwise property of the form

$$\forall_{s\in S}\exists_{t\in T}P(s,t) \tag{2}$$

actually holds uniformly in the form

$$\exists_{t\in T}\forall_{s\in S}P(s,t). \tag{3}$$

Our strategy for attacking this problem is two-fold. First, given a set S of the appropriate sort, we construct a ?-subset B of 2^* such that

▷ if (2) holds, then B is a bar, and
▷ if B is a uniform bar, then (3) holds.

This, though, is only half of the solution. To prove that the implication from (3) to (2) implies **FT**$_?$, we consider a ?-subset B of 2^* and construct a corresponding set S such that

◁ if B is a bar, then (2) holds, and
◁ if (3) holds, then B is a uniform bar.

Our first explicit example of this proof-strategy will be the results of Julian and Richman (Julian and Richman 1984). In this case, the 'appropriate sort' of set S is the set of values of a given uniformly continuous mapping $f : [0,1] \to \mathbb{R}$, T is the set of positive real numbers, and

$$P(s,t) \equiv (s \geqslant t)\,.$$

Unwrapping these ideas, we see that we are dealing with the pointwise property

$$\forall_{x\in[0,1]}\exists_{t>0}\,(f(x) \geqslant t)$$

and the uniform version

$$\exists_{t>0}\forall_{x\in[0,1]}\,(f(x) \geqslant t)\,.$$

Here are the Julian-Richman results, which are exactly of the type our proof-strategy seeks.

Theorem 4. *Let $f : [0,1] \to \mathbb{R}$ be uniformly continuous. Then there exists a detachable subset B of 2^* such that*

▷ *if $f(x) > 0$ for each $x \in [0,1]$, then B is a bar, and*
▷ *if B is a uniform bar, then* $\inf f > 0$.

Theorem 5. *Let B be a detachable bar of 2^*. Then there exists a uniformly continuous $f : [0,1] \to \mathbb{R}$ such that*

◁ *if B is a bar, then $f(x) > 0$ for each $x \in [0,1]$, and*
◁ *if* $\inf f > 0$, *then B is a uniform bar.*

The proofs of these two theorems are subtle and tricky; see either (Julian and Richman 1984) or (Bridges and Richman 1987, ch.6) for details. New, but equally complex, proofs are to appear shortly (Berger and Bridges, forthcoming). Taken together, Theorems 4 and 5 show us that, relative to **BISH**, the fan theorem for detachable bars, $\mathbf{FT}_D$, is equivalent to the statement:

> **POS**: *Each positive-valued, uniformly continuous function of* $[0,1]$ *has positive infimum.*

For if, starting with a positive-valued, uniformly continuous function f on $[0,1]$, we construct B as in Theorem 4, then we see from that theorem that B is a detachable bar; so if $\mathbf{FT}_D$ holds, then the second conclusion of that theorem tells us that $\inf f > 0$. On the other hand, if, starting with a detachable bar B of 2^*, we construct a uniformly continuous function $f : [0,1] \to \mathbb{R}$ as in Theorem 5, then we see from that theorem that f is positive-valued; so if **POS** holds, then the second conclusion of Theorem 5 tells us that B is a uniform bar.

It follows that **POS** is derivable in **INT**, in which the full fan theorem, not just $\mathbf{FT}_D$, is a standard principle. The situation is quite the opposite in **RUSS**: in that model of **BISH**, there exist both a detachable bar of 2^* that is not uniform and a positive-valued, uniformly continuous function on $[0,1]$ that has infimum equal to 0; see Chapters 5 and 6 of (Bridges and Richman 1987).

For the second illustration of our proof-strategy, we need yet more definitions. A sequence $(x_n)_{n\geqslant 1}$ in $\mathbb{R}$ is

- **eventually bounded away from the point** $x \in \mathbb{R}$ if there exist $\delta > 0$ and N such that $|x - x_n| > \delta$ for all $n \geqslant N$;
- **eventually bounded away from the set** $S \subset \mathbb{R}$ if there exist $\delta > 0$ and N such that $|x - x_n| > \delta$ for all $n \geqslant N$ and all $x \in S$.

Specker's theorem in **RUSS** says that there exists a sequence in $[0,1]$ that is eventually bounded away from each point of that interval; see (Specker 1949) or (Bridges and Richman 1987, ch.3).

In classical analysis, the **anti-Specker property** — equivalent to the Bolzano-Weierstrass property — holds: if a sequence of real numbers is eventually bounded away from each point of $[0,1]$, then it is eventually bounded away from the interval itself. Now, the anti-Specker property embodies passage from the pointwise property

$$\forall_{x\in[0,1]}\exists_{\delta>0}\exists_N\forall_{n\geqslant N}\left(|x-x_n|>\delta\right)$$

to the corresponding uniform property

$$\exists_{\delta>0}\exists_N\forall_{x\in[0,1]}\forall_{n\geqslant N}\left(|x-x_n|>\delta\right).$$

In the notation of our proof-strategy, S is the interval $[0,1]$, T is the set of all ordered pairs (δ, N) with $\delta > 0$ and N a positive integer, and

$$P(x,(\delta,N)) \equiv \forall_{n\geqslant N}\left(|x-x_n|>\delta\right).$$

This time, it is the fan theorem for c-bars that comes into play: we have the following two theorems of Berger and Bridges (Berger and Bridges 2007):

Theorem 6. *Let $(x_n)_{n\geqslant 1}$ be a sequence of rational numbers. Then there exists a c-subset B of 2^* such that*

- ▷ *if $(x_n)_{n\geqslant 1}$ is eventually bounded away from each element of $[0,1]$, then B is a bar, and*
- ▷ *if B is a uniform bar and $(x_n)_{n\geqslant 1}$ is eventually bounded away from each point of $[0,1]$, then $(x_n)_{n\geqslant 1}$ is eventually bounded away from the interval $[0,1]$.*

Theorem 7. *Let B be a c-subset of 2^*. Then there exists a sequence $(x_n)_{n\geqslant 1}$ of rational numbers such that*

◁ *if B is a bar, then $(x_n)_{n\geqslant 1}$ is eventually bounded away from each element of* $[0,1]$, *and*
◁ *if $(x_n)_{n\geqslant 1}$ is eventually bounded away from the interval* $[0,1]$, *then B is a uniform bar.*

It follows from Theorems 6 and 7 that, relative to **BISH**, the fan theorem for c-bars, $\mathbf{FT}_c$, is equivalent to the statement:

> *Every sequence of rational numbers that is eventually bounded away from each point of* $[0,1]$ *is eventually bounded away from the interval* $[0,1]$.

A little more work enables us to remove the word *rational* from this last sentence. Thus, relative to **BISH**, $\mathbf{FT}_c$ is equivalent to the anti-Specker property. Since the anti-Specker property is a classical equivalent of the Bolzano-Weierstrass property and seems to have the same relative ease of constructive application that the latter has classically, there is a strong incentive for exploring the system **BISH** + $\mathbf{FT}_c$ as an extended framework for constructive mathematics.

That incentive would be even stronger if we knew that $\mathbf{FT}_c$ was equivalent, over **BISH**, to the uniform continuity theorem for $[0,1]$,

> $\mathbf{UCT}_{[0,1]}$: *Every pointwise continuous mapping of* $[0,1]$ *into* $\mathbb{R}$ *is uniformly continuous,*

the proposition for which Brouwer originally developed his proof of the fan theorem. It follows from results of Berger (Berger 2006) that

$$\mathbf{BISH} \vdash \mathbf{UCT} \Rightarrow \mathbf{FT}_c.$$

Also, Diener and Loeb (Diener and Loeb 2006) have proved that

$$\mathbf{BISH} \vdash \mathbf{FT}_{\Pi^1_0} \Rightarrow \mathbf{UCT}.$$

We do not know if either of these implications can be replaced by a bi-implication. Perhaps **UCT** is equivalent, relative to **BISH**, to some natural, but as yet unidentified, version of the fan theorem. This identification problem remains open. Nevertheless, we can say something about a form of weak uniform continuity (introduced in the next definition) that is implied by, and classically equivalent to, uniform continuity.

We say that a function $f : [0,1] \to \mathbb{R}$ is **uniformly sequentially continuous** if the following holds:

> **USC:** If $(x_n)_{n\geqslant 1}$ and $(x'_n)_{n\geqslant 1}$ are sequences in $[0,1]$ such that $|x_n - x'_n| \to 0$, then $|f(x_n) - f(x'_n)| \to 0$.

In the special case where the sequence $(x'_n)_{n\geqslant 1}$ has the constant value $a \in [0,1]$ for all n, **USC** yields the sequential continuity of f at a.

Theorem 8. *The following are equivalent.*

(i) $\boldsymbol{FT}_c$.
(ii) *Every pointwise continuous, real-valued mapping on* $[0,1]$ *is uniformly sequentially continuous.*
(iii) *Every pointwise continuous mapping of a compact (that is, complete and totally bounded) metric space into a metric space is uniformly sequentially continuous.*

This follows from (Bridges 2006, Corollary 6), with reference to the equivalence of $\mathbf{FT}_c$ and the anti-Specker property.

In order to pass from uniform sequential continuity to uniform continuity, we introduce a definition and a simple principle about positive integers. We call a set S of positive integers **pseudobounded** if

$$\lim_{n\to\infty} \frac{s_n}{n} = 0$$

for each sequence $(s_n)_{n\geqslant 1}$ in S. This leads us to a principle, due to Ishihara (Ishihara 1992), that has several interesting applications in CRM:

> **BD-**$\mathbb{N}$: *Every countable*[3] *pseudobounded set of positive integers is bounded.*

The principle **BD-**$\mathbb{N}$ holds in **INT**, **RUSS**, and **CLASS**, but, as has been shown by Lietz (Lietz 2004), there is a model of **BISH** in which **BD-**$\mathbb{N}$ is false.

Theorem 9. *The following are equivalent.*

(i) ***BD***-$\mathbb{N}$.
(ii) *Every uniformly sequentially continuous mapping of* $[0,1]$ *into* $\mathbb{R}$ *is uniformly continuous.*

This is a special case of (Bridges 2006, Theorem 7).

It follows from Theorems 8 and 9 that

$$\mathbf{BISH} \vdash (\mathbf{FT}_c + \mathbf{BD}\text{-}\mathbb{N}) \Rightarrow \mathbf{UCT},$$

and therefore that

$$\mathbf{BISH} \vdash (\mathbf{AS} + \mathbf{BD}\text{-}\mathbb{N}) \Rightarrow \mathbf{UCT}.$$

Theorems 8 and 9 are part of a pattern in which $\mathbf{FT}_c$ (or, equivalently, the anti-Specker property) is equivalent to the passage from a pointwise property to a sequential one that lies between it and a uniform property, and **BD-**$\mathbb{N}$ is equivalent to the passage from the sequential property to that uniform one; see also (Baroni and Bridges, forthcoming) and (Bridges 2008).

We end with some known equivalents of the uniform continuity theorem which, though falling short of our desired equivalence with a fan theorem, hold considerable interest in their own right.

[3] The word *countable* is not redundant here.

Theorem 10. *The following are equivalent relative to* ***BISH:***

(i) *Every pointwise continuous mapping of* $[0,1]$ into $\mathbb{R}$ *is bounded.*
(ii) *Every pointwise continuous mapping of* $[0,1]$ *into* $\mathbb{R}$ *is uniformly continuous.*
(iii) *Every pointwise continuous mapping of* $[0,1]$ *into* $\mathbb{R}$ *is Lebesgue integrable.*
(iv) *Every pointwise continuous mapping of a compact (that is, complete and totally bounded) metric space into a metric space is uniformly continuous.*

See (Bridges and Diener 2007); a result of this type, but with 'Riemann' replacing 'Lebesgue' in (iii), was given by Loeb (Loeb 2005), working with a stronger notion of continuity.[4]

In the traditional form of **INT**, with the full fan theorem, the statements (i)–(iv) of Theorem 10 are all valid. As you might expect, exactly the opposite holds in **RUSS**, where the existence of so-called *singular covers* leads to that of a bounded, pointwise continuous mapping $f : [0,1] \to \mathbb{R}$ that fails to be Lebesgue integrable and hence cannot be uniformly continuous (Bridges and Demuth 1991).

For long periods of the century following the publication of Brouwer's doctoral thesis (Brouwer 1907), and of the 80-odd years since his first papers on the fan theorem (Brouwer 1924D2, Brouwer 1924G2), it looked as if his deep philosophical and mathematical insights were 'no doubt destined to remain only as an historical curiosity'.[5] All this changed, at first slowly but in the last quarter-century more rapidly, with the revival of interest in constructive issues that was catalyzed by Bishop's book (Bishop 1967). We hope that the work we have outlined in this paper will help convince you that the work of Brouwer goes on, even if the type of foundational investigations that CRM represents might not have been altogether to his taste.

Acknowledgements

Thanks are due to Iris Loeb, for her very helpful comments on the draft of this paper; to the Marsden Fund of the Royal Society of New Zealand, for supporting much of the research reported therein; and to the University of Canterbury Erskine Fund, for supporting my participation at the meeting *1907–2007: One Hundred Years of Intuitionism* held in Cerisy in June 2007.

---

[4] The proof of Theorem 10 in (Bridges and Diener 2007) depends on an obscure result that previously had seemed a mere curiosity without any potential for application. There is a moral here: never disdain your early results, as you may find that they are exactly what you need thirty years after their derivation!

[5] cf. Bourbaki, in (Bourbaki 1960): 'L'école intuitionniste, dont le souvenir n'est sans doute destiné à subsister qu'à titre de curiosité historique'.

Some applications of Brouwer's Thesis on Bars

Wim Veldman

συνχώρησον ἐξ ὑποθέσεως αὐτὸ σκοπεῖσθαι
συνχώρησον ἐξ ὑποθέ ως αὐτὸ σκοπε α

Plato, *Meno*, 86e[1]

1 Introduction

One may distinguish two components in Brouwer's proposal for an intuitionistic reform of the existing mathematical practice.

In the first place, as is well known, he advocated a constructive interpretation of the logical constants. For instance, a proof of a disjunctive statement $A \vee B$ must supply one with a proof of A or with a proof of B and, therefore, the principle of the excluded middle $A \vee \neg A$ has to be rejected. Similarly, a proof of an existential statement $\exists x \in V[P(x)]$, must provide one with an element v from the set V and a proof of the corresponding statement $P(v)$. This constructive interpretation of the disjunction and the existential quantifier is the natural one if one, like Brouwer, starts from the idea that the meaning of a mathematical statement has to be found in its proof.

Secondly, as is somewhat less well known, Brouwer proposed several new axioms. The most important two later have been called *Brouwer's Continuity Principle* and *Brouwer's Thesis on Bars.*

The two components of Brouwer's proposal are closely related. Both are the fruit of a single attempt to handle the concept of the continuum in a meaningful way. It is nonsense to treat infinitary objects like real numbers or infinite sequences of natural numbers as if they were finite objects. Classical mathematics sometimes does so and then becomes difficult to understand.

In this paper, we want to explore the meaning of the second one of the two important assumptions made by Brouwer: his Thesis on Bars. There are at least two ways to do so. One may study the argument Brouwer gave for

---

[1] 'allow me to examine this starting from an hypothesis'.

defending his thesis and weigh its merits. Or, as is our main intention now, one may try to see what the consequences of the thesis are, that is, one may accept the thesis as a hypothesis, and wait and see what happens. We shall see that Brouwer's Thesis on Bars is a very fruitful hypothesis, shedding much light in various dark corners. As K. Gödel once suggested, discussing the possible truth of axioms in set theory, one may perhaps count this fact as positive evidence for Brouwer's Thesis.

Apart from this introduction, the paper contains five sections.

1. The canonical proof that a bar is a bar.
2. The Bar Theorem and the Fan Theorem.
3. On well-founded subsets of $\mathbb{Q}$.
4. The Lusin Separation Theorem.
5. Ramsey's Theorem.

2 The canonical proof that a bar is a bar

We let $\mathbb{N}$ denote the set of the natural numbers. $\mathbb{N}^*$ is the set of all finite sequences of natural numbers. We let $\langle\ \ \rangle$ be a fixed bijective mapping from $\mathbb{N}^*$ onto $\mathbb{N}$. Such a function is called a *coding* of the set of finite sequences of natural numbers: $\langle a_0, a_1, \ldots, a_{k-1}\rangle$ is the code number of the finite sequence $(a_0, a_1, \ldots, a_{k-1})$. We assume that the empty sequence is coded by the number 0 and that for each finite sequence $(a_0, a_1, \ldots, a_{k-1})$, for every $i < k$, the code number $\langle a_0, a_1, \ldots, a_{k-1}\rangle$ is greater than a_i. We let length be the function from $\mathbb{N}$ to $\mathbb{N}$ that associates to any natural number a the *length* of the finite sequence coded by a. We also assume that there is a function $a, i \mapsto a(i)$ from $\mathbb{N} \times \mathbb{N}$ to $\mathbb{N}$, such that, for every k, for every a, if $\text{length}(a) = k$, then $a = \langle a(0), a(1), \ldots, a(k-1)\rangle$.

We let $*$ denote concatenation: $*$ is a function from $\mathbb{N} \times \mathbb{N}$ to $\mathbb{N}$ such that, for all m, n, $m*n$ is the code number of the finite sequence obtained by putting the sequence coded by n behind the sequence coded by m.

For all m, n, m *is an initial part of* n, notation: $m \sqsubseteq n$, if and only if there exists p such that $n = m * p$; and n *is an immediate successor of* m if and only if there exists p such that $n = m * \langle p\rangle$.

For all m, n, m, n are *incompatible*, or: m *is perpendicular to* n, notation: $m \perp n$, if and only if neither $m \sqsubseteq n$ nor $n \sqsubseteq m$.

We let *Baire space* $\mathcal{N}$ be the set of all infinite sequences $\alpha = \alpha(0), \alpha(1), \alpha(2), \ldots$ of natural numbers.

We let *Cantor space* $\mathcal{C}$ be the set of all α in $\mathcal{N}$ assuming no other values than $0, 1$.

We define, given any α and any n, $\overline{\alpha}(n) := \langle \alpha(0), \alpha(1), \ldots, \alpha(n-1)\rangle$.

If confusion is unlikely to arise, we sometimes write $\overline{\alpha}n$ for $\overline{\alpha}(n)$.

We also define, given any α and any s, *s is an initial part of α*, or: *α passes through s*, or: *s contains α*, if and only if, for some n, $\overline{\alpha}n = s$.

Let B be a subset of $\mathbb{N}$. *B is a bar in $\mathcal{N}$* if and only if, for each α in $\mathcal{N}$, there exists n in $\mathbb{N}$ such that $\overline{\alpha}n$ belongs to B.

Let B be a subset of $\mathbb{N}$ and let s belong to $\mathbb{N}$.
B is a bar in $\mathcal{N}$ above s if and only if, for each α in $\mathcal{N}$ *passing through s*, there exists n in $\mathbb{N}$ such that $\overline{\alpha}n$ belongs to B.
Note that *B is a bar in $\mathcal{N}$* is equivalent to *B is a bar (in $\mathcal{N}$) above $\langle\,\rangle$*, where $\langle\,\rangle$ is (the code number of) the empty sequence.

Brouwer asked the following question:

> Suppose B is a subset of $\mathbb{N}$ and a bar in $\mathcal{N}$. How can I ever be sure that B is indeed a bar in $\mathcal{N}$?

Brouwer gave to this question the following answer (*'Brouwer's Thesis'*):

> If B is a bar in $\mathcal{N}$, there must be a *canonical proof* of the fact that B is a bar in $\mathcal{N}$.

We have of course to explain what is understood by a canonical proof of '*B is a bar in $\mathcal{N}$*'.

Such a proof consists of infinitely many statements of the form: '*B is a bar above s*'.

The *conclusion of the proof* is the statement: '*B is a bar above $\langle\,\rangle$*'.

The *starting points of the proof* have the form: '*s belongs to B, therefore: B is a bar above s*'.

There are two kinds of *reasoning steps*, *forward reasoning steps* and *backward reasoning steps*.

A forward reasoning steps has *infinitely many premises*:

B is a bar above $s * \langle 0\rangle$, B is a bar above $s * \langle 1\rangle$, B is a bar above $s * \langle 2\rangle$, ...

and a *single conclusion*:

therefore: B is a bar above s

A backward reasoning step has a *single premiss*:

B is a bar above s

and also a *single conclusion*:

therefore: B is a bar above $s * \langle n\rangle$

(the number n may be chosen freely).

As Brouwer emphasizes, the canonical proof is an (infinite) mental construction. One should not identify proofs with (finite) texts on paper describing them. The function of the text is to help us in communicating and memorizing the proofs.

Brouwer's Thesis appeared in (Brouwer 1924D2), (Brouwer 1927B) and (Brouwer 1954A) and is discussed thoroughly in (Kleene and Vesley 1965). The above formulation is the one used in (Veldman 2006b).

3 The Fan Theorem and the Bar Theorem

Let s in $\mathbb{N}$ code a binary sequence, that is, for each $i <$ length(s), $s(i) < 2$. Let B be a subset of $\mathbb{N}$. *B is a bar in $\mathcal{C}$ above s* if and only if every α in $\mathcal{C}$ passing through s has an initial part in B. *B is a bar in $\mathcal{C}$* if and only if every α in $\mathcal{C}$ has an initial part in B.

For Brouwer, the most useful conclusion from his thesis: *'given any bar in $\mathcal{N}$, one must have a canonical proof that the bar is a bar'* is the famous:

Theorem 1. (Fan Theorem)
Let B be a subset of $\mathbb{N}$ that is a bar in $\mathcal{C}$. There exists a finite subset B' of B that is a bar in $\mathcal{C}$.

Proof: Let B be a subset of $\mathbb{N}$ that is a bar in $\mathcal{C}$. Relativizing Brouwer's thesis on the existence of canonical proofs we find that there is a proof with the conclusion: '*B is a bar in $\mathcal{C}$ above $\langle\,\rangle$*' and with starting points: '*s belongs to B, therefore: B is a bar above s*' that uses only reasoning steps of one of the following three kinds: '*B is a bar in $\mathcal{C}$ above $s * \langle 0\rangle$ and B is a bar in $\mathcal{C}$ above $s * \langle 1\rangle$,* **therefore:** *B is a bar in $\mathcal{C}$ above s*', '*B is a bar in $\mathcal{C}$ above s,* **therefore:** *B is a bar in $\mathcal{C}$ above $s * \langle 0\rangle$*', and: '*B is a bar in $\mathcal{C}$ above s,* **therefore:** *B is a bar in $\mathcal{C}$ above $s * \langle 1\rangle$*'. Any such canonical proof contains finitely many statements of the form: '*s belongs to B*'. Inspecting them, we construct a finite subset B' of B that is a bar in $\mathcal{C}$. □

For each α in $\mathcal{N}$, for each n in $\mathbb{N}$, we define α^n in $\mathcal{N}$ by: for each m, $\alpha^n(m) = \alpha(\langle n\rangle * m)$. We call α^n *the n-th subsequence of α*.

We now introduce the set of *stumps*. We borrowed the word '*stump*' from (Brouwer 1954A) but are giving it a different meaning. Stumps are the characteristic functions of certain decidable subsets of $\mathbb{N}$, that is, they are elements of *Cantor space $\mathcal{C}$*. We define the set of the stumps inductively, as follows:

(i) $\underline{0}$ is a stump. Note that $\underline{0}$ is the characteristic function of the empty set. We sometimes call $\underline{0}$ the *empty stump*.
(ii) For every α in $\mathcal{C}$, if $\alpha(\langle\,\rangle) = 1$ and, for each n, α^n is a stump, then

α is a stump.
(iii) Every stump is obtained from $\underline{0}$ by the repeated application of (ii).

We shall use $\sigma, \tau, \ldots$ as variables over stumps. The set of the stumps will be called **Stump**.

If one accepts an inductive definition of a set like the definition of the set of stumps, one will also accept the corresponding induction principle:

Principle of induction on the set of stumps:
Every subset P of the set of stumps such that
(i) the empty stump $\underline{0}$ belongs to P, and
(ii) for every non-empty stump σ, if every immediate substump σ^n of σ belongs to P, then σ itself belongs to P,
coincides with the set of stumps.

Note that, for every stump σ, *either* $\sigma(\langle\ \rangle) = 0$ and σ coincides with the empty stump $\underline{0}$, *or* $\sigma(\langle\ \rangle) = 1$ and σ does not coincide with the empty stump $\underline{0}$.

For every β in $\mathcal{C}$, we let D_β be the set of all n in $\mathbb{N}$ such that $\beta(n) = 1$. We sometimes call D_β the subset of $\mathbb{N}$ that is *decided by* β.

Theorem 2. (Bar Theorem)
Let B be a subset of $\mathbb{N}$ that is a bar in $\mathcal{N}$. There exists a stump σ such that $B \cap D_\sigma$ is a bar in $\mathcal{N}$.

Let B be a subset of $\mathbb{N}$ that is a bar in $\mathcal{N}$ and let σ be a stump. We define: σ *secures* B if and only if $B \cap D_\sigma$ is a bar in $\mathcal{N}$. The Bar Theorem may be summarized in the statement: if B is a bar in $\mathcal{N}$, then there exists a stump securing B.

The Fan Theorem is the dividing point between intuitionistic mathematicians and constructive mathematicians who want to develop analysis starting from the idea that every infinite sequence of natural numbers is given by an algorithm. A number of equivalents of the Fan Theorem may be found in (Veldman 2005).

The above formulation of the Bar Theorem is the one proposed in (Veldman 2006b), where the Bar Theorem is derived from Brouwer's Thesis, and some difficulties concerning the appropriate formulation of the Bar Theorem are discussed.

4 On well-founded subsets of $\mathbb{Q}$

We let $\mathbb{Q}$ be the set of the rational numbers, and we let $q_0, q_1, q_2, \ldots$ be an enumeration without repetitions of the set $\mathbb{Q}$. We also use $<$ to denote the usual ordering on $\mathbb{Q}$.

Let γ belong to $\mathcal{N}$. We let E_γ be the set of all numbers m, such that, for some n, $\gamma(n) = m + 1$. We call E_γ *the subset of* $\mathbb{N}$ *enumerated by* γ.

The following axiom is a plausible axiom for the intuitionistic mathematician who allows himself to define infinite sequences of natural numbers step-by-step.

> *Second Axiom of Countable Choice:*
> For every binary relation R on $\mathbb{N} \times \mathcal{N}$, if, for every m, there exists α such that $mR\alpha$, then there exists α such that, for every m, $mR\alpha^m$.

Using the Second Axiom of Countable Choice, one may prove that, for every infinite sequence $X_0, X_1, X_2, \ldots$ of enumerable subsets of $\mathbb{N}$, the set $\bigcup_{n \in \mathbb{N}} X_n$ is also an enumerable subset of $\mathbb{N}$.

Let A be a subset of $\mathbb{Q}$. We let A_γ be the set of all numbers q in $\mathbb{Q}$ such that, for some n, for some m, $q = q_m$ and $\gamma(n) = m+1$. We call A_γ *the subset of* $\mathbb{Q}$ *enumerated by* γ.

A subset A of $\mathbb{Q}$ is called *enumerable* if and only if, for some γ, A coincides with A_γ.

Using the Second Axiom of Countable Choice, one may prove that, for every infinite sequence $X_0, X_1, X_2, \ldots$ of enumerable subsets of $\mathbb{Q}$, the set $\bigcup_{n \in \mathbb{N}} X_n$ is also an enumerable subset of $\mathbb{Q}$.

Let A be a subset of $\mathbb{Q}$. A is called *explicitly well-founded* if and only if, for each α in $\mathcal{N}$, such that, for each n, $q_{\alpha(n)}$ belongs to A, there exists n such that $q_{\alpha(n)} \leq q_{\alpha(n+1)}$.

Let A be an enumerable subset of $\mathbb{Q}$ and let γ in $\mathcal{N}$ satisfy $A = A_\gamma$. Note that A_γ is explicitly well-founded if and only if, for each α in $\mathcal{N}$, there exists n in $\mathbb{N}$ such that *either* $\gamma(\alpha(n)) = 0$ *or* $\big(\gamma(\alpha(n)) > 0$ and $\gamma(\alpha(n+1)) > 0$ and $q_{\gamma(\alpha(n))-1} \leq q_{\gamma(\alpha(n+1))-1}\big)$.

Now let σ be a stump, and let γ belong to $\mathcal{N}$. We define: σ *secures that* γ *enumerates an explicitly well-founded subset of* $\mathbb{Q}$ if and only if, for each α in $\mathcal{N}$, there exists k in $\mathbb{N}$ such that $\overline{\alpha}k$ belongs to D_σ, and there exists n in $\mathbb{N}$ such that *either* $n < k$ and $\gamma(\alpha(n)) = 0$ *or* $\big(n+1 < k$ and $\gamma(\alpha(n)) > 0$ and $\gamma(\alpha(n+1)) > 0$ and $q_{\gamma(\alpha(n))-1} \leq q_{\gamma(\alpha(n+1))-1}\big)$.

Brouwer's Thesis implies that, for each γ in $\mathcal{N}$, if γ enumerates an explicitly well-founded subset of $\mathbb{Q}$, then there exists a stump σ securing that γ enumerates an explicitly well-founded subset of $\mathbb{Q}$.

For all subsets X, Y of $\mathbb{Q}$, we define: $X < Y$ if and only if, for all p in X, for all q in Y, $p < q$.

We now define the class of $\mathcal{IWF}$ of the *inductively well-founded subsets of* $\mathbb{Q}$ by means of the following inductive definition:

(i) The empty set $\emptyset$ belongs to $\mathcal{IWF}$ and, for every p in $\mathbb{Q}$, the singleton $\{p\}$ belongs to $\mathcal{IWF}$.

(ii) For every sequence $X_0, X_1, X_2, \ldots$ of subsets of $\mathbb{Q}$, if, for each n, $X_n < X_{n+1}$ and X_n belongs to $\mathcal{IWF}$, then $\bigcup_{n\in\mathbb{N}} X_n$ belongs to $\mathcal{IWF}$.

(iii) Every element of $\mathcal{IWF}$ is obtained from sets mentioned under (i) by the repeated application of the operation mentioned under (ii).

(We are using a terminology different from the one used in (Veldman 2006b). We used the term *'explicitly well-founded'* for sets that are now called *'inductively well-founded'*. Our present use of the words is in keeping with Kleene's choice of terms in his treatment of Brouwer's Bar Theorem in (Kleene and Vesley 1965).)

We accept the principle of induction on the set $\mathcal{IWF}$ that belongs to the definition of the set $\mathcal{IWF}$ and, without spelling it out, we use it in the next theorem:

Theorem 3. *Every inductively well-founded subset of $\mathbb{Q}$ is an enumerable and explicitly well-founded subset of $\mathbb{Q}$.*

Proof: Using the Second Axiom of Countable Choice, one may prove, by induction on $\mathcal{IWF}$, that every inductively well-founded subset of $\mathbb{Q}$ is an enumerable subset of $\mathbb{Q}$.

In order to see that every inductively well-founded subset of $\mathbb{Q}$ is explicitly well-founded, note first that the empty set $\emptyset$ is explicitly well-founded and that, for every p in $\mathbb{Q}$, the singleton $\{p\}$ is explicitly well-founded. Now assume that $X_0, X_1, X_2, \ldots$ is an infinite sequence of subsets of $\mathbb{Q}$, and, for each n, $X_n < X_{n+1}$ and X_n is explicitly well-founded, and let X be the set $\bigcup_{n\in\mathbb{N}} X_n$. We prove, by course-of-values induction on $\mathbb{N}$, that, for every n, for every α, if, for every i, $q_{\alpha(i)}$ belongs to X and $q_{\alpha(0)}$ belongs to X_n, then there exists m such that $q_{\alpha(m)} \leq q_{\alpha(m+1)}$. So assume that n belongs to $\mathbb{N}$, and, for every $k < n$, for every α, if for every i, $q_{\alpha(i)}$ belongs to X and $q_{\alpha(0)}$ belongs to X_k, then there exists m such that $q_{\alpha(m)} \leq q_{\alpha(m+1)}$. Let α be an element of $\mathcal{N}$ such that for every i, $q_{\alpha(i)}$ belongs to X and $q_{\alpha(0)}$ belongs to X_n. Note that, for every q in X, there exists exactly one i such that q belongs to X_i, and, therefore, one may decide if q belongs to X_n or not. Define α' in $\mathcal{N}$, as follows. $\alpha'(0) = \alpha(0)$ and, for each $i > 0$, if $\alpha(i)$ belongs to X_n, then $\alpha'(i) = \alpha(i)$, and, if $\alpha(i)$ does not belong to X_n, then $\alpha'(i) = \alpha'(i-1)$. Note that, for each i, $q_{\alpha'(i)}$ belongs to X_n and find m such that $q_{\alpha'(m)} \leq q_{\alpha'(m+1)}$. Now distinguish two cases.

Case (i). $\alpha'(m) = \alpha(m)$ and $\alpha'(m+1) = \alpha(m+1)$. Then $q_{\alpha(m)} \leq q_{\alpha(m+1)}$.

Case (ii). We find i in $\{m, m+1\}$ such that $\alpha'(i) \neq \alpha(i)$. Find k such that $q_{\alpha(i)}$ belongs to X_k. Then k will be different from n. If $n < k$, then $q_{\alpha(0)} < q_{\alpha(i)}$, and there will exist $m < i$ such that $q_{\alpha(m)} \leq q_{\alpha(m+1)}$. If $k < n$ we may apply the induction hypothesis to the sequence $\alpha(i), \alpha(i+1), \alpha(i+2), \ldots$, and we find m such that $q_{\alpha(m)} \leq q_{\alpha(m+1)}$. □

Theorem 4. *Every enumerable and explicitly well-founded subset of* $\mathbb{Q}$ *is an inductively well-founded subset of* $\mathbb{Q}$.

Proof: It suffices to show: every stump σ *has the property* P, that is: for every γ in $\mathcal{N}$, if σ secures that γ enumerates an explicitly well-founded subset of $\mathbb{Q}$, then the subset of $\mathbb{Q}$ enumerated by γ is an inductively well-founded subset of $\mathbb{Q}$. We prove by induction on the set of stumps that every stump has the property P.

Note that the empty stump $\underline{0}$ has the property P. Now assume that σ is a non-empty stump and that every immediate substump σ^n of σ has the property P. Assume that γ belongs to $\mathcal{N}$ and that σ secures that γ enumerates an explicitly well-founded subset of $\mathbb{Q}$. We now define, by induction, an infinite sequence $X_0, X_1, X_2, \ldots$ of subsets of X. $X_0 := \emptyset$, and, for each n, *if* $\gamma(n) > 0$ and, for each $i < n$, if $\gamma(i) > 0$, then $q_{\gamma(i)-1} < q_{\gamma(n)-1}$, *then* we let X_n be the set of all q in X such that, for all $i < n$, $q \notin X_i$, and $q \leq q_{\gamma(n)-1}$, and *if not*, then we let X_n be the empty set $\emptyset$. Note that for each n, $X_n < X_{n+1}$ and that A_γ coincides with $\bigcup_{n \in \mathbb{N}} X_n$. Also observe that, for each q in X, for each n in $\mathbb{N}$, one may decide if q belongs to X_n or not. We intend to prove that, for each n, X_n is an inductively well-founded subset of $\mathbb{Q}$.

Assume that n is a natural number and that X_n is not the empty set. Define γ' in $\mathcal{N}$ such that, for all i, *if* $\gamma(i) > 0$ and $q_{\gamma(i)-1}$ belongs to X_n and $q_{\gamma(i)-1} < q_{\gamma(n)-1}$, then $\gamma'(i) = \gamma(i)$, and, *if not*, then $\gamma'(i) = 0$. Note that γ' enumerates $X_n \setminus \{q_{\gamma(n)-1}\}$ and that, because of the fact that $\gamma(n) > 0$, $\sigma^{\gamma(n)}$ is not the empty stump and $\sigma^{\gamma(n)}$ secures that γ' enumerates an explicitly well-founded subset of $\mathbb{Q}$: the set $X_n \setminus \{q_{\gamma(n)-1}\}$. It follows that also X_n itself is an inductively well-founded subset of $\mathbb{Q}$.

As, for each n, X_n is an inductively well-founded subset of $\mathbb{Q}$, also A_γ, coinciding with $\bigcup_{n \in \mathbb{N}} X_n$, is an inductively well-founded subset of $\mathbb{Q}$.

We may conclude that σ has the property P. □

The relation between explicitly and inductively well-founded countable linear orderings and its connection with the Bar Theorem has been studied by J. Gielen and W. Gielen in their student days. Theorem 4 as formulated above occurs in (Veldman 1985). The theorem is related to results in (Howard and Kreisel 1966).

5 The Lusin Separation Theorem

C. Parsons, in (Parsons 1967), has called attention to the fact that Brouwer's ideas on bars may have influenced developments in descriptive set theory. The theorem we are going to treat now illustrates this fact.

We want to study subsets of Baire space $\mathcal{N}$. A subset X of Baire space $\mathcal{N}$ is called *basic open* if and only if either X is empty or there exists s such that

X is the set all α in $\mathcal{N}$ passing through s. A subset X of $\mathcal{N}$ is *open* if and only if there exists an infinite sequence $X_0, X_1, X_2, \ldots$ of basic open subsets of $\mathcal{N}$ such that $X = \bigcup_{n \in \mathbb{N}} X_n$. A subset X of $\mathcal{N}$ is *closed* if and only if there exists an infinite sequence $X_0, X_1, X_2, \ldots$ of basic open subsets of $\mathcal{N}$ such that $X = \bigcap_{n \in \mathbb{N}} X_n$.

We introduce the class of the *positively Borel subsets of* $\mathcal{N}$ by means of the following inductive definition:

(i) Every basic open set is a positively Borel subset of $\mathcal{N}$.
(ii) For every sequence $X_0, X_1, X_2, \ldots$ of subsets of $\mathcal{N}$, if, for each n, X_n is positively Borel, then also $\bigcup_{n \in \mathbb{N}} X_n$ and $\bigcap_{n \in \mathbb{N}} X_n$ are positively Borel.
(iii) Every positively Borel subset of $\mathcal{N}$ is obtained from sets mentioned under (i) by the repeated application of the operations mentioned under (ii).

Let γ be an element of $\mathcal{N}$. γ is said to be *the code of a continuous function from* $\mathcal{N}$ *to* $\mathcal{N}$ if and only if $\gamma(\langle\ \rangle) = 0$ and, for each α in $\mathcal{N}$, there exists n such that $\gamma(\overline{\alpha}n) \neq 0$. For every γ in $\mathcal{N}$ coding a continuous function from $\mathcal{N}$ to $\mathcal{N}$, for every α in $\mathcal{N}$, we define a sequence $\gamma|\alpha$, (*the result of applying* γ *to* α), as follows:

For each n, p in $\mathbb{N}$, $(\gamma|\alpha)(n) = p$ if and only if there exists m in $\mathbb{N}$ such that $\gamma(\langle n \rangle * \overline{\alpha}m) = p + 1$, while, for every $i < m$, $\gamma(\langle n \rangle * \overline{\alpha}i) = 0$.

A subset X of $\mathcal{N}$ is called *strictly analytic* if and only if there exists γ in $\mathcal{N}$ coding a continuous function from $\mathcal{N}$ to $\mathcal{N}$ such that X *coincides with the range of* γ, that is, for every α, α belongs to X if and only if, for some β, α coincides with $\gamma|\beta$.

For all α, β in $\mathcal{N}$ we let $\langle \alpha, \beta \rangle$ be the element of $\mathcal{N}$ satisfying: for each n, $\langle \alpha, \beta \rangle(2n) = \alpha(n)$ and $\langle \alpha, \beta \rangle(2n+1) = \beta(n)$.

A subset X of $\mathcal{N}$ is called *analytic* if and only if there exists a closed subset Y of $\mathcal{N}$ such that X coincides with the set of all α such that, for some β, $\langle \alpha, \beta \rangle$ belongs to Y. (X thus is *the projection* of a closed subset Y of $\mathcal{N}$.)

Every strictly analytic subset of $\mathcal{N}$ is also an analytic subset of $\mathcal{N}$. Every positively Borel subset of $\mathcal{N}$ is an analytic subset of $\mathcal{N}$, but there are inhabited and closed subsets of $\mathcal{N}$ that fail to be strictly analytic, see (Veldman 2006a). (A subset X of $\mathcal{N}$ is *inhabited* if there exists an element of $\mathcal{N}$ belonging to X.) It follows that there are inhabited analytic subsets of $\mathcal{N}$ that fail to be strictly analytic. This fact diminishes the value but not the beauty of the Lusin Separation Theorem as we intend to prove it.

For all α, β in $\mathcal{N}$, we define: α *is apart from* β, notation: $\alpha \mathrel{\#} \beta$, if and only if, for some n, $\alpha(n) \neq \beta(n)$. This well-known *constructive inequality* or *apartness* relation is *co-transitive*, that is, for all α, β, if $\alpha \# \beta$, then, for every γ, either $\alpha \# \gamma$ or $\gamma \# \beta$.

Let γ, δ be elements of $\mathcal{N}$ coding continuous functions from $\mathcal{N}$ to $\mathcal{N}$. We say that γ, δ *have separate ranges*, if and only if, for all α, for all β, $\gamma|\alpha \# \delta|\beta$.

Let γ, δ be elements of $\mathcal{N}$ coding continuous functions from $\mathcal{N}$ to $\mathcal{N}$ with separate ranges.

Note that, for every α, β there exists n such that $(\gamma|\alpha)(n) \neq (\delta|\beta)(n)$. Therefore, for every α, β there exist n, p, q in $\mathbb{N}$ such that $\gamma(\langle n \rangle * \overline{\alpha}p) \neq 0$, while, for every $i < p$, $\gamma(\langle n \rangle * \overline{\alpha}i) = 0$, and $\delta(\langle n \rangle * \overline{\beta}q) \neq 0$, while, for every $i < q$, $\delta(\langle n \rangle * \overline{\beta}i) = 0$, and $\gamma(\langle n \rangle * \overline{\alpha}p) \neq \delta(\langle n \rangle * \overline{\beta}q)$.

Let s belong to $\mathbb{N}$. We say that γ, δ *separate at* s if and only if there exist u, v such that $u \perp v$ (that is, neither $u \sqsubseteq v$ nor $v \sqsubseteq u$), and, for all α, β such that $\langle \alpha, \beta \rangle$ passes through s, $\gamma|\alpha$ passes through u and $\delta|\beta$ passes through v.

(Recall that, for all α, β in $\mathcal{N}$, $\langle \alpha, \beta \rangle$ is the element of $\mathcal{N}$ satisfying: for each n, $\langle \alpha, \beta \rangle(2n) = \alpha(n)$ and $\langle \alpha, \beta \rangle(2n+1) = \beta(n)$.)

Let B be the set of all s such that γ, δ separate at s. Note that B is a bar in $\mathcal{N}$.

Let σ be a stump and let γ, δ be elements of $\mathcal{N}$ coding continuous functions from $\mathcal{N}$ to $\mathcal{N}$. We say that σ *secures that* γ, δ *have separate ranges* if and only if, for each α, there exists n such that γ, δ separate at $\overline{\alpha}n$ and $\overline{\alpha}n$ belongs to D_σ.

Brouwer's Thesis implies that, for every γ, δ in $\mathcal{N}$, if γ and δ are coding continuous functions from $\mathcal{N}$ to $\mathcal{N}$ with separate ranges, then there exists a stump σ securing that γ, δ have separate ranges.

Theorem 5. (Lusin Separation Theorem)
Let γ, δ be elements of $\mathcal{N}$, both of them coding a continuous function from $\mathcal{N}$ to $\mathcal{N}$, such that, for all α, for all β, $\gamma|\alpha \# \delta|\beta$. There exist positively Borel subsets C, D of $\mathcal{N}$ such that (i) for all α, $\gamma|\alpha \in C$, (ii) for all β, $\delta|\beta \in D$ and (iii) for all α in C, for all β in D, $\alpha \# \beta$.

Proof: It suffices to show: every stump σ *has the property* P, that is, for all γ, δ in $\mathcal{N}$ coding continuous functions from $\mathcal{N}$ to $\mathcal{N}$, if σ is non-empty and secures that γ, δ have separate ranges, then there exist a pair C, D of positively Borel subsets of $\mathcal{N}$ *separating the range of* γ *from the range of* δ, that is, such that (i) for all α, $\gamma|\alpha \in C$, (ii) for all β, $\delta|\beta \in D$ and (iii) for all α in C, for all β in D, $\alpha \# \beta$.

We do so by induction on the set of stumps.

Note that the empty stump $\underline{0}$ has the property P.

Now let σ be a non-empty stump and suppose that every immediate sub-stump σ^n of σ has the property P. Let γ, δ be elements of $\mathcal{N}$ coding continuous functions from $\mathcal{N}$ to $\mathcal{N}$ such that σ secures that γ, δ have separate ranges. Let ε be an element of $\mathcal{N}$ such that, for each n, ε^n codes a continuous function from $\mathcal{N}$ to $\mathcal{N}$, and, for all α, $\varepsilon^n|\alpha$ coincides with $\gamma|(\langle n\rangle * \alpha)$. Note that, for each β in $\mathcal{N}$, β belongs to the range of γ if and only if, for some n, β belongs to the range of ε^n. Also note that, for each n, one may distinguish two cases. *Either* $\sigma^n = \underline{0}$, and γ, δ separate at $\langle\ \rangle$, (and, therefore, also ε^n, δ separate at $\langle\ \rangle$), *or* $\sigma^n \neq \underline{0}$ and σ^n secures that ε^n, δ have separate ranges. In both cases, using the induction hypothesis in the second case, we find positively Borel sets C, D separating the range of ε^n from the range of δ. Applying an axiom of countable choice, we find an infinite sequence $C_0, D_0, C_1, D_1, C_2, D_2, \ldots$ such that, for each n, the pair (C_n, D_n) separates the range of ε^n from the range of δ. Note that the pair $(\bigcup_{n\in\mathbb{N}} C_n, \bigcap_{n\in\mathbb{N}} D_n)$ separates the range of γ from the range of δ. □

We may arrange the above proof in such a way that the countable axiom of choice we need is in fact the Second Axiom of Countable Choice mentioned in Section 3. To this end, we define, by induction on the set **Stump** of stumps, a function $\mathbb{B}$ that associates to every element of the set $\{0,1\} \times \mathbf{Stump} \times \mathcal{N}$ a positively Borel subset of $\mathcal{N}$, as follows:

(i) for each α, $\mathbb{B}(0, \underline{0}, \alpha) = \mathbb{B}(1, \underline{0}, \alpha)$ is the set of all β passing through $\alpha(0)$, and
(ii)for each non-empty stump σ, for each α,

$$\mathbb{B}(0, \sigma, \alpha) = \bigcup_{n\in\mathbb{N}} \mathbb{B}(1, \sigma^n, \alpha^n), \text{ and}$$

$$\mathbb{B}(1, \sigma, \alpha) = \bigcap_{n\in\mathbb{N}} \mathbb{B}(0, \sigma^n, \alpha^n).$$

Given any Borel subset X of $\mathcal{N}$, one may find i in $\{0,1\}$, σ in **Stump**, and α in $\mathcal{N}$ such that the set $\mathbb{B}(i, \sigma, \alpha)$ coincides with the set X. One might call the triple $\langle i, \sigma, \alpha\rangle$ a *code* of the Borel set X. Theorem 5 may be reformulated in such a way that Borel sets are replaced by codes of Borel sets.

The Lusin separation theorem, in its classical form, dates from 1917, see (Lusin 1930), (Moschovakis 1980) and (Kechris 1996). The intuitionistic version occurs in (Veldman 1981) and (Veldman 2006a). A closely related version was obtained by P. Aczel, see (Aczel, forthcoming).

6 Ramsey's Theorem

Let A be an *at-most-unary* relation on $\mathbb{N}$, that is, A is a subset of $\{\langle\ \rangle\} \cup \{\langle n\rangle | n \in \mathbb{N}\}$.

A is *almost full on* $\mathbb{N}$ if and only if, for every α in $\mathcal{N}$, α *meets* A, that is: either $\langle\ \rangle$ belongs to A or there exists n in $\mathbb{N}$ such that $\langle\alpha(n)\rangle$ belongs to A or there exist m, n in $\mathbb{N}$ such that $m < n$ and $\alpha(m) \geq \alpha(n)$.

Let σ be a stump. We say that σ *secures that A is almost full on* $\mathbb{N}$ if and only if there exists i in $\mathbb{N}$ such that, for every α in $\mathcal{N}$ with the property: $i \leq \alpha(0)$ and, for all n, $\alpha(n) < \alpha(n+1)$, there exists l such that $\overline{\alpha}l$ belongs to D_σ and either $\langle\ \rangle$ belongs to A or there exists $n < l$ in $\mathbb{N}$ such that $\langle\alpha(n)\rangle$ belongs to A.

Brouwer's Thesis implies that, if A is almost full on $\mathbb{N}$, then there exists a stump σ securing that A is almost full on $\mathbb{N}$.

Let A and B be at-most-unary relations on $\mathbb{N}$. We let $A \sqcap^* B$ be the union of the set of all u in A such that either u belongs to B or $\langle\ \rangle$ belongs to B, and the set of all t in B such that either t belongs to A or $\langle\ \rangle$ belongs to A.

Theorem 6. (Intuitionistic Pigeonhole Principle)
For all at-most-unary relations A, B, on $\mathbb{N}$, if both A and B are almost full on $\mathbb{N}$, then $A \sqcap^* B$ is almost full on $\mathbb{N}$.

Proof: It suffices to show: every stump σ *has the property P*, that is, for every at-most-unary relation A on $\mathbb{N}$, if σ secures that A is almost full on $\mathbb{N}$, then for every binary relation B on $\mathbb{N}$, if B is almost full on $\mathbb{N}$, then $A \sqcap^* B$ is almost full on $\mathbb{N}$. We do so by induction on the set of stumps.

Note that the empty stump $\underline{0}$ has the property P.

Now let σ be a non-empty stump and suppose that every immediate sub-stump σ^n of σ has the property P. Let A be an at-most-unary relation on $\mathbb{N}$ such that σ secures that A is almost full on $\mathbb{N}$. Let i be a natural number such that for every α in $\mathcal{N}$ with the property: $i \le \alpha(0)$ and, for all n, $\alpha(n) < \alpha(n+1)$, there exists l such that $\overline{\alpha}l$ belongs to D_σ and either $\langle\ \rangle \in A$ or there exists $n < l$ in $\mathbb{N}$ such that $\langle\alpha(n)\rangle \in A$.

We have to prove: for every at-most-unary relation B on $\mathbb{N}$, if B is almost full on $\mathbb{N}$, then $A \sqcap^* B$ is almost full on $\mathbb{N}$.

Let B be an at-most-unary relation on $\mathbb{N}$ that is almost full on $\mathbb{N}$.

Let α be an element of $\mathcal{N}$.

We define a statement QED, *quod est demonstrandum, what has to be proved*:

$$QED := \alpha \text{ meets } A \sqcap^* B.$$

Let n be a natural number. Note that the stump $\sigma^{\alpha(n)}$ secures that the set $A \cup \{u | u = \langle\ \rangle \wedge \langle\alpha(n)\rangle \in A\}$ is almost full on $\mathbb{N}$: for every β in $\mathcal{N}$ with the property: $\alpha(n) + i + 1 \le \beta(0)$ and, for all j, $\beta(j) < \beta(j+1)$, there exists l such that $\overline{\beta}l$ belongs to $D_{\sigma^{\alpha(n)}}$ and either $\langle\ \rangle$ belongs to A or there exists $j < l$ in $\mathbb{N}$ such that $\langle\beta(j)\rangle$ belongs to A. Applying the induction hypothesis, we find that the set $(A \cup \{u | u = \langle\ \rangle \wedge \langle\alpha(n)\rangle \in A\}) \sqcap^* B$ is almost full on $\mathbb{N}$. In particular, the infinite sequence α meets this set. Therefore: either QED or $\langle\alpha(n)\rangle$ belongs to A.

Now find n such that either $\langle\ \rangle$ belongs to B or $\langle\alpha(n)\rangle$ belongs to B.

We then have: *either* QED, *or* ($\langle\alpha(n)\rangle$ belongs to A and $\langle\ \rangle$ belongs to B) *or* ($\langle\alpha(n)\rangle$ belongs to A and $\langle\alpha(n)\rangle$ belongs to B), and in any of these three cases we have: QED.

We thus see that every α in $\mathcal{N}$ meets $A \sqcap^* B$.

We may conclude that σ has the property P. □

There is an alternative proof of Theorem 6, one that does not use Brouwer's Thesis, but has recourse to an axiom of countable choice. The above proof has been chosen with a view to the next theorem.

Let R be an *at-most-binary relation on* $\mathbb{N}$, that is, R is a subset of $\{\langle\,\rangle\} \cup \{\langle n\rangle | n \in \mathbb{N}\} \cup \{\langle m,n\rangle | m,n \in \mathbb{N}\}$.

We say that the at-most-binary relation R is *almost full on* $\mathbb{N}$ if and only if every α in $\mathcal{N}$ *meets* R, that is: either $\langle\,\rangle$ belongs to R or there exists n in $\mathbb{N}$ such that $\langle\alpha(n)\rangle$ belongs to R or there exist m,n in $\mathbb{N}$ such that $m < n$ and either $\langle\alpha(m),\alpha(n)\rangle$ belongs to R or $\alpha(m) \geq \alpha(n)$.

Let σ be a stump. We say that σ *secures that* R *is almost full on* $\mathbb{N}$ if and only if there exists i in $\mathbb{N}$ such that, for every α in $\mathcal{N}$ with the property: $i \leq \alpha(0)$ and, for all n, $\alpha(n) < \alpha(n+1)$, there exists l in $\mathbb{N}$ such that $\overline{\alpha}l$ belongs to D_σ and *either* $\langle\,\rangle$ belongs to R *or* there exists $n < l$ such that $\langle\alpha(n)\rangle$ belongs to R *or* there exist m,n in $\mathbb{N}$ such that $m < n < l$ and $\langle\alpha(m),\alpha(n)\rangle$ belongs to R.

Brouwer's Thesis implies that, if R is almost full on $\mathbb{N}$, then there exists a stump σ securing that R is almost full on $\mathbb{N}$.

Let R, T be subsets of $\mathbb{N}$. We let $R \sqcap^* T$ be the union of the set of all u in R such that there exists t in T with the property $t \sqsubseteq u$ and the set of all t in T such that there exists u in R with the property $u \sqsubseteq t$.

For all α,β in $\mathcal{N}$, we let $\alpha \circ \beta$ be the element of $\mathcal{N}$ that is the *composition* of α and β, that is, for each n, $\alpha \circ \beta(n) = \alpha(\beta(n))$.

Theorem 7. (Ramsey's Theorem)
For all at-most-binary relations R, T *on* $\mathbb{N}$, *if both* R *and* T *are almost full on* $\mathbb{N}$, *then* $R \sqcap^* T$ *is almost full on* $\mathbb{N}$.

Proof: It suffices to show: every stump σ *has the property* P, that is, for every at-most-binary relation R on $\mathbb{N}$, if σ secures that R is almost full on $\mathbb{N}$, then for every binary relation T on $\mathbb{N}$, if T is almost full on $\mathbb{N}$, then $R \sqcap^* T$ is almost full on $\mathbb{N}$. We do so by induction on the set of stumps.

Note that the empty stump $\underline{0}$ has the property P.

Now let σ be a non-empty stump and suppose that every immediate substump σ^n of σ has the property P. Let R be an at-most-binary relation on $\mathbb{N}$ such that σ secures that R is almost full on $\mathbb{N}$. Let i be a natural number such that, for every α in $\mathcal{N}$ with the property: $i \leq \alpha(0)$ and, for all j, $\alpha(j) < \alpha(j+1)$, there exists l in $\mathbb{N}$ such that $\overline{\alpha}l$ belongs to D_σ and *either* $\langle\,\rangle$ belongs to R *or* there exists $n < l$ such that $\langle\alpha(n)\rangle$ belongs to R *or* there exist m,n in $\mathbb{N}$ such that $m < n < l$ and $\langle\alpha(m),\alpha(n)\rangle$ belongs to R.

We have to prove: for every at-most-binary relation T on $\mathbb{N}$, if T is almost full on $\mathbb{N}$, then $R \sqcap^* T$ is almost full on $\mathbb{N}$.

It suffices to show: every stump τ *has the property* Q, that is, for all stumps τ, for every at-most-binary relation T on $\mathbb{N}$, if τ secures that T is almost full on $\mathbb{N}$, then $R \sqcap^* T$ is almost full on $\mathbb{N}$.

We do so by induction on the set of stumps.

Note that the empty stump has the property Q.

Now let τ be a non-empty stump and suppose that every immediate sub-stump τ^n of τ has the property Q. Let T be a binary relation on $\mathbb{N}$ such that τ secures that T is almost full on $\mathbb{N}$. Let j be a natural number such that, for every α in $\mathcal{N}$ with the property: $j \le \alpha(0)$ and, for all n, $\alpha(n) < \alpha(n+1)$, there exists l in $\mathbb{N}$ such that $\overline{\alpha}l$ belongs to D_τ and *either* $\langle\ \rangle$ belongs to T *or* there exists $n < l$ such that $\langle \alpha(n) \rangle$ belongs to T *or* there exist m, n in $\mathbb{N}$ such that $m < n < l$ and $\langle \alpha(m), \alpha(n) \rangle$ belongs to T.

We now have to prove: $R \sqcap^* T$ is almost full on $\mathbb{N}$.

We start with a preliminary remark.

Let n be a natural number.

Let $R \upharpoonright n$ be the set of all u in $\mathbb{N}$ such that $\langle n \rangle * u$ belongs to R. Note that σ^n secures that $R \cup (R \upharpoonright n)$ is almost full on $\mathbb{N}$: for every α in $\mathcal{N}$ with the property: $i + n + 1 \le \alpha(0)$ and, for all j, $\alpha(j) < \alpha(j+1)$, there exists l in $\mathbb{N}$ such that $\overline{\alpha}l$ belongs to D_{σ^n} and *either* $\langle\ \rangle$ belongs to $R \cup (R \upharpoonright n)$ *or* there exists $m < l$ such that $\langle \alpha(m) \rangle$ belongs to $R \cup (R \upharpoonright n)$ *or* there exist m, p in $\mathbb{N}$ such that $m < p < l$ and $\langle \alpha(m), \alpha(p) \rangle$ belongs to R. Using the first induction hypothesis, we may conclude: $(R \cup (R \upharpoonright n)) \sqcap^* T$ is almost full on $\mathbb{N}$.

Let $T \upharpoonright n$ be the set of all u in $\mathbb{N}$ such that $\langle n \rangle * u$ belongs to T. Reasoning as in the previous paragraph, we find that τ^n secures that $T \cup (T \upharpoonright n)$ is almost full on $\mathbb{N}$. Using the second induction hypothesis, we may conclude: $R \sqcap^* (T \cup (T \upharpoonright n))$ is almost full on $\mathbb{N}$.

Now, let α be an element of $\mathcal{N}$.

We define a statement QED, *quod est demonstrandum, what has to be proved*:

$QED := \alpha$ meets $R \sqcap^* T$.

Note that, for every β in $\mathcal{N}$, $\alpha \circ \beta$ meets $(R \cup (R \upharpoonright \alpha(0))) \sqcap^* T$, and therefore, *either* QED, *or* $\alpha \circ \beta$ meets $R \upharpoonright \alpha(0)$, that is *either* $\langle \alpha(0) \rangle$ belongs to R, *or*, for some p, $\langle \alpha(0), \alpha(\beta(p)) \rangle$ belongs to R.

Also note that, for every β in $\mathcal{N}$, $\alpha \circ \beta$ meets $R \sqcap^* (T \cup (T \upharpoonright \alpha(0)))$, and therefore, *either* QED, *or* $\alpha \circ \beta$ meets $T \upharpoonright \alpha(0)$, that is *either* $\langle \alpha(0) \rangle$ belongs to T, *or*, for some p, $\langle \alpha(0), \alpha(\beta(p)) \rangle$ belongs to T.

Let A be the set of all u in $\mathbb{N}$ such that *either* $u = \langle\ \rangle$ and $\langle \alpha(0) \rangle$ belongs to R or QED *or* there exists p such that $u = \langle p \rangle$ and $\langle \alpha(0), \alpha(p) \rangle$ belongs to R or QED.

Similarly, let B be the set of all u in $\mathbb{N}$ such that *either* $u = \langle\ \rangle$ and $\langle \alpha(0) \rangle$ belongs to T or QED *or* there exists p such that $u = \langle p \rangle$ and $\langle \alpha(0), \alpha(p) \rangle$ belongs to T or QED.

Note that A, B are both at-most-unary relations on $\mathbb{N}$ that are almost full on $\mathbb{N}$. Using Theorem 6, the Intuitionistic Pigeonhole Principle, we conclude that $A \sqcap^* B$ is almost full on $\mathbb{N}$.

Let S be the function from $\mathbb{N}$ to $\mathbb{N}$ such that, for all n in $\mathbb{N}$, $S(n) = n + 1$.

We may conclude: S meets $A \sqcap^* B$. So either QED, *or* $\langle \alpha(0) \rangle$ belongs to both R and T, *or* $\langle \alpha(0) \rangle$ belongs to R and there exists p such that $\langle \alpha(0), \alpha(p) \rangle$ belongs to T, *or* $\langle \alpha(0) \rangle$ belongs to T and there exists p such that $\langle \alpha(0), \alpha(p) \rangle$ belongs to R, *or* there exists p such that $\langle \alpha(0), \alpha(p) \rangle$ belongs to both R and T. In all five cases, we have: QED.

We thus see that every α in $\mathcal{N}$ meets $R \sqcap^* T$.

We may conclude that τ has the property Q.

We may conclude that σ has the property P.

□

The intuitionistic version of the Infinite Ramsey Theorem from (Ramsey 1928) was proven in (Veldman and Bezem 1993). A different approach to the theorem may be found in (Coquand 1994). The above proof is similar to the proof in (Veldman 2004). It may be extended to a proof of an intuitionistic version of the so-called Clopen Ramsey Theorem.

The classical mathematician may obtain Ramsey's classical result from Theorem 7 by contraposition, as follows. Let R be a binary relation on $\mathbb{N}$ and let $T := (\mathbb{N} \times \mathbb{N}) \setminus R$ be its complement. As $R \cap T = R \sqcap^* T$ is empty, either R or T is not almost full on $\mathbb{N}$, that is, there exists an infinite strictly increasing sequence α in $\mathcal{N}$ such that *either*, for all m, n, if $m < n$, then $\langle m, n \rangle \in R$ *or*, for all m, n, if $m < n$, then $\langle m, n \rangle \notin R$.

Concluding remarks at the Cerisy conference

Michael Dummett

The great philosopher Ludwig Wittgenstein once wrote, 'Philosophy leaves everything as it is'. That is, philosophy does not change anything, and can never change anything. We may think — I indeed think — that that was a foolish remark; doubtless all philosophers, even great ones, make foolish remarks sometimes. There are many people who would vehemently repudiate Wittgenstein's observation — followers of Karl Marx, for example. We know, and have been celebrating, how greatly Brouwer's philosophy changed mathematics — not, alas, for all mathematicians, but for a select band, so well represented here, able to understand and appreciate Brouwer's insights.

Brouwer's intuitionism was, of course, a rejection of the realist philosophy of mathematics, generally known as Platonism, which, rather than formalism, is the true philosophical basis of classical mathematics. There are several philosophical doctrines that reject realist views of various subject-matters: phenomenalism rejects realism about the material universe, behaviourism rejects realism about the mental, instrumentalism rejects realism about scientific theories. But, historically, these have all concentrated on what constitute the objects of the sphere within which they oppose realism — material objects, mental events and states, the theoretical entities of science: they may be described crudely as denying the common-sense conception of what these objects are made of. Brouwer, too, held views of this kind: he denied that mathematical objects are abstract entities existing eternally and independently of us, but viewed them as products of our intellectual activity. But his genius enabled him to perceive, what the opponents of other varieties of realism had failed to realise, that the improved conception must be applied, first of all, to our understanding of propositions about the subject-matter in question, in Brouwer's case to our understanding of mathematical propositions. Only when this is done does it become apparent that the new non-realist conception demands a change in the logic with which we operate. Brouwer of course developed intuitionistic mathematics not only by restricting proofs to the modified logic his conception validated, but by introducing new notions and new principles governing them. With these a beautiful new version of mathematics was born, whose centenary we have joined in celebrating.

We are mathematicians and we are philosophers, all of us logicians of one sort or another. I have never before participated in so successful a joint endeavour by people from different disciplines. We owe a deep debt of gratitude to the founder of intuitionism, L.E.J. Brouwer. And each of us owes a more immediate debt of gratitude to the organisers of this conference, the local organisers at the Centre Culturel International in Cerisy, and to all our fellow participants. Thank you all very much.

A bibliography of L.E.J. Brouwer

Dirk van Dalen

[updated April 14, 2008]

The list below contains not only the scientific papers that appeared in journals or in book form, but also his contributions to newspapers, magazines, etc.

I have adopted the convention of dating a paper according to its year of publication. For that reason the codes differ here and there from the codes of the collected works. These codes have been indicated as 'Freudenthal-Heyting Verzeichnis' (FHV).

The data of submission has been adopted from the paper as published, but also from Brouwer's correspondence, when available. The publication year is that of the journal volume; whenever available more details are given, based on reprints. In the case of series published by academies or scientific organizations, the submission is usually made by a member, and the date may be recorded. In addition there may be information available about the submission of the author to a member of the organization. Before Brouwer joined the Royal Dutch Academy of Science, his papers were submitted to the proceedings of the Academy by Korteweg, often jointly with Schoute. Once he was a member, he could submit his own papers.

No doubt, this list is incomplete. Comments and supplements are welcome.

Abbreviations:

CR Paris	Comptes Rendus des Séances de l'Académie des Sciences. Paris
CW I, CW II	Collected Works, Volume 1 and 2
Jahresb. D.M.V.	Jahresbericht der Deutschen Mathematiker Vereinigung
JFM	Jahrbuch über die Fortschritte der Mathematik
KNAW Proc.	Koninklijke Akademie van Wetenschappen. Proceedings
KNAW Verh.	Verhandelingen der Koninklijke Akademie van Wetenschappen te Amsterdam
KNAW Versl.	Koninklijke Akademie van Wetenschappen. Verslagen
MA	Mathematische Annalen
Math. Rev.	Mathematical Reviews
Zbl.	Zentralblatt für Mathematik

1904A1 [**FHV**: 1904 A1]

Over de splitsing van de continue beweging om een vast punt O van R_4 in twee continue bewegingen om O van R_3's. *KNAW Versl.* 12, p.819–839.
submitted: 27.02.1904. **published**: 09.03.1904.
submitted by: D.J. Korteweg.
reviews: JFM 35.0696.04 (publ.), p. . Reviewer: Lampe.

1904A2 [**FHV**: 1904 A2]

On a decomposition of a continuous motion about a fixed point O of S_4 into two continuous motions about O of S_3's. *KNAW Proc.* 6, p.716–735.
also in: *CW II*, p.3–22.
submitted: 27.02.1904 D.J. Korteweg. **published**: –.–.1904.

1904B1 [**FHV**: 1904 B1]

Over symmetrische transformatie van R_4 in verband met R_r en R_l. *KNAW Versl.* 12, p.926–928.
submitted: 23.04.1904. **published**: 05.05.1904.
submitted by: D.J. Korteweg, P.H. Schoute.
reviews: JFM 35.0572.07, 1904 (publ. 1906), p. . Reviewer: Lampe.

1904B2 [**FHV**: 1904 B2]

On symmetric transformation of S_4 in connection with S_r and S_l. *KNAW Proc.* 6, p.785–787.
also in: *CW II*, p.25–27
submitted: 23.04.1904. **published**: –.–.1904.
submitted by: D.J. Korteweg.

1904C1 [**FHV**: 1904 C1]

Algebraïsche afleiding van de splitsbaarheid der continue beweging om een vast punt van R_4 in die van twee R_3's. *KNAW Versl.* 12, p.941–947.
submitted: 23.04.1904. **published**: 05.04.1904.
submitted by: D.J. Korteweg, P.H. Schoute.
reviews: JFM 35.0697.02, 1904 (publ. 1906), p. . Reviewer: Lampe.

1904C2 [**FHV**: 1904 C2]

Algebraic deduction of the decomposability of the continuous motion about a fixed point of S_4 into those of two S_3's. *KNAW Proc.* 6, p.832–838.
also in: *CW II*, p.28–34.
submitted: 23.04.1904 D.J. Korteweg. **published**: –.–.1904.

1904D

Een Blijde Wereld? *Studentenweekblad, Delft,* 6.10.1904. p.8.
remarks: pseudonym Lau van der Zee.
published: 06.10.1904.

1904E

untitled. *Propria Cures,* XVI, 5. 15.10.1904.
remarks: Announcement of Bolland lectures. In Brouwer's style. No signature.
published: 15.10.1904.

1904F

Bolland in Delft. *Studentenweekblad, Delft,* 17.11.1904. p.58.
remarks: pseudonym Lau van der Zee.
published: 17.11.1904.

1904G

Over Moraal. *Propria Cures,* XVI, 10. 19.11.1904, p.110.
remarks: signed by Lau van der Zee.
published: 19.11.1904.

1904H

Troostgronden. *Propria Cures,* XVI, 14. 31.12.1904, p.157.
remarks: draft under heading 'Bollandiana' in Archive. Signed 'L.v.d.Z.'
published: 31.12.1904.

1905A [**FHV**: 1905]
Leven, Kunst en Mystiek. Waltman, Delft 1905, 99 p.
also in: Selections in English translation: *CW I*, p.1–9.
remarks: Life, Art and Mysticism. Advertised June 8. Translated in [Brouwer 1996].
published: –.06.1905.

1905B
Over Moraal (Excerpt). *Studentenweekblad, Delft*, 9.2.1905, p.137–138.
remarks: signed, Lau van der Zee. Copy of Propria Cures, XVI, 10. 19.11.1904, p.110.
published: 09.02.1905.

1906A1 [**FHV**: 1906 A1]
Meerdimensionale vectordistributies. *KNAW Versl.* 15, p.14–26.
submitted: 26.05.1906. **published**: 06.06.1906.
submitted by: D.J. Korteweg & P.H. Schoute.
reviews: JFM 37.0587.01, 1906 (publ. 1909), p.587. Reviewer: Wölfing.

1906A2 [**FHV**: 1906 A2]
Polydimensional vector distributions. *KNAW Proc.* 9, p.66–78;
also in: *CW II*, p.36–48.
remarks: corrected version of 1906 A1.
submitted: 30.06.1906. **published**: –.–.1906.

1906B1 [**FHV**: 1906 B1]
Het krachtveld der niet-Euclidische, negatief gekromde ruimten. *KNAW Versl.* 15, p.75–94.
submitted: 30.06.1906. **published**: 20.07.1906.
submitted by: D.J. Korteweg & P.H. Schoute.
reviews: JFM 37.0587.02, 1906 (publ. 1909), p.587. Reviewer: Wölfing.

1906B2 [**FHV**: 1906 B2]
The force field of the non-Euclidean spaces with negative curvature. *KNAW Proc.* 9, p.116–133.
also in: *CW II*, p.54–71.
submitted: 30.06.1906. **published**: –.–.1906.
submitted by: D.J. Korteweg.

1906C1 [**FHV**: 1906 C1]

Het krachtveld der niet-Euclidische positief gekromde ruimten. *KNAW Versl.* 15, p.293–310.

submitted: 29.09.1906. **published**: 10.10.1906.

submitted by: D.J. Korteweg & P.H. Schoute.

reviews: JFM 37.0587.03, 1906 (publ. 1909). Reviewer: Wölfing.

1906C2 [**FHV**: 1906 C2]

The force field of the non-Euclidean spaces with positive curvature. *KNAW Proc.* 9, p.250–266.

also in: *CW II*, p.73–89.

submitted: 29.09.1906. **published**: 27.10.1906.

1907 [**FHV**: 1907]

Over de Grondslagen der Wiskunde. Maas & van Suchtelen, Amsterdam, 1907, 183p.

also in: [Brouwer 1981C]; [Brouwer 2001]. English translation 'On the Foundations of Mathematics' in CW I, p.11–101.

remarks: Dissertation. Published as book without the Theses by Noordhoff, Groningen.

published: 19.02.1907.

reviews: JFM 38.0081.04, 1907 (publ. 1910), p.81–84. Reviewers: Barrau, Schuh.

Nieuw Archief voor Wiskunde 8, 1909, p.175–180. Reviewer: Mannoury.

De Beweging, 3, 1907, p.241–249. Reviewer: Mannoury.

Wiskundig Tijdschrift 4, 1907, p.33–35. Reviewer: Mannoury.

1908A [**FHV**: 1908 A]

Die mögliche Mächtigkeiten. *Atti IV Congr. Intern. Mat. Roma III,* Ed. G. Castelnuovo. Ac. dei Lincei Roma. p.569–571.

also in: *CW I*, p.102–104.

published: –.–.1908.

reviews: JFM 40.0099.04, 1909 (publ. 1912), p.99. Reviewer: Vivanti.

1908B [**FHV**: 1908 B]

Over de grondslagen der wiskunde. *Nieuw Archief voor Wiskunde (II)* 8, p.326–328.

remarks: translated in *CW I*, p.105–106 'On the foundations of mathematics'.

published: –.–.1908.

reviews: JFM 1908, 39.0081.02 (publ. 1911), p.81. Reviewer: Lampe.

1908C [**FHV**: 1908 C]
De onbetrouwbaarheid der logische principes. *Tijdschrift voor Wijsbegeerte.* 2, p.152–158
remarks: accepted 3.1.1908, letter Kohnstamm.
also in: [Brouwer 1919B]. Translated in *CW I*, p. 107–111. 'The unreliability of the logical principles'. French translation in [Largeault 1992].
published: –.–.1908.

1908D1 [**FHV**: 1908 D1]
Over differentiequotiënten en differentiaalquotiënten. *KNAW Versl.* 17, p.38–45.
submitted: 30.05.1908. **published**: 12.06.1908.
submitted by: D.J. Korteweg & P.H. Schoute.
reviews: JFM 39.0349.01, 1908 (publ. 1911), p. . Reviewer: Kowalewski.

1908D2 [**FHV**: 1908 D2]
About difference quotients and differential quotients. *KNAW Proc.* 11, p.59–66.
also in: *CW II*, p.93–100.
submitted: 30.05.1908. **published**: 27.06.1908

1909A [**FHV**: 1909 A]
Het Wezen der Meetkunde. Clausen, Amsterdam.
also in: *CW I*, p.112–120, 'The Nature of Geometry'. [Brouwer 1919 B]. French translation in [Largeault 1992].
remarks: Openbare Les privaatdocent 12.10.1909. (inaugural lecture).
published: –.10.1909.
reviews: JFM 40.0520.01, 1909. (publ. 1912), p.529. Reviewer: Schuh.

1909B [**FHV**: 1909 B]
Die Theorie der endlichen continuierlichen Gruppen, unabhängig von den Axiomen von Lie. *Atti IV Congr. Intern. Mat. Roma II,* Ed. G. Castelnuovo. Ac. dei Lincei. Roma. p.296–303.
also in: *CW II*, p.109–116.
remarks: conference talk 9.4.1908.
published: –.–.1909.
reviews: JFM 40.0194.02, 1909 (publ. 1912). Reviewer: Vivanti.

1909C [**FHV**: 1909 C]
Die Theorie der endlichen kontinuierlichen Gruppen, unabhängig von den Axiomen von Lie, I. *MA* 67, p.246–267.
also in: *CW II*, p.118–139.
remarks: 'Berichtigung', see [Brouwer 1910G]. Letter to Hilbert 14.5.1909.
published: –.–.1909.
reviews: JFM 40.0194.01, 1909, publ. 1912. p.194. Reviewer: Engel.

1909D [**FHV**: 1909 D]
Over de niet-Euclidische meetkunde. *Nieuw Archief voor Wiskunde (II)* 9, p.72–74.
remarks: comments on a paper of Bouman, NAW 9, 483–495. Bouman's reaction NAW 10, 75–77.
submitted: Brouwer to Kluyver — 15.04.1909; to NAW — 30.05.1908.
published: –.–.1909.

1909E [**FHV**: 1909 E]
Karakteriseering der Euclidische en niet-Euclidische bewegingsgroepen in R_n. *Handelingen van het Nederlandsch Natuur- en Geneeskundig Congres* 12, p.189–199.
also in: translated in *CW II*, p.185–192, 'Characterization of the Euclidean and non-Euclidean motion groups in R_n.'
remarks: corrections p.11.
published: –.–.1909.
reviews: JFM 40.0526.02, 1909 (publ. 1912). Reviewer: Schuh.

1909F1 [**FHV**: 1909 F1]
Over één-éénduidige, continue transformaties van oppervlakken in zichzelf. *KNAW Versl.*, 17, p.741–752.
submitted: 27.02.1909. **published**: 11.03.1909.
submitted by: D.J. Korteweg & P.H. Schoute.
reviews: JFM 40.0692.01, 1909 (publ. 1912). Reviewer: Lampe.

1909F2 [**FHV**: 1909 F2]
Continuous one-one transformations of surfaces in themselves. *KNAW Proc.* 11, p.788–798.
submitted: 27.02.1909. **published**: 27.03.1909.
submitted by: D.J. Korteweg.

1909G1 [**FHV**: 1909 G1]

Over continue vectordistributies op oppervlakken. *KNAW Versl.* 17, p.896–904.

submitted: Brouwer to Korteweg — 16.03.1909; to KNAW — 27.03.1909.
published: 07.04.1909.
submitted by: D.J. Korteweg & P.H. Schoute.

reviews: JFM 40.0372.03, 1909 (publ. 1912), p.372. Reviewer: Wallenberg.

1909G2 [**FHV**: 1909 G2]

On continuous vector distributions on surfaces. *KNAW Proc. 11,* p.850–858.
also in: *CW II*, p.273–281.

submitted: 27.03.1909. **published**: 23.04.1909.
submitted by: D.J. Korteweg.

1909H1 [**FHV**: 1909 H1]

Over één-éénduidige, continue transformaties van oppervlakken in zichzelf, II. *KNAW Versl.* 18, p.106–117.

submitted: Brouwer to Korteweg — 18.06.1909; to KNAW — 26.06.1909.
published: 13.07.1909.
submitted by: D.J. Korteweg & P.H. Schoute.

reviews: JFM 40.0691.02, 1909 (publ. 1912). Reviewer: Lampe.

1909H2 [**FHV**: 1909 H2]

Continuous one-one transformations of surfaces in themselves, II. *KNAW Proc.* 12, p.286–297.
also in: *CW II*, p.207–218.
remarks: contains erratum to 1906B3.

submitted: 26.06.1909. **published**: 30.10.1909.
submitted by: D.J. Korteweg.

1910A1 [**FHV**: 1910 A1]

Over continue vectordistributies op oppervlakken, II. *KNAW Versl.* 18, p.702 - 721.
remarks: Contains corrections to part I, [Brouwer 1909F1].

submitted: 26.02.1910. **published**: 08.03.1910.
submitted by: D.J. Korteweg & P.H. Schoute.

reviews: JFM 41.0543.04, 1910 (publ. 1913), p.543. Reviewer: Lampe. JFM 41.0704.04, 1910 (publ. 1913), p.704. Reviewer: Brouwer.

1910A2 [**FHV**: 1910 A2]

On continuous vector distributions on surfaces, II. *KNAW Proc.* 12, p.716–734.

also in: *CW II*, p.283–302.

submitted: 26.02.1910 **published**: –.–.1910

submitted by: D.J. Korteweg

1910B1 [**FHV**: 1910 B1]

Over de structuur der perfekte puntverzamelingen. *KNAW Versl.* 18, p.833–842

submitted: 26.03.1910 **published**: 11.04.1910

submitted by: D.J. Korteweg & P.H. Schoute

reviews: JFM 41.0103.03, 1910 (publ. 1913). Reviewer: Vivanti.

1910B2 [**FHV**: 1910 B2]

On the structure of perfect sets of points. *KNAW Proc.* 12, p.785–794.

also in: *CW II*, p.341–350.

submitted: 26.03.1909 **published**: –.–.1910

submitted by: D.J. Korteweg

1910C [**FHV**: 1910 C]

Zur Analysis Situs. *MA* 68, p. 422–434.

also in: *CW II*, p. 352–366.

submitted: 14.05.1909 **published**: –.–.1910, joint reprint with A. Schoenflies — *Bemerkung zu dem vorstehenden Aufsatz des Herrn L.E.J. Brouwer*

1910D1 [**FHV**: 1910 D1]

Over continue vectordistributies op oppervlakken, III. *KNAW Versl.* 19, p.36–51

submitted: 28.05.1910 **published**: 09.07.1910

submitted by: D.J. Korteweg & P.H. Schoute

reviews: JFM 41.0543.04, 1910 (publ. 1913), p.543. Reviewer: Lampe. JFM 41.0704.04, 1910 (publ. 1913), p.704. Reviewer: Brouwer.

1910D2 [**FHV**: 1910 D2]

On continuous vector distributions on surfaces, III. *KNAW Proc.* 12, p. 171–186.

also in: *CW II*, p. 303–318.

submitted: 28.05.1910 **published**: –.–.1910

submitted by: D.J. Korteweg

1910E [**FHV**: 1910 E]
Beweis des Jordanschen Kurvensatzes. *MA* 69, p.169–175.
also in: *CW II*, p.377–383.
published: –.–.1910
reviews: JFM 41.0544.02, 1910 (publ. 1913), p.544. Reviewer: Dehn.

1910F [**FHV**: 1910 F]
Über eineindeutige, stetige Transformationen von Flächen in sich. *MA* 69, p.176–180.
also in: *CW II*, p.244–248.
remarks: Berichtigung, see [Brouwer 1910L] and [Brouwer 1919F].
submitted: –.08.1909 **published**: –.–.1910
reviews: JFM 41.0544.01, 1910 (publ. 1913), p.544. Reviewer: Dehn.

1910G [**FHV**: 1910 G]
Berichtigung. *MA* 69, p.180.
remarks: refers to [Brouwer 1909C].
published: –.–.1910

1910H [**FHV**: 1910 H]
Die Theorie der endlichen kontinuierlichen Gruppen, unabhängig von den Axiomen von Lie, II. *MA* 69, p. 181–203.
also in: *CW II*, p.156–178.
published: –.–.1910
reviews: JFM 41.0181.03, 1910 (publ. 1913). Reviewer: Engel.

1910J [**FHV**: 1910 J]
Sur les continus irréductibles de M. Zoretti. *Annales Éc. Norm. Sup.* 27, p.565–566
also in: *CW II*, p. 373–374.
submitted: 05.06.1910 **published**: –.–.1910

1910L [**FHV**: 1910 L]
Berichtigung. *MA* 69, p.592.
remarks: refers to [Brouwer 1910F].
published: –.–.1910

1910M
G. Mannoury, Methodologisches und Philosophisches. *De Nieuwe Amsterdammer,* 28.8.1910
remarks: review
published: 28.08.1910

1911A [**FHV**: 1911 A]
Mannoury, Methodologisches und Philosophisches zur Elementar-Mathematik, Haarlem 1909. *Nieuw Archief voor Wiskunde (II)* 9, p.199–201.
also in: Translation in *CW I*, p.121–122,
remarks: Review
published: –.–.1911

1911B1 [**FHV**: 1911 B1]
Over de structuur der perfecte puntverzamelingen, II. *KNAW Versl.* 19, p.1416–1426.
submitted: 28.04.1911 **published**: 10.05.1911
submitted by: D.J. Korteweg & P.H. Schoute
reviews: JFM 42.0089.13, 1911 (publ. 1914), p.89. Reviewer: Vivanti.

1911B2 [**FHV**: 1911 B2]
On the structure of perfect sets of points, II. *KNAW Proc.* 14, p.137–147.
also in: *CW II*, p.384–394.
submitted: 28.04.1911 **published**: –.–.1911
submitted by: D.J. Korteweg

1911C [**FHV**: 1911 C]
Beweis der Invarianz der Dimensionenzahl. *MA* 70, p.161–165.
also in: *CW II*, p.430–434.
submitted: –.06.1910 **published**: 14.02.1911
reviews: JFM 42.0416.02, 1911 (publ. 1914), p.416–417. Reviewer: Blaschke.

1911D [**FHV**: 1911 D]
Über Abbildung von Mannigfaltigkeiten. *MA* 71, p.97–115.
also in: *CW II*, p.454–472.
remarks: The reprints carry a sticker with a simplification of the proof of thm.3. Berichtigung, see [Brouwer 1911 M] and [Brouwer 1921E].
submitted: –.07.1910 **published**: –.–.25.7.1911
reviews: JFM 42.0417.01, 1911 (publ. 1914), p.417–418. Reviewer: Blaschke.

1911E [**FHV**: 1911 E]
Beweis der Invarianz des n–dimensionalen Gebiets. *MA* 71, p.305–313.
also in: *CW II*, p.477–485.
remarks: Berichtigung, see [Brouwer 1921]. Sent to Hilbert 14.7.1911
submitted: 14.07.1911; printer's proof 20.10.1911
published: 16.11.1911
reviews: JFM 42.0418.01, 1911 (publ. 1914), p.418. Reviewer: Blaschke.

1911F [**FHV**: 1911 F]
Beweis des Jordanschen Satzes für den n–dimensionalen Raum. *MA* 71, p.314–319.
also in: *CW II*, p. 489–494.
published: 16.11.1911
reviews: JFM 42.0418.02, 1911 (publ. 1914), p.418. Reviewer: Blaschke.

1911G [**FHV**: 1911 G]
Über Jordansche Mannigfaltigkeiten. *MA* 71, p.320–327.
also in: *CW II*, p.498–505.
remarks: Bemerkung, see [Brouwer 1911 N]
published: 16.11.1911
reviews: JFM 42.0418.03, 1911 (publ. 1914), p.418–419. Reviewer: Blaschke.

1911H1 [**FHV**: 1911 H1]
Over één-éénduidige, continue transformaties van oppervlakken in zichzelf, III. *KNAW Versl.* 19, p.737–747.
submitted: 24.12.1910 **published**: 11.01.1911
submitted by: D.J. Korteweg & P.H. Schoute
remarks: This and/or the next paper was submitted by Brouwer to Korteweg on 30.11.1910
reviews: JFM 42.0707.02, 1911 (publ. 1914), p.707. Reviewer: Engel.

1911H2 [**FHV**: 1911 H2]
Continuous one-one transformations of surfaces in themselves, III. *KNAW Proc.* 13, p. 767–777.
also in: *CW II*, p.221–231.
submitted: 24.12.1910 **published**: –.–.1911
submitted by: D.J. Korteweg

1911J1 [**FHV**: 1911 J1]
Over één-éénduidige, continue transformaties van oppervlakken in zichzelf, IV. *KNAW Versl.* 20, p.24–34.
submitted: 27.05.1911 **published**: 08.06.1911
submitted by: D.J. Korteweg & P.H. Schoute
reviews: JFM 42.0707.02, 1911 (publ. 1914), p.707. Reviewer: Engel.

1911J2 [**FHV**: 1911 J2]
Continuous one-one transformations of surfaces in themselves, IV. *KNAW Proc.* 14, p.300–310.
also in: *CW II*, p.233–243.
submitted: 27.05.1911 **published**: 29.10.1911
submitted by: D.J. Korteweg

1911K [**FHV**: 1911 K]
Sur une théorie de la mesure. A propos d'un article de M.G. Combebiac. *L'enseignement Math.* 13, p.377–380.
also in: *CW II*, p.181–183.
published: –.–.1911
reviews: JFM 42.0446.01, 1911 (publ. 1914), p.446. Reviewer: Oppenheim.

1911L [**FHV**: 1911 L]
Sur le théorème de M. Jordan dans l'espace à n dimensions. *CR Paris* 153, p.542–543.
also in: *CW II*, p.496–497.
submitted: 11.09.1911 **published**: –.–.1911
reviews: JFM 42.0507.03, 1911 (publ. 1914), p.507. Reviewer: Lichtenstein.

1911M [**FHV**: 1911 M]
Berichtigung. *MA* 71, p.598.
also in: *CW II*, p.474.
remarks: refers to [Brouwer 1911D]
published: –.–.1911

1911N [**FHV**: 1911 N]
Bemerkung zu dem Aufsatz von L.E.J. Brouwer: 'Über Jordansche Mannigfaltigkeiten'. *MA* 71, p.598.
also in: *CW II*, p.506.
remarks: refers to [Brouwer 1911 G]
published: –.–.1911
reviews: JFM 42.0419.01, 1911 (publ. 1914), p.419. Reviewer: Blaschke.

1912A [**FHV**: 1912 A1]
Intuïtionisme en Formalisme. Clausen, Amsterdam.
also in: Wiskundig Tijdschrift 9, p.180–211; [Brouwer 1919B]. English translation [Brouwer 1913C]. Polish translation [Brouwer 1986]. French translation in [Largeault 1992].
remarks: Inaugural Lecture, 14.10.1912. A commercial edition was published by Noordhoff, Groningen (1912).
published: –.–.1912
reviews: JFM 43.0111.01, 1912 (publ. 1915), p.111. Reviewer: Barrau.
Tijdschrift voor Wijsbegeerte 7, (1913), p.237–239.

1912B [**FHV**: 1912 B]
Beweis des ebenen Translationssatzes. *MA* 72, p.37–54.
also in: *CW II*, p.250–267.
published: –.–.1912
reviews: JFM 43.0569.02, 1912 (publ. 1915), p.569. Reviewer: Engel.

1912C [**FHV**: 1912 C]
Zur Invarianz des n–dimensionalen Gebiets. 1912 *MA* 72, p.55–56.
also in: *CW II*, p.509–510.
published: 15.03.1912
reviews: JFM 43.0479.03, 1912 (publ. 1915), p.479. Reviewer: Engel

1912D [**FHV**: 1912 D]
Über die topologischen Schwierigkeiten des Kontinuitätsbeweises der Existenztheoreme eindeutig umkehrbarer polymorpher Funktionen auf Riemannschen Flächen (Auszug aus einem Brief an R. Fricke). *Nachr. Göttingen* 1912, p.603–606.
also in: *CW II*, p. 577–580.
remarks: The letter to Fricke is dated 22.12.1911
submitted: 13.01.1912 **published**: –.06.1912
submitted by: F. Klein
reviews: JFM 43.0527.02, 1912 (publ. 1915), p.527. Reviewer: Weyl.

1912E1 [**FHV**: 1912 E1]
Over omloopcoëfficiënten. *KNAW verslagen* 20, p. 1049–1057.
submitted: 24.02.1912 **published**: 08.03.1912
submitted by: L.E.J. Brouwer & P.H. Schoute

1912E2 [**FHV**: 1912 E2]
On looping coefficients. *KNAW Proc.* 15, p.113–122.
also in: *CW II*, p.511–520.
submitted: 24.02.1912 **published**: –.–.1912
submitted by: D.J. Korteweg

1912F [**FHV**: 1912 F]
Sur l'invariance de la courbe fermée. *CR Paris* 154, p.862–863.
also in: *CW II*, p.521– 522.
submitted: 01.04.1912 **published**: 19.07.1912
reviews: JFM 43.0569.04, 1912 (publ. 1915). Reviewer: Lampe.

1912G [**FHV**: 1912 G]

Über die Singularitätenfreiheit der Modulmannigfaltigkeit. *Nachr. Göttingen.* 1912, p.803–806

also in: *CW II*, p.587–590.

remarks: The argument was communicated to Fricke in a letter of 30.12.1911.

submitted: submitted to Hilbert 12.06.1912 **published**: –.–.1912

submitted by: D. Hilbert in the meeting of 22.06.1912

reviews: JFM 43.0527.02, 1912 (publ. 1915). Reviewer: Weyl.

1912H [**FHV**: 1912 H]

Über den Kontinuitätsbeweis für das Fundamentaltheorem der automorphen Funktionen im Grenzkreisfalle. *Jahresb. D.M.V.* 21, p.154–157.

also in: *CW II*, p.568– 571.

remarks: The reprints carry a sticker added by Brouwer, dated 01.05.1913.

published: 06.08.1912 (date on reprint)

reviews: JFM 43.0525.02, 1912 (publ. 1915). A global review of the Karlsruhe symposium on automorphic functions. Reviewer: Koebe.

1912K1 [**FHV**: 1912 K1]

Over één-éénduidige, continue transformaties van oppervlakken in zichzelf, V. *KNAW Versl.* 21, p.300–309.

submitted: 28.09.1912 **published**: 12.10.1912

submitted by: L.E.J. Brouwer

reviews: JFM 43.0692.01, 1912 (publ. 1915), p.692. Reviewer: Engel.

1912K2 [**FHV**: 1912 K2]

Continuous one-one transformations of surfaces in themselves, V. *KNAW Proc.* 15, p.352–360.

also in: *CW II*, p.527–536.

remarks: cf. [Brouwer 1920 G2]

submitted: 28.09.1912 **published**: –.–.1912

submitted by: L.E.J. Brouwer

1912L [**FHV**: 1912 L]

Beweis der Invarianz der geschlossenen Kurve. *MA* 72, p.422–425.

also in: *CW II*, p.523–526.

published: –.–.1912

1912M **[FHV**: 1912 M]

Sur la notion de 'classe' de transformations d'une multiplicité. *Proc. V Int. Congr. Math. Cambridge 1912, II,* Eds. E.W. Hobson, A.F.H. Love. Cambridge University Press, Cambridge. p.9–10.
also in: *CW II*, p.538–539
published: –.–.1912
reviews: JFM 44.0555.02, 1913 (publ. 1918), p.555. Reviewer: Lampe.

1913A **[FHV**: 1913 A]

Über den natürlichen Dimensionsbegriff. *Journ. r. u. angew. Math.* 142, p.146–152.
also in: *CW II*, p. 540–546.
remarks: Berichtigung, see [Brouwer 1923D] and [Brouwer 1924M]
published: 27.01.1913 (date received by Un. Library Amsterdam)
reviews: JFM 44.0555.01, 1913 (publ. 1916), p.555. Reviewer: Blaschke. The Berichtigung was reviewed by Fraenkel — JFM 50.0128.01, 1924 (publ. 1928/29)

1913B1 **[FHV**: 1913 B1]

Eenige opmerkingen over het samenhangstype η. *KNAW Versl.* 21, 1412–1419
also in: *CW II*, p.396–403.
submitted: 25.04.1913 **published**: 09.05.1913
submitted by: L.E.J. Brouwer
reviews: JFM 44.0556.01, 1913 (publ. 1918). Reviewer: Brouwer.

1913B2 **[FHV**: 1913 B2]

Some remarks on the coherence type η. *Proc. KNAW* 15, p. 1256–1263.
also in: CW II, 396–404
remarks: Autoreferat JFM 44.0556.01, 1913 (publ. 1918)
submitted: 25.04.1913 **published**: –.–.1913
submitted by: L.E.J. Brouwer

1913C **[FHV**: 1912 A2]

Intuitionism and Formalism. *Bull. Am. Math. Soc.*, 20: 81–96.
also in: CW II, 123–138; Benacerraf and Putnam, 1964.
remarks: translated by A. Dresden.

1914 **[FHV**: 1914]

A. Schoenflies und H. Hahn. Die Entwickelung der Mengenlehre und ihrer Anwendungen, Leipzig und Berlin 1913. *Jahresb. D.M.V.*23, p.78–83.
also in: *CW I*, p.139–144
remarks: Review
published: –.–.1914

1915A1 [**FHV**: 1915 A1]
Opmerking over inwendige grensverzamelingen. *KNAW Versl.* 23, p.1325–1326.
submitted: 23.04.1915 **published**: 14.05.1915
submitted by: L.E.J. Brouwer
reviews: JFM 45.0128.03, 1914/15 (publ. 1922). Reviewer: Lampe.

1915A2 [**FHV**: 1915 A2]
Remark on inner limiting sets. *KNAW Proc.* 18, p.48–49.
also in: *CW II*, p.412–413.
remarks: 'Inner limiting set' means G_δ
submitted: 23.04.1915 **published**: 14.05.1915
submitted by: L.E.J. Brouwer

1915B [**FHV**: 1915 B]
Over de loodrechte trajectoriën der baankrommen eener vlakke eenledige projectieve groep. *Nieuw Archief voor Wiskunde II,* 11 p.265–290.
remarks: [On the orthogonal trajectories of the orbits of a one-parameter plane projective group]. English translation in *CW II*, p.319–337.
published: –.–.1915
reviews: JFM 45.0938.01, 1914/15 (publ. 1922). Reviewer: Engel.

1915C
Meetkunde en Mystiek. *De Nieuwe Amsterdammer,* 04.12.1915
remarks: review Dr. Naber
published: 04.12.1915

1916A [**FHV**: 1916 A]
L.E.J. Brouwer, Eenige opmerkingen over het samenhangstypen η. JFM 44.0566.01, p.556, 1914/15 (publ. 1918) .
also in: *CW II*, p.405
remarks: Author's review of [Brouwer 1913 B1].
published: –.–.1916

1916B [**FHV**: 1916 B]
É. Borel, Les ensembles de mesure nulle. JFM 44.0556.02, 1914/15 p.556–557 (publ. 1918).
also in: *CW II*, p.406–407.
remarks: Review
published: –.–.1916

1916C

J.I. de Haan, Rechtskundige significa en haar toepassing op de begrippen: 'aansprakelijk', 'verantwoordelijk', 'toerekeningsvatbaar'. *Groot Nederland,* September 1916, 14, 333–336.
remarks: Review. Legal significs and its application to the notions of 'liability', 'responsibility', 'accountability'.
published: –.–.1916

1916D

Mr. Israël de Haan's Rechtkundige Significa. *De Nieuwe Amsterdammer,* 9.9.1916
published: 09.09.1916

1916E

Een vraag aan de Belgische Regeering. *De Nieuwe Amsterdammer,* 16.9.1916
published: 16.09.1916

1916F

Luchtvaart en photogrammetrie. *Avia* (6), p. 29–30, 122–124, 223–225
published: –.–.1916

1916G

Letter to the editor. *De Nieuwe Amsterdammer* 23.9.1916.
published: 23.09.1916.

1917A1 [**FHV**: 1917]

Addenda en corrigenda over de grondslagen der wiskunde. *KNAW Versl.* 25, p.1418–1423;
also, with minor revisions, in: Nieuw Archief voor Wiskunde II, 12, p. 439–445. Translation 'Addenda and corrigenda to On the Foundations of Mathematics', *CW I,* p.145–149.
remarks: Separate sheet with corrections inserted. The 'Nieuw Archief'-version is complete.
submitted: 27.04.1917 **published**: 23.07.1917
submitted by: L.E.J. Brouwer
reviews: JFM 46.0069.04, 1916/18 (publ. 1923/24), p.69. Reviewer: Szegö.

1917A2 [**FHV**: 1917 A]

Addenda en corrigenda over de grondslagen der wiskunde. *Nieuw Archief voor Wiskunde II,* 12, p.439–445

also in: English translation 'Addenda and corrigenda to On the Foundations of Mathematics', *CW I,* p.145–149.

remarks: Corrections to KNAW Proc. paper. Refers to Verslagen Ak. 27.4.1917

submitted: 27.04.1917 **published**: –.–.1917

submitted by: L.E.J. Brouwer

1917B1 [**FHV**: 1917B1]

Over lineaire inwendige grensverzamelingen. *KNAW Versl.* 25, p. 1424–1426.

submitted: 27.04.1917 **published**: 23.07.1917

submitted by: L.E.J. Brouwer

reviews: JFM 46.0293.01, 1916/1918 (publ. 1923/24). Reviewer: Haalmeijer.

1917B2 [**FHV**: 1917 B2]

On linear inner limiting sets. *KNAW Proc.* 20, p.1192–1194.

also in: *CW II,* p.414–415.

submitted: 27.04.1917 **published**: –.–.1917

submitted by: L.E.J. Brouwer

1917C

Luchtvaart en photogrammetrie. *Het Vliegveld,* 142–144, 165–167.

published: –.–.1917

1917D

Anti-nationale Literatuur. *De Nieuwe Amsterdammer,* 3.2.1917

remarks: review of articles of F. van Eeden and C. van Bruggen

published: 03.02.1917

1918A [**FHV**: 1918 A]

Voorbereidend Manifest. *Mededeelingen van het Internationaal Instituut voor Wijsbegeerte* I, p.3–12.

remarks: Preparatory Manifesto, with French, German and English translations. With H.P.J. Bloemers, H. Borel and F. van Eeden,

published: –.–.1918

1918B [**FHV**: 1918 B]
Begründung der Mengenlehre unabhängig vom logischen Satz vom ausgeschlossenen Dritten. Erster Teil, Allgemeine Mengenlehre. *KNAW Verh. 1e sectie,* deel XII, no.5, p.1–43.
also in: *CW I*, p.151–190.
submitted: 24.11.1917 **published**: –.03.1918
reviews: JFM 46.0310.06, 1916/1918 (publ. 1923/24). Reviewer: Szegö.

1918C [**FHV**: 1918 C]
Über die Erweiterung des Definitionsbereichs einer stetigen Funktion. *MA.* 79, p. 209–211.
also in: *CW II*, p.591–593
remarks: Bemerkung, see [Brouwer 1919E]
published: –.–.1918

1918D [**FHV**: 1918 D]
Lebesguesches Mass und Analysis Situs. *MA* 79, p.212–222.
also in: *CW II*, p.595–605
published: –.–.1918
reviews: JFM 46.0292.02, 1916/18 (publ. 1923/24). Reviewer: Rosenthal.

1918E [**FHV**: 1918 E]
Beweging van een materieel punt op den bodem eener draaiende vaas onder den invloed der zwaartekracht. *Nieuw Archief voor Wiskunde II* 12, p.407–419
also in: *CW II*, p.665–675. Translated as 'Motion of a particle on the bottom of a rotating vessel under the influence of gravity.'
published: –.–.1918
reviews: JFM 46.1184.03, 1916/1918 (publ. 1923/24). Reviewer: Schouten.

1918F
Berichtigung. *MA*, 79, 403. (re 1910F).
also in: *CW II*, 449.
reviews: JFM 46.1184.03, 1916/18 (publ. 1923/24). Reviewer: Schouten.

1918G
Voorbereiding van Hoogleeraarsbenoemingen. *De Nieuwe Amsterdammer*, 27.7.1918
published: 27.07.1918

1918H
Intuïtieve significa. *Propria Cures* 9.9.1918
remarks: announcement lecture F. van Eeden
published: –.–.1918

1918J

Aan de Ned. Burgerij. *De Nieuwe Amsterdammer*, 30.11.1918
remarks: Open letter of Committee concerning Troelstra's revolution. Signed by Brouwer and others.
published: 30.11.1918

1919A [**FHV**: 1919 A]

Begründung der Mengenlehre unabhängig vom logischen Satz vom ausgeschlossenen Dritten. Zweiter Teil, Theorie der Punktmengen, *KNAW Verh. 1e sectie,* deel XII, no.7, p.1–33.
also in: *CW I*, p.191–221.
submitted: 26.10.1918 **published**: –.–.1919

1919B [**FHV**: 1919 B]

Wiskunde, Waarheid, Werkelijkheid. Noordhoff, Groningen
remarks: *Mathematics, Truth, Reality.* Reprint of [Brouwer 1908C], [Brouwer 1909A] and [Brouwer 1912A].
published: –.–.1919

1919C [**FHV**: 1919 C]

Significh Taalonderzoek. Onderscheid der taaltrappen ten aanzien van de sociale verstandhouding. *Mededeelingen van het Intern. Inst. voor Wijsbegeerte* II, p.5–32.
also in: Translation 'Signifische Sprachforschung. Vom Unterschied der Sprachstufen in Bezug auf die soziale Verständigung' is published in *CW I*, p.222–229
remarks: With G. Mannoury, H. Borel and F. van Eeden. German and French translations.
published: –.–.1919

1919D [**FHV**: 1919 D1]

Intuitionistische Mengenlehre. *Jahresb. D.M.V.* 28, p. 203–208.
also in: English translation in [Mancosu 1998] p.23–27, French translation in [Largeault 1993] p.183–189.
remarks: dated 1920 on reprint
published: –.–.1920
reviews: JFM 47.0171.04, 1919/20 (publ. 1924/26). Reviewer: Rosenthal.

1919E **[FHV**: 1919 E]

Nachträgliche Bemerkung über die Erweiterung des Definitionsbereiches einer stetigen Funktion. *MA* 79, p.403.
also in: *CW II*, p.593.
remarks: refers to [Brouwer 1918C]
published: –.–.1919
reviews: JFM 46.0399.03, 1916/1918 (publ. 1923/24), p.399. Reviewer: Rosenthal.

1919F **[FHV**: 1919 F]

Berichtigung. *MA* 79, p.403.
also in: *CW II*, p.249
remarks: corrections to [Brouwer 1910 F]
published: –.–.1919

1919G **[FHV**: 1919 G]

Énumération des surfaces de Riemann regulières de genre un. *CR Paris* 168, p. 677–678, 832
also in: *CW II*, p.631–632.
submitted: 31.03.1919 **published**: –.–.1919
reviews: JFM 47.0355.03, 1919/20 (publ. 1924/26). Reviewer: Helmer.

1919G*

Errata. *CR Paris* 168, p. 832
also in: *CW II*, p.633.
remarks: corrections to 1919G **published**: –.–.1919

1919H **[FHV**: 1919 H]

Énumération des groupes finis de transformations topologiques du tore. *CR Paris* 168, p. 845–848, 1168 and 169, p.552.
also in: *CW II*, p.634–638.
submitted: 28.04.1919 **published**: –.–.1919
reviews: JFM 1919/20 47.0524.04 (publ. 1924/26). Reviewer: Helmer.

1919H*

Errata *CR Paris* 169, p.552, 1168.
remarks: corrections to 1919H.
submitted: 28.04.1919 **published**: –.–.1919

1919J [**FHV**: 1919 J]
Sur les points invariants des transformations topologiques des surfaces. *CR Paris* 168, p. 1042–1044, 169, p.552.
also in: *CW II*, p.639–641.
remarks: Errata: CR Paris 169, p.552.
submitted: 26.05.1919 **published**: –.–.1919
reviews: JFM 47.0525.02 1919/20 (publ. 1924/26). Reviewer: Helmer.

1919J*
Errata. *CR Paris,* 169, p. 552
remarks: Corrections to 1919J.
submitted: 26.05.1919 **published**: –.–.1919

1919K [**FHV**: 1919 K]
Sur la classification des ensembles fermés situés sur une surface. *CR Paris* 169, p. 953–954.
submitted: 24.11.1919 **published**: –.–.1919
submitted by: P. Appell

1919L1 [**FHV**: 1919 L1]
Over één-éénduidige, continue transformaties van oppervlakken in zichzelf, VI. *KNAW Versl.* 27, p. 609–612.
submitted: 30.11.1918 **published**: –.–.1919
submitted by: L.E.J. Brouwer
reviews: JFM 47.0525.03 1919/20 (publ. 1924/26). Reviewer: Helmer.

1919L2 [**FHV**: 1919 L2]
Über eineindeutige stetige Transformationen von Flächen in sich, VI. *KNAW Proc.* 21, p.707–710.
also in: *CW II*, p.611–615
submitted: 30.11.1918 **published**: –.–.1919

1919M1 [**FHV**: 1919 M1]
Opmerking over de vlakke translatiestelling. *KNAW Versl.* 27, p.840–841
submitted: 28.12.1918 **published**: –.–.1919
submitted by: L.E.J. Brouwer
reviews: JFM 47.0524.02 1919/20 (publ. 1924/26). Reviewer: Helmer.

1919M2 [**FHV**: 1919 M2]
Remark on the plane translation theorem. *KNAW Proc.* 21, p.935–936.
also in: *CW II*, p.269–270.
submitted: 28.12.1928 **published**: –.–.1919
submitted by: L.E.J. Brouwer

1919N1 [**FHV**: 1919 N1]
Over topologische involuties. *KNAW Versl.* 27, p.1202–1203.
submitted: 29.03.1919 **published**: –.–.1919
submitted by: L.E.J. Brouwer
reviews: JFM 47.0524.03 1919/20 (publ. 1924/26). Reviewer: Helmer.

1919N2 [**FHV**: 1919 N2]
Über topologische Involutionen. *KNAW Proc.* 21, p.1143–1145.
also in: *CW II*, p. 623–625.
submitted: 29.03.1919 **published**: –.–.1919
submitted by: L.E.J. Brouwer

1919P1 [**FHV**: 1919 P1]
Opsomming der periodieke transformaties van de torus. *KNAW Versl.* 27, p. 1363–1367.
submitted: 03.05.1919 **published**: –.–.1919
submitted by: L.E.J. Brouwer
reviews: JFM 47.0525.01 1919/20 (publ. 1924/26). Reviewer: Helmer.

1919P2 [**FHV**: 1919 P2]
Aufzählung der periodischen Transformationen des Torus. *KNAW Proc.* 21, p. 1352–1356.
also in: *CW II*, p.626–630.
submitted: 03.05.1919 **published**: –.–.1919
submitted by: L.E.J. Brouwer

1919Q1 [**FHV**: 1919 Q1]
Opmerking over meervoudige integralen. *KNAW Versl.* 28, p.116–120.
submitted: 28.06.1919 **published**: –.–.1919
submitted by: L.E.J. Brouwer
reviews: JFM 47.0906.03 1919/20 (publ. 1924/26). Reviewer: Müntz.

1919Q2 [**FHV**: 1919 Q2]
Remark on multiple integrals. *KNAW Proc.* 22, p.150–154.
also in: *CW II*, p.49–53.
submitted: 28.06.1919 **published**: –.–.1919

1919R **[FHV**: 1919 R]
Luchtvaart en photogrammetrie, I. *Nieuw Tijdschrift voor Wiskunde* 7, p.311–331.
remarks: based on texts from Avia (1916) and Het Vliegveld (1917) Drawings by Wijdenes.
published: –.–.1919
reviews: JFM 47.0547.02, 1919/20 (publ. 1924/26). Reviewer: Van der Woude.

1919S **[FHV**: 1919 S]
Über die periodischen Transformationen der Kugel. *MA* 80, p.39–41.
also in: *CW II*, p.620–622.
remarks: sequel to paper of Nielsen, MA 80, 36–38. Proofs received from Klein 12.09.1919
published: –.–.1919
reviews: JFM 47.0527.01 (publ. 1924/26). Reviewer: Tietze.

1919T
Hoogleeraarsbenoemingen. *De Nieuwe Amsterdammer*, 1.2.1919
published: 01.02.1919

1919U
Corrections. *Bull. Am. Math. Soc.*, 135
remarks: refers to 'Intuitionism and Formalism' [Brouwer 1912A]
published: –.–.1919

1919V
Antwort von L.E.J. Brouwer. *Mededeelingen van het Internationaal Instituut voor Wijsbegeerte* 2. p.33–34.
remarks: Reaction to letter of Buber.
published: –.03.19

1920A1 **[FHV**: 1920 A1]
Over de structuur van perfecte puntverzamelingen, III. *KNAW Versl.* 28, p. 373–375.
remarks: handwritten corrections, p.374
submitted: 25.10.1919 **published**: –.–.1920
submitted by: L.E.J. Brouwer
reviews: JFM 47.0178.04, 1919/1920 (publ. 1924/1926). Reviewer: Helmer.

1920A2 [**FHV**: 1920 A2]
Über die Struktur der perfekten Punktmengen, III. + Erratum. *KNAW Proc.* 22, p.471–474, 974.
also in: *CW II*, p.643–646.
submitted: 25.10.1919 **published**: –.–.1920
submitted by: L.E.J. Brouwer

1920A2*
Erratum. *Proc. KNAW* 22, p.974
remarks: correction to 1920A2
published: –.–.1920

1920B1 [**FHV**: 1920 B1]
Over één-éénduidige, continue transformaties van oppervlakken in zichzelf, VII.
KNAW Versl. 28, p.1186–1190.
remarks: Erroneously numbered VI. Erratum in same volume.
submitted: 27.03.1920 **published**: –.–.1920
submitted by: L.E.J. Brouwer
reviews: JFM 47.0525.05, 1919/1920 (publ. 1924/1926). Reviewer: Helmer.

1920B2 [**FHV**: 1920 B2]
Über eineindeutige, stetige Transformationen von Flächen in sich, VII. *KNAW Proc.* 22, p.811–814.
also in: *CW II*, p.647–651
remarks: Erroneously numbered VI
submitted: 26.6.1920 **published**: 16.12.1920
submitted by: Brouwer
reviews: JFM 47.0526.02, 1919/20 (publ. 1924/26). Reviewer: Helmer.

1920C [**FHV**: 1920 C]
Enumération des classes de transformations du plan projectif. *CR Paris* 170, p. 834–835, 1295.
remarks: corrections in Brouwer's copy
submitted: 06.04.1920 **published**: –.–.1920
submitted by: P. Appell
reviews: JFM 47.0525.04, 1919/1920 (publ. 1924/1926). Reviewer: Helmer.

1920D [**FHV**: 1920 D]

Enumération de classes de représentations d'une surface sur une autre surface. *CR Paris* 171, p.89–91, 830.

submitted: 12.07.1920 **published**: –.–.1920

submitted by: P. Appell

reviews: JFM 47.0526.01, 1919/20 (publ. 1924/26). Reviewer: Helmer.

1920E [**FHV**: 1920 E]

Über die Minimalzahl der Fixpunkte bei den Klassen von eindeutigen stetigen Transformationen der Ringflächen. *MA* 82, p.94–96.

also in: *CW II*, p.652–654.

submitted: 20.01.1920 **published**: –.–.1920

reviews: JFM 47.0528.01, 1919/20 (publ. 1924/26). Reviewer: Tietze.

1920F [**FHV**: 1920 F]

Luchtvaart en photogrammetrie, II. *Nieuw Tijdschrift voor Wiskunde* 8, p.300–307.

remarks: sequel to 1919R

published: –.–.1920

reviews: JFM 48.0678.02, 1921/22 (publ. 1925/28). Reviewer: Van der Woude.

1920G1 [**FHV**: 1920 G1]

Over één-éénduidige, continue transformaties van oppervlakken in zichzelf, VIII. *KNAW Versl.* 29, p.640–642.

remarks: erroneously numbered VII

published: –.–.1920

reviews: JFM 47.0526.02, 1919/1920 (publ. 1924/1926). Reviewer: Helmer.

1920G2 [**FHV**: 1920 G2]

Über eineindeutige, stetige Transformationen von Flächen in sich, VIII. *KNAW Proc.* 23, p.232–234.

remarks: erroneously numbered VII; cf. *CW II*, p.536

submitted: 26.06.1920 **published**: –.–.1920

submitted by: L.E.J. Brouwer

1920H

De nieuwe Amsterdammer. *De Nieuwe Amsterdammer*, 1920

1920K

Letter to the members of the Mathematics and Physics section of the KNAW.

remarks: defence of candidacy of Denjoy for KNAW.

published: 27.3.1920

1921A **[FHV**: 1921 A]

Besitzt jede reelle Zahl eine Dezimalbruchentwickelung? *MA* 83; p.201–210.
remarks: Talk at Naturforscherversammlung Nauheim 22.9.1920
submitted: 10.12.1920.
also in: *KNAW Versl.*(submitted 18.12.1920) 29, p.803–812; *KNAW Proc.* 23, p.955–964; *CW I*, p.236–245. English translation in [Mancosu 1998] p.28–35.
published: –.–.1921
reviews: JFM 48.0062.01 (publ. 1925/28). Reviewer: Rosenthal.

1921C **[FHV**: 1921 C]

Opmerking over de bepaling van alle complexe functies, voor welke $f(z) = \varphi(|z|)$ is. *Christiaan Huygens* 1, p.352–354.
published: –.–.1921
reviews: JFM 48.0397.02, 1921/22 (publ. 1925/28). Reviewer: Neder.

1921D **[FHV**: 1921 D]

Aufzählung der Abbildungsklassen endlichfach zusammenhängender Flächen. *MA* 82, p. 280–286.
also in: *CW II*, p.655–661.
submitted: 03.07.1920 **published**: –.–.1921
reviews: JFM 48.0649.01, 1921/22 (publ. 1925/28). Reviewer: Tietze.

1921E **[FHV**: 1921 E]

Berichtigung. *MA* 82, p.286.
also in: *CW II*, p.474
remarks: refers to [Brouwer 1911D]
published: –.–.1921

1921G

Sovjet wetten. *Nieuwe Kroniek* 1 (1921), p.2–3
remarks: review
published: –.–.1921

1921H

Wis- en Natuurkunde en Wijsbegeerte. *Nieuwe Kroniek*, 18.6.1921, p.1–2
published: 18.6.1921

1921J **[FHV**: 1919 D2]

Intuïtionistische verzamelingsleer. *KNAW Versl.* 29 (1921), p. 797–802.
submitted: 18.12.1920 **published**: –.–.1921
submitted by: L.E.J. Brouwer
reviews: JFM 48.0202.01, 1921/22 (publ. 1925/28). Reviewer: Rosenthal.

1922A [**FHV**: 1919 D1]
Intuitionistische Mengenlehre. *KNAW Proc.* 23, p.949–954.
also in: *CW I*, p.230–235.
submitted: 18.12.1920 **published**: –.–.1922

1922B
Wis- en Natuurkunde en Wijsbegeerte. *Nieuwe Kroniek*, 11.2.1922, p. 3–5
published: 11.2.1922

1922C
Wetenschap, humaniteit, Frankrijk en Nederland *?*
published: –.–.1922

1922D
Schrijven van Dr. L.E.J. Brouwer aan Zijne Excellentie den Minister van Onderwijs, Kunsten en Wetenschappen. Open letter to minister of education.
remarks: on position of Denjoy. Contains documentation.
published: 27.09.1922

1923A [**FHV**: 1923 A]
Begründung der Funktionenlehre unabhängig vom logischen Satz vom ausgeschlossenen Dritten. Erster Teil, Stetigkeit, Messbarkeit, Derivierbarkeit. *KNAW Verh. 1e sectie,* deel XIII, no.2, p.1–24.
also in: *CW I*, p.246–267.
remarks: proofs 12.11.1923. **published**: –.–.1923

1923B [**FHV**: 1923 B1]
Over de rol van het principium tertii exclusi in the wiskunde, in het bijzonder in de functietheorie. *Wis- en Natuurkundig Tijdschrift* 2, p.1–7.
remarks: lecture at XXII Vlaamsch Natuur- en Geneeskundig Congres, Antwerpen 1923
published: –.–.1923

1923B3 [**FHV**: 1923 B3]
Die Rolle des Satzes vom ausgeschlossenen Dritten in der Mathematik. *Jahresb. D.M.V.* 32, p.67.
remarks: Summary of [Brouwer 1924N]. Talk at Marburg DMV meeting at Marburg, 21.9.1923
published: –.–.1923

1923C1 [**FHV**: 1923 C1]
Intuïtionistische splitsing van mathematische grondbegrippen. *KNAW Versl.* 32, p.877–880.
also in: English translation in [Mancosu 1998] p.286–289.
submitted: 24.11.1923 **published**: –.–.1923
reviews: JFM 49.0030.02, 1923 (publ. 1927/28). Reviewer: Fraenkel.

1923D1 [**FHV**: 1923 D1]
Over het natuurlijke dimensiebegrip. *KNAW Versl.* 32, p. 881–886
remarks: translation of 1913 plus additional correction (Urysohn's remark) and footnote
submitted: 24.11.1923 **published**: –.–.1923
reviews: JFM 49.0140.04, 1923 (publ. 1927/28), p.140–141. Reviewer: Fraenkel.

1923D2 [**FHV**: 1923 D2]
Über den natürlichen Dimensionsbegriff. *KNAW Proc.* 26, p.795–800.
remarks: Revised edition of [Brouwer 1913A]
published: –.–.1923

1923E
Unberücksichtigt gebliebenes Schreiben von Dr. L.E.J. Brouwer an Seine Exzellenz den Minster für Unterricht Kunst und Wissenschaft vom 27. September 1922. Symptomatisches zu einer Gefährdung der niederländischen Staatshoheit. 24 pages.
remarks: German version of [Brouwer 1922D] with a four-page appendix.
published: –.–.1923

1924A1 [**FHV**: 1924 A1]
Over de toelating van oneindige waarden voor het functiebegrip. *KNAW Versl.* 33, p.41.
published: –.–.1924
reviews: JFM 50.0146.03, 1924 (publ. 1928/29). Reviewer: Van der Corput.

1924A2 [**FHV**: 1924 A2]
Ueber die Zulassung unendlicher Werte für den Funktionsbegriff. *KNAW Proc.* 27, p.248.
also in: *CW I*, p.281.
submitted: 26.01.1924 **published**: –.–.1924

1924B1 [**FHV**: 1924 B1]
Perfecte puntverzamelingen met positief irrationale afstanden. *KNAW Versl.* 33, p.81.
published: –.–.1924

1924B2 [**FHV**: 1924 B2]
Perfect sets of points with positive-irrational distances. *KNAW Proc.* 27, p. 487.
also in: *CW I*, p.282.
submitted: 23.02.1924 **published**: –.–.1924
reviews: JFM 50.0132.03,1924 (publ. 1928/29). Reviewer: Van der Corput.

1924C1 [**FHV**: 1924 C1]
(With B. de Loor) Intuïtionistisch bewijs van de hoofdstelling der algebra. *KNAW Versl.* 33, p. 82–84.
submitted: 23.02.1924 **published**: –.–.1924
reviews: JFM 50.0037.01, 1924 (publ. 1928/29), p.30–31. Reviewer: Fraenkel.

1924C2 [**FHV**: 1924 C2]
(With B. de Loor) Intuitionistischer Beweis des Fundamentalsatzes der Algebra. *KNAW Proc.* 27, 186–188.
also in: *CW I*, p. 283–285.
submitted: 23.02.1924 **published**: –.–.1924

1924D1 [**FHV**: 1924 D1]
Bewijs dat iedere volle functie gelijkmatig continu is. *KNAW Versl.* 33, p. 189–193
published: –.–.1924

1924D2 [**FHV**: 1924 D2]
Beweis dass jede volle Funktion gleichmässig stetig ist. *KNAW Proc.* 27, p. 189–193
also in: *CW I*, p.286–290. English translation in [Mancosu 1998] p.36–39. French translation in [Largeault 1992]
submitted: 29.03.1924 **published**: –.–.1924

1924E1 [**FHV**: 1924 E1]
Intuïtionistische aanvulling van de hoofdstelling der algebra. *KNAW Versl.* 33, p.459–462.
published: –.–.1924

1924E2 [**FHV**: 1924 E2]
Intuitionistische Ergänzung des Fundamentalsatzes der Algebra. *KNAW Proc.* 27 p.631–634.
remarks: *CW I*, p.291–294.
submitted: 31.03.1924 **published**: –.–.1924

1924F1 [**FHV**: 1924 F1]
Bewijs van de onafhankelijkheid van de onttrekkingsrelatie van de versmeltingsrelatie. *KNAW Versl.* 33, p.479–480.
submitted: 28.06.1924 **published**: –.–.1924
reviews: JFM 50.0127.02, 1924 (publ. 1928/29), p.127. Reviewer: Fraenkel.

1924G1 [**FHV**: 1924 G1]
Opmerkingen aangaande het bewijs der gelijkmatige continuïteit van volle functies. *KNAW Versl.* 33, p.646–648.
published: –.–.1924
reviews: JFM 50.0147.02,1924 (publ. 1928/29). Reviewer: Van der Corput.

1924G2 [**FHV**: 1924 G2]
Bemerkungen zum Beweise der gleichmässigen Stetigkeit voller Funktionen. *KNAW Proc.* 27, p.644–646.
also in: *CW I*, p.298–300. French translation in [Largeault 1992]
submitted: 27.09.1924 **published**: –.–.1924

1924J1 [**FHV**: 1924 J1]
Opmerkingen over het natuurlijk dimensiebegrip. *KNAW Versl.* 33, p.476–478.
published: –.–.1924
reviews: JFM 50.0127.03, 1924 (publ. 1928/29), p.127–128. Reviewer: Fraenkel.

1924J2 [**FHV**: 1924 J2]
Bemerkungen zum natürlichen Dimensionsbegriff. *KNAW Proc.* 27, p.635–638
also in: *CW II*, p. 554–557.
remarks: Immediately followed by Menger's *Ueber die Dimension von Punktmengen*; jointly printed in one preprint. Also a single preprint of Brouwer's paper. **submitted**: 28.06.1924
published: –.–.1924
reviews: JFM 50.0127.03, 1924 (publ. 1928/29), p.127–128. Reviewer: Fraenkel.

1924K1 [**FHV**: 1924 K1]
Over de n–dimensionale simplex ster in R_n. *KNAW Versl.* 33, p.1008–1010.
submitted: 27.12.1924 **published**: –.–.1924
submitted by: L.E.J. Brouwer

1924K2 [**FHV**: 1924 K2]

On the n–dimensional simplex star in R_n. *KNAW Proc.* 27, p.778–780.
also in: *CW II*, 606–608.
published: 17.12.1924

1924L [**FHV**: 1924 L]

Zum natürlichen Dimensionsbegriff. *Mathematische Zeitschrift* 21, p.312–314.
submitted: 16.06.1924
published: –.–.1924
reviews: JFM 50.0128.02, 1924 (publ. 1928/29), p.128. Reviewer: Fraenkel.

1924M [**FHV**: 1924 M]

Berichtigung. *Journ. r. u. ang. Math.* 153, p.253.
also in: *CW II*. p. 553
remarks: refers to [Brouwer 1913A].
published: 06.03.1924
reviews: JFM 50.0128.02, 1924 (publ. 1928/29), p.128. Reviewer: Fraenkel.

1924N [**FHV**: 1923 B2]

Über die Bedeutung des Satzes vom ausgeschlossenen Dritten in der Mathematik insbesondere in der Funktionentheorie. *Journ. r. u. angew. Mat.* 154, p.1–7.
also in: *CW I*, p.268–274. English translation, 'On the Significance of the principle of the excluded middle in mathematics, especially in function theory', in [Heijenoort 1967], p.334–341. French translation in [Largeault 1992].
remarks: lecture at Annual meeting DMV, September 1923.
published: –.–.1923
reviews: JFM 50.0146.02, 1924 (publ. 1928/29), p.146. Reviewer: Fraenkel.

1925A [**FHV**: 1925 A]

Zur Begründung der intuitionistischen Mathematik, I. *MA* 93, p. 244–257.
also in: *CW I* p.301–314.
remarks: correction in 1926A. Reprints with extensive handwritten emendations.
submitted: 20.06.1924 **published**: –.–.1925
reviews: JFM 51.0164.02, 1925 (publ. 1932). Reviewer: Rosenthal.

1925B1 **[FHV**: 1925 B1]

Intuitionistischer Beweis des Jordanschen Kurvensatzes. *KNAW Proc.* 28, p.503–508.
also in *CW I*, p.315–320.
submitted: 27.06.1925
published: –.–.1925
reviews: JFM 51.0459.02, 1925 (publ. 1932), p.459. Reviewer: Freudenthal.

1925B2 **[FHV**: 1925 B2]

Intuïtionistisch bewijs van de krommenstelling van Jordan. *KNAW Versl.* 34, p.657.
remarks: Summary of [Brouwer 1925B]
published: –.–.1925

1925C **[FHV**: 1925 C]

Rapport over de verhandeling van P. Urysohn 'Mémoire sur les multiplicités cantoriennes. Deuxième partie : les lignes cantoriennes'. [With R. Weitzenböck and H. de Vries]. *KNAW Versl.* 34, p.516–517
published: –.–.1925
reviews: JFM 51.0451.04, 1925 (publ. 1932). Reviewer: Van der Corput.

1925D **[FHV**: 1924 H]

Zuschrift an dem Herausgeber. *Jahresb. D.M.V.*33, p.124
remarks: remark on report of Behmann's talk at Marburg meeting DMV
published: –.–.1924
reviews: JFM 51.0045.01, 1925 (publ. 1932), p.45. Reviewer: Freudenthal.

1925E **[FHV**: 1923 C2]

Intuitionistische Zerlegung mathematischer Grundbegriffe. *Jahresb. D.M.V.* 33 (1925), p.251–256
also in English translation in [Mancosu 1998] p.290–292, *CW I*, p.275–280. French translation in [Largeault 1992]
submitted: 06.02.1924 **published**: –.–.1925
remarks: content of talk at KNAW 24.11.1923.
reviews: JFM 51.0047.06, 1925 (publ. 1932). Reviewer: Rosenthal.

1926A **[FHV**: 1926 A]

Zur Begründung der intuitionistischen Mathematik, II. *MA* 95, p. 453–472.
also in: *CW I*, p.321–340.
remarks: reprint dated 1925. Contains Berichtigung to part I
submitted: 14.03.1925 **published**: –.–.1925
reviews: JFM 51.0164.02, 1925 (publ. 1932). Reviewer: Rosenthal.

1926B1 [**FHV**: 1926 B1]

Intuïtionistische invoering van het dimensiebegrip. *KNAW Versl.* 35, p.634–642.

published: –.–.1926

reviews: JFM 52.0589.02 (publ. 1935), p.589–590. Reviewer: Freudenthal.

1926B2 [**FHV**: 1926 B2]

Intuitionistische Einführung des Dimensionsbegriffes. *KNAW Proc.* 29, p.855–873.

also in: *CW I*, p.341–349.

submitted: 29.05.1926 **published**: –.–.1926

reviews: JFM 52.0589.02 (publ. 1935), p.589–590. Reviewer: Freudenthal.

1926C1 [**FHV**: 1926 C1]

De intuïtionistische vorm van het theorema van Heine-Borel. *KNAW Versl.* 35, p.667–668.

published: –.–.1926

reviews: JFM 52.0198.03, 1926 (publ. 1935). Reviewer: Freudenthal.

1926C2 [**FHV**: 1926 C2]

Die intuitionistische Form des Heine-Borelschen Theorems. *KNAW Proc.* 29, p.866–867.

also in: *CW I*, p.350–351.

submitted: 26.06.1926

reviews: JFM 52.0198.03, 1926 (publ. 1935). Reviewer: Freudenthal.

1926D1 [**FHV**: 1926 D1]

Over transformaties van projectieve ruimten. *KNAW Versl.* 35, p.643–644.

submitted: 29.05.1926

published: –.–.1926

reviews: JFM 52.0574.03, 1926 (publ. 1935), p.574. Reviewer: Grunsky.

1926D2 [**FHV**: 1926 D2]

On transformations of projective spaces. *KNAW Proc.* 29, p.864–865.

also in: *CW II*, p.475–476.

submitted: 29.05.1926 **published**: –.–.1926

reviews: JFM 52.0574.03, 1926 (publ. 1935), p.574. Reviewer: Grunsky.

1926E

Rapport omtrent de ter opneming in de werken der Akademie aangeboden verhandeling van P. Alexandroff en P. Urysohn 'Mémoire sur les espaces topologiques compacts'. [With R. Weitzenböck and H. de Vries] *KNAW Versl.* 35.

published: –.–.1926

1927A **[FHV**: 1927 A]
Zur Begründung der intuitionistischen Mathematik, III. *MA* 96, p.451–488.
also in: *CW I*, p.352–389.
remarks: reprint dated 1926
submitted: 28.11.1925
reviews: JFM 52.0193.01, 1926 (publ. 1935), p.45. Reviewer: Grunsky.

1927B **[FHV**: 1927 B]
Über Definitionsbereiche von Funktionen. *MA* 97, p.60–75.
also in: *CW I*, p.390–405. [English translation 'On the domains of definition of functions' (§1–3) in [Heijenoort 1967], p.446–463]. French translation in [Largeault 1992]
remarks: reprint dated 1926. Brouwer lecture at Düsseldorf meeting DMV, 23.9.1926, 'Ueber Definitionsbereiche von Funktionen'
submitted: 28.04.1926 **published**: –.–.1927
reviews: JFM 52.0239.01, 1926 (publ. 1935). Reviewer: Freudenthal.

1927C **[FHV**: 1927 C]
Virtuelle Ordnung und unerweiterbare Ordnung. *Journ. r. u. angew. Math.* 157, p. 255–257.
also in CW1, p.406–408.
published: –.–.1927
reviews: JFM 53.0040.01, 1927 (publ. 1931), p.40. Reviewer: Freudenthal.

1927D **[FHV**: 1924 F2]
Zur intuitionistischen Zerlegung mathematischer Grundbegriffe. *Jahresb. D.M.V.* 36 (1927), p.127–129.
also in: English translation in [Mancosu 1998] p.293–295, *CW I* p.295–297.
submitted: 18.10.1926 **published**: –.–.1927
reviews: 53.0039.02 (publ. 1931). Reviewer: Freudenthal.

1928A1 **[FHV**: 1928 A1]
Intuitionistische Betrachtungen über den Formalismus. *KNAW Versl.* 36, p.1189.
remarks: Summary of [Brouwer 1928 A2]
submitted: 16.02.1928 **published**: –.–.1928

1928A2 [**FHV**: 1928 A2]

Intuitionistische Betrachtungen über den Formalismus. *KNAW Proc.* 31, p.374–379

also in: English translation in [Mancosu 1998] p.40–44, of §1 in [Heijenoort 1967], p.490–492. Italian translation in [Mangione-Franchella 1993], p.123–126. French translation in [Largeault 1992]

submitted: 17.12.1927 **published**: –.–.1928

reviews: JFM 54.0053.01, 1928 (publ. 1932). Reviewer: Freudenthal.

1928A3 [**FHV**: 1928 A2]

Intuitionistische Betrachtungen über den Formalismus. *Sitzungsberichte der Preuszischen Akademie der Wissenschaften zu Berlin* 1928, p.48–52.

submitted: 16.02.1928 **published**: 05.05.1928

submitted by: L. Bieberbach

reviews: JFM 54.0053.01, 1928 (publ. 1932). Reviewer: Freudenthal.

1928B1 [**FHV**: 1928 B1]

Beweis dass jede Menge in einer individualisierten Menge enthalten ist. *KNAW Versl.* 36, p.1189.

remarks: Summary of [Brouwer 1928 B2]

published: –.–.1928

reviews: JFM 54.0053.02, 1928 (publ. 1932), p.53. Reviewer: Freudenthal.

1928B2 [**FHV**: 1928 B2]

Beweis dass jede Menge in einer individualisierten Menge enthalten ist. *KNAW Proc.* 31, p.380–381

also in: *CW I*, p.415–416.

submitted: 17.12.1927 **published**: –.–.1928

1928C1 [**FHV**: 1928 C1]

Zur Geschichtschreibung der Dimensionstheorie. *KNAW Versl.* 37, p.626.

remarks: Summary and announcement of [Brouwer 1928 C2]

published: –.–.1928

1928C2 [**FHV**: 1928 C2]

Zur Geschichtschreibung der Dimensionstheorie. *KNAW Proc.* 31, p. 953–957.

also in: *CW II*, p.559–563.

submitted: 29.10.1928 **remarks**: Corrigendum in *KNAW Proc.* 32.

published: –.–.1928

reviews: JFM 54.0007.03, 1928 (publ. 1932). Reviewer: Feigl.

1929A [**FHV**: 1929]
Mathematatik, Wissenschaft und Sprache. *Monatshefte für Mathematik und Physik* 36, p.153–164.
also in: English translation in [Mancosu 1998] p.45–53, *CW I*, p.417–428. French translation in [Largeault 1992]
remarks: talk 10.3.1928 Vienna
published: –.–.1929
reviews: JFM 55.0028.04, 1929 (publ. 1932), p.28. Reviewer: Freudenthal.

1929B
Herinnering aan C.S. Adama van Scheltema. *Ter herdenking van C.S. Adama van Scheltema.* Em. Querido, Amsterdam, p.69
published: –.–.1929

1929C
Erratum. *KNAW Proc.* 32, p. 1022
remarks: corrections to [Brouwer 1928 C2]
published: –.–.1929

1930A [**FHV**: 1930 A]
Die Struktur des Kontinuums. Wien 1930 [Sonderabdruck].
also in: English translation in [Mancosu 1998] p.54–64, *CW I*, p.429–440. French translation in [Largeault 1992]
remarks: talk 14.3.1928 Vienna
remarks: Lecture in Vienna 14.3.1928.
published: –.10.1930
reviews: JFM 56.0083.01, 1930 (publ. 1933). Reviewer: Freudenthal.

1930B [**FHV**: 1930 B]
A. Fraenkel, Zehn Vorlesungen über die Grundlegung der Mengenlehre. *Jahresb. D.M.V.* 39, p.10–11.
also in: *CW I*, p.441–442.
remarks: Review
published: –.–.1930

1931 [**FHV**: 1931]
Ueber freie Umschliessungen im Raume. *KNAW Proc.* 34, p.100–101
also in: *CW II*, p.507–508.
submitted: 31.01.1931 **published**: –.–.1931
reviews: JFM 57.0745.03, 1931 (publ. 1938/39). Reviewer: Feigl; *Zbl.* 0001.29304 (1931), p.293. Reviewer: Grell; *Math. Rev.* 1–46. Reviewer: Hurewicz.

1933A1 [**FHV**: 1933]
Willen, Weten, Spreken. *Euclides* 9, p.177–193.
also in:1933B. English translation 'Will, Knowledge and Speech' in [Van Stigt 1990], appendix 5. Partial translation è Volition, Knowledge, Language' in *CW I*, p.443–446.
remarks: Lecture at Amsterdam University 12.12.1932
published: –.–.1933
reviews: JFM 59.0055.08, 1933 (publ. 1934), p.55. Reviewer: Feigl.

1933A2 [**FHV**: 1933]
Willen, Weten, Spreken. In *De uitdrukkingswijze der wetenschap, kennistheoretische voordrachten, gehouden aan de Universiteit van Amsterdam.* Eds. L.E.J. Brouwer, J. Clay, et al. Amsterdam, 1933, p.43–63.
remarks: Lecture at Amsterdam University 12.12.1932
published: –.–.1933

1934 [**FHV**: 1934]
Erratum. *Euclides,* 11
remarks: re 'Willen, Weten, Spreken'
published: –.–.1934

1937 [**FHV**: 1937]
Significhe dialogen [With F. van Eeden, J. van Ginneken and G. Mannoury]. *Synthese* 2, p.168–174, p.261–268 and p. 316–324.
also in: Significhe Dialogen. Erven Bijleveld, Utrecht, 1939, p.65. Translation in CW I, p.447–452. Selections in *CW I*, p.447–452.
published: –.–.1937

1939A [**FHV**: 1939 A]
Zum Triangulationsproblem. *Indag. Math.* I, p. 248–253.
also in: KNAW Proc. 42, p.701–706. *CW I*, p.453–458.
remarks: talk at WG 24.4.1937
submitted: 30.09.1939 **published**: –.–.1939
reviews: JFM 65.0866.02, 1939 (publ. 1941), p.867. Reviewer: Kneser; *Zbl.* 0022.17201 (1940), p.172. Reviewer: Alexits.

1939B
Theodor Vahlen zum 70. Geburtstag. *Forschungen und Fortschritte* 15, p.238–239.
published: –.–.1939

1941A
D.J. Korteweg† *Euclides*, 17, 266–267

1941B

D.J. Korteweg [Obituary]. *KNAW Verh.* p.266–267.
published: –.–.1941

1942A **[FHV**: 1942 A]

Zum freien Werden von Mengen und Funktionen. *Indag. Math.* 4, p.107–108.
also in: KNAW Proc. 45, p.322–323. *CW I*, p.459–460.
submitted: 28.03.1942 **published**: –.–.1942
reviews: JFM 68.0101.0, 1942, (publ. 1944). Reviewer: Schmidt; *Zbl.* 0027.04906 (1943), p.49. Reviewer: Schmidt; *Math. Rev.* 6–31. Reviewer: Weyl.

1942B **[FHV**: 1942 B]

Die repräsentierende Menge der stetigen Funktionen des Einheitskontinuums. *Indag. Math.* 4, p.154.
also in: KNAW Proc. 45, p.443. *CW I*, p.461.
submitted: 25.04.1942 **published**: –.–.1942
reviews: JFM 68.0101.03, 1942 (publ. 1944). Reviewer: Hornich; *Zbl.* 0027.04907 (1943), p.49. Reviewer: Schmidt; *Math. Rev.* 6–31. Reviewer: Weyl.

1942C **[FHV**: 1942 C]

Beweis dass der Begriff der Mengen höherer Ordnung nicht als Grundbegriff der intuitionistischen Mathematik in Betracht kommt. *Indag. Math.* 4, p.274–276.
also in: KNAW Proc. 45, p.791–793. *CW I*, p.462–464.
submitted: 26.09.1942 **published**: –.–.1942
reviews: JFM 68.0101.02, 1942 (publ. 1944). Reviewer: Schmidt; *Zbl.* 0027.30001 (1943), p.300. Reviewer: Schmidt; *Math. Rev.* 6–31. Reviewer: Weyl.

1946A **[FHV**: 1946 A]

Synopsis of the Signific Movement in the Netherlands. *Synthese* 5, p.201–208.
also in: *CW I*, p.465–471.
published: –.–.1946

1946B1

Toespraak van Prof.Dr. L.E.J. Brouwer en antwoord van Prof.Dr. G. Mannoury *Jaarboek der Universiteit van Amsterdam 1946–1947.*
remarks: Address 16.9.1946
published: –.–.1946

1947A **[FHV**: 1946 B]

Address delivered on September 16th, 1946, on the conferment upon Professor G. Mannoury of the honorary degree of Doctor of Science. *Jaarboek der Universiteit van Amsterdam 1946–1947.*
also in *CW I*, 472–476.
published: –.03.1947

1947B **[FHV**: 1947]

Richtlijnen der intuïtionistische wiskunde. *Indag. Math.* 9, p.197.
also in KNAW Proc. 50, p.339. English translation 'Guidelines of intuitionistic mathematics' in *CW I*, p.477. French translation in [Largeault 1992].
submitted: 29.03.1947 **published**: –.–.1947
reviews: *Zbl.* 0035.14813 (1950), p.148. Reviewer: Schmidt; *Math. Rev.* 8–430.

1948A **[FHV**: 1948 A]

Essentieel negatieve eigenschappen. *Indag. Math.* 10, p.322–323.
also in KNAW Proc. 51, p.963–964. English translation 'Essentially negative properties' in *CW I*, p.478–479. French translation in [Largeault 1992].
submitted: 25.09.1948 **published**: –.–.1948
reviews: *Math. Rev.* 10–421. Reviewer: Heyting.

1948B **[FHV**: 1948 B]

Opmerkingen over het beginsel van het uitgesloten derde en over negatieve asserties. *Indag. Math.* 10, p.383–387.
also in: KNAW Proc. 51. p.1239–1244.
submitted: 27.11.1948 **published**: –.–.1948
reviews: *Math. Rev.* 10–421. Reviewer: Heyting.

1948C

Discussion. Following 'Les conceptions mathématiques et le reel' van Gonseth. *Symposium de l'Institut des Sciences théoriques. Bruxelles 1947. Actualités scientifiques et Industrielles, Paris. Hermann 1948.* p. 31–60.
remarks: see also p. 56–58
published: –.–.1948

1949A **[FHV**: 1949 A]

De non-aequivalentie van de constructieve en negatieve orderelatie in het continuum. *Indag. Math.* 11, p.37–39.
also in: KNAW Proc. 52, p.122–124. English translation 'The non-equivalence of the constructive and negative order relation on the continuum' in *CW I*, p.495–496.
submitted: 29.01.1949 **published**: –.–.1949
reviews: *Math. Rev.* 10–499. Reviewer: Heyting.

1949B [**FHV**: 1949 B]
Contradictoriteit der elementaire meetkunde. *Indag. Math.* 11, p.89–90.
also in: KNAW Proc. 52, p.315–316. English translation 'Contradictority of elementary geometry' in *CW I*, p. 497–498.
submitted: 26.03.1949 **published**: –.–.1949
reviews: *Zbl.* 0037.21511 (1951), p.215 Reviewer: Bachmann. *Math. Rev.* 11–2. Reviewer: Heyting.

1949C [**FHV**: 1948 C]
Consciousness, Philosophy and Mathematics. *Proceedings of the 10th International Congress of Philosophy, Amsterdam 1948 III*, p.1235–1249.
also in *CW I*, p.480–494. French translation in [Largeault 1992]
published: –.–.1948
reviews: *Math. Rev.* 10–422. Reviewer: Nelson.

1950A [**FHV**: 1950 A]
Remarques sur la notion d'ordre. *CR Paris* 230, p.263–265.
also in: *CW I*, p.499–500.
published: –.–.1950
submitted by: É. Borel
reviews: *Zbl.* 0034.29001 (1950), p.290. Reviewer: Schröter. *Math. Rev.* 11–305. Reviewer: Nelson.

1950B [**FHV**: 1950 B]
Sur la possibilité d'ordonner le continu. *CR Paris* 230, p.349–350.
also in: *CW I*, p.501–502.
submitted: 23.01.1950 **published**: –.–.1950
submitted by: É. Borel
reviews: *Math. Rev.* 11–305. Reviewer: Nelson; *Zbl.* 0041.37503. Reviewer: Schröter/Asser.

1950C [**FHV**: 1950 C]
Discours final. Les méthodes formelles en axiomatique. *Colloques Internationaux du C.N.R.S. Paris 1950*, p.75.
also in *CW I*, p.503.
published: –.–.1950

1951 [**FHV**: 1951]
On order in the continuum, and the relation of truth to non-contradictority. *Indag. Math.* 13, p.357–358.
also in: KNAW Proc. Ser. A 54, p.357–358. *CW I*, p.504–505.
submitted: 24.11.1951 **published**: –.–.1951
reviews: *Zbl.* 0043.25002 (1952), p.250. Reviewer: Müller.

1952A [**FHV**: 1952 A]

An intuitionist correction of the fixed–point theorem on the sphere. *Proceedings of the Royal Society London,* Ser. A 213, p.1–2.
also in: *CW I*, p.506–507.
submitted: 31.12.1951 **published**: –.–.1952
reviews: *Zbl.* 0046.40902 (1953), p.409. Reviewer: Freudenthal. *Math. Rev.* 12–965. Reviewer: Freudenthal.

1952B [**FHV**: 1952 B1]

Historical background, principles and methods of intuitionism. *South African Journal of Science* 49, p.139–146.
also in: *CW I*, p.508–515. French translation in [Largeault 1992].
submitted: 31.12.1951 **published**: –.10/11.1952
remarks: Lecture at South African Ass. for the Advancement of Science, Cape Town, July 1952.
reviews: *Math. Rev.* 14–528. Reviewer: Nelson.

1952C [**FHV**: 1952 C]

Over accumulatiekernen van oneindige kernsoorten. *Indag. Math.* 14, p.439–441.
also in: KNAW Proc. Ser. A 55, p. 439–441. English translation 'On accumulation cores of infinite core species' in *CW I*, p.516–518.
submitted: 27.09.1952 **published**: –.–.1952
reviews: *Zbl.* 0047.05901 (1953), p.59. Reviewer: Freudenthal. *Math. Rev.* 14–441. Reviewer: Heyting.

1952D [**FHV**: 1952 D]

Door klassieke theorema's gesignaleerde pinkernen die onvindbaar zijn. *Indag. Math.* 14, p.443–445.
also in: KNAW Proc. Ser. A 55, p.443–445. English translation: 'Fixed cores which cannot be found, though they are claimed to exist by classical theorems' in *CW I*, p.519–521.
submitted: 25.10.1952 **published**: –.–.1952
reviews: *Zbl.* 0048.028404 (1953), p.284. Reviewer: Freudenthal. *Math. Rev.* 14–783. Reviewer: Heyting.

1954A **[FHV**: 1954 A]

Points and Spaces. *Canadian Journal for Mathematics* 6, p.1–17.
also in: *CW I*, p.522–540. Italian translation in [Mangione-Franchella 1993], p.95–122. French translation in [Largeault 1993], p. 195–220.
remarks: After lectures at the Seminar of the Canadian Mathematical Congress at Kingston, 10–31.8.1953.
submitted: 01.09.1953 **published**: –.–.1954
reviews: *Math. Rev.* 15–593. Reviewer: Heyting; *Zbl.* 0055.04601. Reviewer: Freudenthal.

1954B **[FHV**: 1954 B]

Addenda en corrigenda over de rol van het principium tertii exclusi in de wiskunde. *Indag. Math.* 16, p.104–105.
also in: KNAW Proc. Ser. A 57, p.104–105. English translation in [Heijenoort 1967], p.341–342. *CW I*, p. 539–540.
submitted: 30.01.1954 **published**: –.–.1954
reviews: *Math. Rev.* 15–670. Reviewer: Heyting.

1954C **[FHV**: 1954 C]

Nadere addenda en corrigenda over de rol van het principium tertii exclusi in de wiskunde. *Indag. Math.* 16, p.109–111.
also in KNAW Proc. Ser. A 57, p.109–111. English translastion 'Further addenda and corrigenda on the role of the principle of the excluded middle in mathematics' in [Heijenoort 1967], p.342–345. *CW I*, p. 541–543.
submitted: 20.03.1954 **published**: –.–.1954
reviews: *Zbl.* 0059.01402 (1965/66), p.14. Reviewer: Freudenthal. *Math. Rev.* 15–925. Reviewer: Heyting.

1954D **[FHV**: 1954 D]

Ordnungswechsel in Bezug auf eine coupierbare geschlossene stetige Kurve. *Indag. Math.* 16, p.112–113.
also in KNAW Proc. Ser. A 57, p.112–113. *CW I*, p.544–545.
published: –.–.1954
reviews: *Zbl.* 0055.16801 (1955), p.168. Reviewer: Haupt. *Math. Rev.* 15–88. Reviewer: Heyting.

1954E **[FHV**: 1954 E]

Intuïtionistische differentieerbaarheid. *Indag. Math.* 16, p.201–203.
also in KNAW Proc. Ser. A 57, p. 201–203. English translation 'Intuitionistic differentiability' in *CW I*, p.546–548.
submitted: 24.04.1954 **published**: –.–.1954
reviews: *Math. Rev.* 16–2. Reviewer: Heyting; *Zbl.* 0056.01304. Reviewer: Freudenthal.

1954F **[FHV**: 1954 F]

An example of contradictority in classical theory of functions. *Indag. Math.* 16, p.204–205.
also in: KNAW Proc. Ser. A 57, p.204–205. *CW I*, p. 549–550.
submitted: 24.04.1954 **published**: –.–.1954
reviews: *Zbl.* 0056.01305 (1966), p.13. Reviewer: Freudenthal; *Math. Rev.* 16–2. Reviewer: Heyting.

1955 **[FHV**: 1955]

The effect of intuitionism on classical algebra of logic. *Proceedings of the Royal Irish Academy Sect. A* 57, p.113–116.
also in *CW I*, p. 551–554. French translation in [Largeault 1992].
published: –.–.1955
reviews: *Zbl.* 0066.01101, p.11. Reviewer: Lorenzen.

1956 **[FHV**: 1952 B2]

Voorgeskiedenis, beginsels en metodes van die intuitionisme. *Tydskrif vir Wetenskap en Kuns* 1956, p.186–197.
published: –.–.1952

1975

L.E.J. Brouwer. Collected Works 1. Philosophy and Foundations of Mathematics Ed. A. Heyting. North-Holland Publ. Co. 627p.
reviews: *Zbl.* 0311.01021. Reviewer: Schultz.

1976

L.E.J. Brouwer. Collected Works 2. Geometry, Analysis, Topology and Mechanics Ed. H. Freudenthal. North-Holland Publ. Co. 706p.
reviews: *Zbl.* 0328.01020. Reviewer: Friedrich.

1981A

Brouwer's Cambridge Lectures on Intuitionism. Ed. D. van Dalen. Cambridge University Press. Cambridge. xii + 109p.
also in translation S. Bernini *Lezioni Sull' Intuitionismo.* Boringhieri, Torino. 1983.
reviews: *Zbl.* 0476.03056. Reviewer: Luckhardt.

1981B

Over de Grondslagen der Wiskunde. Ed. D. van Dalen. Mathematisch Centrum. Amsterdam.
267p.
remarks: Dissertation + Correspondence Brouwer–Korteweg + reviews + [Brouwer 1908C]
reviews: *Zbl.* 0484.01009. Reviewer: Bottema.

1983

Lezioni Sull' Intuitionismo Boringhieri. Torino, 113p. Translation of [Brouwer 1981A] by S. Bernini.

1984

L.E.J. Brouwer & C.S. Adama van Scheltema. *Droeve Snaar, Vriend van Mij. Brieven.* Ed. D. van Dalen. Arbeiders Pers. Amsterdam. 173p. (Correspondence Brouwer-Adama van Scheltema + short biography.)

1985

Recensie 'Amsterdam' van C.S. Adama van Scheltema. Unpublished review; published by R. Delvigne in *Juffrouw Ida*, 11 no.3, 41–44, 1985.

1986

Intuigionizm i formalizm. Translation of [Brouwer 1913C] by R. Murawski. In R. Murawski, *Filozofia matematyki. Antologia tekstów klasycznych*, p.263–275. Wydawnictwo Naukowe UAM. Poznan.

1992

Intuitionismus. Bibliographisches Institut, Wissenschaftsverlag. Mannheim. 161p. (Ed. D. van Dalen). Berlin 1927 lectures and parts of an unpublished book on real functions.
reviews: *Zbl.* 0761.03001. Reviewer: Kreisel.

1996

Life, Art and Mysticism. Translation of [Brouwer 1905A] by W.P. van Stigt. *Notre Dame Journal of Formal Logic, 37. 381–429.*

2001

L.E.J. Brouwer en de Grondslagen van de Wiskunde. Ed. D. van Dalen. Contains the correspondence between Korteweg and Brouwer on the dissertation + [Brouwer 1907] with Brouwer's own corrections + reviews + [Brouwer 1908B] + [Brouwer 1908C] + [Brouwer 1912]+ [Brouwer 1918].

First published by the Mathematical Centre, Amsterdam [Brouwer 1981B]. This expanded version was published by Epsilon, Utrecht, 2001, 2005.

The following items have a vaguely Brouwerian ring, but it is more likely that they were not written by him. Possibly his brother could have tried to imitate him. The choice of the name 'J. van de Zee' is suggestive.

1905C

Wat is Wijsheid? *Studentenweekblad, Delft,* 19.10.1905, p. 20
remarks: signed J.v.d.Z. Style reminiscent of Brouwer. Letter to the editor.
published: 19.10.1905

1905D

Over Wiskunde. *Studentenweekblad*, 30.11.1905, p. 75
remarks: signed Jan van der Zee. Style reminiscent of Brouwer.
published: 30.11.1905

Collections

1964. Benacerraf and Putnam

Philosophy of Mathematics. Selected Readings. Eds. P. Benacerraf, H. Putnam. Prentice Hall. Englewood Cliffs. Second edition 1983.

1967. Heijenoort

From Frege to Gödel, Ed. J. van Heijenoort. Harvard University Press. Cambridge, MA.

1992. Largeault

Intuitionisme et théorie de la démonstration. Ed. J. Largeault. Vrin. Paris.

1993. Mangione-Franchella

Letture di Logica. Fondamenti della Matematica. Linee di Ricerca Attuali, Eds. Corrado Mangione, Miriam Franchella. Edizioni Universitarie di Lettere Economia Diritto. Milano.

1993. Largeault

Intuition et Intuitionisme, Ed. J. Largeault. Vrin. Paris. 238p.

1998. Mancosu

From Brouwer to Hilbert. The Debate on the Foundations of Mathematics in the 1920s., Ed. P. Mancosu. Oxford University Press. Oxford.

Dissertations under supervision of Brouwer

Bernardus Petrus Haalmeijer (28.11.1917) *Bijdragen tot de theorie der elementairoppervlakken.*

Maurits Joost Belinfante (12.12.1923) *Over Oneindige Reeksen.*

Barend de Loor (23.2.1925) *Die Hoofstelling van die Algebra van Intuïtionistiese Standpunt.*

Arend Heyting (25.5.1925) *Intuïtionistische Axiomatiek der Projektieve Meetkunde.*

Wilfrid Wilson (10.2.1928) *Afbeeldingen van Ruimten.*

Frans Loonstra (11.7.1941) *Analytische Untersuchungen über bewertete Körper.*

Johanna Adriana Geldof (11.7.1951) *Over de arithmetisering van axiomatische meetkunden.*

References

Works by Brouwer are listed in the Bibliography of L.E.J. Brouwer, *pp.343–390.*

P. Aczel. An introduction to inductive definitions. In J. Barwise, editor, *Handbook of Mathematical Logic*, chapter C.7, pages 739–782. North-Holland, 1977.

P. Aczel. A constructive version of the Lusin separation theorem. In S. Lindström, E. Palmgren, K. Segerberg, and V. Stoltenberg-Hansen, editors, *Logicism, Intuitionism, and Formalism: What Has Become of Them?*, forthcoming.

P. Aczel and M. Rathjen. Notes on Constructive Set Theory. Technical Report 40, Institut Mittag-Leffler, Djursholm, 2001.

Ācariya Anuruddha. *Abhidhammattha Sangaha.* No publisher, ca. 1150.

T. Arai. Introducing the hardline in proof theory. Draft, 1996a.

T. Arai. Proof theory of theories of ordinals I: Reflecting ordinals. Draft, 1996b.

T. Arai. Systems of ordinal diagrams. Draft, 1996c.

T. Arai. A sneak preview of proof theory of ordinals. Revised version of a résumé for a talk at the Kobe Seminar on Logic and Computer Science, Dec 1997, 1997a.

T. Arai. Proof theory of theories of ordinals II: Σ_1 stability. Draft, 1997b.

T. Arai. Proof theory of theories of ordinals III: Π_1 collection. Draft, 1997c.

Asanga. *Abhidharma Samuccaya,* trans. W. Rahula and S. Boin-Webb. Jain, 2001.

M. van Atten. Review of D. Hesseling, *Gnomes in the Fog. The Reception of Brouwer's Intuitionism in the 1920s. Bulletin of Symbolic Logic*, 10(3): 423–427, 2004a.

M. van Atten. *On Brouwer.* Wadsworth, Belmont, 2004b.

M. van Atten. The correspondence between Oskar Becker and Arend Heyting. In V. Peckhaus, editor, *Oskar Becker und die Philosophie der Mathematik*, pages 119–142. Wilhelm Fink, München, 2005.

M. van Atten. *Brouwer meets Husserl. On the Phenomenology of Choice Sequences.* Springer, Dordrecht, 2007.

M. van Atten. The hypothetical judgement in the history of intuitionistic logic. In C. Glymour, W. Wang, and D. Westerstahl, editors, *Logic, Methodology, and Philosophy of Science XIII: Proceedings of the 2007 International Congress in Beijing*, London, 2008. King's College Publications.

M. van Atten and R. Tragesser. Mysticism and mathematics: Brouwer, Gödel, and the Common Core Thesis. In W. Deppert and M. Rahnfeld, editors, *Klarheit in Religionsdingen*, pages 145–160. Leipziger Universitätsverlag, 2003.

M. van Atten, D. van van Dalen, and R. Tieszen. Brouwer and Weyl: the phenomenology and mathematics of the intuitive continuum. *Philosophia Mathematica*, 10(3):203–226, 2002.

M. van van Atten and D. van Dalen. Arguments for the continuity principle. *Bulletin of Symbolic Logic*, 8(3):329–347, 2002.

H. Bachmann. *Transfinite Zahlen.* Springer, Berlin, 1955.

Abu al-Barakāt al Baghdādī. *Kitāb al-Mu'tabar* 3 vols. No publisher, Heydarabad, 1357.

H.P. Barendregt. Buddhist phenomenology, part I. In M. dalla Chiara, editor, *Proceedings of the Conference on Topics and Perspectives of Contemporary Logic and Philosophy of Science (Cesena, Italy, January 7-10, 1987)*, pages 37–55, Bologna, 1988. Clueb. URL: <www.cs.ru.nl/~henk/BP/bp1.html>.

H.P. Barendregt. Mysticism and beyond, Buddhist phenomenology, part II. *The Eastern Buddhist, New Series*, XXIX:262–287, 1996. URL: <www.cs.ru.nl/~henk/BP/bp2.html>.

H.P. Barendregt. The Abhidhamma Model of Consciousness AM_0 and some of its consequences. In K.J. Gergen M.G.T. Kwee and F. Koshikawa, editors, *Buddhist Psychology: Practice, Research & Theory*, chapter 22, pages 331–349. Taos Institute, Taos, New Mexico, 2006. Available at <ftp.cs.kun.nl/pub/CompMath.Found/G.pdf>.

M. Baroni and D. Bridges. Continuity properties of preference relations. *Elec. Notes in Theor. Comp. Science*, forthcoming.

J. Barwise. *Admissible Sets and Structures. An Approach to Definability Theory.* Springer, Berlin, 1975.

J. Barwise. On Moschovakis closure ordinals. *Journal of Symbolic Logic*, 42 (2):292 – 296, June 1977.

O. Becker. Beiträge zur phänomenologischen Begründung der Geometrie und ihrer physikalischen Anwendungen. *Jahrbuch für Philosophie und phänomenologische Forschung*, VI:385–560, 1923.

O. Becker. Mathematische Existenz. Untersuchungen zur Logik und Ontologie mathematischer Phänomene. *Jahrbuch für Philosophie und phänomenologische Forschung*, VIII:439–809, 1927.

M. Beeson. *Foundations of Constructive Mathematics.* Springer, Heidelberg, 1985.

C. Bellaar-Spruyt. *Leerboek der Formeele Logica. bewerkt naar de Dictaten van wijlen Prof.Dr. C.B. Spruyt door M. Honigh.* Vincent Loosjes, Haarlem, 1903.

S. Bellantoni and S. Cook. A new recursion-theoretic chararcterization of the polytime functions. *Computational Complexity*, 2:97–110, 1992.

S. Bellantoni and M. Hofmann. A new 'feasible' arithmetic. *Journal of Symbolic Logic*, pages 104–116, 2002.

S. Bellantoni, K.-H. Niggl, and H. Schwichtenberg. Higher-types recursion, ramification and polynomial time. *Annals of Pure and Applied Logic*, 104: 17–30, 2000.

Y. Ben-Menahem. Convention: Poincaré and some of his critics. *British Journal for the Philosophy of Science*, 52(3):471–513, 2001.

P. Benacerraf and H. Putnam, editors. *Philosophy of Mathematics: Selected Readings*, 2nd ed. Cambridge University Press, Cambridge, 1983.

J. Benoist, J.-Y. Girard, Y. Ogawa, and M. Okada, editors. *Towards New Logic and Semantics.* Keio University Press, Tokyo, 2006.

J. Berger. The logical strength of the uniform continuity theorem. In A. Beckmann, U. Berger, B. Lowe, and J. Tucker, editors, *Logical Approaches to Computational Barriers*, number 3988 in Lecture Notes in Computer Science, pages 35–39. Springer, Berlin, 2006.

J. Berger and D. Bridges. A fan-theoretic equivalent of the antithesis of Specker's theorem. *Indagationes Mathematicae*, 18(2):195–202, 2007.

J. Berger and D. Bridges. The fan theorem and positive-valued uniformly continuous functions on compact intervals. *New Zealand J. Math.*, forthcoming.

P. Bernays. Die philosophie der Mathematik und die Hilbertsche Beweistheorie. *Blätter für deutsche Philosophie*, 4:326–367, 1930.

P. Bernays. Hilberts Untersuchungen über die Grundlagen der Arithmetik, 1935. In (Hilbert 1935, pp.196–216).

P. Bernays. A system of axiomatic set theory. Part I. *Journal of Symbolic Logic*, 2:65–77, 1937.

P. Bernays. Comments on Ludwig Wittgenstein's *Remarks on the Foundations of Mathematics*. *Ratio*, 2:1–22, 1959.

F. Bernstein. *Untersuchungen aus der Mengenlehre.* Waisenhaus, Halle, 1901. Dissertation Göttingen.

F. Bernstein. Zum Kontinuumproblem. *Mathematische Annalen*, 60:463–464, 1905.

E. Beth. Semantic construction of intuitionistic logic. *Med. Nederl. Akad. Wet., Afd. Lett.*, 19(11):357–388, 1956a.

E. Beth. Poincaré et la philosophie. In *Poincaré,* Œuvres, volume 11, pages 232–238. Gauthier-Villars, 1956b.

E. Beth. Remarks on intuitionistic logic. In *Constructivity in Mathematics, Proc. Colloq. Amsterdam 1957*, pages 15–25, 1959.

E. Bishop. *Foundations of Constructive Analysis.* McGraw-Hill, New York, 1967.

E. Bishop and D. Bridges. *Constructive Analysis.* Springer, Heidelberg, 1985.

Bikkhu Bodhi, editor. *A comprehensive manual of Abhidhamma.* Pariyatti, 2000. Annotated translation of (Anuruddha ca. 1150).

B. Bolzano. *Wissenschaftslehre (§§121-163)*, volume 12-1 of *Gesamtausgabe.* Fromann-Holzboog, Stuttgart, 1987.

É. Borel. *Leçons sur les Fonctions Monogènes Uniformes d'une Variable Complexe.* Gauthier-Villars, Paris, 1917.

É. Borel. *Méthodes et Problèmes de la Théorie des Fonctions.* Gauthiers-Villars, Paris, 1922.

É. Borel. *Leçons sur la Théorie des Fonctions* (3e éd.). Gauthier-Villars, Paris, 1928.

É. Borel. *Œuvres de Émile Borel. Tome I.* Centre National de la Recherche Scientifique, Paris, 1972a.

É. Borel. *Œuvres de Émile Borel. Tome IV.* Centre National de la Recherche Scientifique, Paris, 1972b.

N. Bourbaki. *Éléments de mathématique I, première partie : Les structures fondamentales de l'analyse, livre I : Théorie des ensembles (fascicule de résultats).* Number 846 in Actualités Scientifiques et Industrielles. Hermann, Paris, 1939.

N. Bourbaki. *Eléments d'Histoire des Mathématiques.* Hermann, Paris, 1960. English-language edition: *Elements of the History of Mathematics*, trans. J. Meldrum, Springer, Berlin, 1994.

J. Bouveresse. On the meaning of the word 'Platonism' in the expression 'mathematical Platonism'. *Proceedings of the Aristotelian Society*, 105:55–79, 2005.

R. Braithwaite. *Scientific Explanation.* Cambridge University Press, Cambridge, 1953.

D. Bridges. Constructive notions of equicontinuity. Preprint, University of Canterbury, 2006.

D. Bridges. Uniform continuity properties of preference relations. *Notre Dame J. Formal Logic*, 49(1):97–106, 2008.

D. Bridges and O. Demuth. On the Lebesgue measurability of continuous functions in constructive analysis. *Bulletin of the American Mathematical Society*, 24(2):259–276, 1991.

D. Bridges and H. Diener. The pseudocompactness of $[0, 1]$ is equivalent to the uniform continuity theorem. *Journal of Symbolic Logic*, 72(4):1379–1384, 2007.

D. Bridges and F. Richman. *Varieties of Constructive Mathematics.* Cambridge University Press, 1987.

D. Bridges and L. Vîţă. *Techniques of Constructive Analysis.* Springer, New York, 2006.

W. Buchholz. A simplified version of local predicativity. In P. Aczel, H. Simmons, and S. Wainer, editors, *Proof Theory. A selection of papers from the Leeds Proof Theory Programme 1990*, pages 115 – 147. Cambridge University Press, 1992.

W. Buchholz. Explaining Gentzen's proof with infinitary proof theory. In G. Gottlob, A. Leitsch, and D. Mundici, editors, *Computational logic and proof theory. 5th Kurt Gödel Colloquium, KGC '97*, pages 4 – 17. Springer Lecture Notes in Computer Science, 1289, 1997.

W. Buchholz, S. Feferman, W. Pohlers, and W. Sieg. *Iterated Inductive Definitions. Recent Proof Theoretical Studies*, volume 897 of *Springer Lecture Notes in Computer Science.* Springer, 1981.

Bhadantacariya Buddhagosa. *The Path of Purification.* Buddhist Publication Society, 1999. Original 5th century AD. Translated by Bhikkhu Nanamoli.

G. Cantor. Foundations of a general theory of manifolds: A mathematico-philosophical investigation into the theory of the infinite, 1883. In (Ewald 1996, pp.878–920).

R. Carnap. *The Unity of Science.* Kegan Paul, Trench & Trubner, London, 1934.

J. Cassinet and M. Guillemot. *L'Axiome du Choix dans les Mathématiques, de Cauchy (1921) à Gödel (1940).* PhD thesis, Université Paul Sabatier (Toulouse III), 1983.

J. Cavaillès. *Méthode Axiomatique et Formalisme.* Hermann, Paris, 1938.

P. Cohen. The independence of the continuum hypothesis II. *Proceedings of the National Academy of Sciences, U.S.A.*, 51:105–110, 1964.

T. Coquand. An analysis of Girard's Paradox. In *LICS 1986*, pages 227–236, 1986.

T. Coquand. An analysis of Ramsey's Theorem. *Inf. Comp.*, 110:297–304, 1994.

H. Curry. *Outlines of a Formalist Philosophy of Mathematics.* North-Holland, Amsterdam, 1952.

D. van Dalen. Lectures on intuitionism. In H. Rodgers and A.R.D. Mathias, editors, *Cambridge Summer School in Mathematical Logic 1971*, volume 337 of *Lecture Notes in Mathematics*, pages 1–94, Heidelberg, 1973. Springer.

D. van Dalen. Four Letters from Edmund Husserl to Hermann Weyl. *Husserl Studies*, 1(1):1–12, 1984.

D. van Dalen. *Mystic, Geometer, and Intuitionist. The Life of L.E.J. Brouwer. 1: The Dawning Revolution.* Clarendon Press, Oxford, 1999.

D. van Dalen. Brouwer and Fraenkel on Intuitionism. *Bulletin of Symbolic Logic*, 6(3):284–310, 2000.

D. van Dalen. Kolmogorov and Brouwer on constructive implication and the Ex Falso rule. *Russian Math Surveys*, 59:247–257, 2004.

D. van Dalen. *Mystic, Geometer, and Intuitionist. The Life of L.E.J. Brouwer. 2: Hope and Disillusion.* Clarendon Press, Oxford, 2005.

J. Dauben. *Georg Cantor, his Mathematics and Philosophy of the Infinite.* Harvard University Press, Cambridge MA, 1979.

D. Dennett. *Consciousness Explained.* Little, Brown, Boston, 1991.

M. Detlefsen. Poincaré against the logicians. *Synthese*, 90:349–378, 1992.

M. Detlefsen. Logicism and the nature of mathematical reasoning. In A. Irvine and G. Wedekind, editors, *Russell and Analytical Philosophy*, pages 265–292. Toronto University Press, Toronto, 1993.

M. Detlefsen. Contructive existence claims, 1998. In (Schirn 1998), pp.307–335.

H. Diener and I. Loeb. Sequences of real functions on $[0, 1]$ in constructive reverse mathematics. preprint, University of Canterbury, 2006 (available at `http://www.math.canterbury.ac.nz/\symbol{126}i.loeb/SRF.pdf`), 2006.

H. Dingler. [Rezension von] Julius König, Neue Grundlage der Logik, Arithmetik und Mengenlehre. *Archiv der Mathematik und Physik*, 24:152–159, 1915.

P. Du Bois-Reymond. *Die allgemeine Functionentheorie. Erster Theil.* Verlag der H. Laupp'schen Buchhandlung, Tübingen, 1882.

J. Dubucs and S. Lapointe. On Bolzano's alleged explicativism. *Synthese*, 150 (2):229–246, 2006.

M. Dummett. Wittgenstein's philosophy of mathematics. *Philosophical Review*, 68:324–48, 1959.

M. Dummett. The philosophical basis of intuitionistic logic. In H. Rose and J. Shepherdson, editors, *Logic Colloquium '73*, pages 5–40, Amsterdam, 1975. North-Holland. Also in (Dummett 1978), pp.215–247.

M. Dummett. What is a theory of meaning? II. In G. Evans and J. McDowell, editors, *Truth and Meaning*, pages 67–137. Clarendon Press, Oxford, 1976.

M. Dummett. *Elements of Intuitionism.* Oxford University Press, Oxford, 1977.

M. Dummett. *Truth and Other Enigmas.* Harvard University Press, Cambridge MA, 1978.

M. Dummett. Critical notice. L.E.J. Brouwer: *Collected Works*. *Mind*, 89: 605–616, 1980.

M. Dummett. *Frege: Philosophy of Mathematics.* Duckworth, London, 1991a.

M. Dummett. *The Logical Basis of Metaphysics.* Duckworth, London, 1991b.

M. Dummett. Reply to Sundholm. In B. McGuinness and G. Oliveri, editors, *The Philosophy of Michael Dummett*, pages 318–328. Kluwer, 1994.

M. Dummett. Is time a continuum of instants? *Philosophy*, 75:497–515, 2000a.

M. Dummett. *Elements of Intuitionism,* 2nd, rev. ed. Clarendon Press, Oxford, 2000b.

P. Dybjer. A general formulation of simultaneous inductive-recursive definitions in type theory. *Journal of Symbolic Logic*, 65(2):525 – 549, 2000.

P. Dybjer and A. Setzer. A finite axiomatization of inductive-recursive definitions. In J.-Y. Girard, editor, *Typed Lambda Calculi and Applications*, volume 1581 of *Lecture Notes in Computer Science*, pages 129–146, 1999.

P. Dybjer and A. Setzer. Indexed induction-recursion. In R. Kahle, P. Schroeder-Heister, and R. Stärk, editors, *Proof Theory in Computer Science*, Lecture Notes in Computer Science 2183, pages 93 – 113, Berlin, 2001. Springer.

P. Dybjer and A. Setzer. Induction-recursion and initial algebras. *Annals of Pure and Applied Logic*, 124:1 – 47, 2003.

P. Dybjer and A. Setzer. Indexed induction-recursion. *Journal of Logic and Algebraic Programming*, 66:1 – 49, 2006.

H. Edwards. Kronecker's place in history. In W. Aspray and P. Kitcher, editors, *History and Philosophy of Modern Mathematics*, pages 139–144. Minnesota Press, Minneapolis, 1988.

M. Eigen. *The Psychotic Core.* Jason Aronson, Northvale NJ, 1986.

W. Ewald. *From Kant to Hilbert. Readings in the Foundations of Mathematics.* Oxford University Press, Oxford, 1996.

S. Feferman. Systems of predicative analysis. *Journal of Symbolic Logic*, 29: 1–30, 1964.

S. Feferman. A language and axioms for explicit mathematics. In J. Crossley, editor, *Algebra and Logic. Proc. 1974, Monash Univ Australia*, volume 450 of *Springer Lecture Notes in Mathematics*, pages 87 – 139, 1975. doi: 10.1007/BFb0062852.

S. Feferman. Does reductive proof theory have a viable rationale? *Erkenntnis*, 53(1 – 2):63 – 96, 2000.

H. Feigl. The *Wiener Kreis* in America. In *Inquiries and Provocations. Selected Writings 1929–1974,* ed. R. Cohen, pages 57–94. D. Reidel, Dordrecht, 1981.

P. Février. La philosophie mathématique de Poincaré. *Epistemologia*, 4:151–172, 1981.

H. Le Roy Finch. *Wittgenstein. The Later Philosophy.* Humanities Press, Atlantic Highlands, N. J., 1977.

R. Fogelin. Wittgenstein and intuitionism. *American Philosophical Quarterly*, 5:267–274, 1968.

J. Folina. *Poincaré and the Philosophy of Mathematics.* MacMillan, London, 1992.

A. Fraenkel. *Einleitung in die Mengenlehre. Eine elementare Einführung in das Reich der unendlichen Grössen. 2te erweiterte Auflage.* Springer, Berlin, 1923.

A. Fraenkel. *Einleitung in die Mengenlehre. Eine elementare Einführung in das Reich der unendlichen Grössen. 3te umgearbeitete und stark erweiterte Auflage.* Springer, Berlin, 1929.

A. Fraenkel, Y. Bar-Hillel, and A. Lévy. *Foundations of Set theory,* 2nd, rev. edition (with the collaboration of D. van Dalen). North-Holland, Amsterdam, 1973.

M. Franchella. Like a bee on a window pane: Heyting's reflections on solipsism. *Synthese*, 105:207–251, 1995.

M. Franchella. Towards a re-evaluation of Julius König's contribution to logic. *Bulletin of Symbolic Logic*, 6:45–66, 2000.

M. Franchella. Arend Heyting and phenomenology: is the meeting feasible? *Bulletin d'Analyse Phénoménologique*, 3(2):1–21, 2007. Available online at `http://popups.ulg.ac.be/bap/document.php?id=147`.

H. Freudenthal. Zur intuitionistischen Deutung logischer Formeln. *Compositio Mathematica*, 4:112–116, 1936.

H. Friedman. *Subsystems of Analysis*. PhD thesis, MIT, 1968.

H. Friedman. The consistency of classical set theory relative to a set theory with intuitionistic logic. *Journal of Symbolic Logic*, 38(2):315–319, 1973.

H. Friedman. Some systems of second order arithmetic and their use. In *Proceedings of the 1974 International Congress of Mathematicians*, volume 1, pages 235–242, 1975.

H. Friedman. Set theoretic foundations for constructive analysis. *Ann. Math.*, 105(1):1–28, 1977.

P. Geach. Assertion. *Philosophical Review*, 74:449–465, 1965.

G. Gentzen. Untersuchungen über das logische Schließen. *Mathematische Zeitschrift*, 39:176–210, 405–431, 1934.

G. Gentzen. Die Widerspruchsfreiheit der reinen Zahlentheorie. *Mathematische Annalen*, 112:493–565, 1936.

G. Gentzen. Der erste Widerspruchsfreiheitsbeweis für die klassische Zahlentheorie. *Archiv für mathematische Logik und Grundlagenforschung*, 16(3 - 4):97 - 118, 1974.

J.-Y. Girard. Une extension de l'interprétation de Gödel à l'analyse, et son application à l'élimination des coupures dans l'analyse et la théorie des types. In J. Fenstad, editor, *Proceedings of the Second Scandinavian Logic Symposium (Oslo, June 18–20, 1970)*, volume 63 of *Studies in Logic and The Foundations of Mathematics*, pages 63–92, Amsterdam, 1971. North-Holland.

J.-Y. Girard. *Interprétation Fonctionelle et Élimination des Coupures dans l'Arithmétique d'Ordre Supérieure*. PhD thesis, Université Paris 7, 1972.

J.-Y. Girard. Linear logic. *Theoretical Computer Science*, 50(1):102–187, 1987.

J.-Y. Girard. Linear logic: its syntax and semantics. In J.-Y. Girard, Y. Lafont, and L. Regnier, editors, *Advances in Linear Logic*. Cambridge University Press, 1995.

J.-Y. Girard. Light linear logic. *Information and Computation*, 14(3):175–204, 1998.

K. Gödel. Über formal unentscheidbare Sätze der Principia mathematica und verwandter Systeme I. *Monatshefte für Mathematik und Physik*, 38:173 - 198, 1931.

K. Gödel. Zur intuitionistischen Arithmetik und Zahlentheorie. *Ergebnisse eines mathematischen Kolloquiums*, 4:34–38, 1933a.

K. Gödel. Eine Interpretation des intuitionistischen Aussagenkalküls. *Ergebnisse eines mathematischen Kolloquiums*, 4:39–40, 1933b.

K. Gödel. The present situation in the foundations of mathematics, 1933c. Lecture in Cambridge MA (published in (Gödel 1995, pp.45–53)).

K. Gödel. Über die Länge von Beweisen. *Ergebnisse eines mathematischen Kolloquiums*, 7:23–24, 1936.

K. Gödel. Lecture at Zilsel's, 1938. Lecture in Vienna (published in (Gödel 1995, pp.86–113)).

K. Gödel. *The Consistency of the Axiom of Choice and of the Generalized Continuum Hypothesis with the Axioms of Set Theory*, volume 3 of *Annals of Mathematical Studies*. Princeton University Press, Princeton, 1940.

K. Gödel. Über eine bisher noch nicht benutzte Erweiterung des finiten Standpunktes. *Dialectica*, 12:280–287, 1958.

K. Gödel. On an extension of finitary mathematics which has not yet been used. Circulated earlier version of (Gödel 1972), 1970.

K. Gödel. On an extension of finitary mathematics which has not yet been used, 1972. Revised and expanded translation of (Gödel 1958), first published in (Gödel 1990, pp.271–280).

K. Gödel. *Collected Works. II: Publications 1938–1974*. eds. S. Feferman et al. Oxford University Press, Oxford, 1990.

K. Gödel. *Collected Works. III: Unpublished Essays and Lectures*. eds. S. Feferman et al. Oxford University Press, Oxford, 1995.

K. Gödel. *Collected works. IV: Correspondence A-G.*, eds. S. Feferman et al. Oxford University Press, Oxford, 2003.

W. Goldfarb. Poincaré against the logicists. In W. Aspray and P. Kitcher, editors, *History and Philosophy of Modern Mathematics*, pages 61–81. University of Minnesota Press, Minneapolis, 1988.

I. Grattan-Guinness. *The Search for Mathematical Roots, 1870–1940*. Princeton University Press, Princeton, 2000.

J. Gray. *The Hilbert Challenge*. Oxford University Press, Oxford, 2000.

E. Griffor and M. Rathjen. The strength of some Martin-Löf type theories. *Archive for Mathematical Logic*, 33:347 – 385, 1994.

V. Grishin. A nonstandard logic and its application to set theory (Russian). In *Studies in Formalized Languages and Nonclassical Logics*, pages 135–171. Nauka, Moscow, 1974.

V. Grishin. Predicate and set-theoretic calculi based on logic without contractions. *Izvestija AN SSSR. Ser. Mat. 1981*, 45(1):47–68, 1981.

V. Grishin. On the weight of the comprehension axiom in a theory based on logic without contractions. *Mat. Zametki*, 66(5):643–652, 1999.

M. Guillemot. La communication de Gyulia König au Congrès de 1904. *Séminaire d'histoire des mathématiques de Toulouse*, 3:57–82, 1981.

P. Hacker. *Insight & Illusion. Wittgenstein on Philosophy and the Metaphysics of Experience*. Clarendon Press, Oxford, 1972.

P. Hacker. *Insight & Illusion. Themes in the Philosophy of Wittgenstein*, (2nd, rev. ed.). Clarendon Press, Oxford, 1986.

M. Ha'iri Yazdi. *Knowledge by Presence*. Cultural Studies and Research Institute, Tehran, 1982.

L. Hallnäs. Partial inductive definitions. *Theoretical Computer Science*, 87: 115–142, 1991.

L. Hallnäs. On the proof-theoretic foundation of general definition theory. *Synthese*, 148(3):589–602, 2006. Special issue on proof-theoretic semantics, eds. R. Kahle and P. Schroeder-Heister.

L. Hallnäs and P. Schroeder-Heister. A proof-theoretic approach to logic programming. I. Clauses as rules. *Journal of Logic and Computation*, 1: 261–283, 1990.

L. Hallnäs and P. Schroeder-Heister. A proof-theoretic approach to logic programming. II. Programs as definitions. *Journal of Logic and Computation*, 1:635–660, 1991.

F. Hausdorff. Der Potenzbegriff in der Mengenlehre. *Jahresbericht der deutschen Mathematiker-Vereinigung*, 13:569–571, 1904.

J. van Heijenoort, editor. *From Frege to Gödel: A Sourcebook in Mathematical Logic, 1879–1931.* Harvard University Press, Cambridge MA, 1967.

G. Heinzmann. *Poincaré, Russell, Zermelo et Peano. Textes de la Discussion (1906-1912) sur les Fondements des Mathématiques : des Antinomies à la Prédicativité.* Blanchard, Paris, 1986.

H. Hermes. *Semiotik. Eine Theorie der Zeichengestalten als Grundlage für Untersuchungen von formalisierten Sprachen.* Hirzel, Leipzig, 1938.

H. Hermes. Zum Inversionsprinzip der operativen Logik. In A. Heyting, editor, *Constructivity in Mathematics: Proceedings of the Colloquium Held at Amsterdam, 1957*, pages 62–68, Amsterdam, 1959. North-Holland.

H. Hertz. *Principles of Mechanics.* MacMillan, London, 1899.

A. Heyting. Die formalen Regeln der intuitionistischen Logik I. *Sitzungsberichte der Preussischen Akademie der Wissenschaften*, pages 42–56, 1930a. English translation in (Mancosu 1998), pp.311–327.

A. Heyting. Die formalen Regeln der intuitionistischen Logik II. *Sitzungsberichte der Preussischen Akademie der Wissenschaften*, pages 57–71, 1930b.

A. Heyting. Die formalen Regeln der intuitionistischen Logik III. *Sitzungsberichte der Preussischen Akademie der Wissenschaften*, pages 158–169, 1930c.

A. Heyting. Sur la logique intuitionniste. *Académie Royale de Belgique, Bulletin de la Classe des Sciences*, 16:957–963, 1930d. English translation in (Mancosu 1998), pp.306–310.

A. Heyting. Die intuitionistische Grundlegung der Mathematik. *Erkenntnis*, 2: 106–115, 1931. English translation in (Benacerraf and Putnam 1983, pp.52–61).

A. Heyting. *Mathematische Grundlagenforschung, Intuitionismus, Beweistheorie.* Springer, Berlin, 1934.

A. Heyting. Intuïtionistische wiskunde. *Mathematica B*, 4:72–83, 1935.

A. Heyting. Bemerkungen zu dem Aufsatz von Herrn Freudenthal 'Zur intuitionistischen Deutung logischer Formeln'. *Compositio Mathemathica*, 4: 117–118, 1936.

A. Heyting. Logique et intuitionnisme. In *Applications scientifiques de la logique mathématique. (Actes du 2e colloque internationale de logique mathématique, Paris, 1952)*, Paris, 1954. Gauthier-Villars.

A. Heyting. *Intuitionism, an Introduction.* North–Holland, Amsterdam, 1956.

A. Heyting. Intuitionism in mathematics. In R. Klibansky, editor, *La philosophie au milieu du vingtième siècle*, volume 1, pages 101–115. La nuova Italia, Firenze, 1958a.

A. Heyting. Blick von der intuitionistischen Warte. *Dialectica*, 12:332–345, 1958b.

A. Heyting. Remarques sur le constructivisme. *Logique et Analyse*, 3:177–182, 1960.

A. Heyting. Méthodes et problèmes de l'intuitionnisme. In *Actes du Colloque de Mathématiques Réuni à Clermont à l'Occasion du Tricentenaire de la Mort de Blaise Pascal. Tome I. Introduction et Logique Mathématique*, pages 101–105. Faculté des Sciences de l'Université de Clermont, 1962.

A. Heyting. *Intuitionism, an Introduction* 3rd, rev. ed. North-Holland, Amsterdam, 1971.

D. Hilbert. *Grundlagen der Geometrie. Festschrift zur Feier der Enthüllung des Gauss-Weber Denkmals in Göttingen.* Teubner, Leipzig, 1899.

D. Hilbert. Über den Zahlbegriff. *Jahresbericht der deutschen Mathematiker-Vereinigung*, 8:180–194, 1900a.

D. Hilbert. Mathematische Probleme. Vortrag, gehalten auf dem internationalen Mathematiker-Kongress zu Paris 1900. *Archiv der Mathematik und Physik (3)*, 1:44–63, 1900b.

D. Hilbert. Über die Grundlagen der Logik und der Arithmetik. In *Verhandlungen des Dritten Internationalen Mathematiker-Kongresses in Heidelberg vom 8. bis 13. August 1904*, pages 174–185, Leipzig, 1905. Teubner.

D. Hilbert. Axiomatisches Denken. *Mathematische Annalen*, 78:405–415, 1918.

D. Hilbert. Neubegründung der Mathematik (Erste Mitteilung). *Abhandlungen aus dem Mathematischen Seminar der Hamburgischen Universität*, 1: 157–177, 1922.

D. Hilbert. Die logischen Grundlagen der Mathematik. *Mathematische Annalen*, 88:151–165, 1923.

D. Hilbert. Über das Unendliche. *Mathematische Annalen*, 95:161–90, 1926.

D. Hilbert. Die Grundlagen der Mathematik. *Abhandlungen aus dem Mathematischen Seminar der Hamburgischen Universität*, 6:65–85, 1928.

D. Hilbert. Die Grundlegung der elementaren Zahlenlehre. *Mathematische Annalen*, 104:485–494, 1931a.

D. Hilbert. Beweis des Tertium non datur. *Nachrichten von der Gesellschaft der Wissenschaften zu Göttingen, Math.-Phys. Klasse*, I(22):120–125, 1931b.

D. Hilbert. *Gesammelte Abhandlungen*, volume 3. Springer, Berlin, 1935.

J. Hintikka. *Selected Papers. 1: Ludwig Wittgenstein. Half-Truths and One-and-a-Half Truths.* Kluwer, Dordrecht, 1996.

M. Hofmann. A mixed modal/linear lambda calculus with applications to Bellantoni-Cook safe recursion. In *CSL '97*, number 1414 in Springer Lecture Notes in Computer Science, pages 275–294, 1998.

M. Hofmann. Type systems for polynomial-time computation. Technical report, Edinburgh University, 1999a. LCFS Technical reports ECS-LCFS-99-406.

M. Hofmann. Linear types and non-size increasing polynomial time computation. In *Proceedings 14th Symposium in Computer Science, LICS'99*, pages 464–473, 1999b.

M. Hofmann. Programming languages capturing complexity classes, 2000a. Logic column 9, ed. J. Riecke, Bell Laboratories, Lucent Technologies.

M. Hofmann. Safe Recursion with higher types and BCK algebras. *Annals of Pure and Applied Logic*, 104:113–160, 2000b.

R. Hood, B. Hunsberger, R. Gorsuch, and B. Spilka. *The Psychology of Religion: An Empirical Approach.* Guilford, New York, 2003.

J. Hook and D. Howe. Impredicative Strong Existential Equivalent to Type:Type. Technical report, Cornell University, Ithaca, NY, 1986.

W. Howard. The formulae-as-types notion of construction. In J. Seldin and J. Hindley, editors, *To H.B. Curry: Essays on Combinatory Logic, Lambda Calculus and Formalism*, pages 479–490. Academic Press, London, 1980. Circulated from 1969.

W. Howard and G. Kreisel. Transfinite induction and bar induction of types zero and one, and the role of continuity in intuitionistic analysis. *Journal of Symbolic Logic*, 31:325–358, 1966.

E. Husserl. *Zur Phänomenologie des inneren Zeitbewußtseins (1893–1917)*, ed. R. Boehm, volume X of *Husserliana.* Martinus Nijhoff, Den Haag, 1966.

E. Husserl. *Formal and Transcendental Logic* trans. D. Cairns. Martinus Nijhoff, Den Haag, 1969.

E. Husserl. *Logical Investigations,* 2nd ed., trans. J. Findlay. Routledge and Kegan Paul, London, 1973a.

E. Husserl. *Cartesian Meditations,* trans. D. Cairns. Martinus Nijhoff, Den Haag, 1973b.

E. Husserl. *Ideas Pertaining to a Pure Phenomenology and to a Phenomenological Philosophy, First Book: General Introduction to Phenomenology,* trans. F. Kersten, volume II of *Edmund Husserl Collected Works.* Kluwer, Dordrecht, 1983.

E. Husserl. *The Idea of Phenomenology,* trans. L. Hardy, volume VIII of *Edmund Husserl Collected Works.* Kluwer, Dordrecht, 1999.

H. Ishihara. Continuity properties in constructive mathematics. *Journal of Symbolic Logic*, 57(2):557–565, 1992.

H. Ishihara. Constructive reverse mathematics: compactness properties. In L. Crosilla and P. Schuster, editors, *From Sets and Types to Analysis and Topology: Towards Practicable Foundations for Constructive Mathematics*, number 48 in Oxford Logic Guides, pages 245–267. Oxford Univ. Press, 2005.

H. Ishihara. Reverse mathematics in Bishop's constructive mathematics. *Philosophia Scientiae, Cahier Spécial 6*, pages 43–59, 2006.

G. Jäger. *Theories for Admissible Sets: A Unifying Approach to Proof Theory.* Bibliopolis, Naples, 1986.

G. Jäger. First order theories for nonmonotonic inductive definitions: Recursively inaccessible and Mahlo. *Journal of Symbolic Logic*, 66(3):1073–1089, 2001.

W. James. *The Varieties of Religious Experience: A Study in Human Nature.* Simon & Schuster, New York, 1902.

P. Jourdain. On the transfinite cardinal numbers of well-ordered aggregates. *Philosophical Magazine*, 6(6):61–75, 1904.

W. Julian and F. Richman. A uniformly continuous function on [0, 1] that is everywhere different from its infimum. *Pacific J. Math.*, 111:333–340, 1984.

R. Kahle and P. Schroeder-Heister, editors. *Proof-Theoretic Semantics*, 2006. Special issue of *Synthese*, 148(3), pp.503–756. Introduction: pp.503–506.

M. Kanovich, M. Okada, and K. Terui. Intuitionistic phase semantics is almost classical. *Mathematical Structures in Computer Science*, pages 1–20, 2006.

I. Kant. *Critique of Pure Reason* (trans. Norman Kemp Smith). St Martin's Press, New York, 1965.

F. Kaufmann. *Das Unendliche in der Mathematik und seine Ausschaltung. Eine Untersuchung über die Grundlagen der Mathematik.* Franz Deuticke, Leipzig, 1930.

A. Kechris. *Classical Descriptive Set Theory.* Springer, Berlin, 1996.

S. Kleene. On the interpretation of intuitionistic number theory. *Journal of Symbolic Logic*, 50:109–124, 1945.

S. Kleene. *Introduction to Metamathematics.* North-Holland, Amsterdam, 1952.

S. Kleene. Realisability: a retrospective survey. In H. Rodgers and A.R.D. Mathias, editors, *Cambridge Summer School in Mathematical Logic 1971*, volume 337 of *Lecture Notes in Mathematics*, pages 95–112, Heidelberg, 1973. Springer.

S. Kleene and R. Vesley. *The Foundations of Intuitionistic Mathematics, Especially in Relation to Recursive Functions.* North-Holland, Amsterdam, 1965.

F. Klein. *The Evanston Colloquium Lectures on Mathematics.* Macmillan, New York, 1911.

A. Kolmogorov. O principe tertium non datur. *Matematiceskij Sbornik*, 32: 646–667, 1925. English translation in (van Heijenoort 1967), pp.416–437.

A. Kolmogorov. Sovremennye spory o prirode matematiki. *Nauchnoe Slovo*, 6:41–54, 1929. English translation 'Contemporary debates on the nature of mathematics', Problems of Information Transmission 42(4), 2006, pp.129Ũ141.

A. Kolmogorov. Zur Deutung der intuitionistischen Logik. *Mathematische Zeitschrift*, 35:58–65, 1932.

J. König. *Einleitung in die allgemeine Theorie der algebraischen Größen.* Teubner, Leipzig, 1903.

J. König. Zum Kontinuum-Problem. *Mathematische Annalen*, 60:177–180, 1905a. Berichtigung, p.462.

J. König. Über die Grundlagen der Mengenlehre und das Kontinuumproblem. *Mathematische Annalen*, 61:156–160, 1905b.

J. König. Über die Grundlagen der Mengenlehre und das Kontinuumproblem (Zweite Mitteilung). *Mathematische Annalen*, 63:217–221, 1906a.

J. König. Sur la théorie des ensembles. *Comptes Rendus des Séances Hebdomadaires de l'Académie des Sciences de Paris*, 143:110–112, 1906b.

J. König. *Neue Grundlagen der Logik, Arithmetik und Mengenlehre.* Veit, Leipzig, 1914.

G Kowalewski. *Bestand und Wandel.* Oldenburg, München, 1950.

G. Kreisel. Wittgenstein's *Remarks on the Foundations of Mathematics. British Journal for the Philosophy of Science*, 9:135–158, 1958.

G. Kreisel. On weak completeness of intuitionistic predicate logic. *Journal of Symbolic Logic*, 27(2):139–158, 1962a.

G. Kreisel. Foundations of intuitionistic logic. In E. Nagel, P. Suppes, and A. Tarski, editors, *Logic, Methodology and Philosophy of Science, Proc. 1960 Int. Congr.*, pages 198–210, Stanford, 1962b. Stanford University Press.

G. Kreisel. Functions, ordinals, species. In *Logic, Methodology Philos. Sci. III, Proc. 3rd Int. Congr. Amsterdam 1967*, pages 145–159, 1968.

G. Kreisel. Church's Thesis and the ideal of informal rigour. *Notre Dame Journal of Formal Logic*, 28(4):499–519, 1987a.

G. Kreisel. Gödel's excursions into intuitionistic logic. In P. Weingartner and L. Schmetterer, editors, *Gödel remembered*, pages 67–179. Bibliopolis, Napoli, 1987b.

S. Kripke. Semantical analysis of intuitionistic logic I. In M. Dummett and J. Crossley, editors, *Formal Systems and Recursive Functions*, pages 92–130, Amsterdam, 1965. North-Holland.

S. Kripke. *Wittgenstein on Rules and Private Language. An Elementary Exposition.* Harvard University Press, Cambridge MA, 1982.

B. Kushner. *Lectures on Constructive Mathematical Analysis.* Amer. Math. Soc., Providence RI, 1985.

J. Lambek. The mathematics of sentence structure. *American Mathematical Monthly*, 65:154–170, 1958.

J. Largeault. *Intuition et Intuitionisme.* Vrin, Paris, 1993.

H. Lebesgue. *Leçons sur la Théorie de l'Intégration,* 2e éd. Gauthier-Villars, Paris, 1928.

D. Leivant. Stratified functional programs and computational complexity. In *Proceedings 20th IEEE Symposium on Principles of Programming Languages*, 1993.

D. Leivant. Intrinsic theories and computational complexity. In D. Leivant, editor, *Logic and Computational Complexity, International Workshop LCC'94*, volume 960 of *Lecture Notes in Computer Science*, pages 177–194, Berlin, 1995. Springer.

J. Leuba. *The Psychology of Religious Mysticism.* Harcourt, New York, 1925.

E. Levinas. *The Theory of Intuition in Husserl's Phenomenology,* 2nd ed. Northwestern University Press, 1995.

P. Lietz. *From Constructive Mathematics to Computable Analysis via the Realizability Interpretation.* PhD thesis, Universität Darmstadt, 2004.

I. Loeb. Equivalents of the (weak) Fan Theorem. *Ann. Pure and Applied Logic*, 132:51–66, 2005.

K. Lorenz. Logik, operative. In J. Ritter and K. Gründer, editors, *Historisches Wörterbuch der Philosophie*, volume 5, pages 444–452. Schwabe, Basel, 1980.

K. Lorenz. Basic objectives of dialogue logic in historical perspective. *Synthese*, 127:255–263, 2001.

P. Lorenzen. Konstruktive Begründung der Mathematik. *Mathematische Zeitschrift*, 53:162–201, 1950.

P. Lorenzen. *Einführung in die operative Logik und Mathematik.* Springer, Berlin, 1955. 2nd ed. Berlin 1969.

P. Lorenzen. Über die Begriffe 'Beweis' und 'Definition'. In A. Heyting, editor, *Constructivity in Mathematics: Proceedings of the Colloquium Held at Amsterdam, 1957*, pages 169–177, Amsterdam, 1959. North-Holland.

P. Lorenzen. Ein dialogisches Konstruktivitätskriterium. In *Infinitistic Methods: Proceedings of the Symposium on Foundations of Mathematics (Warsaw 1959)*, pages 193–200, Warszawa, 1961. Państwowe Wydawnictwo Naukowe.

P. Lorenzen. Bibliographie der Schriften von Paul Lorenzen. *Journal for General Philosophy of Science*, 27:187–202, 1996.

N. Lusin. *Leçons sur les Ensembles Analytiques et Leurs Applications.* Gauthier-Villars, Paris, 1930. Second edition: Chelsea, New York 1972.

N. Lusin and W. Sierpinski. Sur une propriété du continu. *Comptes Rendus Hebdomadaires des Séances de l'Académie des Sciences de Paris*, 165:498–500, 1917.

E. Mach. *Popular Scientific Lectures.* Open Court, LaSalle IL, 1943.

E. Mach. *The Analysis of Sensations.* Dover, New York, 1959.

M. Maietti and G. Sambin. Toward a minimalist foundation for constructive mathematics. In L. Crosilla and P. Schuster, editors, *From Sets and Types to Topology and Analysis. Towards Practicable Foundations for Constructive Mathematics*, pages 91–114. Clarendon Press, 2005.

M. Maietti and G. Sambin. Choice sequences in a minimalist foundation, in preparation.

M. E. Maietti. Modular correspondence between dependent type theories and categories including pretopoi and topoi. *Mathematical Structures in Computer Science*, 15:1089–1149, 2005.

M. E. Maietti. A minimalist two-level foundation for constructive mathematics, forthcoming.

P. Mancosu. *From Brouwer to Hilbert. The Debate on the Foundations of Mathematics in the 1920s.* Oxford University Press, Oxford, 1998.

P. Mancosu. Between Russell and Hilbert: Behmann on the foundations of mathematics. *Bulletin of Symbolic Logic*, 5:303–330, 1999.

P. Mancosu and T. Ryckman. Mathematics and phenomenology: the correspondence between O. Becker and H. Weyl. *Philosophia Mathematica*, 10: 130–202, 2002.

M. Marion. Wittgenstein and Finitism. *Synthese*, 105:141–176, 1995.

M. Marion. Wittgenstein and Ramsey on identity. In J. Hintikka, editor, *From Dedekind to Gödel*, pages 343–372. Kluwer, 1996.

M. Marion. *Wittgenstein, Finitism, and the Foundations of Mathematics.* Clarendon, Oxford, 1998.

M. Marion. Carnap, lecteur de Wittgenstein ; Wittgenstein, lecteur de Carnap. In *Carnap Aujourd'hui*, pages 87–113. Bellarmin/Vrin, Montréal/Paris, 2002.

M. Marion. Wittgenstein and Brouwer. *Synthese*, 137:103–127, 2003.

J. Maritain. *An Introduction to Logic.* Sheed and Ward, London, 1946.

P. Martin-Löf. *Notes on Constructive Mathematics.* Almqvist & Wiksell, 1970.

P. Martin-Löf. Hauptsatz for the intuitionistic theory of iterated inductive definitions. In J. Fenstad, editor, *Proceedings of the 2nd Scandinavian Logic Symposium (Oslo 1970)*, pages 179–216, Amsterdam, 1971. North Holland.

P. Martin-Löf. *Intuitionistic Type Theory. Notes by G. Sambin of a Series of Lectures Given in Padua, June 1980.* Bibliopolis, Napoli, 1984.

P. Martin-Löf. Truth of a proposition, evidence of a judgment, validity of a proof. *Synthese*, 73:407–420, 1987.

P. Martin-Löf. Analytic and synthetic judgement in type theory. In Paolo Parini, editor, *Kant and Contemporary Epistemology*, pages 87–99. Kluwer, Dordrecht, 1994.

P. Martin-Löf. On the meaning of the logical constants and the justification of the logical laws. *Nordic Journal of Philosophical Logic*, 1(1):11 – 60, May 1996. Lectures in Siena, 6–9 April 1983. Available from `http://www.hf.uio.no/filosofi/njpl/vol1no1/meaning/meaning.html`.

P. Martin-Löf. Truth and knowability: on the principles C and K of Michael Dummett. In H. Dales and G. Oliveri, editors, *Truth in Mathematics*, pages 105–114. Clarendon Press, Oxford, 1998.

E. Martino and P. Giaretta. Brouwer, Dummett, and the Bar Theorem. In *Atti del Congresso Nazionale di Logica, Montecatini Terme, 1–5 Ottobre 1979*, Napoli, 1981.

C. McCarty. Information systems, continuity and realizability. In E. Clark and D. Kozen, editors, *Workshop on the Logic of Programs 1983*, volume 164 of *Springer Lecture Notes in Computer Science*, pages 341–359. Springer, New York, 1984.

C. McCarty. Incompleteness in intuitionistic metamathematics. *Notre Dame Journal of Formal Logic*, 32(3):323–358, 1991.

C. McCarty. The coherence of antirealism. *Mind*, 115(460):947–956, 2006a.

C. McCarty. Thesis and variations. In A. et al. Olszewski, editor, *Church's Thesis After 70 Years*, pages 281–303. Ontos, Frankfurt, 2006b.

B. McGuinness. Wittgenstein and Ramsey. In M.-C. Galavotti, editor, *Cambridge and Vienna: Frank P. Ramsey and the Vienna Circle*, pages 19–28. Springer, Vienna, 2006.

J. McKinsey and A. Tarski. Some theorems about the sentential calculi of Lewis and Heyting. *Journal of Symbolic Logic*, 13:1–15, 1948.

C. McLarty. Poincaré: mathematics & logic & intuition. *Philosophia Mathematica*, 5:97–115, 1997.

K. Menger. *Reminiscences of the Vienna Circle and the Mathematical Colloquium*, eds. L. Golland, B. McGuinness, and A. Sklar. Kluwer, Dordrecht, 1994.

K. Menninger, M. Mayman, and P. Pruyser. *The Vital Balance. The Life Process in Mental Health and Illness.* Viking, New York, 1963.

D. Mirimanoff. Compte rendu de l'ouvrage de J. König, Neue Grundlagen der Logik, Arithmetik und Mengenlehre. *L'Enseignement Mathématique*, 16:399–402, 1914.

J. Mitchell and R. Harper. The essence of ML. In *Proceedings of the 15th ACM SIGPLAN-SIGACT Symposium on Principles of Programming Languages*, pages 28–46, 1988.

R. Monk. *Ludwig Wittgenstein. The Duty of Genius.* Jonathan Cape, London, 1990.

J. Mooij. *La Philosophie des Mathématiques de Henri Poincaré.* Nauwelaerts, Louvain, 1966.

E. Moriconi and L. Tesconi. On inversion principles. *History and Philosophy of Logic*, 29(2):103–113, 2008.

Y. Moschovakis. *Descriptive Set Theory.* North-Holland, Amsterdam, 1980.

J. Myhill. Embedding classical type theory in 'intuitionistic' type theory. In *Axiomatic Set Theory*, volume 13, Part 1 of *Proceedings of Symposia in Pure Mathematics*, pages 267–270, Providence, R.I., 1971. American Mathematical Society.

J. Myhill. Embedding classical type theory in 'intuitionistic' type theory: a correction. In T. Jech, editor, *Axiomatic Set Theory*, volume 13, part 11 of *Proceedings of Symposia in Pure Mathematics*, pages 185–188, Providence, R.I., 1974. American Mathematical Society.

J. Myhill. Constructive set theory. *Journal of Symbolic Logic*, 40(3):347–382, 1975.

P. Nabonnand. La polémique entre Poincaré et Russell au sujet du Statut des axiomes de la Géométrie. *Revue d'Histoire des Mathématiques*, 6:219–269, 2000a.

P. Nabonnand. *La genèse psycho-physiologique de la géométrie selon Poincaré.* Textes du Séminaire 'Histoires de Géométries'. Maison des Sciences de l'Homme, 2000b.

U. Nandamalabhivamsa. Personal communication to H. Barendregt, 2004.

E. Nelson. *Predicative Arithmetic.* Princeton University Press, Princeton, 1986.

J. von Neumann. Eine Axiomatisierung der Mengenlehre. *Journal für die reine and angewandte Mathematik*, 154:199–240, 1925.

J. von Neumann. Zur Hilberschen Beweistheorie. *Mathematische Zeitschrift*, 16:2–46, 1927.

B. Nordström, K. Petersson, and J.Smith. *Programming in Martin-Löf's Type Theory, an Introduction.* Oxford University Press, 1990.

W. Ockham. *Ockham's Theory of Terms. Part 1 of the Summa Logicae* (trans. M. Loux). University of Notre Dame Press, Notre Dame, 1974.

M. Okada. Introduction to linear logic: Expressiveness and phase semantics. In M. Takahashi, M. Okada, and M. Dezani, editors, *Theories of Types and Proofs,* 2nd edition, pages 255–295. Mathematical Society of Japan, 1999.

M. Okada. Linear logic and intuitionistic logic. *Revue Internationale de Philosophie*, 58(4):449–481, 2004.

M. Okada and K. Terui. Simple logic and simple phase semantic completeness. Circulated since 2003; to appear, 2003.

H. Ono and Y. Komori. Logics without the contraction rule. *Journal of Symbolic Logic*, 50(1):169–201, 1985.

E. Palmgren. *On Fixed Point Operators, Inductive Definitions and Universes in Martin-Löf's Type Theories.* PhD thesis, University of Uppsala, 1991. UUDM Report 1991:8.

E. Palmgren. On universes in type theory. In G. Sambin and J. Smith, editors, *Twenty-Five Years of Constructive Type Theory.* Oxford University Press, 1998.

C. Parsons. Introduction to (Brouwer1927B). In van Heijenoort (1967), pages 446–457.

C. Parsons. Intuition in constructive mathematics. In J. Butterfield, editor, *Language, mind and logic*, pages 211–229. Cambridge University Press, Cambridge, 1986.

C. Parsons. On some difficulties concerning intuition and intuitive knowledge. *Mind*, 102:233–246, 1993.

C. Parsons. Mathematical intuition. In W. Hart, editor, *The Philosophy of Mathematics*, pages 95–Ű113. Clarendon Press, Oxford, 1996.

G. Peano. Additione. *Revista de Mathematica*, 8:143–157, 1906.

S. Pines. *Collected Works. 1: Studies in Abu'l-Barakat al-Baghdadi: Physics and Metaphysics.* Brill, Leiden, 1979.

G. Pitcher. *The Philosophy of Wittgenstein.* Prentice-Hall, Englewood Cliffs NJ, 1964.

A. Pitts. Polymorphism is set theoretic, constructively. In D. Pitts et al., editors, *Category Theory and Computer Science*, volume 283 of *Lecture Notes in Computer Science*, pages 12–39. Springer, Berlin, 1987.

T. Placek. *Mathematical Intuitionism and Intersubjectivity. A Critical Exposition of Arguments for Intuitionism.* Kluwer, Dordrecht, 1999.

R. Platek. *Foundations of recursion theory.* PhD thesis and supplement, Stanford, CA: Stanford University, 1966.

Plato. *Philebus,* trans. D. Frede. Hackett, Indianapolis, 1993.

E.M. Podvoll. *Recovering Sanity: A Compassionate Approach to Understanding and Treating Pyschosis.* Shambhala, 2003.

W. Pohlers. Pure Proof Theory. Aims, methods and results. Extended version of talks given at Oberwolfach and Haifa. *Bulletin of Symbolic Logic*, 2(2): 159 – 188, 1996.

W. Pohlers and G. Jäger. Eine Beweistheoretische Untersuchung von $(\Delta^1_2 - Ca) + (BI)$ und verwandter Systeme. In *Sitzungsber., Bayer. Akad. Wiss., Math.-Naturwiss. Kl.*, pages 1–28, 1983.

H. Poincaré. On the foundations of geometry. *The Monist*, 9:1–43, 1898. French translation by L. Rougier, *Des Fondements de la Géométrie*, Paris: Chiron (1921); reprint in (Poincaré 2002, pp.5–46).

H. Poincaré. L'espace et ses trois dimensions. *Revue de Métaphysique et de Morale*, 11:281–301, 407–429, 1903.

H. Poincaré. La terre tourne-t-elle? *Bulletin de la Société Astronomique de France*, 18:216–217, 1904.

H. Poincaré. *Science and Hypothesis.* Walter Scott, London, 1905.

H. Poincaré. Les mathématiques et la logique. *Revue de Métaphysique et de Morale*, 14:294–317, 1906. Reprint in (Heinzmann 1986, pp.79–104).

H. Poincaré. La logique de l'infini. *Scientia*, 12:1–11, 1912.

H. Poincaré. Analyse des travaux scientifiques de Henri Poincaré faite par lui-même. *Acta Mathematica*, 38:3–135, 1921. Reprint in F. Browder (ed.), *The Mathematical Heritage of Henri Poincaré*, vol. II, Providence R.I. 1983 (Proc. Symp. in Pure Math. 39), pp.257–357.

H. Poincaré. *Science and Hypothesis.* Dover, New York, 1952.

H. Poincaré. *The Value of Science.* Dover, New York, 1958.

H. Poincaré. *Mathematics and Science: Last Essays.* Dover, New York, 1963.

H. Poincaré. *L'Opportunisme Scientifique* (comp. L. Rougier, ed. L. Rollet). Birkhäuser, Basel, 2002.

H. Poincaré. *Science and Method.* Thomas Neldon & Sons, London, n.d.

K. Popper. *The Logic of Scientific Discovery.* Harper, New York, 1968.

E. Post. Introduction to a general theory of elementary propositions. *American Journal of Mathematics*, 43:163–185, 1921.

C. Posy. A note on Brouwer's definition of unextendable order. *History and Philosophy of Logic*, 1:129–149, 1980.

C. Posy. Epistemology, ontology and the continuum. In E. Grosholz and H. Breger, editors, *The Growth of Mathematical Knowledge.* Kluwer, Dordrecht, 2000.

D. Prawitz. *Natural Deduction, a Proof-Theoretical Study.* Almqvist & Wiksell, 1965. Reprinted Mineola NY: Dover, 2006.

D. Prawitz. Ideas and results in proof theory. In J. Fenstad, editor, *Proceedings of the 2nd Scandinavian Logic Symposium (Oslo 1970)*, pages 235–308, Amsterdam, 1971. North-Holland.

D. Prawitz. Towards a foundation of a general proof theory. In P. Suppes et al, editor, *Logic, Methodology, and Philosophy of Science IV*, pages 225–250., Amsterdam, 1973. North-Holland.

D. Prawitz. On the idea of a general proof theory. *Synthese*, 27:63–77, 1974.

D. Prawitz. Proofs and the meaning and completeness of the logical constants. In J. Hintikka, editor, *Essays on Mathematical and Philosophical Logic*, pages 25–40. Reidel, Dordrecht, 1978. Revised German translation in *Conceptus* 16 (1982).

D. Prawitz. Remarks on some approaches to the concept of logical consequence. *Synthese*, 62:152–171, 1985.

D. Prawitz. Meaning approached via proofs. *Synthese*, 148(3):507–524, 2006. Special issue on proof-theoretic semantics, eds. R. Kahle and P. Schroeder-Heister.

W. Quine. *Philosophy of Logic.* Prentice-Hall, Englewood Cliffs NJ, 1970.

F. Ramsey. On a problem in formal logic. *Proc. London Math. Soc.*, 30: 264–289, 1928.

F. Ramsey. *Philosophical Papers.* Cambridge University Press, Cambridge, 1990.

A. Ranta. Propositions as games as types. *Synthese*, 76:377–395, 1988.

M. Rathjen. Ordinal notations based on a weakly Mahlo cardinal. *Archive for Mathematical Logic*, 29(4):249–263, 1990.

M. Rathjen. Proof-theoretic analysis of KPM. *Archive for Mathematical Logic*, 30(5–6):377–403, 1991.

M. Rathjen. Recent advances in ordinal analysis: Π^1_2-CA and related systems. *Bulletin of Symbolic Logic*, 1:468 – 485, 1995a.

M. Rathjen. An ordinal representation system for Π^1_2-comprehension and related systems. Preprint, 1995b.

M. Rathjen. The strength of Martin-Löf type theory with a superuniverse. Part I. *Archive for Mathematical Logic*, 39(1):1 – 39, 2000.

M. Rathjen. The strength of Martin-Löf type theory with a super universe, part II. *Archive for Mathematical Logic*, 40:207–233, 2001.

M. Rathjen. The constructive Hilbert program and the limits of Martin-Löf type theory. *Synthese*, 147:81 – 120, 2005a.

M. Rathjen. An ordinal analysis of stability. *Archive for Mathematical Logic*, 44(1):1 – 62, January 2005b.

M. Rathjen. An ordinal analysis of parameter free Π^1_2-comprehension. *Archive for Mathematical Logic*, 44(3):263 – 362, April 2005c.

M. Resnik. On understanding mathematical proofs. In J.-L. Greffe, G. Heinzmann, and K. Lorenz, editors, *Henri Poincaré. Science and Philosophy.* Blanchard, Paris, 1996.

W. Richter. Recursively Mahlo ordinals and inductive definitions. In R. O. Gandy and C. E. M. Yates, editors, *Logic Colloquium '69*, pages 273 – 288. North-Holland, 1971.

R. Rucker. *Infinity and the Mind.* Birkhäuser, Basel, 1983.

B. Russell. *The Principles of Mathematics.* Allen & Unwin, London, 1903.

B. Russell. On some difficulties in the theory of transfinite numbers and order types. In *Proc. London Mathematical Society*, volume 4, pages 29–53, 1906.

B. Russell. Mathematical logic as based on the theory of types. *American Journal of Mathematics*, 30:222–282, 1908.

B. Russell. *Our Knowledge of the External World.* Open Court, London, 1914.

B. Russell. *Our Knowledge of the External World.* Allen & Unwin, London, 1926. Revised edition of (Russell 1914).

B. Russell. *The Problems of Philosophy.* Oxford University Press, Oxford, 1997.

B. Russell. *Autobiography.* Routledge, London, 1998.

B. Russell. *Mysticism and Logic.* Routledge, London, 2004.

E. Saks. *The Center Cannot Hold: My Journey Through Madness.* Harper & Row, New York, 2007.

G. Sambin. Intuitionistic formal spaces. A first communication. In D. Skordev, editor, *Mathematical Logic and its Applications*, pages 187–204. Plenum, New York, 1987.

G. Sambin. Steps towards a dynamic constructivism. In P. Gärdenfors, K. Kijania-Placek, and J. Wolenski, editors, *Proceedings of the XI International Congress of Logic, Methodology and Philosophy of Science, Krakow, August 1999*, pages 261 - 284. Kluwer, 2001.

G. Sambin. *The Basic Picture. A Structural Basis for Constructive Topology.* Oxford University Press, forthcoming a.

G. Sambin. A minimalist foundation at work, forthcoming b. In: *Essays in Honour of John L. Bell*, eds. P. Clark, D. DeVidi, M. Hallett, *Western Ontario Series in the Philosophy of Science*, Springer.

G. Sambin, G. Battilotti, and C. Faggian. Basic logic: reflection, symmetry, visibility. *Journal of Symbolic Logic*, 65(3):979–1013, 2000.

M. Schirn, editor. *The Philosophy of Mathematics Today.* Oxford University Press, 1998.

M. Schlick. *Philosophical Papers. II: 1925–1936.* D. Reidel, Dordrecht, 1979.

A. Schlüter. On provability in set theories with reflection, n.d. Preprint.

A. Schoenflies. Zum Erinnerung an Georg Cantor. *Jahresbericht der deutschen Mathematiker-Vereinigung*, 31:97–106, 1922.

P. Schroeder-Heister. Rules of definitional reflection. In *8th Annual IEEE Symposium on Logic in Computer Science (Montreal 1993)*, pages 222–232, Los Alamitos, 1993. IEEE Computer Society Press.

P. Schroeder-Heister. On the notion of *assumption* in logical systems. In R. Bluhm and C. Nimtz, editors, *Selected Papers Contributed to the Sections of GAP5, Fifth International Congress of the Society for Analytical Philosophy, Bielefeld, 22–26 September 2003*, pages 27–48, Paderborn, 2004. mentis.

P. Schroeder-Heister. Dummett, Michael. In J. Mittelstraß, editor, *Enzyklopädie Philosophie und Wissenschaftstheorie*, volume 2, pages 258–262. Metzler, Stuttgart, 2005.

P. Schroeder-Heister. Validity concepts in proof-theoretic semantics. *Synthese*, 148(3):525–571, 2006. Special issue on proof-theoretic semantics, eds. R. Kahle and P. Schroeder-Heister.

P. Schroeder-Heister. Generalized definitional reflection and the inversion principle. *Logica Universalis*, 1:355–376, 2007.

P. Schroeder-Heister. Lorenzens operative Logik und moderne beweistheoretische Semantik. In J. Mittelstraß, editor, *Der Konstruktivismus in der Philosophie im Ausgang von Wilhelm Kamlah und Paul Lorenzen. Beiträge zum Symposium in Konstanz, September 2005.* mentis, 2008a. In print.

P. Schroeder-Heister. Proof-theoretic vs. model-theoretic consequence. In *Logica 2007 Yearbook.* Czech Academy of Sciences, 2008b. In print.

P. Schroeder-Heister and P. Contu. Folgerung. In W. Spohn, P. Schroeder-Heister, and E. Olsson, editors, *Logik in der Philosophie*, pages 247–276. Synchron Wissenschaftsverlag, Heidelberg, 2005.

K. Schütte. Beweistheoretische Erfassung der unendlichen Induktion in der Zahlentheorie. *Mathematische Annalen*, 122:369Ũ–389, 1951.

K. Schütte. Predicative well-orderings. In *Formal Systems and Recursive Functions, Proc. 8th Logic Colloquium, Oxford 1963*, pages 280–303, 1965.

H. Schwichtenberg. An arithmetic for polynomial-time computation. *Theoretical Computer Science*, 357(1-3):202–214, 2006.

H. Schwichtenberg and S. Bellantoni. Feasible computations with higher types. In H. Schwichtenberg and R. Steinbrüggen, editors, *Proof and System Reliability. Proceedings NATO Advanced Study Institute, Marktoberdorf, 2001*, pages 399–415. Kluwer, 2002.

A. Setzer. *Proof theoretical strength of Martin-Löf Type Theory with W-type and one universe.* PhD thesis, Universität München, 1993.

A. Setzer. A model for a type theory with Mahlo Universe. Unpublished, 1996.

A. Setzer. Well-ordering proofs for Martin-Löf's type theory with W-type and one universe. *Annals of Pure and Applied Logic*, 92:113 – 159, 1998a.

A. Setzer. An introduction to well-ordering proofs in Martin-Löf's type theory. In G. Sambin and J. Smith, editors, *Twenty-five Years of Constructive Type Theory*, pages 245 – 263, Oxford, 1998b. Clarendon Press.

A. Setzer. Ordinal systems. In Cooper B. and J. Truss, editors, *Sets and Proofs*, pages 301 – 331, Cambridge, 1999. Cambridge University Press.

A. Setzer. Ordinal systems part 2: One inaccessible. In S. Buss, P. Hajek, and P. Pudlak, editors, *Logic Colloquium '98*, ASL Lecture Notes in Logic 13, pages 426 – 448, Massachusetts, 2000a. Peters.

A. Setzer. Extending Martin-Löf type theory by one Mahlo-universe. *Archive for Mathematical Logic*, 39:155 – 181, 2000b.

A. Setzer. Proof theory of Martin-Löf Type Theory —An overview. *Mathematiques et Sciences Humaines*, 42(165):59 – 99, 2004.

A. Setzer. Universes in type theory part II: Autonomous Mahlo and Π_3-reflection. Available at `http://www.cs.swan.ac.uk/$\sim$csetzer/`, 2005.

A. Setzer. Review of (Rathjen 2005a). *Mathematical Reviews*, MR2182643 (2006m:03023), 2006.

A. Setzer. Universes in type theory part I — inaccessibles and Mahlo. In A. Andretta, K. Kearnes, and D. Zambella, editors, *Logic Colloquium '04*, pages 123 – 156. Association of Symbolic Logic, Lecture Notes in Logic 29, 2008.

S. Simpson. Which set existence axioms are needed to prove the Cauchy/Peano theorem for ordinary differential equations? *Journal of Symbolic Logic*, 49(3):783–802, 1984.

S. Simpson. *Subsystems of Second Order Arithmetic.* Perspectives in Mathematical Logic. Springer, 1999. Second edition to be published by the Association for Symbolic Logic.

S. Simpson, editor. *Reverse Mathematics 2001.* ASL lecture notes in logic 21, Association for Symbolic Logic and A. Peters, 2005.

Ibn Sīnā. *al-Ishārāt wa al-Tanbihāt,* ed. M. Shahabi. Tehran University Press, Tehran, 1960. In Arabic.

T. Skolem. Begründung der elementaren Arithmetik durch die rekurrierende Denkweise ohne Anwendung scheinbarer Veränderlichen mit unendlichem Ausdehnungsbereich. *Videnskapsselskapets skripfter, I. Matematisk-naturvidenskabelig klasse*, 6, 1923. English translation in (van Heijenoort 1967, pp.302–333).

E. Specker. Nicht konstruktiv beweisbare Sätze der Analysis. *Journal of Symbolic Logic*, 14:145–158, 1949.

C. Spector. Provably recursive functionals of analysis: A consistency proof of analysis by an extension of principles formulated in current intuitionistic mathematics. In *Proc. Sympos. Pure Math.*, volume 5, pages 1–27, 1962.

F. Staal. *Exploring Mysticism. A Methodological Essay.* University of California Press, Berkeley, 1975.

W. Stegmüller. *Probleme und Resultate der Wissenschaftstheorie und Analytischen Philosophie, Band IV: Personelle und Statistische Wahrscheinlichkeit, 1. Halbband: Personelle Wahrscheinlichkeit und Rationale Entscheidung.* Springer, Berlin, 1973.

W. van Stigt. *Brouwer's Intuitionism.* North-Holland, Amsterdam, 1990.

Sŭhrawardī. Ḥikmat al-ishrāq, 2002a. In (Sŭhrawardī 2002), vol.2.

Sŭhrawardī. Kitāb al-talwīḥāt, 2002b. In (Sŭhrawardī 2002), vol.1.

Sŭhrawardī. Kitāb al-mashāri' wa'l-muṭāriḥāt, 2002c. In (Sŭhrawardī 2002), vol.1.

Sŭhrawardī. *Collected works,* eds. H. Corbin and S. Nasr. Institute for Humanities and Cultural Studies, Tehran, 2002.

J.R. Suler. *Contemporary Psychoanalysis and Eastern Thought.* State University of New York Press, 1993.

G. Sundholm. Constructions, proofs and the meaning of logical constants. *Journal of Philosophical Logic*, 12:151–172, 1983.

G. Sundholm. Brouwer's anticipation of the principle of charity. *Proceedings of the Aristotelian Society*, LXXXV:i–xiv, 1985.

G. Sundholm. Questions of proof. *Manuscrito*, XVI(2):47–70, 1993.

G. Sundholm. Implicit epistemic aspects of constructive logic. *Journal of Logic, Language and Information*, 6:191–212, 1997.

H. de Swart. Another intuitionistic completeness proof. *Journal of Symbolic Logic*, 41(3):644–662, 1976.

W. Tait. Intensional interpretations of finite type I. *Journal of Symbolic Logic*, 32:198–212, 1967.

W. Tait. Normal derivability in classical logic. In J. Barwise, editor, *The Syntax and Semantics of Infinitary Languages*, pages 204 – 236. Springer Lecture Notes in Mathematics 72, 1968.

W. Tait. Finitism. *Journal of Philosophy*, 78(9):524 – 546, 1981.

W. Tait. *The Provenance of Reason. Essays in the Philosophy of Mathematics and its History.* Oxford University Press, Oxford, 2005.

W. Tait. Gödel's correspondence on constructive mathematics and proof theory. *Philosophia Mathematica*, 14:76–111, 2006a.

W. Tait. Gödel's interpretation of intuitionism. *Philosophia Mathematica*, 14: 208–228, 2006b.

G. Takeuti. Consistency proofs of subsystems of classical analysis. *Ann. Math (2)*, 86:299–348, 1967.

G. Takeuti. *Proof Theory.* North-Holland, Amsterdam, second edition, 1987.

J. Tannery. Einleitung in die allgemeine Theorie der algebraischen Größen von J. König. *Bulletin des Sciences Mathématiques*, 27:346–355, 1905.

A. Tarski. Der Wahrheitsbegriff in den formalisierten Sprachen. *Studia Philosophica*, 1:261–405, 1935.

K. Terui. Light affine set theory: a naive set theory of polynomial time. *Studia Logica*, 77:9–40, 2004.

C. Thiel. Paul Lorenzen (1915-1994). *Journal for General Philosophy of Science*, 27, 1996.

R. Tieszen. *Mathematical Intuition. Phenomenology and Mathematical Knowledge.* Kluwer, Dordrecht, 1989.

R. Tieszen. Kurt Gödel and phenomenology. *Philosophy of science*, 59:176–194, 1992.

R. Tieszen. What is the philosophical basis of intuitionistic mathematics? In D. Prawitz, B. Skyrms, and D. Westerstahl, editors, *Logic, Methodology, and Philosophy of Science IX*, pages 579–594, Amsterdam, 1994. Elsevier.

R. Tieszen. Intuitionism, meaning theory and cognition. *History and Philosophy of Logic*, 21(3):179–194, 2000a. Reprinted with minor revisions in (Tieszen 2005, pp.227–247).

R. Tieszen. The philosophical background of Weyl's mathematical constructivism. *Philosophia Mathematica*, 3:274–301, 2000b. Reprinted with minor revisions in (Tieszen 2005, pp.248–275).

R. Tieszen. *Phenomenology, Logic, and the Philosophy of Mathematics.* Cambridge University Press, Cambridge, 2005.

R. Tieszen. Mathematical realism and transcendental phenomenological idealism, forthcoming. Forthcoming in a volume of papers from the *Phenomenology and Mathematics* conference in Tampere, Finland, May 2007.

A. Troelstra. *Principles of Intuitionism.* Springer, Berlin, 1969.

A. Troelstra. *Choice Sequences. A Chapter of Intuitionistic Mathematics.* Oxford University Press, Oxford, 1977.

A. Troelstra. Intuitionistic extensions of the reals. *Nieuw Archief voor Wiskunde*, XXVIII:63–113, 1980.

A. Troelstra and H. Schwichtenberg. *Basic Proof Theory,* 2nd ed. Cambridge University Press, Cambridge, 2000.

A. Troelstra and D. van Dalen. *Constructivism in Mathematics,* 2 vols. North-Holland, Amsterdam, 1988.

Vasubandhu. *Abhidharmakosabhasyam,* trans. L. de la Vallee Poussin and L. Pruden. Jain, 1988.

W. Veldman. An intuitionistic completeness theorem for intuitionistic predicate logic. *Journal of Symbolic Logic*, 41(1):159–166, 1976.

W. Veldman. *Investigations in Intuitionistic Hierarchy Theory.* PhD thesis, Catholic University of Nijmegen, 1981.

W. Veldman. Intuïtionistische Wiskunde, 1985. Lecture Notes in Dutch, Department of Mathematics, Katholieke Universiteit Nijmegen.

W. Veldman. An intuitionistic proof of Kruskal's Theorem. *Archive for Mathematical Logic*, 43:215–264, 2004.

W. Veldman. The fan theorem as an axiom and as a contrast to Kleene's alternative. Technical Report 0509, Department of Mathematics, Radboud University Nijmegen, July 2005.

W. Veldman. The Borel hierarchy and the projective hierarchy from Brouwer's intuitionistic perspective. Technical Report 0604, Department of Mathematics, University of Nijmegen, June 2006a.

W. Veldman. Brouwer's real thesis on bars. *Philosophia Scientiae, Cahier Spécial 6*, pages 21–42, 2006b.

W. Veldman and M. Bezem. Ramsey's Theorem and the pigeonhole principle in intuitionistic mathematics. *J. London Math. Soc.*, 47:193–211, 1993.

H. Wang. *Reflections on Kurt Gödel.* MIT Press, Cambridge MA, 1987.

H. Wang. *A Logical Journey. From Gödel to Philosophy.* MIT Press, Cambridge, 1996.

H. Weyl. *Das Kontinuum. Kritische Untersuchungen über die Grundlagen der Analysis.* Veit, Leipzig, 1918.

H. Weyl. Über die neue Grundlagenkrise der Mathematik. *Mathematische Zeitschrift*, 10:39–79, 1921.

H. Weyl. Die heutige Erkenntnislage in der Mathematik. *Symposion*, 1:1–32, 1925.

H. Weyl. *Philosophie der Mathematik und Naturwissenschaft.* Leibniz Verlag, München, 1926. (Weyl 1949) is an expanded English version.

H. Weyl. Diskussionsbemerkungen zu dem zweiten Hilbertschen Vortrag über die Grundlagen der Mathematik. *Abhandlungen aus dem mathematischen Seminar der Hamburgischen Universität*, 6:86–88, 1928.

H. Weyl. *Philosophy of Mathematics and Natural Science.* Princeton University Press, Princeton, 1949.

H. Weyl. *The Continuum. A Critical Examination of the Foundation of Analysis* trans. S. Pollard and T. Bole. Dover, New York, 1994.

L. Wittgenstein. *Philosophical Investigations,* 2nd ed., ed. E. Anscombe and R. Rhees, trans. E. Anscombe. Blackwell, Oxford, 1958.

L. Wittgenstein. *Philosophical Remarks.* Blackwell, Oxford, 1975.

L. Wittgenstein. *Ludwig Wittgenstein and the Vienna Circle. Conversations Recorded by Friedrich Waismann,* ed. B. McGuinness. Blackwell, Oxford, 1979a. Translation of *Wittgenstein und der Wiener Kreis*, Suhrkamp, Frankfurt am Main, 1967.

L. Wittgenstein. *On Certainty.* Blackwell, 1979b.

L. Wittgenstein. *Wittgenstein's Lectures, Cambridge, 1930–1932,* ed. D. Lee. Blackwell, Oxford, 1980.

H. Woodin. The tower of Hanoi. In H. Dales and G. Oliveri, editors, *Truth In Mathematics*, pages 329–351. Clarendon Press, Oxford, 1998.

C. Wright. *Wittgenstein on the Foundations of Mathematics.* Duckworth, London, 1980.

E. Zermelo. Untersuchungen über die Grundlagen der Mengenlehre (Aus einem zu Herrn Hilbert gerichteten Briefe). *Mathematische Annalen*, 65: 514–516, 1908.

Index of citations

SPRINGER NATURE

GPSR Compliance

The European Union's (EU) General Product Safety Regulation (GPSR) is a set of rules that requires consumer products to be safe and our obligations to ensure this.

If you have any concerns about our products, you can contact us on ProductSafety@springernature.com

In case Publisher is established outside the EU, the EU authorized representative is:

Springer Nature Customer Service Center GmbH
Europaplatz 3
69115 Heidelberg, Germany

Kolibri 360 GmbH
Ferdinand-Jühlke-Straße 7
99095 Erfurt, Deutschland
produktsicherheit@kolibri360.de